The Economics of Biotechnology
Volume I

The International Library of Critical Writings in Economics

Series Editor: Mark Blaug

Professor Emeritus, University of London, UK
Professor Emeritus, University of Buckingham, UK
Visiting Professor, University of Amsterdam, The Netherlands

This series is an essential reference source for students, researchers and lecturers in economics. It presents by theme a selection of the most important articles across the entire spectrum of economics. Each volume has been prepared by a leading specialist who has written an authoritative introduction to the literature included.

A full list of published and future titles in this series is printed at the end of this volume.

Wherever possible, the articles in these volumes have been reproduced as originally published using facsimile reproduction, inclusive of footnotes and pagination to facilitate ease of reference.

For a list of all Edward Elgar published titles visit our site on the World Wide Web at
www.e-elgar.com

The Economics of Biotechnology Volume I

Edited by

Maureen McKelvey

Professor of Economics of Innovation
Chalmers University of Technology, Sweden

and

Luigi Orsenigo

Professor of Industrial Organisation, University of Brescia and
CESPRI (Center on the Processes of Innovation and Internationalisation),
Bocconi University, Milan, Italy

THE INTERNATIONAL LIBRARY OF CRITICAL WRITINGS IN ECONOMICS

An Elgar Reference Collection
Cheltenham, UK • Northampton, MA, USA

Published by
Edward Elgar Publishing Limited
Glensanda House
Montpellier Parade
Cheltenham
Glos GL50 1UA
UK

Edward Elgar Publishing, Inc.
136 West Street
Suite 202
Northampton
Massachusetts 01060
USA

A catalogue record for this book is available from the British Library

ISBN-10: 1 84376 776 7 (2 volume set)
ISBN-13: 978 1 87376 776 3 (2 volume set)

Printed and bound in Great Britain by MPG Books Ltd, Bodmin, Cornwall

Contents

Acknowledgements

The editors and publishers wish to thank the authors and the following publishers who have kindly given permission for the use of copyright material.

American Economic Association for article: Lynne G. Zucker, Michael R. Darby and Marilynn B. Brewer (1998), 'Intellectual Human Capital and the Birth of U.S. Biotechnology Enterprises', *American Economic Review*, **88** (1), March, 290–306.

Blackwell Publishing Ltd for article: Iain M. Cockburn and Rebecca M. Henderson (1998), 'Absorptive Capacity, Coauthoring Behavior, and the Organization of Research in Drug Discovery', *Journal of Industrial Economics*, **XLVI** (2), June, 157–82.

Business History Review for article: Louis Galambos and Jeffrey L. Sturchio (1998), 'Pharmaceutical Firms and the Transition to Biotechnology: A Study in Strategic Innovation', *Business History Review*, **72** (2), Summer, 250–71, 274–78.

Cambridge University Press for excerpt: Rebecca Henderson, Luigi Orsenigo and Gary P. Pisano (1999), 'The Pharmaceutical Industry and the Revolution in Molecular Biology: Interactions Among Scientific, Institutional, and Organizational Change', in David C. Mowery and Richard R. Nelson (eds), *Sources of Industrial Leadership: Studies of Seven Industries*, Chapter 7, 267–311.

Elsevier for articles: Martin Kenney (1986), 'Schumpeterian Innovation and Entrepreneurs in Capitalism: A Case Study of the U.S. Biotechnology Industry', *Research Policy*, **15**, 21–31; Ashish Arora and Alfonso Gambardella (1994), 'Evaluating Technological Information and Utilizing It: Scientific Knowledge, Technological Capability, and External Linkages in Biotechnology', *Journal of Economic Behavior and Organization*, **24**, 91–114; Lynne G. Zucker and Michael R. Darby (1997), 'Present at the Biotechnological Revolution: Transformation of Technological Identity for a Large Incumbent Pharmaceutical Firm', *Research Policy*, **26**, 429–46; Shyama V. Ramani (2002), 'Who is Interested in Biotech? R&D Strategies, Knowledge Base and Market Sales of Indian Biopharmaceutical Firms', *Research Policy*, **31** (3), 381–98; Vincent Mangematin, Stéphane Lemarié, Jean-Pierre Boissin, David Catherine, Frédéric Corolleur, Roger Coronini and Michel Trommetter (2003), 'Development of SMEs and Heterogeneity of Trajectories: The Case of Biotechnology in France', *Research Policy*, **32** (4), 621–38; Govindan Parayil (2003), 'Mapping Technological Trajectories of the Green Revolution and the Gene Revolution from Modernization to Globalization', *Research Policy*, **32**, June, 971–90; Joanna Chataway, Joyce Tait and David Wield (2004), 'Understanding Company R&D Strategies in Agro-Biotechnology: Trajectories and Blind Spots', *Research Policy*, **33** (6/7), 1041–1057.

Institute for Operations Research and the Management Sciences (INFORMS) for article: Michelle Gittelman and Bruce Kogut (2003), 'Does Good Science Lead to Valuable Knowledge? Biotechnology Firms and the Evolutionary Logic of Citation Patterns', *Management Science*, **49** (4), April, 366–82.

Macmillan Magazines Ltd for article: Hannah E. Kettler and Sonja Marjanovic (2004), 'Engaging Biotechnology Companies in the Development of Innovative Solutions for Diseases of Poverty', *Nature Reviews: Drug Discovery*, **3**, February, 171–6.

Merck and Co., Inc. for two photographs included in article: Louis Galambos and Jeffrey L. Sturchio (1998), 'Pharmaceutical Firms and the Transition to Biotechnology: A Study in Strategic Innovation', *Business History Review*, **72** (2), Summer, 250–78.

Oxford University Press for excerpt and article: Maureen D. McKelvey (1996), 'Introduction' and 'Conclusions for Science and Technology', in *Evolutionary Innovations: The Business of Biotechnology*, Chapters 1 and 10, 1–15, 280–99, references; Paul Nightingale (2000), 'Economies of Scale in Experimentation: Knowledge and Technology in Pharmaceutical R&D', *Industrial and Corporate Change*, **9** (2), June, 315–59.

Springer Science and Business Media for article: David B. Audretsch (2001), 'The Role of Small Firms in U.S. Biotechnology Clusters', *Small Business Economics*, **17**, 3–15.

John Wiley and Sons Ltd for article: Gary P. Pisano (1994), 'Knowledge, Integration, and the Locus of Learning: An Empirical Analysis of Process Development', *Strategic Management Journal*, **15**, Winter, 85–100.

Every effort has been made to trace all the copyright holders but if any have been inadvertently overlooked the publishers will be pleased to make the necessary arrangement at the first opportunity.

In addition the publishers wish to thank the Marshall Library of Economics, University of Cambridge, UK, the Library at the University of Warwick, UK, for their assistance in obtaining these articles.

Preface

Maureen McKelvey and Luigi Orsenigo

This reference collection has developed out of a long-term collaboration between Maureen McKelvey and Luigi Orsenigo. In the context of several EU projects, conferences and personal friendship, we have written together (but also debated with each other) about the economics of biotechnology. The topics found here represent our way of 'making sense' of how and why the overlap of biotech and pharmaceuticals is an emergent, interesting and extreme example of wider issues about knowledge and the economy. Selecting themes and published pieces to include in this collection has given us the opportunity to continue that debate, and to structure our interpretation in a way accessible to other readers who share an interest in these topics.

Actually delivering this collection has involved finding, reading and rereading a much larger literature, in order to select our eight themes in the pieces reproduced here. We should like to thank all our research colleagues working on biotechnology, especially those who answered our call to suggest references and those debates that they considered key to the area. Some colleagues are part of our local research environments at, respectively, the Department of Technology, Management and Economics and RIDE Research Center at Chalmers University of Technology, at Brescia University and at CESPRI, Bocconi University. Other colleagues are located at universities and research institutes throughout the globe. The usual caveats apply, for despite all the advice and discussions, we have gone on to select and interpret the economics of biotechnology in our own way.

Finally, we should like to acknowledge sources of financing for our ongoing research, in addition to our workplaces. Maureen McKelvey acknowledges two projects that she leads at Chalmers University of Technology, namely strategic projects on 'IT and Biotechnology' and on the 'Economics of Innovation'. She also thanks the Rueben Rausing Foundation, the Institute for Management of Innovation and Technology (IMIT) and the University of Queensland, Australia (UQ). In the context of a meeting for a projected OUP title on 'Flexibility and Stability in the Innovating Economy', Maureen arranged a book meeting, financed by the Rausing Foundation and managed by IMIT. As editors, we spent several pleasant and useful days at UQ, and also working on our reference collection. Thanks especially are due to Mark Dodgson and John Foster for helping to organize our visits to UQ.

Introduction

Maureen McKelvey and Luigi Orsenigo

This reference collection *The Economics of Biotechnology*, includes our selection of published articles and book chapters in a two-volume work. This collection covers the economics and business side of the social scientific debate about the economics of what is generally called 'modern biotechnology' or 'the biotechnology industry'.

Biotechnology, which is here defined as a set of technologies and knowledge and the economic activities that have been developing around them, has attracted an enormous interest, both within economics and other social sciences as well as with the public at large. Research has spawned work on a variety of theoretical issues concerning economic dynamics and innovation systems, and about what might be called – in the current jargon – the modern 'learning economy' and innovation economy. More generally, biotechnology is often perceived as one of the most important and cutting-edge technologies of the contemporary era. As such, biotech is often considered as one (albeit extreme) example for studying the current and future trends of technological change, as well as the study of the related consequences, controversies and problems. In public opinion, biotechnology has spurred complex and often bitter debates, both in terms of moral and policy issues, and of the related interpretations about use and the implications for society.

The purpose of this Introduction, and of the reference collection as a whole, is to provide the reader with an accessible and structured understanding of the main issues which have characterized the economic debate. Thus the present two volumes mainly concern the economics of innovation and technological change, and as such, the work is addressed to economists, managers, professionals, policy-makers and students, and to those readers who are interested in the specific case of biotechnology. Here we also introduce ideas about how biotech has informed the broader conceptual issues in the field of economics of innovation and technological change.

In what follows, we shall try to provide a roadmap for the reader, by illustrating the major issues and debates over the past few decades, based on a few select references, as developed by researchers in the business, economics and broader social science communities. Obviously, the process of selection of the material here presented has entailed many trade-offs and difficult choices, given the large amount of work which has been done on modern biotechnology. So, clearly this collection reflects our own perspectives, and the selection is the result of our specific editorial choices. Thus the collection provides only a limited and focused selection of published articles and book chapters.

The following four criteria have been used to guide our work as editors. They have influenced our choice of articles and chapters, as well as our conceptual framework for interpreting the debates:

1. The work selected should, in its totality, provide insights into biotech in different industries. This includes biotechnology as linked to the pharmaceutical industry, as well as the use of biotechnology techniques, knowledge and equipment within many other types of businesses and sectors. This criterion is perhaps more difficult to meet than might be apparent, since, until recently, almost all research was focused on small, dedicated biotech firms and pharmaceuticals.
2. The work selected should, in its totality, represent a mix of early pieces which have framed the debates as well as pieces which are more recent and/or less well known. This criterion reflects our understanding of research into the economics of biotechnology having developed along a series of debates, with new questions and debates arising over time, even if the conclusions from previous research has remained controversial to many authors.
3. The work should deal with aspects of the business and economics of biotechnology from a social scientific perspective. This excludes consultant reports, technical overviews, ethics or other topics not addressing business and economics *per se*. The one exception, found in Volume I, Chapter 1, of the collection, is the OTA (1984) report.[1]
4. The work should be scholarly, published papers. Again, this criterion excludes technical reports and the so-called 'grey literature' (which includes, however, much extremely informative, rigorous and challenging material), as well as unpublished conference and working papers.

These four criteria have been applied, in choosing published work to reflect the different categories of debate as defined in this volume. Note that these criteria do not restrict our choice to papers written within any given or strict disciplinary approach. The reason that here we have expanded beyond economics as a discipline, *per se*, will be clear in the subsequent sections, given the importance of knowledge, social networks, ethics, and the like, to our story. Various debates have spawned an enormous literature, which includes a variety of approaches and methodologies, ranging from case-studies, and comparative case studies to econometric analyses and theoretical models, and having either a formalised or appreciative structure. Hence, our view is that an understanding of the economics of biotechnology requires insights into the contributions from a variety of social scientists.

Here we present our interpretation, or the conceptual framework, which has helped us interpret the debates and guided the organization of this book. This is followed by a short introduction to each of the selected articles. The theoretical and empirical questions in this reference collection are framed in terms of the following eight issues, which reflect the eight parts of this two-volume work:

1. Overview;
2. Science and innovation;
3. New specialised biotechnology firms;
4. Reaction and adaptation of large incumbent companies;
5. Division of labour in innovative activities and networks of innovators;
6. Geographical agglomeration;
7. Institutions supporting the biotechnology industry;
8. Intellectual property.

Below we present a brief overview of modern biotechnology as a business and economic phenomenon. We then provide our interpretation of the main conceptual issues, which we consider necessary in understanding biotechnology but which are also part of the wider debates about the economics of innovation and technological change. In particular, we address the relationships between science, technology and innovation; the role and nature of the new biotechnology firms; the behaviour of large incumbent companies; the economic meaning of collaborative relationships and of networks of interactions among differentiated agents; the geography of innovation in biotechnology and the emergence and development of clusters of innovation; and the institutional variables affecting the development of biotechnology and the co-evolution between technology, markets and institutions. These issues help define the themes within which the selected articles and chapters are grouped. We then introduce the material contained in the reference collection, briefly summarising and contextualising each individual contribution and, finally, offer brief remarks about future research and the emerging unsolved problems and methodological issues associated with the study of the economics of biotechnology.

As editors, we hope that this book proves timely for readers who have a particular interest in the business and economics of modern biotechnology.

An Overview

This part of the collection provides a brief historical overview of biotechnology during the past three decades or so. This rather superficial sweep of history is followed by the more nuanced discussion of scientific publications and debates, both within subsequent sections and subsequent work.

As is well known, 'biotechnology' emerged in the mid-1970s not only as an immediate consequence of two key scientific and technological discoveries – restriction enzymes (better known as recombinant DNA) and monoclonal antibodies – but also as the outcome of a much longer process of dramatic advances in fundamental biomedical sciences, especially molecular biology.

These fundamental scientific discoveries took place in universities, but were quickly used to set the basis for commercial uses. Genentech was the first company to be started, with the specific strategy to use r-DNA techniques to produce drugs. Genentech was based on the crucial support and initiative of venture capitalists and of basic scientists. This was a pattern often followed in later years, where the creation of new, technology-based firms, usually spin-offs from academic labs, became one of the distinguishing features of the emerging 'industry'. However, rates of entry of DBFs (dedicated biotech firms) fluctuated wildly over time, following both the vagaries of the stock market and the appearance of new waves of technological innovations. Examples of waves of technological innovations range from r-DNA and monoclonal antibodies, to the so-called 'platform' technologies, like combinatorial chemistry and high throughput screening, all the way down to genomics, proteinomics, etc. Simultaneously, universities became increasingly involved in the commercialization of their research, through patents as well as spin-offs.

In its early stages, biotechnology was conceived as a sort of 'general purpose' technology, whose potential applications could span to an enormous variety of sectors, ranging from pharmaceuticals, agriculture, food and chemicals to the environment (or what might now be called the life-science industries). Hence, in its constituent phase, it was often thought that biotechnology could become the basis for powerful opportunities of technological convergence and of diversification of companies (OTA, 1984; Orsenigo, 1989). However, this promise did not materialise. Some companies did move in this direction, with the case of Monsanto being an extreme example of a large company betting on transforming itself into a fully diversified and integrated life-science firm. Yet, history took a somewhat different path, such that economic applications of biotech concentrated mainly on the pharmaceutical and, to a lesser extent, the agro-food sectors. So, despite a few such examples of general-purpose applications, the economic exploitation of scientific progress has followed quite distinct and specialised trajectories within different sectors.

As to firms in the pharmaceutical and agro-food industries, large established firms approached the new scientific and technical developments mainly from a different perspective from the DBFs – i.e. as research tools in the discovery process. This still required, however, a very substantial extension of the range of scientific skills employed by the firm and deep organisational changes in the way innovative activities were conducted. The adoption of biotechnology by established pharmaceutical companies was, in general, a slow, painful and difficult process. Moreover, there was enormous variation across firms in the speed with which the new techniques were adopted. This internal reorganization occurred in parallel with the collaboration with small firms.

The performance and role of spin-off firms have also been re-evaluated over time. The tumultuous rate of scientific and technical progress did not automatically translate into commercial growth and success. Only a few of the DBFs were making profits, and only a handful of them were really able to transform themselves from being research companies to integrated corporations. Instead, DBFs survived and, in some cases, grew through the establishment of collaborative agreements with large corporations. Collaboration provided the financial resources necessary to fund R&D as well as the access to organizational capabilities in product development and marketing. Established companies faced the opposite problem. While they had the experience and the structures necessary to control testing, production and marketing of pharmaceuticals, they could use collaboration to explore, acquire, absorb and develop the new knowledge.

The intensity of collaboration among DBFs and large firms was so strong that – according to many analysts – a new form of organization of the innovation process emerged. Various concepts have been used, including innovation systems and networks of innovators. The underlying idea has been that innovation can no longer be conceptualised as the result of the efforts of any individual firm, but as the outcome of the interactions. Such interactions exist through market-mediated exchanges, as well as other complex and socially embedded forms of exchange and collective learning, and interaction occurs among different types of firms and a wide range of institutions – e.g. universities, public research laboratories, venture capitalists, etc. The biotech firm is thus strongly linked to other actors and its socio-political context.

A related feature of the development of biotechnology is its strong geographical concentration, both within countries and regions. One aspect is that, ever since its inception, biotechnology has been largely an American phenomenon, both in terms of the rates of innovation and commercialization on the one hand, and the forms of organisation of innovative activities on the other hand. European countries (the UK being only a partial exception) and

Japan lagged behind the USA. Over time, despite some recent signals of vitality in the Nordic countries, Germany and France, the gap does not appear to have become smaller. One could argue that this gap helps contribute to the declining competitiveness of traditional European strongholds such as pharmaceuticals and the agro-food industries. Similarly, one observes increasing innovative activities in countries like Israel, in Singapore and East Asia. Yet despite international attention and investment in biotech, the US leadership remains unchallenged and, if anything, it may be still increasing.

Geographical concentration in terms of linking the biotech firm and industry to its socio-political context goes well beyond the level of national economies. In fact, one observes a strong concentration of innovative activities within countries in a few local clusters, typically centred around top universities, such as Boston, the Bay Area and San Diego in the USA, and Cambridge in the UK. Such clusters strongly overlap with the networks of collaboration discussed earlier, but the latter are not univocally defined in terms of geography. If anything, it would appear that the more successful regional clusters are characterised both by dense internal relations, but also by a high degree of openness and a strong degree of interaction with agents located outside the cluster itself.

Taken together, concepts such as networks, clusters and the geographical distribution of innovative activities highlight that the development of biotechnology as an economic activity is embedded in and deeply affected by broader institutional and social variables. For example, the dominance of the USA has been explained by a mix of factors, ranging from the sheer level of public expenditure in fundamental biomedical research, and the resulting size and quality of academic research, to the institutional structure of universities and the ethos of academic scientists, which create conditions and incentives for researchers to engage in business activities. Such explanations go far beyond the traditional boundaries of 'economic' analysis.

Similarly, there is no question that the very existence of American DBFs was made possible by the existence and further development of a strong venture capital industry, as coupled with the establishment of legislation and courts' decisions that have promoted extremely strong patent protection. Patent protection is important both in relation to biotechnological innovations and to the commercialization of the results of publicly funded research by universities. In the USA, the Bayh-Dole Act was passed in 1982 and introduced incentives and facilitations for the patenting of academic research results and their commercial exploitation. Many other countries have attempted to implement institutional changes to mirror their interpretation of these 'lessons'. Last but not least, public attitudes against agro-food applications of biotechnology in Europe are commonly cited as a major obstacle to the development of the European industry in this segment. However, and interestingly enough, one observes an inversion of public attitudes about research on stem cells, with the USA adopting a much more restrictive stance as compared to Europe.

In summary, this brief overview has indicated that the biotech industry is driven by a combination of factors. Individual firms, whether DBFs or large firms, as well as the biotechnology industry as a whole, are clearly changing, along with their environments. As argued above, analyses of networks and clusters consistently emphasize the role played by institutional and social factors in shaping the evolution and the differential success of individual firms and of the biotechnology industry. More provocatively, there is the observation that technology, institutions, markets and the industry have co-evolved by mutually influencing one another, sometimes in a purposive way and sometimes as the unintended or unexpected outcomes of actions taken for different motivations. So, even though each piece of research addresses a specific question, an interpretation of the overall economics of biotechnology is dependent upon a multidisciplinary understanding of many underlying processes. This suggests a large set of theoretical and empirical issues which can be studied, and which are also of great relevance in terms of strategy and policy-making.

Theoretical and Empirical Questions

This part of the collection provides a broader literature review of the debates and core questions, and the following section describes the specific contribution of each work reproduced in this collection. Many of these questions have attracted the interest of economists and other social scientists. Here we have chosen to concentrate on the eight themes specified above, which we consider constitute the 'hard core' of the economics of biotechnology. These themes represent core theoretical and empirical questions to interpret the economics of biotechnology, as discussed below in relation to selected debates in the broader literature. Some of these publications were important in framing the early debates, such as the OTA (1984) report. Others provided a broad sweep of analysis, such as the overview of pharmaceuticals, while others have represented areas of research which as yet have stimulated too little academic interest; they represent interesting and important areas for future research and debate.

Science, Technology and Industrial Innovation

A second set of issues in this part of the collection concerns 'Science and innovation', especially the relationships between science, technology and industrial innovation. A crucial feature of biotechnology, as a technology and as an economic activity, is that it has been strongly science based. In contrast to most other technologies and industries – including pharmaceuticals at least up until the early 1970s – basic scientific research has an almost immediate and direct impact on biotechnology and on its industrial applications. Indeed, in many cases, the knowledges useful for science, technology and industrial innovation are practically indistinguishable – even if over time, more and more industrial innovation using biotech knowledge differs in significant ways in technological and business activities from that of the scientific activities. But how does science actually influence technology and industrial innovation? And how do science, technology and economic forces interact cognitively, organizationally and institutionally?

The case of biotechnology confirms that the relationship between science, technology and the economy is much more complex than the 'traditional' perspective has framed the problem. If we oversimply a traditional view, then the relationship between scientific research and industrial innovation is relatively straightforward. Scientific research is considered as generating information – i.e. abstract and codified knowledge (Nelson, 1959; Arrow, 1962; Arora and Gambardella, 1994) – aiming at understanding phenomena. Moreover, science is to be considered as a public good: its use is non-rival and non-excludable. It is non-rival because scientific knowledge is costly to produce, but it can be indefinitely reused at negligible marginal costs. It is non-excludable because scientific research is (or should be) governed by the principle of 'open science' – i.e. it is freely made available to anybody through publication. Conversely, technical knowledge has a largely practical nature, which aims at producing 'things that work'. Technology belongs to the private domain, since its development is largely motivated in modern economies by the perspective and circulation of supra-normal profits. The diffusion of such technological knowledge is restricted by intellectual property rights and other mechanisms that shield technological innovations from imitation.

This distinction between science, technology and industrial innovation becomes even sharper if technical knowledge is conceived as having a strongly tacit, local and specific nature (Nelson and Winter, 1982; Pavitt, 2001); that is to say, technical knowledge is not completely codified in instructions, recipes, handbooks, etc. Therefore it cannot be easily and costlessly transferred and acquired. Rather, mastery requires practice and direct experience, which lead to different abilities, skills and outcomes. The (successful) use and application of such knowledge is therefore limited to specific domains, as well as to specific individuals and teams, as a result of the specific competencies accumulated over time. In this approach, science contributes to increasing the productivity of technological research essentially by informing researchers about which among the possible directions of inquiry are more or less promising, or simply wrong. Thus the search space becomes 'simpler' and more focused.

Science and technology, then, can be differentiated from a cognitive perspective (understanding versus functioning; abstract and codified versus tacit and specific) and from an institutional perspective (public versus private; opportunities versus incentives). However, the empirical case of the economics of biotechnology raises a few paradoxes and problems. First of all, why did incumbent large corporations find it so difficult to absorb and master the new scientific knowledge base relevant for innovation? After all, if scientific knowledge is public and codified in nature, in principle, it should be available to anybody and everywhere. Second, why did the process of translation of scientific knowledge into new products and processes require the development of such complex forms of organization of innovative activities, like the DBFs, innovation systems, networks of collaboration and markets for knowledge? Do these new forms constitute an organizational innovation in themselves, destined to supplant the more traditional, vertically integrated model of R&D in science-based industries? There are questions about the trade-offs involved in different ways of organizing innovation activities. Finally, why is it that despite the public nature of scientific knowledge, industrial innovative activities are so unevenly distributed, organizationally and geographically? What do firms do to develop technological knowledge that differs from science? As has been persuasively argued in various ways over the past two decades, science and technology interact in subtle – and, in most cases, certainly non-linear – ways in biotechnology.

Cognitively, this interaction involves feedback from technology to science (Kline and Rosenberg, 1986; Pavitt, 2001); and scientific discoveries do not only 'simplify' the search space for researchers, but also increase and deform it, by providing new opportunities to be explored and new ways of defining and framing problems, and new problems which were previously not even conceived (Dosi, 1982). Moreover, science, like technology, is not completely codified: its production and use requires, in any case, pre-existing knowledge, direct experience and the development of protocols and procedures which are themselves at least partly tacit.

Organisationally, the interaction between science and innovation requires the development of interfaces and specific capabilities by individuals, firms and other organizations (Dasgupta and David, 1994). For example, the actors need capabilities in order to explore new technological opportunities and to survey the progress achieved by others, to absorb new knowledge and to integrate it into existing competences and organisational structures. In turn, this often requires deep transformations into extant organisational architectures and the development of new forms of organization and of interaction.

Institutionally, a closer relationship between science and innovation implies that the traditional boundaries of actors and of activities become blurred. New forms of division of labour and channels of interaction between the public and the private domain have to be created and managed. In any case, science, technology and innovation – when they are so strictly intertwined – are subject to an overlap between multiple and different selection and incentive mechanisms (McKelvey, 1996). How can they be reconciled and how do agents – individuals, firms and institutions – react? To what extent it is possible for individuals to balance the demands of both selection environments? And to what extent do commercial pressures tend to dominate other selection principles? Thus the empirical case of biotechnology raises many questions about our interpretation of science, technology and industrial innovation.

Creative Destruction or Creative Accumulation?

The issue of 'creative destruction or creative accumulation' within our broader field refers to the third and fourth parts of this book (see above), namely 'New specialised biotechnology firms' and 'Reactions and adaptation of large incumbent companies'. This leads us into research questions about company strategy as well as about the relative advantage of each type of organization in relation to the dynamic effects of new technologies and industries. However, the case of biotechnology challenges any standard, simplified accounts of the patterns of industrial evolution and of their life cycles, and it does not unequivocally support any of the extreme interpretations.

These two issues refer to an important research question in the whole history of the economics of innovation, and more broadly of industrial organization and its dynamics, namely the debate on the so-called 'Schumpeterian hypotheses' on the relationships between innovation, firms' size and age, and market concentration. Is innovation mainly generated by small, new firms or large incumbent corporations? Are higher rates of technological change linked to a high degree of monopolistic power or innovation best promoted by vibrant competitive industries? One aspect is the extreme interpretation, that there are absolute, relative advantages of new/small and incumbent/large firms. Another aspect is whether, in a dynamic perspective, these questions can be reframed in terms of both the industry and the technology. In one influential interpretation, new industries typically follow a distinct life cycle, marked by high entry and turbulence in the early stages, followed by a shake-out in the number of firms and a period of consolidation, with the emergence of a few dominant companies (typically early entrants) (Abernathy and Utterback, 1978; Utterback, 1994; Klepper, 1996).

Indeed, in biotechnology one observes a series of waves of entry of new firms in relation to the appearance of new technologies (e.g. from r-DNA and monoclonal antibodies, to the so-called 'platform' technologies, like combinatorial chemistry and high throughput screening, and all the way down to genomics, proteinomics, etc.). There is not one technology and industry, but several. Yet, only very few DBFs have been able to gain significant market shares. Indeed, incumbents have not been swept away by these new entrants. They have been able to maintain the leadership of the industry and consolidated their position.

DBFs were primarily university spin-offs and usually formed through collaboration between scientists and professional managers, backed by venture capital. Their specific skills resided in the knowledge of the new techniques and in the research capabilities in that area. This type of NBF has mobilized fundamental knowledge created in universities and transformed it into potentially commercially useful techniques and products. Their aim consisted in applying the new scientific discoveries to commercial applications, in particular, drug development. Their internal organizational structure reflected their origin and competencies. They were organized very much like academic units, and they deeply embodied some fundamental academic principles like the importance attributed to publication and to work at the frontier of knowledge. However, these organizational principles (in terms of norms, incentives, practices) had to be made consistent with their commercial nature too. Thus secrecy and the search for broad property rights became crucial features of these new firms. Moreover, financial constraints coupled with high burn rates have made 'time to patent' a characteristic feature of the research style of these companies.

The growth of NBFs was constrained by various factors. First, many of the early trajectories of research proved to be dead-ends and/or much more difficult to develop than expected.[2] Second, the large majority of these new companies never managed to become fully integrated producers, since they lacked competencies in other different crucial aspects of the innovative process: in particular (and specifically in the case of pharmaceuticals), they lacked on the one hand knowledge and experience of clinical testing and other procedures related to product approval, and marketing on the other hand. Thus, with few exceptions most of these NBFs exploited their basic competence and acted primarily as research companies and specialized suppliers of high-technology intermediate products, performing contract research for and in collaboration with established pharmaceutical corporations.

Third, even remaining at the level of pre-clinical R&D, most NBFs lacked crucial competencies in a rather different way. In fact, many individual NBFs were actually started on the basis of a specific hypothesis or technique, following the processes of growth of knowledge in the field. Successive generations of NBFs were increasingly specialised in particular fields and techniques and, with few exceptions, they were stuck in specific cognitive/research niches, with limited opportunities to exploit the advantages stemming from the possibility to pursue alternative routes to discovery, the cognitive complementarities among different techniques and bodies of knowledge, and the realization and exploitation of economies of scope. Moreover, the process of discovery and development of new products – especially in the case of drugs – still requires a broader and more 'general' perspective, which integrates several differentiated fragments of knowledge.

Indeed, later generations of NBFs were largely created on the basis of specialization into radically different new technologies like genomics, gene therapy, combinatorial chemistry and so-called 'platform technologies'. These technologies are essentially research tools, and the companies essentially provide services to the corporations involved in product discovery and development.

Collaboration allowed NBFs to survive and, in some cases, to pave the way for subsequent growth in many respects. Basically, collaboration with large companies provided the financial resources necessary to fund R&D, and provided the access to organizational capabilities in product development and marketing. Established companies faced the opposite problem. While they needed to explore, acquire and develop the new knowledge, they had the experience and the structures necessary to control testing, production and marketing. Both companies also wanted collaboration with the relevant basic scientific communities, in order to gain access to new sources of knowledge.

Large incumbent firms were able to preserve their leadership by gradually learning and absorbing the new technologies, controlling key complementary assets and establishing a dense and complex web both of collaborative and competitive relations with the entrants. However, the continuing dominance of large corporations has required long and painful processes of learning and strategic as well as organizational adaptation. These processes have entailed outright failures and losers, and still go on over time, also through processes of mergers and acquisitions.

An enormous literature has studied the conditions in which technological discontinuities can be destructive for incumbents. These explanations range from the consideration of strategic considerations in game theoretic settings to a variety of variables which ultimately have to do with the nature of the competencies possessed by firms *vis-à-vis* those required in the new technological environment and the inherent inertia in corporate behaviour (Henderson and Clark, 1990; Christensen and Rosenbloom, 1995; Christensen, 1997). A review of this literature goes well beyond the scope of this Introduction; but as mentioned in the previous paragraph, in the case of biotechnology, established corporations had to get access to an entirely new knowledge base, and to absorb and integrate it into their strategies and organisation. This task turned out to be relatively difficult.

The transition to biotechnology required a substantial extension of the range of scientific skills employed by the firm, a scientific workforce that was tightly connected to the larger scientific community and an organizational structure that supported a rich and rapid exchange of scientific knowledge across the firm (Henderson and Cockburn, 1996; Gambardella, 1995). The new techniques also significantly increased returns to the scope of the research effort (Henderson and Cockburn, 1996). In turn, this required the recruitment of star scientists and the adoption of organizational practices and incentive structures which in some way attempted to replicate some of the typical characteristics of an academic environment, like the adoption of 'pro-publication' incentives (Cockburn, Henderson and Stern, 2000).

More generally, the molecular biology revolution made innovative capabilities critically dependent on publicly generated scientific research. Far from being a costless and direct process, this implied that companies had to establish much closer and tighter linkages with the scientific community, in various forms, including research contracts, long-run funding agreements to particular teams or institutions, etc. In turn, this required firm investment to monitor and maintain networks for potential sources of information. A necessary condition for getting access to such knowledge, both from a cognitive and a sociological perspective, was that companies had to become active players in the scientific arena and not simply passive observers and users. In facts, the relation between firms and public research is 'very much a bidirectional one, characterized by the rich exchange of information in both directions' (Cockburn and Henderson, 1996). In other words, companies had to build in-house competencies for at least three reasons. First, they needed to do so in order to develop the 'absorptive capabilities' necessary to understand the scientific progresses taking place in academia and in the NBFs. Second, they needed to do so in order to get the 'ticket of admission' to the scientific community. Third, because the development of new products required not simply the availability of specific techniques, but the evaluation and testing of alternative approaches and the integration of different techniques, scientific disciplines, etc.

Clearly, there was enormous variation across sectors and firms in the modalities through which the transition to biotechnology was experienced. For example, in the case of pharmaceuticals, biotechnology did not imply the appearance of totally new markets and categories of users. The main difficulties were related to the absorption and integration of biological sciences – as contrasted to chemistry – in the process of drug discovery and development, with the related organizational problems of establishing and maintaining relationships with the scientific community. In agro-food, an additional critical issue was the hostility of public opinion and the evolution of related legislature. At the firm level, the transition to biotechnology was influenced by conventional variables like their size, but also by their past history, in terms of the extent of their relationships with science, strategies and organizational structures.

Towards New Forms of Organization of R&D?

The first part of Volume II in this reference collection addresses 'Division of labour in innovative activities and networks of innovators'. Indeed, much effort has gone into conceptualising and analysing the development of new forms of organization of R&D, including collaborative relationships, in biotechnology. The term 'network' has increasingly become widespread in the economic literature, and the concept 'innovation system' more apparent in the policy domain.

As mentioned above, the distinctions between the nature of science and technology as economic goods have important implications for understanding how and why such knowledge can be transferred and exchanged. To the extent that knowledge is considered as information, and therefore codified, standard economic theory of information would suggest that significant market failures characterise these exchanges. A potential buyer cannot properly evaluate the 'quality' of the information to be exchanged and the seller cannot disclose such information, because this would imply actually transferring the good for free. Moreover, in conditions of information asymmetry, which are intrinsic in markets for information, mutually beneficial transactions may be forgone as a consequence of opportunistic behaviour and free-riding, leading to the underdevelopment or even to the absence of markets (Akerlof 1970; Mowery, 1981; Arrow, 1983; Arora, Fosfuri and Gambardella, 2001).

Similarly, there are debates about how and why difficulties in the process of buying and selling information may entail other serious inefficiencies, especially in matching the production of new knowledge to its industrial uses. Under-investment in the production of information (including R&D and technologies) is likely to arise if the markets for such goods do not work efficiently. One possible solution to this problem resides in vertical integration (Mowery, 1981; Arrow, 1983), where production and further exploitation of knowledge are integrated in the same organisation. This solution to the fundamental problem of under-investment in the production of knowledge is sometimes considered to be one of the main reasons explaining the establishment of R&D labs within firms, as opposed to alternative arrangements. For example, an alternative also debated in the literature is the role of interactions and 'markets for technology', where specialised producers of knowledge sell their output to other agents who, in turn, use such knowledge as an input in the production of their products.

Yet, many debates remain. These arguments would be further strengthened if technical knowledge were considered as tacit, as discussed above. In this case, vertical integration facilitates communication, face-to-face contacts, exposure to similar learning processes among individuals and teams sharing the same (firm-specific) procedures, routines and organisational practices (Nelson and Winter, 1982; Mowery, 1981). This, then, is similar to some of the debates about the geographical agglomeration of biotechnology. Another issue is that vertical integration of this kind may imply inefficiencies, as linked to the lower power incentives typical of an organisation as compared to a market and as linked to the failure to exploit potential advantages of specialisation and division of labour.

Indeed, biotechnology represents an interesting and important empirical area to discuss many of these issues. On the one hand, the model of vertical, integrated R&D has been of one the major organisational revolutions in the history of firms' organisation and it has remained for several decades the dominant mode of organising corporate research. On the other hand, the broader management and economic literature has stressed many new ways to organize R&D. Indeed, the spreading of collaborative research and of more decentralised and flexible forms of organisation of innovative activities, based on 'loosely coupled' and network-like structures, is by no means a phenomenon limited to biotechnology. But biotechnology has been the forum for much research and debate. Within pharmaceuticals, it is clear that increasingly large firms are complemented by a dynamic population of smaller companies sustained by venture capital and outsource R&D to specialist private-sector research houses. Similarly, the role of academic research as a source of innovation, and of universities as active agents in the process of economic exploitation of knowledge, has been emphasized, which also links to the series of debates about what type of science is most relevant for technology and industrial innovation.

One finds in the literature widely different interpretations of the nature, motivations, structure and functions of these networks, ranging from more sociologically orientated approaches to economic explanations based on (various mixes of) alternative theoretical backgrounds – e.g. transaction costs, contract theories, game theory and competence-based accounts of firms' organization. As further developed in the summary of the work in this collection, there are different interpretations of this phenomenon.

According to one influential interpretation, collaborations represent a new form of organization of innovative activities, which are emerging in response to the increasingly codified and abstract nature of the knowledge bases on which innovations draw (Arora and Gambardella, 1994; Gambardella, 1995). These properties of knowledge, coupled with the establishment of property rights, would make it possible, in principle, to separate the innovative process in different vertical stages. Thus, in this view, the innovative process can be adequately represented as a sequence going downstream from science to marketing, in which division of labour can occur at any stage of the process. Different types of institutions tend to specialize in the stage of the innovative process at which they are more efficient: universities in the first stage, small firms in the second and big, established firms in the third (see also Arrow, 1983). Thus a network of ties between these actors can provide the necessary coordination of the innovative process. Collaborations are likely to be a permanent feature of the industry, generating an intricate network within which each subject specializes in particular technological areas or stages of the innovative process and getting benefits from an increasing division of innovative labour.

According to a competing interpretation, collaborative relations are instead considered as a transient phenomenon, bound to decrease in scale and scope as the technology matures. When knowledge is used in applications closer to the actual realization of products, and often becomes 'more tacit' in industrial innovation, this implies that the industrial structure will have higher degrees of vertical integration (see Pisano, 1991; Volume II, Chapter 1). There are debates here, for example, about the relative advantage of each strategy. Empirical studies on the rates of success and failures of projects carried on entirely in-house, as compared to projects involving the acquisition of licenses from third parties, show that indeed licensed projects have higher probability of success (see Arora, Gambardella, Pammolli and Riccaboni, 2000; and for conflicting evidence see Pisano, 1997).

Finally, another view is that each actor holds unique resources and competencies to solve problems of a scientific, technological and business nature. In other words, the complex and interdisciplinary nature of relevant knowledge bases in R&D tend to make technological innovations the outcome of interactions and cooperation among different types of agents commanding complementary resources and competencies (Sharp, 1985; Orsenigo, 1989; McKelvey, 1996; Pammolli, 1997). In more radical interpretations within this perspective, it has also been suggested that the locus of innovation (and the proper unit of analysis) is no longer a firm, but a network of differentiated agents (see Powell, Koput and Smith-Doerr, 1996; Volume II, Chapter 3). In this case, the direction of causation is reversed: it is the structure of the network and the position of agents within it that fundamentally determines agents' access to relevant sources of scientific and technological knowledge and therefore innovative activities and performances (see also Kogut, Shan and Walker, 1992). There is little question that the ability of firms to access and make efficient use of such network of collaborative relations and of the underlying market for technology has become an important source of competitiveness for the firm.

Thus, in relation to the development of new ways of organizing R&D, the empirical studies around biotechnology have generated a series of debates related to both the type of knowledge involved and the relative advantages and disadvantages of the firm versus networks and innovation systems. Much of this research has implications for understanding the dynamics of the industrial structure, especially in pharmaceuticals, where these issues have been most studied. First of all, it is worth noting that collaboration does not simply involve the transfer of knowledge from a NBF lacking complementary assets to an established corporation that uses such knowledge to develop and market the final product. In pharmaceuticals, for example, collaboration takes place mainly in the pre-clinical stage rather than in the marketing stage and usually an established corporation strikes different agreements with different DBFs at the same time and within the same project. Even more important, the innovative process involves the effective integration of a wide range of pieces of knowledge and activities, which are not ordered in a linear way and that may not be easily separated and codified (Orsenigo, 1989). And substantial costs still remain in transferring knowledge across different organizations. Thus, one does not observe simply processes of specialisation and division of labour. Indeed, what we instead observe as a distinct feature is the hybridisation of different organizational forms and the coexistence of deepening division of innovative labour with concomitant processes of new forms of vertical and horizontal integration.

Clusters of Innovation

Clusters relate to the second part in Volume II, 'Geographical agglomeration'. Biotechnology, once again, represents an ideal case for studying and questioning how and why innovative activities tend to cluster in specific areas, as well as examining the characteristics of those areas. And again, results obtained in the case of biotechnology are having an impact on the more abstract literature that goes beyond the specifics of the industry.

In the broad literature on clusters, there seems to be a wide consensus that clustering in innovative activities cannot be simply explained by some sort of given and immobile 'endowments' that provide comparative advantage to any one area. Instead, powerful agglomeration forces must be at work in order to produce geographical concentration. In fact, literature to develop the geography of innovation has concentrated around explanations to questions which are essentially various reformulations of the fundamental sources of agglomeration externalities originally suggested by Marshall (see Henderson, 1986; Krugman, 1991, among others). This includes:

1. Economies of intra-industry specialisation: a localised industry can support a greater number of specialised local suppliers of industry-specific intermediate inputs and services, thus obtaining a greater variety at a lower cost.
2. Labour market economies: a localised industry attracts and creates a pool of workers with similar skills, which benefits both the workers and their employers.
3. Ease of communication among firms: information about new technologies, goods and processes seems to flow more easily among agents located within the same area, owing to social bonds that foster reciprocal trust and frequent face-to-face contacts. Therefore adoption, diffusion and innovation seem faster and more intense in geographical clusters than in scattered locations. That is, some 'knowledge spillovers' exist, which are geographically bounded.

Of course, within this literature, there are debates over the exact variables or 'forces' and their relative explanatory role.

This debate about different types of agglomeration externalities also overlaps with the debate within economics and other fields about 'knowledge spillovers'. The notion of knowledge spillovers (in its various formulations) has taken a central role not only in scientific analysis, but also in policy-making. Based on these arguments, strong emphasis has been attributed to the idea that strategic and public policy interventions should be directed primarily towards the attempt at facilitating such knowledge flows and spillovers, through, for example, offices for technology transfer, measures for strengthening university–industry relations, etc.

Yet, it has become increasingly acknowledged that the evidence supporting the role of knowledge spillovers is largely indirect and that it is quite difficult to clearly separate knowledge spillovers from other types of pecuniary externalities. More generally, it is difficult to separate Marshallian externalities and the more classic urbanisation externalities or even natural endowments (Glaeser *et al.*, 1992; Ellison and Glaeser, 1999). Similarly, there is conflicting evidence about the relative role of specialisation versus diversity in the economic structure of clusters, that is Marshallian versus Jacobian externalities.[3] Moreover, as forcefully argued by Breschi and Lissoni (2001), it has proven hard to show precisely how knowledge spillovers actually work and even whether legitimately they can be interpreted as spillovers. This has implications, for example, whether we emphasize 'spillover' mechanisms or the role of skilled individuals. For instance, diverse studies show that even the pool of knowledge that should constitute the very origin of knowledge spillovers seems to be embodied in specific people and/or in the pool of a specialised workforce, as argued, for example, by Almeida and Kogut (1999) and, with specific reference to biotechnology, by Zucker, Darby and Brewer (1997).

Another way of viewing clustering and regional agglomeration is by stressing the development and flows of knowledge. Hence, in some views, clusters are often associated with cooperation in innovative activities and interactive learning (Maskell, 2001; Cooke, 2002). According to this view, firms within innovative clusters learn through a variety of types of interactions, ranging from user–producer relationships, formal and informal collaborations, interfirm mobility of skilled workers and the spin-off of new firms from existing firms, universities and public research centres. Local firms are embedded in a thick network of knowledge sharing, supported by close social interactions and by (formal and informal) institutions that promote the development of trust among participants in the network.

As should be evident, to a considerable extent, this debate links with the issue of the new, network-like, forms of organization of innovative activities. Thus, for example, Annalee Saxenian (Saxenian, 1994) proposed the extremely influential argument that the superior performance of Silicon Valley was related to the particular form of organization of innovative activities that had been developing over time, based indeed on network-like structures of interactions among entrepreneurial agents. On the contrary, Route 128 fared worse as a consequence of the dominance of a more traditional organisational structure, based on large, vertically integrated firms. Against this background, one can understand why many of the conceptual building-blocks of the discussion about collaborative R&D and networks are so prominent in the debate about the geographical distribution (concentration) of economic activities in biotechnology.

Indeed, explanations of why innovative activities tend to concentrate in specific geographical areas, and the characteristics of such areas, rely, in many cases, on the identification of a few crucial ingredients in the case of biotechnology. These include: the strength of the scientific base, the propensity towards and the existence of institutions, policies and other infrastructures that support and promote entrepreneurship and the availability of a vibrant venture capital industry. There appears to be a dominant and robust result, namely that excellence in scientific research spanning a differentiated spectrum of areas as well as an integration along the horizontal and vertical dimensions of the innovative process are the crucial ingredients for the development of biotechnology. The role of the other factors appears to be ancillary or irrelevant, absent these capabilities. However, in most cases, it is stressed that there are processes of collective learning, exchange of information and tacit knowledge among agents, and establishment of a dense network of interactions at the core of clusters.

The differing debates about clusters and geographical agglomerations open up for many interesting questions, in the current debates and in the future. For example, biotechnology can be clearly interpreted as a case where knowledge within a biotech cluster does not appear to simply 'spill over'. Rather, access to such knowledge seems to require deep involvement in the research process and bench-level scientific collaboration, as well as the conscious investment of resources not simply to search for new knowledge, but to build the competencies to absorb the knowledge developed by others. This is linked to mechanisms for coordination and selection. In many cases, knowledge flows occur via (localised) mobility of researchers and of the workforce. These 'flows' are mediated by market transactions and other institutionalised or quasi-institutionalised mechanisms involving not simply mutual trust and face-to-face contacts, but also highly complex economic and social structures.

Indeed, in contrast to predications from the knowledge spill-over literature, biotech knowledge tends to remain sticky within the cluster for rather different reasons, and that may be interpreted as linked to attempts at privately appropriating knowledge and at restricting its circulation. Moreover, one should ask whether the relevance of 'soft' institutions might have been overestimated, given that knowledge flows in the biotechnology industry appear to be channelled significantly through market transactions and interorganizational rules. In this perspective, biotechnology would look much more similar to a market for technology than to the classical notion of the industrial district. In general, flows of knowledge occurring in a cluster are structured and mediated by a host of different mechanisms involving different flows of knowledge. If anything, as mentioned previously, successful clusters tend to be more open towards external interactions.

A further set of issues concerns the specific nature of the dynamic processes leading to clusterisation of innovative activities.[4] Indeed, very little is known about the genesis of clusters (Braunerhjelm and Feldman, forthcoming) – i.e. how clusters initially get formed and how they develop over time. The studies within biotechnology suggest that the basic ingredients which are usually associated with successful clusters were not necessarily simultaneously in place at the beginning – i.e. at the time of the genesis of the cluster. On the one hand, certainly, those locations where most of these conditions are (or were) available enjoy (or enjoyed) significant advantages. But on the other hand, their existence was neither a necessary nor a sufficient condition. Nor do the factors that promote the genesis of a cluster coincide with those that sustain its growth over time. Rather, in almost all cases, the process of construction of such ingredients would seem to be a crucial part of the story: in a sense, the process is a fundamental ingredient itself. As indicated, the current debates about clusters and geographical agglomeration open many fundamental questions in the economics of biotechnology, such as how these processes can be analysed and formalised.

Technology, Innovation, Institutions and their Co-evolution

Finally, the third and fourth parts of Volume II address 'Institutions supporting the biotechnology industry' and 'Intellectual property'. Within our co-evolutionary perspective, the development of institutional settings includes broader institutions in society which support the economy, as well as the specific institution of intellectual property rights. Institutions have played, and will continue to play, a crucial role in the evolution of biotechnology.

Institutions are defined in a broad sense, and refer not only to the 'rules of the game', but especially, how different elements of innovation systems are structured and affect the development of biotechnology. Research is obviously important, given the science-based nature of much industrial innovation. Research systems, in particular, universities, have not only provided the fundamental opportunities for innovation, but their structure and organisation have also shaped the capabilities and the incentives for engaging academic researchers in commercial activities. Thus some research has focused on the different structure of the American university system *vis-à-vis* the European system (notwithstanding enormous differences within European countries), and this is always indicated as one of the main motivations for the US supremacy.

Another area of debate about institutions refers to the structure of the financial system, especially as it concerns venture capital. Venture capital is commonly thought of as a key factor in sustaining the creation of DBFs in the USA and, conversely, in hindering their development elsewhere. One issue in this debate is whether one should emphasise only venture capital, or also take into account the many other sources of funding. For example, other sources of capital were available in Europe, the USA and other countries for prospective start-ups, especially but not exclusively through government programs. Thus scarcity of venture capital might have played a role less for purely financial reasons than for the contribution that these institutions provide in supporting the management of start-ups (Orsenigo, 1989). One could reverse the direction of causality, and thereby argue that the dearth of a sufficiently large number of good projects has slowed down the growth of venture capital in many countries.

A third important area for debate about institutions is public opinion and public policies. Public attitudes towards genetic engineering and the legislative frameworks regulating genetic experimentation are further elements of the institutional landscape influencing biotechnology. For example, despite early opposition, the attitudes of both the public and the regulators have always been much more permissive in the USA as compared to Europe. Public policy to support research and companies within biotech is also acknowledged to have played an important role, although this has varied considerably across countries and over time.

Comparative studies suggest that there are differing ways of organizing public policy, which can be analysed with reference, for example, to what type of government agencies are involved, the relative degree of associative governance as opposed to 'traditional' top-down policy-making, the diversity of programs, and so on. For example, the BioRegion program launched by the German federal government has attracted the interest of analysts and policy-makers in other countries. The program consisted of a competition between projects submitted by local clusters rather than by individual firms or research institutions. The outcomes of this program have so far been mixed. Yet, it has represented a significant innovation in European policy-making towards high-tech industries, and it probably has been quite important in the development of biotechnology, at least in areas like Munich and Heidelberg. However, as with the debates over regional agglomeration, the problem in evaluating public policy is often how to evaluate the relative importance of specific variables, such as strong science, the degree of regional interaction, openness towards the global system, etc., when comparing and contrasting examples. In what ways and why do different aspects of public policy contribute to the genesis and dynamics of both successful and less successful cases of biotechnology concentration?

Another way of conceptualizing institutions is in terms of how they structure the 'workings' of capitalism (Hall and Soskice, 2001). Hence, such a definition of institutions implies that the influence of institutions goes much beyond specific examples of institutions or elements of the innovation system, such as venture capital and the regulatory framework. Broader and deeper institutions, such as company law, the structure of the labour market for skilled workers and researchers, the forms of organisation of large corporations and the structure of the research systems, probably play a less direct and less purposive but more important role in affecting incentives and capabilities for innovation in an area like biotechnology.

Such institutional contexts are deeply ingrained in each country (and even each region), and much empirical research suggests that they are extremely resilient to abrupt changes. Thus past history, as crystallised in institutional settings, delimits what individuals and organisations (and governments) can do and even conceive as doable. Moreover, it is increasingly recognised that innovation emerges not simply from specific agents or particular institutions, but from the entire institutional architecture that impinges on the multifaceted activities that lead to innovation itself.

This, then, leads to debates about how to conceptualize the institutional architecture under-pinning innovation in biotechnology. For example, technological and industrial change are now frequently conceptualised as taking place and being shaped by systems of innovation, which involve directly or indirectly a large variety of actors and relations among them All these actors are, in many senses, different. They know different things, they have different incentives and motivations, they have different rules of action. The actors are linked together through a web of different relationships: they include almost pure market transactions, command and control, competition, collaboration and cooperation and all sorts of 'intermediate forms'. Such innova-tion systems have overlapped national, regional and sectoral dimensions (Edquist, 1977; Nelson, 1982; Edquist and McKelvey, 2000; Malerba, 2005); they are typically fraught with tensions and contradictions, as they have evolved over time in response to quite different and specific problems, interests and circumstances, not as a result of a 'grand', conscious design.

Very important research questions center on the issue of how and why these systems change over time. Such change results from different sources, where the overlap of sectoral systems of innovation in pharmaceuticals and biotechnology provides rich empirical material to test and develop concepts (McKelvey, Orsenigo and Pammolli 2005). Such systemic dynamics are spurred by the disequilibria and imbalances that connotate the system, and are driven by ex-ternal shocks, both 'small' ones and 'big' ones (like the emergence of a new technological paradigm). The process of change is driven also (and mainly) by the interaction of endogenous learning and selection processes. Agents learn how to improve their position by developing new techniques, products and marketing strategies. They improve also their ability to use such products and techniques. They learn how to compete *vis-à-vis* their old and new competitors. They adapt and sometimes try to change to new forms of regulation and forms of market or-ganization. Mechanisms of selection themselves change. Changes in regulation are just one obvious example.

Even more interesting from a research perspective, this implies that different selection mechanisms coexist, influence each other and sometimes mix together, and the principles of selection become themselves partly endogenous. In fact, they result from the interaction of dif-ferent mechanisms, from the purposive actions of agents who actively try to change the 'rules of the game' and from the disequilibria that at any point in time characterize the system.

The case of intellectual property rights is a good case in point. In biotechnology one observes, for example, changes in the intellectual property rights regime that have greatly influenced the subsequent evolution of the industry, but many of the major institutional changes were not made having in mind the specific case of biotech. Later on, conscious attempts have been made to further modify such regimes which have in turn resulted in other, unintended, consequences.

For example, it is well known that the establishment of clearly defined property rights indeed played an important role in making possible the explosion of new firm foundings in the USA, since the new firms, by definition, had few complementary assets that would have enabled them to appropriate returns from the new science in the absence of strong patent rights (Teece, 1986). In the early years of 'biotechnology' considerable confusion surrounded the conditions under which patents could be obtained. In the first place, research in genetic engineering was on the borderline between basic and applied science. Much of it was conducted in universities or otherwise publicly funded, and the degree to which it was appropriate to patent the results of such research became almost immediately the subject of bitter debate.[5] Similarly, a growing tension emerged between publishing research results versus patenting them. While the norms of the scientific community and the search for professional recognition had long stressed rapid publication, patent laws prohibited the granting of a patent to an already published discovery (Merton, 1973; Kenney, 1986). In the second place, the law surrounding the possibility of patenting life-formats and procedures relating to the modification of life forms was not defined. This issue involved a variety of problems (see OTA, 1984), but it essentially boiled down first to the question of whether living things could be patented at all, and second to the scope of the claims that could be granted to such a patent (Merges and Nelson, 1994; Mazzoleni and Nelson, 1998).

In fact, these trends were partly spurred by a growing concern about how to exploit more efficiently academic research and by the need to put some order in the system that governed the conditions at which universities could obtain patents, and therefore income, on the results of publicly funded research. The Bayh-Dole Act in 1980 sanctioned these attitudes by greatly facilitating university patenting and licensing. But as Mowery *et al.* (2001) have shown, the emergence of the 'industry–university complex' (Kenney, 1986) and of the entrepreneurial university predates Bayh-Dole and depends critically on the rise of the two main technological revolutions of the second half of the century, namely micro-electronics and, especially, biotechnology.

Parallel to Bayh-Dole, a series of judicial and Congress decisions further strengthened the appropriability regime of the emerging sectoral system. In 1980, the US Supreme Court ruled in favour of granting patent protection to living things (*Diamond* v. *Chakrabarty*), and in the same year the second reformulation of the Cohen and Boyer patent for the r-DNA process was approved. In subsequent years, a number of patents were granted establishing the right for very broad claims (Merges and Nelson, 1994). Finally, a one-year grace period was introduced for filing a patent after the publication of the invention.

These developments led to an increasing relevance of court decisions upon the fate of individual firms, and of the industry in general. Litigation appears to be a distinct feature of the new biotechnology sectoral system and IPR experts have become crucial components of firms' human resources and competencies.

Increasingly, in the USA doubts are voiced by economists, lawyers and industry analysts that the diffusion of an excessively permissive attitude towards the granting of broad claims on patents might actually slow down the process of diffusion and circulation of knowledge and hence the future rate of technological advance. Similarly, fragmentation of broad property rights among a large number of patent-holders could create an 'anticommons problem' whereby researchers might be impeded in their projects by having to gain the rights to use all the needed patents, and blockaded by strategic considerations of one particular agent (see Heller and Eisenberg, 1998; Volume II, Chapter 18). It is also important to notice that the rationale for stronger protection to intellectual property in biomedical research is not based on the traditional argument that the concession of broad property rights is an incentive to the production of knowledge. Rather, the argument is based on the assumption that property rights would favour the creation of markets for technology and hence a faster and more ordered diffusion and use of knowledge (Merges and Nelson, 1994).

Understanding these processes of co-evolution between institutions, technology and industrial change presents an important challenge to the economics of biotechnology and to economic analysis more generally.

Structure and Content of the Reference Collection

This section describes the structure and content of the individual contributions presented in this two-volume reference collection. They have been organised along the conceptual framework of the debate, as described above. This section provides a brief description of each article or book chapter, grouped within the eight themes we have outlined.

Volume I, Part I: Overview

This part includes contributions meant to provide a general overview of the development of biotechnology in various fields. The first document is the executive summary of the report prepared by the US Office of Technology Assessment (Volume I, Chapter 1) on the early developments and future prospects of biotechnology back in 1984. This report has been extremely important, in that it marks the first official recognition of biotechnology as a set of technologies and as a nascent industry, and it provides a view that has deeply influenced analysts, businesspeople and policy-makers. As such, this represents an important document, useful for understanding and putting in perspective the ways in which biotechnology has been conceptualised and changed over time.

The paper by Henderson, Orsenigo and Pisano (Volume I, Chapter 2) provides an overview of the evolution of the pharamaceutical industry in the era of molecular biology. The authors emphasise how biotechnology actually can be understood as a chapter in a longer process of transformation of the relevant knowledge base of pharmaceuticals, from chemistry and random screening to molecular biology and the attempt to scientifically design new drugs. The argument goes on to discuss how these changes in the underlying knowledge base have triggered profound transformations in the industry – simultaneously, through the emergence of the DBFs, the processes of adjustment within incumbent companies and the development of complex forms of interactions between the firms, universities and other agents. Finally, the paper discusses the broader institutional factors that might have led – jointly with firm-specific characteristics – to different patterns of development in the USA, Europe and Japan.

The article by Parayil (Volume I, Chapter 3) discusses the impact and the patterns of diffusion of biotechnology in agriculture. In a provocative and controversial interpretation, the author makes the point that the 'Gene Revolution' marks a sharp discontinuity with the 'Green Revolution', mainly because the latter is shaped, in his view, by different social, political and economic contexts, that is mainly by the logics of the large agro-biotech corporations.

Finally, in this part, the article by Kettler and Marjanovic (Volume I, Chapter 4) opens a window on a further (often neglected by the scholarly literature) issue, namely the potential effect of biotechnology on developing countries. In particular, the authors look at how biotechnology companies can contribute to alleviate the health needs of people in the developing world and conclude that, despite some encouraging progress, major global policy interventions are needed for scientific and technological change to be more directly relevant to the improvement of health conditions in poor countries.

Volume I, Part II: Science and Innovation

Part II of Volume I is devoted to the tangled relationships between science and innovation in biotechnology. Arora and Gambardella (Volume I, Chapter 5) develop a model that distinguishes between the ability to evaluate and the ability to utilise scientific and technological information. The former allows firms to evaluate better ex-ante the value of an innovation project before conducting applied R&D. The latter allows firms to extract higher value from a given technology through better experimentation and development capabilities. The model shows that greater capabilities to utilize implies that firms will carry out projects which would seem a priori less promising: through better experimentation and development capabilities, a firm can derive an acceptable pay-off even from projects with relatively poor ex-ante expectations. Conversely, better predictive capabilities lead firms to concentrate on better projects, discarding with greater confidence the less appealing ones. The relationships between these variables and some important predictions of the model are then explored econometrically in a sample of large US chemical and pharmaceutical companies active in the biotechnology sector in the 1980s.

In a similar vein, Pisano (Volume I, Chapter 6) presents a framework which links approaches to experimentation and the structure of underlying knowledge. The framework suggests that where underlying knowledge is sufficiently strong, effective learning may take place outside the final use environment in laboratories (i.e. 'learning-before-doing'). This proposition is tested by comparing how an emphasis on laboratory experimentation impacts process development lead times in two different technological environments: traditional chemical-based pharmaceuticals, and new biotechnology-based pharmaceuticals. Results show that learning-by-doing is essential for efficient development in an environment like biotechnology, where underlying theoretical and practical knowledge, and thus the possibility of learning-*before*-doing, is relatively thin. In contrast, the need for learning-by-doing is far lower in environments like chemical synthesis, where underlying theoretical and practical knowledge is deep enough to enable the design of laboratory experiments that effectively model future production experience.

Nightingale (Volume I, Chapter 7) partially adds to these insights, but at the same time introduces a different perspective and implications, as compared to the two previous papers. He claims that science does not produce technology directly, but instead gradually reduces the costs of solving complex technical problems. He notes that traditional craft-based, sequential experimentation in chemistry and biology has been complemented by, first, the automated, mass-production analysis of populations, and second by, 'in silico' experimentation using simulations and databases. Nightingale then argues that the use of genetic, database, high throughput screening and bioinformatics technologies has allowed pharmaceutical firms to exploit economies of scale in experimentation, concluding that the organizational and technological features of modern drug discovery may favour large firms with large R&D departments.

Gittelman and Kogut (Volume I, Chapter 8) challenge the conventional view that science leads to more technology. By analysing the publications and patents of 116 biotechnology firms during the period 1988–95, they conclude scientific ideas are not simply inputs into inventions: important scientific ideas and influential patents follow different and conflicting selection logics. In particular, important scientific papers are negatively associated to high-impact innovations. These results point to the key relevance of specific individuals or organisations (like the DBFs) that perform the function of effectively integrating science with innovation.

McKelvey (Volume I, Chapter 9) analyses the different selection conditions for science and technology, and stresses the difficulties involved in turning 'scientific results' into workable commercial innovations. The study is based on a detailed analysis of different firms and actors involved in the development of genetic engineering techniques to produce human growth hormone, one of the first uses of biotech in the pharmaceutical industry. The study includes the major scientific developments at UCSF and elsewhere, the commercial development in Genentech and Kabi (later part of Pharmacia concern) and the role of medical science in greatly expanding the market. The analysis shows that the search for innovations involves competing trajectories and choices, but the outcome depends upon a combination of incentives, skills and available knowledge to solve different types of problems within universities and firms.

Volume I, Part III: New Specialised Biotechnology Firms

Part III of Volume I focuses on the nature, roles and characteristics of the DBFs – i.e. one of the most spectacular features of the biotechnology industry. The article by Kenney (Volume I, Chapter 10) provides an interpretation of the first ten years in the history of the US biotechnology industry in a Schumpeterian theoretical perspective. It is argued that the work of the early Schumpeter (*A Theory of Economic Development*, 1912) more accurately describes the US biotechnology industry in 1985 than his later work (*Capitalism, Socialism and Democracy*, 1942). Kenney discusses the role of large established firms and of small entrepreneurial firms, highlighting the tensions inherent in the 'cooperative' arrangements between these two types of business enterprises. The role of the small firms in reducing the innovations to practice and in their ability to continue to grow demonstrates, in Kenney's view, that the independent entrepreneur recognized by the early Schumpeter has played a key role in the biotechnology industry.

Audretsch (Volume I, Chapter 11) emphasises that innovation in biotechnology is predominantly produced by new start-ups whose core mission consists in moving basic research to commercialisation through technological innovation. Furthermore, he highlights the complementary nature of biotechnology firms and established firms, particularly in the pharmaceutical industry. Audretsch also insists upon the high degree of geographical concentration of the biotechnology industry, suggesting that this phenomenon reflects the concentration of the most important input in this industry – i.e. specialised knowledge – in few locations.

Zucker, Darby and Brewer (Volume I, Chapter 12) provide strong empirical evidence for these assertions. They examine the relationship between the intellectual capital of scientists making frontier discoveries, the presence of great university bioscience programs, the presence of venture capital firms, other economic variables and the founding of US biotechnology enterprises during 1976–89. Their findings show that the timing and location of the birth of biotechnology enterprises is determined primarily by intellectual capital measures. In other words, timing and the location of new biotechnology firms is primarily explained by the presence at a particular time and place of scientists who are actively contributing to the basic science.

Mangematin, Lemarié, Boissin, Catherine, Corolleur, Coronini and Trommetter (Volume I, Chapter 13) examine the patterns of development of biotechnology firms in France. They remark that the conventional model of the development of high-tech SMEs, while appealing, is simply one of the possible models of biotechnology development. According to this classical model, entrepreneurs rely on growth forecasts to persuade capital investors (business angels and venture capitalists) to invest in a radical innovation project. Firms aim for a world market to industrialise their innovation, and initial public offering (IPO) enables investors to make profits that offset risky initial investment. However, they show that some firms are not designed to experience exponential growth, and choose to target instead local markets. Moreover, not all firms have the ambition of being listed on the stock exchange. The authors define the development trajectories of each of these models. They share common characteristics: science base, leading role of SMEs and resource acquisitions through alliances. However, when an SME focuses on a market niche and conducts small research programmes, it will experience steady growth if it is able to reach financial equilibrium fairly quickly. By contrast, when SMEs embark on large research programmes in partnership or competition with major companies in the sector, development is possible only with outside capital and the participation of venture capital firms. The founding members' experience is then a key factor if the SME is to enter into certain partnerships. The authors conclude that certain business models that are less risky, but probably make less use of the founders' knowledge, can appear as gateways in the establishment of a final business model.

Volume I, Part IV: Reaction and Adaptation of Large Incumbent Companies

Part IV of Volume I examines how large incumbent companies have reacted to and eventually adopted biotechnology. Galambos and Sturchio (Volume I, Chapter 14) consider the pharmaceutical industry. They contextualise the biotechnology revolution within broader and longer-term processes of technological discontinuities along the history of the industry and show how large firms adopted one of two strategic pathways through the transition. The first (the more common one) was to start developing highly specific capabilities and the attempt to generalise them across a range of therapeutic categories. The second strategy consisted of attempting to acquire and build upon general capabilities very early, also by establishing various forms of relationships with the biotech firms. In both cases, companies experienced substantial transition costs and adjustment was slow. However, the authors conclude that by the mid-1990s a number of the leading pharmaceutical firms had managed to make their way through the transition, establishing themselves once again as dominant players in the process of innovation in pharmaceuticals.

Chataway, Tait and Wield (Volume I, Chapter 15) examine the strategies of large firms in the agro-food segment. They claim that, in many instances, managers did not sufficiently recognise the importance of the complex interactions between public policies and public opinion, thus failing to incorporate public policy into strategic R&D decision-making. This blind spot compounded initial difficulties in bringing products to market and had a significant impact on the rate and direction of innovation, including contributions to the demise of the idea of an integration of agro and health sectors based on the life sciences.

Cockburn and Henderson (Volume I, Chapter 16) look in more detail into the reorganisation of the process of drug discovery within large corporations. In particular, they examine the interface between for-profit and publicly funded research. They show that firms access basic research through in-house R&D investment in absorptive capacities and the introduction of 'pro-publication' internal incentives. Some firms also maintain extensive connections to the wider scientific community, which is measured by data on co-authorship of scientific papers between corporate scientists and publicly funded researchers. They find that connectedness to public research is significantly correlated with firms' internal organisation and performance in drug discovery.

Zucker and Darby (Volume I, Chapter 17) develop these results by showing that the transition to 'biotechnology' was championed in large incumbent firms by sophisticated senior management, and it was achieved mainly through hiring new scientists and the adoption of pro-publication incentive schemes. Collaborations with academics are found to be ubiquitous and often non-public. Collaborations with new biotechnology firms are used primarily to substitute for developing internal expertise judged of marginal value.

Ramani (Volume I, Chapter 18) tackles this issue from the perspective of developing countries, specifically India. She looks at the nature and impact of the R&D strategies of Indian firms active in the biopharmaceutical sector, and shows that market performance is positively correlated with the knowledge base of the firm as embodied in its qualified labour outside of the R&D department. Interestingly, internal R&D and foreign collaborations appear to be strategic substitutes, while patents and publications are strategic complements. Finally, these firms are likely to be younger and implementing more aggressive learning strategies

Volume II, Part I: Division of Labour in Innovative Activities and Networks of Innovators

Part I of Volume II collects contributions to the analysis of the processes of division of innovative labour, collaborative relations among firms and the resulting networks of innovators. The paper by Pisano (Volume II, Chapter 1) analyses and compares the various governance structures used by firms involved in biotechnology in the early stages (the first 15 years) of the evolution of the biotechnology industry. Pisano finds a clear trend towards vertical integration: as firms have gone further along the product development trajectory, much more of the know-how generated is idiosyncratic to the product and firm-specific and therefore difficult to transfer. The transaction costs of such arrangements provide a rationale for many of the more mature NBFs to build 'downstream' competencies needed to commercialize the results of their R&D and for established enterprises to conduct biotechnology R&D on their own. However, Pisano concludes that is unlikely that vertical integration will completely replace all alternative organizational forms very soon, given limits on the rate at which NBFs and established enterprises can expand their boundaries. Similarly, rapid internalization of biotechnology R&D through acquisition of NBFs is unlikely to become a model for organizational change, since the strategic risks for the downstream sponsor may be much greater than the direct relationship. A second reason why collaborative or market forms of organization will survive in biotechnology is, according to Pisano, that not all segments will be characterized by high transaction costs. Vertical integration is more likely to predominate where the innovation chain is characterized by uncertain property rights, transaction-specific assets and complex technology transfer. Where these conditions are not present, such as in research products, functionally specialized firms, dealing with each other through various contractual relations, should thrive.

The article by Arora and Gambardella (Volume II, Chapter 2) takes a somewhat different view. They test the hypothesis that the strategies of external linkage of the large firms with other parties (universities and small/medium sized research-intensive firms) are complementary to one another. They show also that if any two strategies are complementary (i.e. undertaking more of one strategy raises the marginal value of the other), then they are positively correlated. Using data for a sample of large US, European and Japanese chemical and pharmaceutical producers, it is found that the strategies are positively correlated even after controlling for firm-specific characteristics.

Powell, Koput and Smith-Doerr (Volume II, Chapter 3) introduce forcefully the notion that networks, rather than individual firms, are to be considered as the relevant locus of innovation when the knowledge base of an industry is both complex and expanding and the sources of expertise are widely dispersed. The authors develop a network approach to organizational learning and derive firm-level, longitudinal hypotheses that link R&D alliances, experience with managing interfirm relationships, network position, rates of growth and portfolios of collaborative activities. The hypotheses are then tested on a sample of dedicated biotechnology firms in the years 1990–94.

In a similar vein, Liebeskind, Oliver, Zucker and Brewer (Volume II, Chapter 4) examine how two highly successful, new biotechnology firms source their most critical input – scientific knowledge. They find that firm scientists enter into large numbers of collaborative research efforts with scientists at other organizations, especially universities. Formal market contracts are rarely used to govern these exchanges of scientific knowledge. The authors suggest that the use of boundary-spanning social networks by the two firms increases both their learning and their flexibility in ways that would not be possible within a self-contained hierarchical organization.

Walker, Kogut and Shan (Volume II, Chapter 5) proceed to provide a strong theoretical foundation to the concept of a network of innovators. They claim that theories of social capital and structural holes have fundamentally different implications for network formation. Their paper investigates these theories by examining empirically the formation of the interorganizational network among biotechnology firms. It is proposed that network structure determines the frequency with which an NBF establishes new relationships. A critical test of the theories is whether new relationships reproduce or alter the inherited network structure. Strong support is found for the power of social capital in reproducing the network over time.

Powell, White, Koput and Owen-Smith (Volume II, Chapter 6) explore the logics which dynamically drive the development of interorganizational collaboration in the field of biotechnology. Specifically, they identify four alternative logics: attachment and accumulative advantage, homophily, follow-the-trend and multiconnectivity. The authors map the network dynamics of the field over the period 1988–99. Using multiple novel methods, they demonstrate how a preference for diversity shapes network evolution. Collaborative strategies pursued by early commercial entrants are supplanted by activities influenced more by universities, research institutes, venture capital and small firms. As organizations increase both the number of activities on which they collaborate and the diversity of organizations with which they are linked, cohesive sub-networks form that are characterized by multiple, independent pathways. These structural components, in turn, condition the choices and opportunities available to members of a field, thereby reinforcing an attachment logic based on connection to partners that are diversely and differently linked.

Finally, in this part, Orsenigo, Pammolli and Riccaboni (Volume II, Chapter 7) introduce a somewhat different perspective, investigating how underlying relevant technological conditions induce distinguishable patterns of change in industry structure and evolution. A mapping is detected between the specific nature and growth of the relevant knowledge base at the micro level and patterns of structural evolution at the macro level of the industry network. Graph-theoretic techniques are used to map major technological discontinuities on changes observed at the level of dominant organization forms.

Volume II, Part II: Geographical Agglomeration

This part includes papers which describe and explain the patterns of geographical agglomeration in the biotechnology industry as well as the characteristics of innovative clusters. The seminal paper by Audretsch and Stephan (Volume II, Chapter 8) explores why geography matters more in certain economic relationships than in others by focusing on the locational incidence of contacts between firms in the biotechnology industry and university-based scientists affiliated with these firms. The authors suggest, in particular, that the specific role played by the scientist shapes the importance of geographic proximity in the firm–scientist link.

Feldman (Volume II, Chapter 9) poses the question about the relationships between academic research, new biotechnology forms and regional development in a policy-oriented framework. The article explores the policy issues related to the commercialization of biotechnology, its role as an engine of economic development and the appropriate public policy responses. Feldman argues that standard locational models used in policy-making are based on the premise of the footloose actor surveying the landscape for the optimal location, and she claims that policy based on a model of endogenous growth would be more appropriate. Under this model, state and local government policy has an important role in creating an environment that will be conducive to creating and retaining science-based firms. According to Feldman, policy should be predicated on an understanding of the area's scientific resources and industrial mix, reflect the region's emerging expertise and be patient. Developing, or fine-tuning, the role of state government in growing a technology-intensive industry requires an understanding of how firms develop. Different technology-intensive industries have unique patterns of development; however, there are some underlying similarities.

Most notably, industries appear to cluster geographically in a few sites in the early stages of development. As the industry matures, the typical pattern is that one or two sites will become dominant. The economic development question is how may policy best anchor an industry in a region. Communities differ widely, however, in their apparent ability to benefit from the science-driven model of economic development. The question of how and why some regions capture the benefits of new technology and thrive while other regions do not is a question of great practical importance. Moreover, there appear to be important distinctions between a region's ability to generate commercial start-ups and to grow them into larger, more successful companies. In addition, there is a distinction between growing a company and growing an industry. While the success, or failure, of an individual company is easier to track, the best result for economic development would be a well-developed local industry.

Stuart and Sorensen (Volume II, Chapter 10) develop an explanation for firm co-location in high-technology industries that draws upon a relational account of new venture creation. They argue that industries cluster because entrepreneurs find it difficult to leverage the social ties necessary to mobilize essential resources when they reside far from those resources. Therefore opportunities for high-tech entrepreneurship mirror the distribution of critical resources. The same factors that enable high-tech entrepreneurship, however, do not necessarily promote firm performance. In the empirical analyses, Stuart and Sorensen investigate the effects of geographic proximity to established biotechnology firms, sources of biotechnology expertise (highly skilled labor) and venture capitalists on the location-specific founding rates and performance of biotechnology firms. The article finds that the local conditions that promote new venture creation differ from those that maximize the performance of recently established companies.

The article by Cooke (Volume II, Chapter 11) focuses on the concept of interactive innovation and Regional Innovative Systems. It starts by operationalizing regional innovation systems in the context of multi-level governance. It shows how regional and external innovation interaction among firms and other innovation organizations is important for regional innovation potential. Cooke argues, in particular, that the ability to access and use funding for innovation support for regional firms and organizations is crucial for regional innovation promotion. Conversely, equity investment funding is found to be more important than public funding, which tends to be cautious and otherwise risk avoiding. The key message is that regional systems of innovation are broader than single sectors or clusters, but some of these will be strategically privileged recipients of policy support because of their growth performance or potential, rather than, as in the past, their uncompetitiveness. The paper devotes space to exploring biotechnology clustering from a regional innovation systems viewpoint, as an instance of rather strong, sectoral innovation systems capabilities, though integrated also to global knowledge supply and markets. Illustration is provided of the way such sectoral innovation systems work at local regional level by reference to cases from Cambridge, Massachusetts, and Cambridge, England.

Niosi and Bas (Volume II, Chapter 12) focus on the Canadian experience in order to examine how Canada hosts two major diversified biotechnology regional systems of innovation in its two largest cities. Similar in many respects – for example, research universities and hospitals, science-based firms, venture capital, etc. – these two regional systems each includes two somewhat separate subsystems. One subsystem is focused on science-based firms, universities and venture capital around 'biotech', whereas the other focuses on pharmaceutical companies and contract research organizations around 'pharma'. The authors analyse how these networks differ, both between and across the subsystems in Toronto and Montreal. The authors review the main theories on regional innovation systems and innovative clusters and proceed to analyse these two regional systems before concluding on the usefulness of combining several theories to study biotechnology regional innovation systems.

Volume II, Part III: Institutions Supporting the Biotechnology Industry

Part III of Volume II focuses on the role and nature of institutions that bear relevance to the development of biotechnology. Prevezer (Volume II, Chapter 13) explores the ingredients that stimulated the development of the biotechnology industry in the USA and contrasts conditions with those in Europe. The main findings are that: (1) the funding of the medical science research base has been substantially more generous in the US than Europe. It is the funding of the science base rather than of the biotechnology industry directly that has provided the foundations for start-ups to be created out of the science base. (2) It has been easier for US academics to found start-ups, close to their research establishment, and to retain their academic posts and status, as well as being involved in a commercial enterprise. (3) Start-ups have been concentrated in the therapeutics and agricultural fields, with strong scientific research inputs into their commercialization, in contrast to other sectors where downstream processing innovations have been more important, and which have been undertaken in-house by the large incumbent companies. (4) Financing and managerial conditions have been significantly easier in the US for start-ups, in terms of access to venture capital specializing in high technology, ability to use the stock market to raise capital and access to people able to forge links between scientists and entrepreneurs, and introducing managerial expertise into new companies. (5) There has been a greater facility in the US for alliances to be formed between incumbent companies and indigenous US start-ups; European start-ups have not found similar backing from European incumbent companies.

Casper and Kettler (Volume II, Chapter 14) move the analysis forward, adopting and developing the 'national institutional framework' approach. They begin from the following puzzle: why, given what institutional scholars have described as an inhospitable institutional climate for entrepreneurial business, has the German biotechnology industry suddenly taken off, while in the UK, where according to the same thinkers the 'correct' institutional architecture exists, the industry has shown signs of stagnation? In seeking to explain these seemingly contradictory trends, a firm-centred and dynamic approach is taken to conceptualizing the link between institutional arrangements and the competencies developed by firms. This dynamic view recognizes the ability of firms to work with institutional frameworks – often with help from public policies – to create new business strategies. Such processes are associated with the 'hybridization' of business strategies at the micro level, combined with the generation of new constellations of particular institutional frameworks within what are seen as relatively stable national models.

The paper by Lehrer and Akasawa (Volume II, Chapter 15) shifts attention to the role and nature of public policies in two extremely interesting cases, namely in Japan and Germany. Moreover, the paper provides a suggestive case study about the co-evolution of technology and institutions. Partially moving within the same approach used in the preceding paper by Casper and Kettler, the authors emphasise that public policies to promote biotechnology are motors of broader R&D reforms in these two countries. The idiosyncracies of biotechnology commercialisation could not be readily accommodated by these two countries' traditional technology policies, thus prompting reforms in institutions governing the commercialisation of basic scientific research. The paper also draws attention to variables different from those that have been conventionally used for explaining the development of biotechnology. The authors emphasise the importance of the level of public governance within national research institutions, the autonomy of the university/research sector and the historical role of the public sector generally.

Owen-Smith, Riccaboni, Pammolli and Powell (Volume II, Chapter 16) focus attention on the structure and institutional organization of upstream life-science research across the USA and Europe. Network visualization methods and correspondence analyses are used to demonstrate that innovative research in biomedicine has its origins in regional clusters in the USA and in European nations. But the scientific and organizational composition of these regions varies in consequential ways. In the USA, public research organizations and small firms conduct R&D across multiple therapeutic areas and stages of the development process. Ties within and across these regions link small firms and diverse public institutions, contributing to the development of a robust national network. In contrast, the European story is one of regional specialization with a less diverse group of public research organizations working in a smaller number of therapeutic areas. European institutes develop local connections to small firms working on similar scientific problems, while cross-national linkages of European regional clusters typically involve large pharmaceutical corporations. The authors show that the roles of large and small firms differ in the US and Europe, arguing that the greater heterogeneity of the US system is based on much closer integration of basic science and clinical development (Gittelman, 1999).

The paper by Baum and Silverman (Volume II, Chapter 17) provides finally in this part a look at venture capital – a financial institution that has been almost invariably indicated as one of the key elements of the American model of development of biotechnology. The authors emphasise that venture capital firms (VCs) are perhaps the dominant source of selection shaping the environment within which new ventures evolve. VCs affect selection both by acting as a 'scout' able to identify future potential and as a 'coach' that can help realize it. Baum and Silverman claim that despite the large literature on the role of VCs in encouraging start-ups, it is generally taken for granted that VCs are expert scouts and coaches, and so the ways in which VCs actually enhance start-up performance are not well understood. In this study, we examine whether VCs emphasize picking winners or building them by comparing the effects of start-ups' alliance, intellectual and human capital characteristics on VCs' decisions to finance them with the effects of the same characteristics on future start-up performance. The findings point to a joint logic that combines the roles: VCs finance start-ups that have strong technology, but are at risk of failure in the short run, and so in need of management expertise. The results of the paper support the belief in VC expertise, but only up to a point. VCs also appear to make a common attribution error in overemphasizing start-up companies' human capital when making their investment decisions.

Volume II, Part IV: Intellectual Property

This final part concentrates on one specific, but extremely important, institutional aspect that has shaped the evolution of biotechnology, namely intellectual property rights. This debate has also important broader implications for the future development of science, technology and innovation even beyond the specific instance of the life sciences.

In this extremely influential and controversial paper, Heller and Eisenberg (Volume II, Chapter 18) cast serious doubts about the wisdom of establishing a tight appropriability regime in biomedical research. In standard economics, the 'tragedy of the commons' metaphor helps explain why people overuse shared resources. However, the recent proliferation of intellectual property rights in biomedical research suggests a different tragedy, an 'anticommons' in which people underuse scarce resources because too many owners can block one another. In the authors' view, privatization of biomedical research must be more carefully deployed to sustain both upstream research and downstream product development. Otherwise, more intellectual property rights may lead paradoxically to fewer useful products for improving human health.

In the same vein, Mazzoleni and Nelson (Volume II, Chapter 19) challenge the conventional wisdom that strong property rights are unequivocally conducive to innovation and economic progress. This paper reviews the main theories on the benefits of patents, stressing also that they imply significant economic costs. Moreover, the authors clarify the point that arguments in favour of strong and broad patents are not based on the conventional arguments that patents induce innovation, but on a rather different one: that is, that patents favour the commercialisation and diffusion of inventions. The authors conclude that there is reason for concern that the present movement towards stronger patent protection may hinder rather than stimulate technological and economic progress.

The paper by Walsh, Arora and Cohen (Volume II, Chapter 20) is one of the few empirical analyses on the effects of the changing patenting regime in biomedical research. Using interviews and archival data, the authors find that there has in fact been an increase in patents on the inputs to drug discovery ('research tools'). However, they also find that drug discovery and university research have not been substantially impeded by these changes. Restrictions on the use of patented genetic diagnostics are an important exception, where evidence is observed of patents interfering with university research. There is also some evidence of delays associated with negotiating access to patented research tools, and there are areas in which patents over targets limit access and where access to foundational discoveries can be restricted. There are also cases in which research is redirected to areas with more intellectual property (IP) freedom. The authors advance the view that we do not observe as much breakdown or even restricted access to research tools as might be expected because firms and universities have been able to develop 'working solutions' that allow their research to proceed. The paper ends with a discussion of the potential social welfare effects of these changes in the industry and the adoption of working solutions for dealing with a complex patent landscape.

Conclusion

We hope that this reference collection provides the interested reader with insights into a variety of debates and direct access to high-quality material which can stimulate thought and discussion over a wide range of important issues, be they theoretical, empirical or, in a broader perspective, 'political'. We have refrained from taking an obviously idiosyncratic perspective in interpreting this subject. Hence, rather than trying to spell out and impose our own views on these areas, we have attempted to keep as much an 'objective' and open approach as possible in our role as editors, by presenting different ideas, approaches and methodologies. Clearly, our own ideas cannot be entirely repressed or hidden, and they show up both in this Introduction and in the selection of articles presented.

As we have stated, the purpose of this reference collection is to provide the reader with an accessible and structured understanding of the main issues which have characterized the economic debates about biotech. This implies that we, as editors, have had to make difficult choices in relation to defining how and why to study biotechnology. Many of these difficulties are related to the evolving, changing nature of relevant knowledge and industrial activities within biotechnology, and given that these will continue in the foreseeable future, researchers will open up many avenues of future research. We wish now to point to five areas which may be seen as difficulties, or as opportunities.

First, 'biotechnology' is itself an ill-defined and plastic concept, which stretches and shrinks to meet the purposes to which it is applied, and which has been constantly changing over time. In particular, the concept 'biotech' may be delineated by the type of firm and the type of sector, and/or by scientific and technology knowledge. Hence, in one report, biotech may be defined only as encompassing dedicated biotech firms (DBFs), many of which are small and young, thereby limiting the concept to a particular organizational form. In another report, biotech may refer to the pharmaceutical sector – whether only to the DBFs or also the large pharmaceutical firms – thereby limiting the concept to a particular industrial sector. In yet another report, biotech may refer to the whole body of relevant science and technology, such as the knowledge, equipment and techniques that are used and which affect many sectors and many different types of firms, thereby widening the concept to a broad range of possible knowledge and industrial applications. This plasticity of definition implies that there is a continuing need to redefine and to refine the definitions of biotechnology – precisely, in order to understand their economic dimensions.

Second, the economic analysis of biotechnology is characterised by some intriguing and curious biases and paradoxes, which arise partly as a consequence of the vagueness and changing nature of the definition (see McKelvey, Rickne and Laage-Hellman, 2005, pp. 3–5). In particular, most of the available literature is disproportionately focused on empirical material describing the American case, describing dedicated biotechnology firms (DBFs) and describing also the pharmaceuticals applications of biotechnology. In some senses, this bias should not perhaps come as too big a surprise. After all, modern biotechnology has seemed for several decades to be primarily a US phenomenon, with the rest of the world lagging behind. Similarly, the proliferation of specialised research companies, which are usually university spin-offs, is certainly one of the most prominent characteristics of the development of biotechnology, particularly in the USA. Finally, pharmaceuticals have been by far the most dynamic field of application of the new techniques and scientific advances. Still, this bias implies that other business and economic aspects of biotechnology are missing from the debates.

Indeed, biotechnology is evolving in many other countries and types of firms as well as in many other industrial and service sectors. Agricultural and food applications are a crucially important case in point. Biotechnology keeps changing, due to discoveries, applications and further developments within a large number and variety of different knowledge fields. Yet, relatively little is known about these other developments, thereby pointing as well to the usefulness of research in developing theoretical concepts in connection with empirical studies of these other areas.

Third, the growth of modern biotechnology is only partially the result of interactions between scientific and technical progress in the life sciences, with attempts to apply and use such knowledge for economic and commercial purposes. Its growth is also the result of, and profoundly influenced by, broader social and institutional processes. Thus the economic analysis of biotechnology implies, almost inevitably, that one is forced to take a rather broad and perhaps unconventional approach to economic analysis. In particular, it becomes necessary to take a multidisciplinary stance, considering and indeed emphasising the inherent connections and overlapping with other social sciences, primarily (but not only) with management and sociology.

Fourth, the economic analysis of biotechnology has to rely, to a considerable extent, on an explicitly dynamic and historical perspective. One reason for this is the combination of the above two characteristics of the constantly changing nature of biotech as well as its embeddedness in transforming institutional and social forces. Hence, on the one hand, biotechnology can be analysed through abstract and theoretical concepts. On the other hand, the business and economic debates about biotechnology have been significantly changing over time, with the emergence of new issues, techniques of analysis and interpretations.

Interestingly, the dynamics and evolution of networks appear to be strongly influenced by logics that do not correspond to economic variables narrowly defined, but involve fundamental social processes. If anything, one of the more interesting and recent developments in the fields of the economics of innovation studies and industrial dynamics is the increasing overlap and exchange between economic and sociological concepts and methods. Network analysis, for example, is not only rapidly becoming a standard tool for economists, but also a general language for dealing with interactions among agents and, in this sense, having a similar (prospective) status as game theory. This is, in our view, perhaps one of the most fruitful contributions of the economics of biotechnology to economic analysis more generally.

In summary, the economics of biotechnology is an interesting empirical area of study, because it can raise rich, complex and intrinsically difficult questions and present the opportunity to develop methodologies. As such, the analysis of the emergence and development of biotechnology poses, in itself, some fundamental challenges to analysts. In fact, it could be argued that studies of biotechnology occasionally pioneered (and in any case, contributed heavily to) the development of new concepts and methodologies that are now becoming state-of-the-art in the analysis of economics of innovation and industrial dynamics. Moreover, biotechnology poses difficult and often dramatic policy and moral dilemmas that are likely to emerge in different fashion time and again in the future. This phenomenon of the emergence of new dilemmas is something with which democratic societies are still grappling.

So, obviously, given the wealth of published material available, it is clear that many other fundamental issues and good papers have by necessity been left out of this particular reference collection. The reader, however, should by now have sufficient insights into significant issues and debates, in order to search for additional material. There are also missing subjects which have been touched upon here but not been satisfactorily covered, because the available material was not rich and developed enough to provide a coherent and articulated presentation of the relevant issues. One important issue which was only marginally touched upon despite its immense relevance relates to the development and use of biotechnology in developing countries. Another important issue concerns the use of data and the technical (analytical) treatment of data in order to understand the impact and problems raised by the new emerging technological trajectories in biotechnology.

In the future, the ongoing evolution of biotechnology will certainly generate new questions and approaches that are not even conceivable at the present moment. One of the messages of this reference collection is that this future evolution of the area will open up interesting new research areas.

Notes

1. The reason for this exception is that the OTA report of 1984 has been extremely influential in setting the scene for a large fraction of the early research, by providing definitions and by highlighting some of the most prominent issues characterizing biotechnology as an industrial activity.
2. The first biotechnology product, human insulin, was approved in 1982 and, between 1982 and 1992, 16 biotechnology drugs were approved for the US market. As is the case for small molecular weight drugs, the distribution of sales of biotechnology products is highly skewed. Three products were

major commercial successes: insulin (Genentech and Eli Lilly), tPA (Genentech in 1987) and erythropoietin (Amgen and Ortho in 1989). By 1991, there were over 100 biotechnology drugs in clinical development and 21 biotechnology drugs with submitted applications to the FDA (Grabowski and Vernon, 1994): this was roughly one third of all drugs in clinical trials. Sales of biotechnology-derived therapeutic drugs and vaccines had reached $2 billion, and two new biotechnology firms (Genentech and Amgen) have entered the club of the top eight major pharmaceutical innovators (Grabowski and Vernon, 1994).

3.   Thus, Feldman and Audretsch (1999), for instance, support the view that diversity matters more than specialisation, finding that the number of innovations in sector states owes more to the presence in the state of other industries whose science base is related to that of industry, rather than to the specialisation in sector. In a similar vein, Bresnahan, Gambardella and Saxenian (2001) suggest that clusters tend to form taking advantage of new technological and market opportunities that have not already been exploited.

4.   Most of the literature on clusters relies on some form of static and/or dynamic externalities as the key factors of agglomeration (e.g. combinations of knowledge spillovers, attraction of entrepreneurs and innovators from outside the clusters, increasing returns stemming from proximity and network externalities (Arthur, 1994). Other accounts instead stress the role of processes of spin-offs of new firms by incumbent companies and, more generally, on some form of cumulativeness of innovation within specific firms and/or institutions not strictly reducible to externalities (Kenney and von Burg, 2001; Klepper, 2002). In some cases, these dynamic processes are conceived as self-reinforcing and path-dependent.

5.   Millstein and Kohler's groundbreaking discovery – hybridoma technology – was never patented, while Stanford University filed a patent for Boyer and Cohen's process in 1974. Boyer and Cohen renounced their own rights to the patent but nevertheless they were strongly criticized for having been instrumental in patenting what was considered to be a basic technology.

References

Abernathy, W. and Utterback, J. (1978), 'Patterns of Industrial Innovation', *Technology Review*, **50**, 40–47.

Akerlof, G. (1970), 'The Market for Lemons: Uncertainty and the Market Mechanisms', *Quarterly Journal of Economics*, **84**, 488–500.

Almeida, P. and Kogut, B. (1999), 'Localisation of Knowledge and the Mobility of Engineers in Regional Networks', *Management Science*, **45**, 905–17.

Arora, A., Fosfuri, A. and Gambardella, A. (2001), *Markets for Technology*. Cambridge, MA: MIT Press.

Arora, A. and Gambardella, A. (1994), 'The Changing Technology of Technical Change: General and Abstract Knowledge and the Division of Innovative Labor', *Research Policy*, **23**, 5, 523–32.

Arora, A., Gambardella, A., Pammolli, F. and Riccaboni, M. (2000), 'Advantage Lost? On the Leveling Effect of the Market for Technology in Biopharmaceuticals', paper prepared for the International Conference on Technological Policy and Innovation, Paris, 20–22 November.

Arrow, K.J. (1962), 'Economic Welfare and the Allocation of Resources for Invention', in Nelson, R.R. (ed.), *The Rate and Direction of Inventive Activity:Economic and Social Factors*, Princeton, NJ: Princeton University Press, 609–25.

Arrow, K.J. (1983), 'Innovation in Large and Small Firms', in Ronen, J. (ed.), *Entrepreneurship*. Lexington, MA: Lexington Books.

Arthur, B.A. (1994), *Increasing Returns and Path-Dependency in Economics*. Ann Arbor, MI: University of Michigan Press.

Braunerhjelm, P. and Feldman, M. (forthcoming), *The Genesis of Clusters*.

Breschi, S. and Lissoni, F. (2001), 'Knowledge Spillovers and Local Innovation Systems: A Critical Survey, *Industrial and Corporate Change*, **10**, 4, 975–1006.

Bresnahan, T., Gambardella, A. and Saxenian, A. (2001), 'Old Economy Inputs for "New Economy"

Outcomes: Cluster Formation in the New Silicon Valleys', *Industrial and Corporate Change*, **10**, 4, 836–60.

Christensen, C.M. (1997), *The Innovator's Dilemma: When New Technologies Cause Great Firms to Fail*. Cambridge, MA: Harvard Business School Press.

Christensen, C.M. and Rosenbloom, R.S. (1995), 'Explaining the Attacker's Advantage: Technological Paradigms, Organizational Dynamics and the Value Network', *Research Policy*, **24**, 233–57.

Cockburn, I. and Henderson, R. (1996), 'Public–private interaction in pharmaceutical research', *Proceedings of the National Academy of Sciences*, **93**, 23, 12, 725–30.

Cockburn, I., Henderson, R. and Stern, S. (2000), 'Untangling the Origins of Competitive Advantage', *Strategic Management Journal*, **21**,123–45.

Cooke, P. (2002), *Knowledge Economies: Clusters, Learning & Cooperative Advantages*. London: Routledge.

Dasgupta, P. and David, P.A. (1994), 'The New Economics of Science', *Research Policy*, **23**, 5, 487–521.

Dosi, G. (1982), 'Technological Paradigms and Technological Trajectories: A Suggested Interpretation of the Determinants of Technical Change', *Research Policy*, **11**, 3, 147–62.

Edquist, C. (ed.) (1977), *Systems of Innovation: Technologies, Institutions and Organisations*. London: Pinter.

Edquist, C. and McKelvey, M. (eds) (2000), *Systems of Innovation: Growth, Competitiveness and Employment*, Cheltenham: Edward Elgar.

Ellison, G. and Glaeser, E.L. (1999), 'The Geographic Concentration of Industry: Does Natural Advantage Explain Agglomeration?', *American Economic Review*, **89**, 2, 311–16.

Feldman, M.P and Audretsch, D.B. (1999), 'Innovation in Cities: Science-based Diversity, Specialization and Localized Competition', *American Economic Review*, **43** (Special Issue), 409–29.

Gambardella, A. (1995), *Science and Innovation – the US Pharmaceutical Industry during the 1980s*. Cambridge: Cambridge University Press.

Gittelman, M. (1999), 'Knowledge as Property: Innovation in Biotechnology in the United States and France', Best Papers and Proceedings of the Academy of Management 1999 Annual Meeting.

Glaeser, E.L., Kallal, H.D., Scheinkman, J.A. and Shleifer, A. (1992), 'Growth of cities', *Journal of Political Economy*, **100**, 1126–52.

Grabowski, H.G. and Vernon, J.M. (1994), 'Innovation and Structural Change in Pharmaceuticals and Biotechnology', *Industrial and Corporate Change*, **3**, 2, 435–49.

Hall, P. and Soskice, D. (2001), *Varieties of Capitalisms. The Institutional Foundations of Comparative Advantage*. Oxford: Oxford University Press.

Henderson, J.V. (1986), 'The Efficiency of Resource Usage and City Size', *Journal of Urban Economics*, **19**, 47–70.

Henderson, R. and Clark, K.B. (1990), 'Architectural Innovation: The Reconfiguration of Existing Product Technologies and the Failure of Established Firms', *Administrative Science Quarterly*, **35**, 9–30.

Henderson, R. and Cockburn, I. (1996), 'Scale, Scope and Spillovers: The Determinants of Research Productivity in Drug Discovery', *RAND Journal of Economics*, **27**, 1, 32–59.

Kenney, M. (1986), *Biotechnology: The Industry–University Complex*. Ithaca, NY: Cornell University Press.

Kenney, M. and von Burg, U. (2001), 'Technology, Entrepreneurship and Path-dependence: Industrial Clustering in Silicon Valley and Route 128', *Industrial and Corporate Change*, **10**, 67–104.

Klepper, S. (1996), 'Entry, Exit, Growth and Innovation over the Product Life Cycle', *American Economic Review*, **86**, 3, 562–83.

Klepper, S. (2002), 'The Capabilities of New Firms and the Evolution of the US Automobile Industry', *Industrial and Corporate Change*, **11**, 645–66.

Kline, S. and Rosenberg, N. (1986), 'An Overview of Innovation', in Landau, R. and Rosenberg, N. (eds), *The Positive Sum Strategy: Harnessing Technology for Economic Growth*. Washington, DC: The National Academy Press, pp. 275–305.

Kogut, B., Shan, W. and Walker, G. (1992), 'The make or cooperate decision in the context of an industry network', in Nohria, N. and Eccles, R.G. (eds), *Networks and Organizations*. Cambridge, MA: Harvard Business School Press, 348–65.

Krugman, P. (1991), *Geography and Trade*. Cambridge, MA: MIT Press.

Malerba, F. (ed.) (2005), *Sectoral Systems of Innovation*. Cambridge: Cambridge University Press.

Maskell, P. (2001), 'Towards a Knowledge-based Theory of the Geographical Cluster', *Industrial and Corporate Change*, **10**, 4, 921–44.

Mazzoleni, R. and Nelson, R.R. (1998), 'Economic Theories about the Benefits and Costs of Patents', *Journal of Economic Issues*, **32**, 1031–52.

Merges, R. and Nelson, R.R. (1994), 'On Limiting or Encouraging Rivalry in Technical Progress: The Effect of Patent Scope Decisions', *Journal of Economic Behavior and Organization*, **25**, 1–24.

McKelvey, M. (1996), *Evolutionary Innovations: The Business of Biotechnology*. Oxford: Oxford University Press.

McKelvey, M., Orsenigo, L. and Pammolli, F. (2005), 'Pharmaceuticals Analysed through the Lens of a Sectoral Innovation System', in Malerba, F. (ed.), *Sectoral Systems of Innovation*, Cambridge: Cambridge University Press.

McKelvey, M., Rickne, A. and Laage-Hellman, J. (2005), *The Economic Dynamics of Modern Biotechnology*. Cheltenham: Edward Elgar Publishers.

Merton, D. (1973), *The Sociology of Science: Theoretical and Empirical Investigation*, Chicago: University of Chicago Press.

Mowery, D. (1981), 'The Emergence and Growth of Industrial Research in American Manufacturing, 1899–1945', PhD thesis, Stanford University, Stanford, CA.

Mowery, D.C., Nelson, R.R., Sampat, B. and Ziedonis, A. (2001), 'The Growth of Patenting and Licensing by US Universities: An Assessment of the Effects of the Bayh-Dole Act of 1980', *Research Policy*, **30**, 99–119.

Nelson, R.R. (1959), 'The Simple Economics of Basic Scientific Research', *Journal of Political Economy*, **67**, 297–306.

Nelson, R.R. (1982), *National Systems of Innovation*. Oxford: Oxford University Press.

Nelson, R.R. and Winter, S. (1982), *An Evolutionary Theory of Economic Change*. Cambridge, MA: The Bellknap Press of Harvard University Press.

Office of Technology Assessment (1984), *Commercial Biotechnology: An International Analysis*. Washington, DC: US Government Printing Office.

Orsenigo, L. (1989), *The Emergence of Biotechnology*. London: Frances Pinter.

Pavitt, K. (2001), 'Public Policies to Support Basic Research: What Can the Rest of the World Learn from US Theory and Practice? (and What They Should Not Learn)', *Industrial and Corporate Change*, **10**, 3, 761–80.

Pammolli, F. (1997), *Innovation and Industry Structure: The International Pharmaceutical Industry during the Nineties*. Milan: Guerini.

Pisano, G. (1997), *R&D Performance, Collaborative Arrangements and the Market for Knowhow: A Test of the 'Lemons' Hypothesis in Biotechnology* (Working paper). Cambridge, MA: Harvard Business School Press.

Saxenian, A. (1994), *Regional Advantage, Culture and Competition in Silicon Valley and Route 128*. Cambridge, MA: Harvard University Press.

Sharp, M. (1985), *The New Biotechnology: European Governments in Search of a Strategy* (Sussex European Papers 115). Brighton: University of Sussex.

Teece, D.J. (1986), 'Profiting from Technological Innovation: Implications for Integration, Collaboration, Licensing and Public Policy', *Research Policy*, **15**, 185–219.

Utterback, J. (1994), *Mastering the Dynamics of Innovation*. Cambridge, MA: Harvard Business School Press.

Zucker, L., Darby, M. and Brewer, M. (1997), *Intellectual Human Capital and the Birth of US Biotechnology Enterprises* (Working Paper No. 4,653), Boston, MA:National Bureau of Economic Research.

Part I
Overview

[1]

Summary

Introduction

In the past 10 years, dramatic new developments in the ability to select and manipulate genetic material have sparked unprecedented interest in the industrial uses of living organisms. Following the first successful directed insertion of foreign DNA in a host microorganism in 1973, scientific researchers in the United States and other countries began to recognize the potential for directing the cellular machinery to develop new and improved products and processes in a wide diversity of industrial sectors. Potential industrial applications of those novel genetic techniques include the production of new drugs, food, and chemicals, the degradation of toxic wastes, and the improvement of agricultural products. Thus, these new techniques could have a major economic impact on industries throughout the world.

Beginning around 1976, many small entrepreneurial firms were formed in the United States specifically to build on the growing body of fundamental knowledge in molecular biology and to exploit it to a profitable end. Furthermore, large established American, Japanese, and European companies in a spectrum of industrial sectors expanded their research and development (R&D) programs to include the new genetic techniques. In the United States, private sector investments to commercialize these new techniques exceeded $1 billion in 1983.

This report assesses the competitive position of the United States with respect to Japan and four European countries-the Federal Republic of Germany, the United Kingdom, Switzerland, and France—believed to be the major competitors in the commercial development of "new biotechnology," as defined below. Although the United States is currently the world leader in both basic science and commercial development of new biotechnology, continuation of the initial preeminence of American companies in the commercialization of new biotechnology is not assured. Japan and other countries have

identified new biotechnology as a promising area for economic growth and have therefore invested quite heavily in R&D in this field, Congressional policy options for improving U.S. competitiveness in new biotechnology are identified in this report.

Definitions

Biotechnology, broadly defined, includes any technique that uses living organisms (or parts of organisms) to make or modify products, to improve plants or animals, or to develop microorganisms for specific uses. Biological processes and organisms have been used with great success throughout history and have become increasingly sophisticated over the years. Since the dawn of civilization, people have deliberately selected organisms that improved agriculture, animal husbandry, baking, and brewing. More recently, a better understanding of genetics has led to more effective applications of traditional genetics in such areas as antibiotic and chemical production.

This report focuses on the industrial use of recombinant DNA (rDNA, cell fusion and novel bioprocessing techniques To differentiate between biotechnology using these novel techniques and the more traditional forms of biotechnology, this report uses the terms '(new biotechnology" and "old biotechnology)" respectively. Thus, for example, traditional wine production is old biotechnology, but the use of yeast modified with rDNA techniques to produce wine with a higher alcohol content is new biotechnology. Where no specific distinction is made, the term biotechnology alone henceforth refers to new biotechnology.

Biotechnology is the most recent phase in a historical continuum of the use of biological organisms for practical purposes. Furthermore, developments arising from existing technologies are providing a base from which other technologies will emerge, and new technologies can make even the most potentially useful current technology obsolete in a short time. Of necessity, this assessment describes the development of biotechnology at a particular point in time, but it is important to emphasize that dynamic and progressive change has characterized biotechnology for the last decade. Figure 1 shows some prominent events that illustrate the rapid progress made in the development of biotechnology over the last decade. This pace is likely to continue into the 21st century.

The technologies

The novel techniques used in biotechnology are extremely powerful because they allow a large amount of control over biological systems *Recombinant DNA technology,* one of the new techniques, allows direct manipulation of the genetic material of individual cells. The ability to

direct which genes are used by cells permits more control over the production of biological molecules than ever before. Recombinant DNA technology can be used in a wide range of industrial sectors to develop micro-organisms that produce new products, existing products more efficiently, or large quantities of otherwise scarce products. This technology can also be used to develop organisms that themselves are useful, such as microorganisms that degrade toxic wastes or new strains of agriculturally important plants.

Cell *fusion,* the artificial joining of cells, combines the desirable characteristics of different types of cells into one cell. This technique has been used recently to incorporate in one cell the traits for immortality and rapid proliferation from certain cancer cells and the ability to produce useful antibodies from specialized cells of the immune system. The cell line resulting from such

Figure 1.—Major Events in the Commercialization of Biotechnology

1973	**First gene cloned.**
1974	First expression of a gene cloned from a different species in bacteria. Recombinant DNA (rDNA) experiments first discussed in a public forum (Gordon Conference).
7975	U.S. guidelines for rDNA research outlined (Asilomar Conference). First hybridoma created.
T976	First firm to exploit rDNA technology founded in the United States (Genentech). Genetic Manipulation Advisory Group (U. K.) started in the United Kingdom.
1980	*Diamond v. Chakrabarty*—U.S. Supreme Court rules that micro-organisms can be patented under existing law. Cohen/Boyer patent issued on the technique for the construction of rDNA. United Kingdom targets biotechnology (Spinks' report). Federal Republic of Germany targets biotechnology (Leistungsplan). Initial public offering by Genentech sets Wall Street record for fastest price per share increase ($35 to $89 in 20 minutes).
T981	First monoclinal antibody diagnostic kits approved for use in the United States. First automated gene synthesizer marketed. Japan targets biotechnology (Ministry of International Trade and Technology declares 1981 "The Year of Biotechnology"). France targets biotechnology (Pelissolo report). Hoescht/Massachusetts General Hospital agreement. Initial public offering by Cetus sets Wall Street record for the largest amount of money raised in an initial public offering ($1 15 million). Industrial Biotechnology Association founded. DuPont commits $120 million for life sciences R&D. Over 80 NBFs had been formed by the end of the year.
1982	First rDNA animal vaccine (for colibacillosis) approved for use in Europe. First rDNA pharmaceutical product (human insulin) approved for use in the United States and the United Kingdom. First R&D limited partnership formed for the funding of clinical trials.
383	First plant gene expressed in a plant of a different species. **$500 million** raised in U.S. public markets by NBFs.

SOURCE: Office of Technology Assessment

a fusion, known as a hybridoma, produces large quantities of *monoclinal antibodies (MAbs)*, so called because they are produced by the progeny, or clones, of a single hybridoma cell. MAbs can potentially be used for many purposes, including the diagnosis and treatment of disease and the purification of proteins.

The commercial success of specific industrial applications of rDNA and cell fusion techniques will hinge on advances in bioprocess engineering. *Bioprocess technology*, though not a novel genetic technique, allows the adaptation of biological methods of production to large-scale industrial use. Most industrial biological syntheses at present are carried out in single batches, and a small amount of product is recovered from large quantities of cellular components, nutrients, wastes, and water. Recent improvements in techniques for immobilizing cells or enzymes and in bioreactor design, for example, are helping to increase production and facilitate recovery of many substances. Additionally, new genetic techniques can aid in the design of more efficient bioreac - tors, sensors, and recovery systems. In the next decade, competitive advantage in areas related to biotechnology may depend as much on developments in bioprocess engineering as on innovations in genetics, immunology, and other areas of basic science.

The same technologies that yield commercial products will also provide new research tools. The new genetic technologies described above have ignited an explosion of fundamental knowledge. The widespread use of rDNA and cell fusion techniques in the investigation of a wide variety of biological phenomena in plants, animals, microorganisms, and viruses highlights the impact of these technologies on basic science research and the advances in fundamental knowledge that they make possible. This new knowledge, in turn, may reveal new commercial opportunities.

Industrial development

Biotechnology could potentially affect any current industrial biological process or any process in which a biological catalyst could replace a

chemical one. As discussed in this report, industrial applications of biotechnology will be found in several industrial sectors, including pharmaceuticals, animal and plant agriculture, specialty chemicals and food additives, environmental areas, commodity chemicals and energy production, and bioelectronics.

The industrial sector in which the earliest applications of new biotechnology have occurred is the pharmaceutical sector. Reasons for the rapid diffusion of the new techniques into the pharmaceutical sector include the following:

• Recombinant DNA and MAb technologies were developed with public funds directed toward biomedical research. The first biotechnology products, such as rDNA-produced human insulin, interferon, and MAb diagnostic kits, are a direct result of the biomedical nature of the basic research that led to these new technologies.

• Pharmaceutical companies have had years of experience with biological production methods, and this experience has enabled them to take advantage of the new technologies.

• Pharmaceutical products are high value-added and can be priced to recover costs incurred during R&D, so the pharmaceutical sector is a good place to begin the costly process of developing a new technology.

Because of the rapid diffusion of the new genetic techniques into pharmaceutical R&D programs, the pharmaceutical sector is currently most active in commercializing biotechnology. For this reason, it serves as a model for the industrial development of biotechnology in much of this report. It is important to recognize, however, that the development of biotechnology in other industrial sectors will differ from its development in the pharmaceutical sector. Regulatory and trade barriers and a marketing and distribution system unique to the pharmaceutical sector limit *its* usefulness as a model. Furthermore, the techniques may not diffuse as rapidly into other industrial sectors, such as the chemical industry, because of difficulties companies may have in recovering investments in R&D and physical plants required to convert to biological methods of production.

Findings

Industrial applications of biotechnology

The earliest industrial applications of biotechnology (i.e., during the next 5 to 10 years) are likely to occur in pharmaceuticals, animal agriculture, and specialty chemicals. Applications of biotechnology to pharmaceuticals being pursued at present are in the production of proteins such as insulin, interferon, and human serum albumin; antibiotics; MAb diagnostics; and vaccines for viral, bacterial, and parasitic diseases. As more is learned about hormone growth factors, immune regulators, and neurological peptides, their importance in the treatment of disease may increase dramatically. Eventually, the production of such regulatory proteins may turn out to be the largest application of biotechnology in the pharmaceutical industry. U.S. companies pursuing biotechnological applications in pharmaceuticals include many of the established pharmaceutical companies* and a large number of small, entrepreneurial new biotechnology firms (NBFs). * * Additionally, many established companies in other sectors are using biotechnology as a way to diversify into pharmaceuticals.

In animal agriculture, biotechnology is being used to develop products similar to those being developed in the pharmaceutical industry. However, since animal producers cannot afford to purchase expensive products made with new technology, biotechnologically produced products may initially be limited to products for "high value" animals such as pets and breeding stock. The most important products are likely to be vaccines and growth promotants.

Unlike the production of pharmaceuticals, the production of animal health products using traditional technologies is not dominated by a few large companies. Additionally, the animal agriculture industry differs from the pharmaceutical industry in that the regulatory requirements for animal health products, especially for vaccines and diagnostics, are significantly less stringent than for human health products; markets for animal products are smaller and more accessible; and the distribution and delivery systems are different. Because of these features, many NBFs are finding animal agriculture an attractive field for the application of biotechnology.

The potential applications of biotechnology are probably more varied for specialty chemicals (i.e., chemicals costing more than $1flb) and food additives* than for any other industrial sector at the present time. Possible applications include improvements in existing bioprocesses, such as in the production of amino acids. Other products, such as vitamins and steroid compounds, are currently made in multistep production processes involving chemical syntheses. Biotechnology could provide one or more enzymatic conversion process to increase the specificity of currently used chemical conversions. Generally, complex products, such as enzymes and some polysaccharides, can only be made economically using bioprocesses. The production of specialty chemicals represents one of the largest opportunities for the application of biotechnology because of the diversity of potential applications. Several companies in the United States are pursuing biological production of specialty chemicals, but most specialty chemicals currently produced biologically are made almost exclusively in Japan and Europe, and these countries intend to pursue new applications for specialty chemical production.

Applications of rDNA technology to plant agriculture are proceeding faster than anyone anticipated 3 to 4 years ago. Some important traits of plants, including stress-, herbicide-, and pest-resistances, appear to be rather simple genetically, and it may be possible to transfer these traits to important crop species in the next few years. Other traits, such as increased growth rate, increased photosynthetic ability, and the stimula-

---

*Established companies pursuing applications of biotechnology are generally process-oriented, multiproduct companies in traditional industrial sectors such as pharmaceuticals, energy, chemicals, and food processing.

● *NBFs, as defined in this report, are entrepreneurial ventures started specifically to pursue applications of biotechnology.

---

*Food additives are considered together with specialty chemicals because many (though not all) food additives are also specialty chemicals, e.g., amino acids and vitamins.

tion of nitrogen fixation, are genetically complex, and it is likely to be several years before plants with these characteristics developed with rDNA technology will be ready for field testing. Microorganisms that interact with plants offer possibilities for genetic manipulation that may be more near-term. For instance, it may be possible to manipulate micro-organisms to produce pesticides or inhibit frost formation. Companies pursuing these applications include many NBFs and established companies in agricultural chemicals and seed production.

Environmental applications of biotechnology include mineral leaching and metal concentration, pollution control and toxic waste degradation, and enhanced oil recovery. These applications may take longer to reach the market, because little is known of the genetics of the most potentially useful micro-organisms. Additionally, regulation is expected to be a major factor influencing development of this area because these applications use microorganisms that are deliberately released into the environment. The nature and extent of this regulation remains uncertain, and this uncertainty may deter some firms from entering the field, thus slowing development.

Commodity chemicals, which are now produced from petroleum feedstocks, could be produced biologically from biomass feedstocks such as cornstarch and lignocellulose. Commodity chemical production from cornstarch will probably occur before production from lignocellulose because of the high energy inputs necessary for the solubilization of lignocellulose. Although the technology exists now for the cost+ ffective biological production of some commodity chemicals such as ethanol, the complex infrastructure of the commodity chemical industry will prevent the replacement of a large amount of commodity chemical production using biotechnology for at least 20 years. This distant time horizon is due more to the integrated structure of the chemical industry, its reliance on petroleum feedstocks, and its low profit margins than to technical problems in the application of the biotechnology.

In the area of bioelectronics, biotechnology could be used to develop improved biosensors or new conducting devices called biochips. Sensors that use enzymes for detecting specific substances are available now. However, their use is limited by the narrow range of substances they detect and by their temperature instability. Biotechnology could be instrumental in the development *of* more versatile sensors that use enzymes or MAbs. Better sensors would be especially useful in the control of industrial bioprocesses. Biotechnology may also make it possible to construct devices that use proteins as a framework for molecules that act as semiconductors. The anticipated advantages of these biochips are their small size, reliability, and the potential for self assembly. The production of biochips, however, is one of the most distant applications of biotechnology.

The U.S. competitive position

A well-developed life science base, the availability of financing for high-risk ventures, and an entrepreneurial spirit have led the United States to the forefront in the commercialization of biotechnology. For the most part, the laws and policies of this country have made it possible for industrialists and scientists to capitalize rapidly on the results of basic research in biotechnology conducted in the university system and government laboratories. The relative freedom of U.S. industry to pursue a variety of courses in the development of products has also given the United States a comparative advantage. The flexibility of the U.S. industrial system and the plurality of approaches taken by entrepreneurial NBFs and established companies in the development of products have facilitated the rapid development of biotechnology in the United States.

Japan is likely to be the leading competitor of the United States for two reasons. First, Japanese companies in a broad range of industrial sectors have extensive experience in bioprocess technology. Japan does not have superior bioprocess technology, but it does have relatively more industrial experience using old biotechnology, more established bioprocessing plants, and more bioprocess engineers than the United States. Second, the Japanese Government has targeted biotechnology as a key technology of the future, is funding its commercial development, and is coordinating interactions among representatives from industry, universities, and government. The United States may compete very favorably with Japan if it can direct more attention to research problems associated with the scaling-up of bioprocesses for production

The European countries are not moving as rapidly toward commercialization of biotechnology as either the United States or Japan, in part because the large established pharmaceutical and chemical companies in Europe have hesitated to invest in biotechnology and in part because of cultural and legal traditions that tend not to promote venture capital formation and, consequently, risk-taking ventures. Nevertheless, several of the large pharmaceutical and chemical houses in the United Kingdom, the Federal Republic of Germany, Switzerland, and France will surely be competitors in selected product areas in the future because of their prominent position in world sales of biologically derived products. Additionally, the increased interest shown recently by the British Government in biotechnology may speed its development in the United Kingdom.

The United States could have difficulty maintaining its competitive position in the future if several issues are not addressed. If U.S. Government funding for basic life science research continues its decline, the science base, which is the source of innovation in bio technology as well as in other fields, may be eroded. U.S. Government funding of generic applied research, * especially in the areas of bioprocess engineering and applied micro biology, is currently insufficient to support rapid commercialization U.S. Government funding for personnel training in these areas may also be insufficient. Additionally, clarification and modification of certain aspects of U.S. health, safety, and environmental regulation and intellectual property law may be necessary for the maintenance of a strong U.S. competitive position in biotechnology.

● Generic applied research, which is nonproprietary and bridges the gap between basic research and applied research, is aimed at the solution of generic problems that are associated with the use of a technology by industry.

Analysis of international competitiveness in biotechnology

Often international competitiveness is defined as the relative ability of firms based in one country to develop, produce, and market equivalent goods or services at lower costs than firms in other countries. Competitiveness is a matter of relative prices, and these usually reflect relative costs of developing, producing, and distributing goods and services. In the case of biotechnology, two factors preclude a traditional analysis of international competitiveness. First, standard analyses of competitiveness examine the marketing of products, but as of the end of 1983, only a few products of new biotechnology had reached the marketplace—notably human insulin, some MAb diagnostic kits, and some animal vaccines. Most of these products are substitutes for already existing products, and the markets are well defined

and relatively limited, Furthermore, even the markets for some new animal vaccines are quite small when compared to potential markets for applications of biotechnology in the production of some chemicals or new crop plants. Thus, the biotechnology products that have reached the market to date may be inaccurate indicators of the potential commercial success in world markets of the much larger number of biotechnology products and processes still in R&D stages. Which of the biotechnology products and processes in development are likely to be marketed and when cannot be accurately predicted. Second, even with many more products on the market, a traditional competitive analysis might not be appropriate because an economic analysis of competitiveness usually addresses a specific industrial sector. The set of techniques that constitute biotechnology, however, are potentially applicable to many industrial sectors.

Since the technologies are still emerging and most biotechnology products and processes are in early development, most of this report focuses on *potential* rather than actual products and processes. In the case of biotechnology, knowledge about market size, distribution systems, customers, production* processes, and learning curve economies is lacking. Thus, traditional parameters of competitiveness are difficult or impossible to estimate. Instead of examining the classical measures of competitiveness, this analysis of international competitiveness in biotechnology examines the aggregate industrial activity in biotechnology in both domestic and foreign firms and 10 factors that *might* be influential in determining the competitive position of the United States and other countries with respect to the commercialization of biotechnology.

In investigating competitiveness in biotechnology, this report analyzes the commercialization efforts of five countries in addition to the United States: Japan, the Federal Republic of Germany, the United Kingdom, Switzerland, and France. Although companies from many countries will have biotechnology products in world markets, these five countries were selected because of their research capabilities in biology and their existing capabilities in old biotechnology and because, as a whole, their companies are most likely to reach world markets first with biotechnology-produced products. Japan leads the world both in the microbial production of amino acids and in large-scale plant cell culture, and it has a strong position in new antibiotic markets. Japan is also the world leader in traditional bioprocess engineering. Furthermore, the Ministry of International Trade and Industry (MITI) in Japan has designated biotechnology for industrial development. The European pharmaceutical houses, notably in the United Kingdom, France, the Federal Republic of Germany, and Switzerland, lead the world in pharmaceutical sales. Like Japan, three of these European countries, the Federal Republic of Germany, the United Kingdom, and France, have national plans for the promotion of biotechnology. The Federal Republic of Germany and the United King-

dom have good basic biology research and especially good bioprocess engineering research.

The first step in the analysis of international competitiveness in biotechnology was to consider the aggregate level of industrial activity and the number and kinds of firms commercializing biotechnology in the competitor countries. OTA'S industrial analysis, presented in *Chapter 4: Firms Commercializing Biotechnolo~*, was approached from three perspectives:

- the number and kinds of companies commercializing biotechnology,
- the markets targeted by industrial biotechnology R&D, and
- the interrelationships among companies applying biotechnology and the overall organization of the commercial effort.

The analysis began with the United States and comparisons were then made with other countries.

The second step in providing an overall picture of competitiveness in biotechnology involved the evaluation of the following 10 factors identified as potentially important in determiningg the future position of the United States and other countries in the commercialization of biotechnology:

- financing and tax incentives for firms;
- government funding 'of basic and applied research;
- personnel availability and training;
- health, safety, and environmental regulation;
- intellectual property law;
- university/industry relationships;
- antitrust law;
- international technology transfer, investment, and trade;
- government targeting policies in biotechnology; and
- public perception.

The relative importance of each of the factors was first evaluated-to determine their importance to competitiveness today (see fig. 2) and which ones could be important as the technology matures and more products reach the marketplace. Then, each of the factors was analyzed for each of the six competitor countries: the United States, Japan, the Federal Republic of Germany, the United

Figure 2.—The Relative Importance of Factors Affecting the Commercialization of Biotechnology

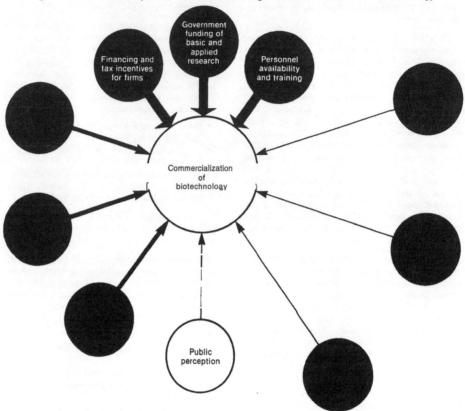

SOURCE Office of Technology Assessment

Kingdom, Switzerland, and France. Since the importance to competitiveness of any given factor is not necessarily the same for every industrial sector in which applications are being pursued—for instance, a country's intellectual property laws may protect pharmaceuticals better than plants—the importance of each factor was evaluated for different industrial sectors.

Additional considerations taken into account in the analysis are historical patterns of industrial

commercialization, the lack or abundance of particular natural resources, and the tendency toward risk taking in each country. These other considerations were used as modifiers of the results of the analysis.

OTA'S principal findings with respect to the types and activities of firms commercializing biotechnology, the factors potentially important to international competitiveness in biotechnology, and the other considerations just mentioned are presented below.

The importance of established and new firms in the commercialization of biotechnology

U.S. and foreign efforts to develop and commercialize biotechnology differ substantially in character and structure. In the United States, two distinct sets of firms are pursuing commercial applications of biotechnology -NBFs and established companies. Because NBFs were founded specifically to exploit perceived research advantages, they are providing the United States with a commercial edge in the current research-intensive phase of biotechnology's development. Through their R&D efforts, NBFs are contributing to innovation, expansion of the U.S. research base, technology diffusion, and encouragement of technical advances through the increased domestic competition they create. All of these contributions provide the United States with a competitive advantage.

Although NBFs have assumed much of the risk for biotechnology's early development in the United States, established U.S. companies are making substantial contributions to the U.S. commercialization effort. Through equity investments and licensing and contract research agreements with NBFs, established U.S. companies are providing many NBFs with the necessary financial resources to remain solvent. Through joint development agreements with NBFs, many established companies will also provide the necessary production and marketing resources to bring many NBF products to world markets. These resources could help to sustain the rapid pace of technical advance spurred by NBFs. Recently, more and more established U.S. companies have been investing in their own research and production facilities, so the role of established companies in the U.S. biotechnology effort is expanding.

U.S. efforts to commercialize biotechnology are currently the strongest in the world. The strength of U.S. efforts is in part derived from the unique complementarily and competition that exists between NBFs and established U.S. companies in developing biotechnology for wider commercial application. At present, most NBFs are still specializing in research-oriented phases of development, precisely the commercial stage where

they excel. The established companies, on the other hand, have assumed a major share of the responsibility for production and marketing of, and, when necessary, obtaining regulatory approval for, many of the earliest biotechnology products— the commercial stages where their resources are strongest. Since established companies control the later stages of commercialization for many new products being developed through production and marketing agreements with NBFs, they will also have considerable control over the pace at which these new products reach the market. Whether the dynamism arising fmm the competition and complementarily between NBFs and estab lished companies will continue giving the United States a comparative advantage in the context of product introduction remains unclear. Some established companies, for example, might have disincentives to market new products because the new products might compete with products they already have on the market.

In Japan, the Federal Republic of Germany, the United Kingdom, Switzerland, and France, biotechnology is being commercialized almost exclusively by established companies. The Japanese consider biotechnology to be the last major technological revolution of this century, and the commercialization of biotechnology is accelerating *over* a broad range of industries, many of which have extensive bioprocessing experience. The general chemical and petroleum companies especially are leaning strongly toward biotechnology, and some of them are making rapid advances in R&D through their efforts to make biotechnology a key technology for the future. In Europe, large pharmaceutical and chemical companies, many of which already have significant strength in biologically produced product markets, are the major developers of biotechnology. Their inherent financial, production, and marketing strengths will be important factors as the technology continues to emerge internationally.

The commercial objectives of biotechnology R&D vary across national boundaries. In the United States, commercial research projects appear primarily focused on pharmaceutical and plant and animal agriculture, and American com-

petitive vigor in these application areas is cor-
respondingly strong. Much of the investment in
animal agriculture has been made by NBFs
whereas much of the investment in plant agriculture has been made by major U.S. agrichemical
companies.

In Japan, a competitive drive has been launched
to enter international pharmaceutical markets.
Furthermore, Japanese companies are world leaders in large-scale plant tissue culture, and MITI
has identified secondary compound synthesis
from plants as a major area for commercialization. Unlike the United States, Japanese companies
appear to be dedicating a great deal of biotechnology R&D to specialty chemical production, an
area where they are already internationally
prominent.

To the extent that large companies in Europe
began their commercialization efforts later than
U.S. companies and may also lack the dynamism
and flexibility to compete with the combined efforts of NBFs and established companies in the
United States, European companies could initially be at a competitive disadvantage. The United
Kingdom's major pharmaceutical companies are
among the leading producers of biologically produced products, however, and their expertise in
bioprocessing is impressive. Furthermore, the
United Kingdom possesses some of the strongest
basic research in interdisciplinary plant sciences.
Whether or not the basic research will be commercialized successfully is difficult to predict.

U..S. competitive strength in biotechnology will
be tested when large-scale production begins and
bioprocessing problems are addressed. Pharmaceutical markets will be the first proving ground
for U.S. competitive strength. The Japanese have
extensive experience in bioprocess technology,
and dozens of strong "old biotechnology" companies from several industrial sectors in Japan are
using new biotechnology as a lever to enter profitable and expanding pharmaceutical markets. In
addition to competing against Japanese companies, U.S. pharmaceutical and chemical companies will be competing against pharmaceutical and
chemical companies of Western Europe, all of
whom expect to recover their biotechnology investments through extensive international market

penetration. There seem to be fewer European
companies than Japanese companies strong in bio-
technology now, but the competitive strength of
European multinationals such as Hoechst (F. R.G.),
Rhone Poulenc and Elf Aquitaine (France), ICI,
Glaxo, and Wellcome (U.K.), and Hoffmann-La
Roche (Switzerland) in the long run should not
be underestimated.

Factors potentially important to international competitiveness in biotechnology

MOST IMPORTANT FACTORS

The three factors most important to the commercial development of biotechnology are financing and tax incentives for firms, government
funding of basic and applied research, and personnel availability and training.

Financing and Tax Incentives for Firms—
The availability of venture capital to start new
firms and tax incentives provided by the U.S.
Government to encourage capital formation
and stimulate R&D in the private sector are
very important to development of biotechnology in the United States. Since 1976, private venture capital in the United States has funded the
startup of more than 100 NBFs. Many of these
firms have already obtained second- and third-
round financing, while others, still seeking additional funds, are relying heavily on the currently strong stock market, R&D limited partnerships,
and private placements to fund research, production scale-up, clinical trials, and early product
development. Between March and July of 1983,
23 NBFs raised about $450 million. R&D limited
partnerships in biotechnology are expected to
total $500 million in 1983 and $1.5 billion by 1984.
Corporate equity investment in NBFs, although
now diminishing, has also been an important
source of financing for the new firms. From 1977
to August 1983, corporate venture capital sup-
plied over $350 million to NBFs in equity in-
vestments alone.

Current price/earnings ratios* for NBFs appear
high, because most NBFs still have negative earn-

*A price/earnings ratio ($\frac{\text{market price per share}}{\text{company earnings per share}}$) reflects the stock mar-
ket's anticipation of the company's future performance based on
the earnings per share.

ings records. Continued reliance on the stock market and R&D limited partnerships to raise funds will place increased pressure on the new firms to begin showing profits. If NBFs do not begin showing profits within the time frame expected by investors, additional financing from public offerings and R&D limited partnerships may be difficult to obtain.

The future performance of NBFs now extensively using the stock market and R&D limited partnerships for financing may influence the availability of financing for other firms seeking capital in the future. If some of these companies do not begin to manufacture soon in order to generate product revenues, investors may lose confidence in many of the firms' ability to commercialize biotechnology.

In the United States, venture capital is generally more difficult to obtain for later rounds of financing than for initial rounds, in part because venture capitalists are more eager to invest in the earlier rounds to maximize their investment returns. The difficulty in getting subsequent financing for production scale-up may prove to be an insurmountable problem for some NBFs; the ability to self-finance may still be 5 to 10 years away.

Of all the six competitor countries, the United States has the most favorable tax environment for capital formation and financing small firms. Tax incentives, more than government funding, are used in the United States to stimulate business and encourage R&D expenditures. Thus, R&D limited partnerships, low capital gains tax rates, R&D tax credits (due to expire in 1985), and subchapter S provisions all benefit small firms.

In Japan and the European competitor countries, venture capital has played a very minor role in the commercialization of biotechnology, because these countries do not have tax provisions that promote the formation of venture capita-1 and investment in high-risk ventures. As a consequence, few NBFs exist outside the United States. Instead, established foreign companies have initiated efforts to commercialize biotechnology because they generally can finance R&D activities through retained earnings. Established companies also have access to financing from bank loans. Additionally, the governments of Japan, the United

Kingdom, the Federal Republic of Germany, and France have provided the private sector with public funds for biotechnology.

After the United States, Japan has the most financing available for companies using biotechnology. The Japanese Government has made the commercialization of biotechnology a national priority and is financing cooperative interindustry biotechnology projects. Most of the established companies commercializing biotechnology in Japan have at least one bank as a major shareholder that provides the company with low-interest loans for R&D. Wealthy individual investors in Japan, although few in number, have also provided some risk capital for new ventures.

Tax incentives relevant to established companies commercializing biotechnology are those which stimulate R&D investments and those which encourage capital formation. Corporate tax rates are also important. For purposes of international comparisons, the most reliable basis is the overall effective corporate tax rate. Unlike statutory rates, the effective rate takes into account different definitions of taxable income and treatments of depreciation. Available studies suggest that Switzerland, followed by Japan and the United Kingdom, have the lowest effective corporate tax rates. The effective rates in the United States, the Federal Republic of Germany, and France are higher and about equal.

Government Funding of Basic and Applied Research.-The objective of basic research is to gain a better understanding of the fundamental aspects of phenomena without goals toward the development of specific products or processes. Such research is critical to maintaining the science base on which a technology rests and to stimulating advances in a technology. Basic research is usually conducted by academic researchers who receive government funds. The objective of applied research is to gain the knowledge needed to supply a recognized and specific need, through a product or process. Such research is usually funded by industry. Generic applied science can be viewed as bridging a gap between basic science done mostly in universities and applied, proprietary science done in industry for the development of specific products. Such research is aimed at the solution of general problems that are associated with the use of a technology by industry. Generic applied research areas in biotechnology, for instance, include development of bioreactors, screening of microorganisms for potential products, and better understanding of the genetics and biochemistry of industrially important microorganisms. Support of basic science and of generic applied science is generally viewed as the responsibility of government, because it ultimately contributes to the public good and because it is high risk and too expensive for individual firms.

Controversy exists over the relative importance of national support of basic and applied science. Some argue that since the findings of basic research are readily accessible worldwide because they are published in journals with international distribution, strong government support for basic research is therefore not required for the maintenance of a leading position in the development of a technology. Others argue that the development of a technology within a country will progress faster if companies have access to local basic research scientists for consulting and contractual arrangements. Domestic technology transfer can help give industry a lead in innovation.

Of the competitor countries, the United States, both in absolute dollar amounts and in relative terms, has the largest commitment to basic research in biological sciences Like the United States, the Federal Republic of Germany, the United Kingdom, and Switzerland have a strong basic science base. On the other hand, the U.S. Government% commitment to generic ap plied research in biotechnology is relatively small The governments of Japan, the Federal Republic of Germany, and the United Kingdom fund a significant amount of generic applied science in biotechnology.

During the past few decades, the U.S. Government increased its commitment to basic biological sciences, although this commitment has decreased in the last few years. While the Government was increasing its commitment to basic science, there was a concomitant decrease in its commitment to generic applied fields such as bioprocess engineering and applied microbiology.

The rationale for this policy has been that most applied science, regardless how general, is the responsibility of industry. This policy has contributed to a widening scientific gap between purely basic research funded by the U.S. Government and short-term, relatively product-specific applied research funded by private industry. In fiscal year 1983, the Federal Government spent $511 million on basic biotechnology re search • compared to $6.4 million on generic applied research in biotechnology. The relatively low level of U.S. Government funding for generic applied research in biotechnology may cause a bottleneck in this country's bio technology commercialization efforts,

The Japanese Government, in contrast, is devoting proportionately more public funding to the solution of generic applied science problems than to basic research, The pattern of funding in Japan may reflect a policy of placing a greater priority on generic applied research in lieu of basic research because the Japanese may rely on the United States and other countries to prove the early feasibility of new technologies for commercialization. This strategy worked well in the semiconductor industry, and Japan may very well attain a larger market share for biotechnology products than the United States because of its ability to rapidly apply results of basic research available from other countries,

Personnel Availability and Training.—Adequately trained scientific and technical personnel are vital to any country's industrial competitiveness in biotechnology. For the most part, countries with good science funding in a field also have a good supply of well-trained people in that field.

The commercial development of biotechnology will require several specific types of technical personnel. Especially important categories include specialists in rDNA and MAb technology such as molecular biologists and immunologists; specialists in scale-up and downstream processing such as microbiologists, biochemists, and bioprocess engineers; and specialists for all aspects of biotechnology such as enzymologists and cell culture

---

• From $20 million to $30 million of the $511 million may actually be generic applied research, because definitions of biotechnology differ among agencies.

specialists. Scale-up personnel will become more important as companies using biotechnology move into production.

The United States currently has a competitive edge in the supply of molecular biologists and immunologists able to meet corporate needs, in part because the U.S. Government has provided substantial funding since World War 11 for basic life sciences research in U.S. universities The supply of Ph. D. plant molecular biologists and scaleup personnel in the United States, however, may be inadequate. Like the United States, the United Kingdom and Switzerland have funded life sciences well and have a sufficient supply of basic biological scientists. Unlike the United States, Japan, the United Kingdom, and the Federal Republic of Germany maintained a steady supply of both industrial and government funding for generic applied microbiology and bioprocess engineering in the past few decades and have adequate personnel in these fields. In Japan and the Federal Republic of Germany, slight shortages of molecular biologists and immunologists exist; Japanese companies are seeking to train personnel abroad. France appears to have shortages in all types of personnel.

The training of personnel is important to the continuing commercialization of biotechnology. The United States has, for the most part, good training programs for basic scientists. Specialists in plant molecular biology may be in short supply now, but training in this discipline can be readily achieved with interdisciplinary programs in biology departments in universities. On the other hand, the United States does not have more than a handful of training programs for personnel in the more applied aspects of biotechnology, nor does it have Government programs, such as training grants, to support training in these fields. The training of bioprocess engineers and industrial microbiologists will require greater interdisciplinary cooperation between engineering and biology departments within universities.

The United States promotes and funds the training of foreign nationals in laboratories in the United States, yet funds very little training of Americans abroad. Foreign countries have many significant research programs in biotechnology that U.S. researchers could be visiting were funding available.

FACTORS OF MODERATE IMPORTANCE

The three factors found to be of moderate importance to international competitiveness in biotechnology are health, safety, and environmental regulation; intellectual property law; and university/industry relationships.

Health, Safety, and Environmental Regulation.—The analysis of the effect of health, safety, and environmental regulation on competitiveness in biotechnology was made by determining how restrictive a country's laws would be with respect to marketing biotechnology products and whether there were any uncertainties about their application. The analysis focused on the drug laws for humans and animals and, to a lesser extent, on laws governing the production of chemicals and the deliberate release of novel organisms into the environment. In all the competitor countries, there is some uncertainty as to the environmental regulation governing the deliberate release into the environment of genetically manipulated organisms.

The only government controls directed specifically toward biotechnology are the rDNA guidelines adopted by the six competitor countries. They are essentially voluntary and directed primarily at research. Their containment and oversight provisions have been substantially relaxed since they were originally adopted, and this trend is expected to continue. The United States has the most liberal guidelines, whereas Japan has the most stringent.

Since companies generally approach domestic markets first, the countries with the least stringent regulation may have products on the market earlier. Japan has the most stringent health and safety regulation for pharmaceuticals and animal drugs, followed by the United States. Switzerland appears to be the most liberal. Thus, the regulatory environment favors the European companies over those of Japan and the United States reaching their own domestic markets sooner for pharmaceuticals and animal drugA In the United States, the Food and

Drug Administration has taken the position that rDNA products whose active ingredients are identical to ones already approved or to natural substances will still need to go through the new product approval process. However, data requirements may be modified and abbreviated. This appears not to be the situation in the competitor countries, although there have not been definitive pronouncements by their regulatory agencies.

Regulation may also influence where companies locate their production facilities. A country with liberal regulation may attract production facilities and, as a consequence, gain access to technology, Alternatively, companies may set up facilities in the United States and Japan regardless of regulation because of market size and as a way to avoid certain nontariff trade barriers on imports. NBFs may not have the capital to establish foreign subsidiaries in order to avoid regulatory barriers. Thus, they may beat a competitive disadvantage with respect to larger firms for entering world markets.

Countries wishing to market their products abroad will have to abide by the regulations of the countries to which they are exporting. Thus, countries can control access to their domestic markets by the regulations they impose. This is a form of nontariff trade barrier. These barriers are considered further in the discussion of trade policy.

Intellectual Property Law.—The ability to secure property interests in or otherwise protect processes, products, and knowhow will encourage development of biotechnology, because it provides incentives for a private company to invest the time and money for R&D. Without the ability to prevent competitors from taking the results of this effort, many new and risky R&D projects would not be undertaken. Thus, a strong intellectual property law system will enhance a country's competitiveness in biotechnology.

The areas of intellectual property law most relevant to biotechnology are those dealing with patents, trade secrets, and plant breeders' rights. These areas work together as a system; an invention may be protected by one or more of them, and if one has disadvantages, a company can look to another. Thus, to the extent that a country's

intellectual property law provides several alternative ways for companies to protect biotechnological inventions, it is more likely to be competitive in biotechnology.

The patent laws of the competitor countries provide fairly broad protection for biotechnological inventions, but the laws differ to some degree in the types of inventions that are protected, the effect of publication on patent rights, and the requirements regarding public disclosure of the invention, which is the quid pro quo for the grant of the patent. The United States provides the widest coverage, Patents are available for living organisms (including plants and possibly animals), their products, their components, and methods for making or using all of these. In addition, patents can be granted on therapeutic and diagnostic methods, In the United Kingdom, the Federal Republic of Germany, France, Switzerland, and Japan, patent coverage is almost as broad, but patents are not permitted on plants and animals nor on therapeutic and diagnostic methods. In addition, Switzerland does not permit patents on microaganisms. In Japan, the relatively strict guidelines governing rDNA research also may bar patents on those genetically manipulated organisms viewed as hazardous.

With regard to the effect of publication on patent rights, the United. States also has a slight advantage over the other countries analyzed here. The four European countries do not permit a patent to be granted to an inventor who has disclosed his or her invention in a publication before the patent application is filed, assuming the disclosure enables others to make it. This absolute novelty requirement is viewed as impeding the free exchange of scientific information and possibly providing a disincentive for scientists to seek patent rights. The United States, on the other hand, provides a l-year grace period between the date that an inventor publishes an article and the date on which the patent application must be filed. Japan provides a 6-month grace period for certain activities, such as presenting scientific papers. The U.S. advantage is limited, however, because when U.S. inventors wish to secure patents in other countries, they must refrain from publication in order to protect their patent rights in those countries.

The patent law requirement that an invention be described in sufficient detail so that it could be replicated creates unique problems for biological inventions. Since a living organism generally cannot be described in writing with sufficient specificity to allow others to make and use it, granting of patents on such organisms and methods of using them generally is contingent on their deposit in a public depository. However, these deposits, in effect, turn over the factory for making a product to one's competitors, unlike patents in other technologies. The four European countries, and particularly the Federal Republic of Germany, place restrictions on access to such deposits that may be advantageous for their inventors.

Most aspects of biotechnology lend themselves to protection as trade secrets, and owners of such technology may rely on trade secrets when patent rights are uncertain or when they judge trade secrecy to be more advantageous. All of the competitor countries protect trade secrets relating to biotechnology, but the Federal Republic of Germany and, to a lesser extent, Switzerland, provide the greatest degree of protection. Japan appears to provide the least degree of protection.

All of the competitor countries recognize property rights in new varieties of plants, but the United States provides the greatest degree of protection. Protection in the United States is most favorable because the plant breeder has the greatest number of options among which to choose in securing property rights for a new variety of plant, including pursuing a patent under the traditional patent laws.

In the final analysis, the U.S. intellectual property system appears to offer the best protection for biotechnology of any system in the world, thus providing the United States with a competitive advantage with regard to this factor. This advantage results from the fact that the system provides the widest choice of options for protecting biotechnological inventions, the broadest scope of coverage, and some of the best procedural safeguards.

University/Industry Relationships.-A factor that has moderate overall importance is the relationship that exists between universities and industries. Interest in the commercial potential of biotechnology has dramatically increased university/industry interactions, especially in the United States. Established U.S. and foreign companies have invested substantial amounts of money in U.S. universities doing work in biotechnology in order to gain a "window on the technology." Many university/industry agreements in biotechnology focus on research directed toward applications of biotechnology in a specific industrial sector, whereas other university/industry agreements are directed at many applications of biotechnology. The various agreements in the United States appear to be working well and fears concerning conflict of interest and commingling of Government and industry funds have diminished.

The increase of industry funding of university research in the United States in several disciplines came at a time when Federal funding of science was decreasing in constant dollars. Although the infusion of industry funds to the U.S. universities has been substantial, it accounts for only a small fraction (less than 10 percent) of the total funding of university research. In some university departments, however, such as electrical engineering, chemistry, and possibly now molecular biology, industrial funding of university research may exceed 10 percent. Even with the increase in industrial support, industrialists agree that private funding can never replace Federal funding of basic science research if past and current levels of basic research are to continue.

University/industry interactions are a very effective way of transferring technology from a research laboratory to industry. Such interactions promote communication between industrialists and academicians, a two-way interaction that benefits both sides. Industrial scientists learn the latest techniques and research results, while academicians gain increased familiarity with challenges of industrial R&D.

Neither Japan nor the European competitor countries identified in this assessment have as many or as well-funded university/industry relationships as the United States does, but varying degrees of cooperation do exist. In Japan, the ties between university applied research departments and industry have always been close. Additionally, the Japanese Government is implementing new policies to encourage closer ties between basic research scientists and industry. In the Federal Republic of Germany, the Federal Ministry of Science and Technology (BMFT, Bundesministerium fur Forschung und Technologies) has a history of promoting close contact between academia and industry and is cosponsoring with industry many projects important to biotechnology. Switzerland encourages communication between individuals in academia and industry, and relationships are easy to maintain. The universities in both the United Kingdom and France have had very few ties with industry in biotechnology, but the governments of both countries have recently set up programs designed to encourage university/industry relationships.

Industrial funding for research in American universities is helping to promote the transfer of technology. However, the multimillion dollar arrangements that have characterized the initial relationships in biotechnology are most likely short term and will probably become less important as the firms develop in-house expertise and their research becomes more applied. **As in other fields, consulting and contractual research agreements are likely to predominate in university/industry relationships in bio technology in the future.**

LEAST IMPORTANT FACTORS

The least important of the 10 factors analyzed were found to be antitrust law; international technology transfer, investment, and trade; government targeting policies in biotechnology; and public perception. Any of these factors, however, could become important as the technology develops and products reach the marketplace.

Antitrust Law.—Antitrust laws are based on the general economic assumption that competition among a country's industries will result in greater productivity, innovation, and general consumer benefits than will cooperation. Recently there has been much public debate about whether US. antitrust laws have, in fact, accomplished these goals in all cases and whether they place U.S. companies at a competitive disadvantage in the international marketplace when foreign companies face allegedly less restrictive antitrust laws.

The antitrust laws of the United States and the other major competitors in biotechnology are generally similar in that they prohibit restraint of trade and monopolization. However, the foreign laws generally provide for exemptions and vest much discretion with the enforcement authorities, especially in Japan. Thus, in practice, they are often less restrictive than in the United States. In addition, countries differ in the consequences to firms for failure to comply with antitrust laws, In the United States, the consequences of noncompliance can be more severe than in the competitor countries because private, in addition to Government, suits can be brought against alleged antitrust violators, and treble damages are assessed if a violation is found.

U.S. companies commercializing biotechnology face no major antitrust compliance problems, because the lack of concentration and the absence of measurable markets mean that most types of joint research arrangements would not be anticompetitive. Technology licensing agreements can raise antitrust concerns, but these generally are not unique to biotechnology. However, there is some degree of uncertainty about the scope and applicability of the antitrust laws to R&D joint ventures and licensing agreements. This uncertainty, plus the expense of litigation and the threat of treble damages, could deter some activities that might lead to innovation in biotechnology, thus limiting the ability of U.S. companies commercializing biotechnology to exploit their technology. * For these reasons, the current U.S. antitrust laws may have some modest adverse effect on biotechnology.

International Technology Transfer, Investment, and Trade.—Technology transfer across national boundaries can be promoted or inhibited by export control laws and by laws governing international joint ventures and technology licensing. Most export controls are directed at overseeing technology transfer for national security reasons, and the concept of national security is fairly narrowly interpreted in all of the competitor countries except the United States. Therefore export controls may not be very

---

● In addition, the rigid application of certain "per se rules" in the area of licensing may actually lead to anticompetitive results.

important for the international development of biotechnology. However, the export controls of the United States, which are the most restrictive of the competitor countries, include the control of pharmaceuticals and of many microorganisms that potentially could be used in biotechnology product production. These controls may have a slightly adverse affect on the competitive ness of U.S. companies commercializing biotechnology because they could cause delays that result in sales' being lost to foreign competitors. U.S. export control laws may need clarification as biotechnology products proceed to the marketplace because there is some uncertainty as to what products or data will be restricted. In addition, the current U.S. export control law expired in October 1983. While it is virtually certain that a new law will be passed, the form that law will take is still unclear.

The U.S. Government has no laws governing international joint ventures and technology licensing among U.S. and foreign companies. As a consequence, technology can be transferred readily to other countries. The predominance of NBFs in the United States and their need for capital has led to the formation of many transnational joint ventures involving NBFs, Because of this, the United States appears to be transferring more technology outside of its national borders than are other countries at the present time. However, as biotechnology products reach the market, foreign firms will probably set up subsidiaries in the United States in order to have access to U.S. markets. If this happens, the United States could become a net importer of technology.

In contrast with the United States, France and Japan have Government programs for the review of potential transnational agreements, but it is uncertain whether such programs help or hinder the transfer of technology into those countries. As of now, laws governing the transfer of technology are not very important to the U.S. competitive position in biotechnology. However, if other countries establish themselves more favorably in world markets, the current outward flow of technology from the United States may hurt the U.S. competitive position.

Foreign exchange and investment control laws help prevent access to domestic markets and tech-

nology by foreign firms. The United States has the fewest controls, whereas Japan and France have the most control mechanisms. Japanese controls exist in the form of nontariff barriers such as ministerial review and screening of foreign investments and licensing agreements with respect to a number of criteria ranging from national security to competition with other Japanese business. Ministries also have the power to designate specific companies for special controls on foreign ownership. In France, the Government has the ability to object or order alteration of licensing agreements and foreign investments. Foreign direct investment in certain domestic industries is not encouraged. Thus, U.S. **markets are the most accessible to foreign firms and therefore the most vulnerable to foreign competition,** whereas Japanese and French markets are the least accessible and the most protected against foreign competition.

Trade policy was assessed by examining the competitor countries' abilities to protect domestic industries from imports and to control foreign investment in domestic industries. **Trade policy is not important for the commercialization of biotechnology today because of the small number of products that have reached the market and because trade in biotechnologically produced products is not likely to raise any unique trade issues.** However, trade policy will become increasingly important as more products reach the marketplace, especially in the area of pharmaceuticals, where significant nontariff barriers, such as conforming to country standards with appropriate testing data, quality control standards, and packaging requirements exist. Problems with nontariff barriers are now being negotiated with Japan and other countries including the European Economic Community, and it apears as though some trade barriers may become less stringent.

Government Targeting Policies in Biotechnology.-The governments of four of the competitor countries-Japan, the Federal Republic of Germany, the United Kingdom, and France-have instituted comprehensive pro grams to help domestic companies develop certain areas of biotechnology. The targeting policies are intended to reduce economic risk and lessen corporate duplication in biotechnology R&D. A variety of policy measures are used within each country. In Japan and West Germany, the Governments carry out their policies mostly through projects that combine the resources of the Government and private companies to meet specific objectives set by the Government. The United Kingdom and France have adopted a different approach; they support startup of small firms, which are expected to commercialize the results of Government-funded basic and applied research.

At this early stage, any evaluation of the eventual success of foreign targeting pro grams is preliminary. History has shown that even the best thought-out targeting policies do not guarantee competitive success. Whether targeting policies of foreign governments in biotechnology are superior to the U.S. Government policy of funding basic research in the life sciences and encouraging R&D in all industries with tax credits remains to be seen. Though targeting policies are not of great importance when compared to other competitive factors, they could tip the balance of a competitive position in the future.

Public perception.—Public perception of the risks and benefits of biotechnology is of greater importance in countries with representative, democratic forms of government than it is in countries with other forms of government, simply because of the greater attention paid to public opinion in democracies and the independence of the media. Therefore, public perception could influence commercialization of biotechnology in all of the countries examined here. As a factor influencing competitiveness, however, public perception is probably of greater importance in the United States than in the other competitor countries. Historically, the American public has been more involved than the public in Japan or the European countries with issues pertaining to genetic research and technology (e.g., issues regarding the safety of rDNA research).

In all countries, the importance of public perception as a factor influencing competitiveness will be greatly increased in the event of an accident or perceived negative conse quence of biotechnology. Particularly in such a case, the level of scientific and technological literacy in the various competitor countries becomes important, as judgments must be made concerning complex issues. In the United States, survey data show that only a small fraction of the public is fully informed about genetics in general and therefore, probably, about biotechnology in particular. Survey data also suggest that there is public apprehension concerning applied genetics. Thus, an accident associated with biotechnology could arouse strong public reaction in the United States, a reaction that might be greater than in the competitor countries.

Given the lack of public knowledge in the United States, it is particularly important that the media play a responsible role with respect to biotechnology. The role of the media already extends beyond mere reporting of the facts, by virtue of the events and issues the media elect to cover.

At the current time, public perception is not an important factor in the commercialization of biotechnology. However, the volatility of a potential public response must be noted. Were thereto be an accident due to commercial biotechnology, the public's reaction could be extremely important to the future of biotechnology.

Other influences on competitiveness in biotechnology

Three other considerations that should be noted in evaluating competitive positions in the commercialization of biotechnology are, for each country, historical patterns of industrial commercialization, the availability of natural resources, and cultural attitudes toward risk-taking.

Historically, industries in some countries have moved research results into commercialization rapidly, while industries in other countries have moved more slowly. This observation is especially important in this analysis of biotechnology. For instance, the United Kingdom has a good science base, trained personnel, and industries that could be using these new technologies; however, the United Kingdom may not be a major contender in the commercialization of biotechnology mainly because it does not have a history of rapid commercialization. On the other hand, both the United States and Japan historically commercialize scientific advances rapidly.

Another historical consideration is the quantity of sales of specific products in a country. For example, Japan's per capita consumption of pharmaceuticals is significantly higher than that of the other competitor countries; therefore, Japan may have more interest than other countries have in applying biotechnology to the production of pharmaceuticals. In other words, cultural differences will probably play a role in determining the markets each country will attempt to dominate.

The absence or presence of certain natural resources may also determine how quickly a country moves into the commercialization of biotechnology. For instance, Japan does not have domestic petroleum resources. Because biomass can potentially replace petroleum as a feedstock in the chemical industry, Japan may be more in-terested in applying biotechnology in the chemical industry than a country, such as the United Kingdom, which has domestic petroleum resources, The United States, a country that produces excesses of grain each year, may find commercialization of processes that can use grain as a feedstock particularly attractive. However, it is too early to predict the degree to which natural resources will determine the commercial applications of biotechnology a country may undertake.

The United States, as a general rule, is not averse to risk-taking in business. Risk-taking is a part of the American lifestyle. European countries are more risk averse. Since investment in biotechnology is considered risky, countries that are more risk averse are less likely to move rapidly to commercialize biotechnology.

Conclusion

The unique complementarities between established and new firms, the well-developed science base, the availability of finances, and an entrepreneurial spirit have been important in giving the United States its present competitive advantage in the commercialization of biotechnology. In order to maintain this advantage, increased funding of research and training of personnel in basic and generic applied sciences, especially bioprocess engineering and industrial microbiology, may be necessary. The United States may also need to be concerned with the continued availability of finances for NBFs until they are self-supporting. on most of the other factors influencing competitiveness, the United States rates very favorably, although there are changes in laws and policies that could potentially improve or help maintain the U.S. competitive position. These changes include clarification and modification of particular aspects of intellectual property law; health, safety, and environmental regulation; antitrust law; and export control law.

Japan will be the most serious competitor of the United states in the commercialization of biotechnology. Japan has a very strong bioprocess technology base on which to build, and the Japanese Government has specified biotechnology as a national priority. The demonstrated ability of the Japanese to commercialize rapidly developments in technology will surely manifest itself in biotechnology.

The Federal Republic of Germany, the United Kingdom, Switzerland, and France lag behind the United States and Japan in the commercialization of biotechnology. The European countries generally do not promote risk-taking, either industrially or in their government policies. Additionally, they have many fewer companies commercializing biotechnology. Thus, the European countries are not expected to be as strong general competitors in biotechnology as the United States and Japam In markets for specific products, including some pharmaceuticals, specialty chemicals, and animal agriculture products, however, some European companies will undoubtedly be strong international competitors.

Issues and options

Congressional issues and options for improving the competitive position of the United States in biotechnology are presented at the end of most of the chapters in part IV. To improve the competitive position of the United States, legislation could be directed toward any of the 10 factors OTA identified as influencing competitiveness, although coordinated legislation directed toward all of the factors might be more effective in promoting U.S. biotechnology efforts. The chapters in part IV discuss only those options that are specific to the development of biotechnology. Some of the options presented in part IV are limited and straightforward, such as some options concerning health and safety regulation and R&D limited partnerships. Other options are much broader with potentially large political, ethical, and financial considerations. Some examples of the latter include establishing university/industry cooperative research centers, regulating the deliberate release of genetically manipulated organisms into the environment, and changing patterns of research funding. Thus, the adoption of some options may occur more rapidly than others.

Policy options in some areas are not specific to biotechnology but apply to high technology or industry in general. These options are to:

- improve U.S. science and engineering education and the retraining of industrial personnel,
- change U.S. antitrust law to promote more research collaboration among domestic firms,
- regulate imports into the United States to protect domestic industries,
- regulate the transfer of technology from the United States to other countries, and
- target specific industries or technologies for Federal assistance.

There are many arguments for and against these options that are beyond the scope of this report. Because of their broad applicability to industry in general, these options are not discussed in part IV. It is important to note, however, that legislation in any of these areas could affect the development of biotechnology and potentially have a large influence on the U.S. competitive position.

[2]

The Pharmaceutical Industry and the Revolution in Molecular Biology: Interactions Among Scientific, Institutional, and Organizational Change

REBECCA HENDERSON, LUIGI ORSENIGO, AND GARY P. PISANO

I. Introduction

The last 25 years have seen a revolution in the biological sciences that has had several dramatic effects on the global pharmaceutical industry.[1] These effects raise a number of fascinating questions about patterns of industrial evolution and about the interaction of scientific, organizational, and institutional changes. Although a cursory analysis might suggest that the revolution in molecular biology can be interpreted simply as a classic "Schumpeterian" event, in which the early days of the industry were characterized by high rates of entry and incumbents were gradually supplanted by a new breed of innovators, it has several features that make it quite distinctive from the "traditional" model.

First, whereas the traditional model is derived largely from the study of radical shifts in *engineering* knowledge (Abernathy and Utterback, 1978; Tushman and Anderson, 1986), the revolution in molecular biology represented a shift in the *scientific* knowledge base of an industry. Second, despite the sweeping nature of the molecular revolution, incumbent pharmaceuticals companies have *not* been swept away by new entrants. Third, and relatedly, the relationships between incumbents and entrants has entailed not only competition, but also cooperation and the establishment of complex interactions between firms. Finally, and perhaps most importantly, the revolution did not create a monolithic new paradigm of technical development, but instead created two quite

The authors would like to express their appreciation to the project participants and to Louis Galambos, Richard Nelson, Dick Rosenbloom, David Mowery, Anita McGahan, and Sidney Winter in particular for their helpful comments and suggestions on earlier drafts. The usual disclaimers apply.

[1] This revolution includes the dramatic advances in genetics and genetic engineering that popularly go by the name of "biotechnology," as well as important advances in peptide chemistry and in molecular and cell biology. For the purposes of this chapter we define these advances collectively as the "molecular biology revolution."

R. HENDERSON, L. ORSENIGO, AND G.P. PISANO

distinct trajectories of development that have only recently been combined: the use of biotechnology as a tool for the *production* of proteins whose therapeutic properties were already well understood, and the use of biotechnology as a tool in the *search* for entirely new therapies.

The revolution in molecular biology is also intriguing because despite the fact that it is global in nature, and despite the fact that scientific advances are normally thought of as creating a "free good," or as being instantaneously available throughout the world, the revolution is producing quite different changes in industry structure across different regions of the world. In the United States, it has spawned both the emergence of radically new actors – the new specialized biotechnology start-ups – as well as the gradual creation of biotechnology programs within established firms. In Europe, responses have differed dramatically from country to country. Despite a strong research tradition in molecular biology, in general Europe has not witnessed the creation of a specialized biotechnology sector. Several of the leading Swiss and British incumbent firms have attempted to build strong biotechnology capabilities through a combination of internal development and an aggressive program of external acquisition, but the French, German, and Italian firms have been much slower to adopt the new techniques. In Japan, where historically the pharmaceutical industry has been rather less innovative than its Western rivals, most substantial investments in biotechnology have been made by firms with historical strengths in fermentation based industries, and the large pharmaceutical companies have been particularly slow to embrace the new technology.

The case of the molecular biology revolution thus lends itself to a study of the detailed mechanisms of industrial transformation at the firm and industry levels and of the interaction and coevolution of scientific knowledge, on the one hand, and organizational capabilities, industry structure, and institutional context on the other. In this chapter we focus particularly on regional differences in the impact of biotechnology on pharmaceutical industry structure and the nature of competition, and thus hope to gain insight into the role of specific features of the "national systems of innovation" in shaping firm capabilities and the diffusion of new technology (Nelson, 1992).

Section II discusses briefly the structure of the pharmaceutical industry before the advent of the molecular biology revolution, focusing particularly on early patterns of competition, the variety of institutional structures across the world within which innovation was conducted, and the nature of pharmaceutical demand. Section III provides some background on the history and scope of the molecular biology revolution and examines its impact on the process of drug research and development,

The Revolution in Molecular Biology

focusing particularly on its impact on the organizational capabilities fundamental to effective pharmaceutical R&D. Section IV describes the evolution of the structure of the pharmaceutical industry following the molecular revolution across the regions of the developed world. Section V speculates as to the ways in which institutional context has shaped the development of competence and the adoption of the new technologies across firms. Section VI concludes the chapter with implications for our broader understanding of the coevolution of science, national systems of innovation, organizational capabilities, and patterns of competition.

II. Historical Background: The Pharmaceutical Industry Before the Molecular Biology Revolution

The history of the pharmaceutical industry can be usefully divided into three major epochs. The first, corresponding roughly to the period 1850–1945, was one in which little new drug development occurred, and in which the minimal research that was conducted was based on relatively primitive methods. The large-scale development of penicillin during World War II marked the emergence of the second period of the industry's evolution, which we date somewhat arbitrarily as running from 1945 to roughly 1990. This period was characterized by the institution of formalized in-house R&D programs and relatively rapid rates of new drug introduction. During the early part of the period the industry relied largely on so-called random screening as a method for finding new drugs, but in the seventies the industry began a transition to "guided" drug discovery or "drug development by design," a research methodology that drew heavily on advances in molecular biochemistry, pharmacology, and enzymology. The third epoch of the industry, and the one constituting the main focus of this chapter, has its roots in the seventies but did not come to full flower until quite recently as the use of the tools of genetic engineering in the production and discovery of new drugs has come to be more widely dispersed.

In this section, we briefly review the first two of these periods. Understanding the evolution of the industry during these earlier periods is important since they played a critical role in molding not only the industrial and institutional structure of the industry, but also the organizational capabilities of individual firms that continue to influence the industry today. Indeed, we shall argue that the impact of biotechnology on industry structure cannot be understood without an appreciation of the fact that the techniques of guided drug discovery did not diffuse uniformly, and that this variation in diffusion was critically important

R. HENDERSON, L. ORSENIGO, AND G.P. PISANO

in shaping incumbent firms' response to the revolution in molecular biology.

A. Early History

By almost any measure pharmaceuticals is a classic "high-technology" or "science-based" industry.[2] Yet drugs are as old as antiquity. For example, the Ebers Papyrus lists 811 prescriptions used in Egypt in 550 B.C. Eighteenth-century France and Germany had pharmacies where pharmacists working in well-equipped laboratories produced therapeutic ingredients of known identity and purity on a small scale. Mass production of drugs dates back to 1813, when J. B. Trommsdof opened the first specialized pharmaceutical plant in Germany. However, during the first half of the 19th century, there were virtually no standardized medicines for treating specific conditions. A patient instead would be given a customized prescription which would be formulated at the local pharmacy by hand.

The birth of the modern pharmaceutical industry can be traced to the mid-19th century with the emergence of the synthetic dye industry in German and Switzerland. At that time, Switzerland and Germany were the leading centers of the synthetic dye industry (this was due in part to the strength of German universities in organic chemistry, and in part to the fact that Basel was close to the leading silk and textile regions of Germany and France). During the 1880s, the medicinal effects (such as antiception) of dyestuffs and other organic chemicals were discovered. It was thus initially Swiss and German chemical companies such as Ciba, Sandoz, Bayer, and Hoechst, leveraging their technical competencies in organic chemistry and dyestuffs, who began to manufacture drugs (usually based on synthetic dies) later in 19th century. For example, salicytic acid (aspirin) was first produced in 1883 by the German company Bayer.

In the United States and the United Kingdom, mass production of pharmaceuticals also began in the later part of the 19th century. However, the pattern of development in the English-speaking world was quite different from that of Germany and Switzerland. Whereas Swiss and German pharmaceutical activities tended to emerge within larger chemical producing enterprises, the United States and United Kingdom witnessed the birth of specialized pharmaceutical producers such as Wyeth (later American Home Products), Eli Lilly, Pfizer, Warner-

---

[2] Including R&D intensity, innovative output, and use of new scientific concepts.

The Revolution in Molecular Biology

Lambert, and Burroughs-Wellcome. Up until World War I German companies dominated the industry, producing approximately 80% of the world's pharmaceutical output.

In the early years the pharmaceutical industry was not tightly linked to formal science. Until the 1930s, when sulfonamide was discovered, drug companies undertook little formal research. Most new drugs were based on existing organic chemicals or were derived from natural sources (e.g., herbs) and little formal testing was done to ensure either safety or efficacy. Harold Clymer, who joined SmithKline in 1939, noted:

[Y]ou can judge the magnitude of [SmithKline's] R&D at that time by the fact I was told I would have to consider the position temporary since they had already hired two people within the previous year for their laboratory and were not sure that the business would warrant the continued expenditure. (Clymer, 1975)

World War II and wartime needs for antibiotics marked the drug industry's transition to an R&D intensive business. Penicillin and its antibiotic properties were discovered by Alexander Fleming in 1928; however, throughout the 1930s, it was produced only in laboratory-scale quantities and was used almost exclusively for experimental purposes. With the outbreak of World War II, the U.S. government organized a massive research and production effort that focused on commercial production techniques and chemical structure analysis. More than 20 companies, several universities, and the Department of Agriculture took part. Pfizer, which had production experience in fermentation, developed a deep-tank fermentation process for producing large quantities of penicillin. This system led to major gains in productivity and, more important, laid out an architecture for the process and created a framework in which future improvements could took place.

The commercialization of penicillin marked a watershed in the industry's development. Partially as a result of the technical experience and organizational capabilities accumulated through the intense wartime effort to develop penicillin, as well as the recognition that drug development could be highly profitable, pharmaceutical companies embarked on a period of massive investment in R&D and built large-scale internal R&D capabilities. At the same time there was a very significant shift in the institutional structure surrounding the industry. Whereas before the war public support for health related research had been quite modest, after the war it boomed to unprecedented levels, helping to set the stage for a period of great prosperity.

R. HENDERSON, L. ORSENIGO, AND G.P. PISANO

B. Patterns of Competition and Industrial Organization: 1950–1990

The period from 1950 to 1990 was a golden age for the pharmaceutical industry, as the industry in general, and particularly the major U.S. players – firms such as Merck, Eli Lilly, Bristol-Myers, and Pfizer – grew rapidly and profitably. R&D spending literally exploded and with this came a steady flow of new drugs. Drug innovation was a highly profitable activity during most of this period. Statman (1983), for example, estimated that accounting rates of return on new drugs introduced between 1954 and 1978 averaged 20.9% (compared to a cost of capital of 10.7%). Between 1982 and 1992, firms in the industry grew at an average annual rate of 18%. During the early 1980s, double-digit rates of growth in earnings and return on equity were the norm for most pharmaceutical companies and the industry as a whole ranked among the most profitable in the United States.[3]

A number of structural factors supported the industry's high average level of innovation and economic performance. One was the sheer magnitude of both the research opportunities and the unmet needs. In the early postwar years, there were many physical ailments and diseases for which no drugs existed. In every major therapeutic category – from painkillers and antiinflammatories to cardiovascular and central nervous system products – pharmaceutical companies faced an almost completely open field (before the discovery of penicillin, very few drugs effectively *cured* diseases).

Faced with such a "target rich" environment but very little detailed knowledge of the biological underpinnings of specific diseases, pharmaceutical companies invented an approach to research now referred to as "random screening." Under this approach, natural and chemically derived compounds are randomly screened in test tube experiments and laboratory animals for potential therapeutic activity. Pharmaceutical companies maintained enormous "libraries" of chemical compounds and added to their collections by searching for new compounds in places such as swamps, streams, and soil samples. Thousands, if not tens of thousands, of compounds might be subjected to multiple screens before researchers honed in on a promising substance. Serendipity played a key role since in general the "mechanism of action" of most drugs – the specific biochemical and molecular pathways that were responsible for their thera-

---

[3] Note that these figures are based on accounting rates of return. Figures that are recalculated to take account of the fact that the industry spends heavily on advertising and research suggest that rates of return were actually somewhat lower than the accounting figures would suggest (Myers and Howe, 1997).

The Revolution in Molecular Biology

peutic effect – were not well understood. Researchers were generally forced to rely on the use of animal models as screens. For example, researchers injected compounds into hypertensive rats or dogs to explore the degree to which they reduced blood pressure. Under this regime it was not uncommon for companies to discover a drug to treat one disease while searching for a treatment for another.

Although random screening may seem inefficient, it worked extremely well for many years and continues to be widely employed. Several hundred chemical entities were introduced in the 1950s and 1960s and several important classes of drug were discovered in this way, including a number of important diuretics, all of the early vasodilators, and a number of centrally acting agents including reserpine and guanethidine.

Beginning in the early seventies, the industry also began to benefit more directly from the explosion in public funding for health related research that followed the war. Publicly funded research had been important to the industry's health since the war, but it was probably most important as a source of knowledge about the cause of disease. From the middle seventies on, however, substantial advances in physiology, pharmacology, enzymology, and cell biology – the vast majority stemming from publicly funded research – led to enormous progress in the ability to understand the mechanism of action of some existing drugs and the biochemical and molecular roots of many diseases. This new knowledge made it possible to design significantly more sophisticated screens. By 1972, for example, the structure of the renin angiotensive cascade, one of the systems within the body responsible for the regulation of blood pressure, had been clarified by the work of Laragh and his collaborators (Laragh et al., 1972), and by 1975 several companies had drawn on this research in designing screens for hypertensive drugs (Henderson and Cockburn, 1994). These firms could replace ranks of hypertensive rats with precisely defined chemical reactions. In the place of the request "Find me something that will lower blood pressure in rats" pharmacologists could make the request "Find me something that inhibits the action of the angiotensin-2 converting enzyme."

In turn the more sensitive screens made it possible to screen a wider range of compounds. Prior to the late seventies, for example, it was difficult to screen the natural products of fermentation (a potent source of new antibiotics) in whole animal models. The compounds were available in such small quantities, or triggered such complex mixtures of reactions in living animals, that it was difficult to evaluate their effectiveness. The use of enzyme systems as screens made it much easier to screen these kinds of compounds. It also triggered a "virtuous cycle" in that the availability of drugs whose mechanisms of action was well known made

R. HENDERSON, L. ORSENIGO, AND G.P. PISANO

possible significant advances in the medical understanding of the natural history of a number of key diseases, advances which in turn opened up new targets and opportunities for drug therapy (see Gambardella, 1995, and Henderson, 1994, for a fuller discussion of this transition, and Maxwell and Eckhardt, 1990, for a fuller discussion of the role of the public sector in making it possible).

These techniques were *not* uniformly adopted across the industry. For any particular firm, the shift in the technology of drug research from random screening to one of "guided" discovery or "drug discovery by design" was critically dependent on the ability to take advantage of publicly generated knowledge (Gambardella, 1995; Cockburn and Henderson, 1996) and of economies of scope within the firm (Henderson and Cockburn, 1996). Smaller firms, those farther from the centers of public research, and those that were most successful with the older techniques of rational drug discovery appear to have been much slower to adopt the new techniques than their rivals (Henderson and Cockburn, 1994; Gambardella, 1995; Cockburn et al., 1997). There was also significant geographical variation in adoption. Whereas the larger firms in the United States, the United Kingdom, and Switzerland were among the pioneers of the new technology, other European and Japanese firms appear to have been slow in responding to the opportunities afforded by the new science. These differences had significant implications for the industry's later response to the revolution in molecular biology.

As is well known to both practitioners and scholars of innovation, new products do not ensure profits, and the dramatic success of the industry's research efforts need not, in themselves, have generated supranormal returns. Rents from innovation can be competed away unless "isolating mechanisms" (Lippman and Rumelt, 1982) are in place to inhibit imitators and new entrants. For most of the postwar period, pharmaceutical companies (particularly those operating in the United States) had a number of isolating mechanisms working in their favor. Several of these mechanisms, including the strength of intellectual property protection and the nature of the regulatory regime for pharmaceutical products, were institutional in origin and differed significantly across national boundaries. We discuss these types of mechanisms in more detail later. However, it is important to note that the organizational capabilities developed by the larger pharmaceutical firms may also have acted as isolating mechanisms. Consider, for example, the process of random screening itself. As an organizational process, random screening was anything but random. Over time, early entrants into the pharmaceutical industry

274

The Revolution in Molecular Biology

developed highly disciplined processes for carrying out mass screening programs. Because random screening capabilities were based on internal organizational processes and tacit skills, they were difficult for potential entrants to imitate and thus became a source of first-mover advantage. In addition, in the case of random screening, spillovers of knowledge between firms were relatively small since when firms essentially rely on the law of large numbers, there is little to be learned from the competition. These advantages, combined with the presence of scale economies in pharmaceutical research, may help to explain the dearth of new entry prior to the mid-1970s. Until that time, only one company – Syntex, the developer of the oral contraceptive – succeeded in entering the industry. Indeed, many of the leading firms during this period – companies like Roche, Ciba, Hoechst, Merck, Pfizer, and Lilly – had their origins in the "pre-R&D" era of the industry.

The advent of guided drug discovery appears only to have increased the advantages of incumbency. Although it increased the importance of publicly generated knowledge and thus reduced the importance of firm scale, it also appears to have increased returns to scope (Henderson and Cockburn, 1996). Moreover, under both regimes, the organizational capabilities developed to manage the process of drug development and delivery – competencies in the management of large-scale clinical trials, the process of gaining regulatory approval, and marketing and distribution – also appear to have acted as powerful barriers to entry into the industry.

However, significant differences in the competitive and innovative performance at the country level suggest that institutional factors have also played a critical role in generating isolating mechanisms. Although global in nature, the postwar pharmaceutical industry has been dominated by companies from the United States, Switzerland, Germany, and the United Kingdom. French and Italian firms have not played major international roles, and although Japan is the second largest pharmaceutical market in the world and is dominated by local firms (largely for regulatory reasons), Japanese firms have to date been consciously absent from the global industry. Only Takeda, for instance, ranks among the top 20 pharmaceutical firms in the world, and until relatively recently the innovative performance of Japanese pharmaceutical firms has been weak relative to that of their U.S. and European competitors.

We next turn to a discussion of the institutional forces that have shaped the industry. We argue that one of the central factors underlying U.S. success in pharmaceuticals has been a combination of institutional factors that provided powerful inducements to innovation.

R. HENDERSON, L. ORSENIGO, AND G.P. PISANO

C. Institutional Environments

From its inception, the evolution of the pharmaceutical industry has been tightly linked to the structure of national institutions. The pharmaceutical industry emerged in Switzerland and Germany, in part, because of strong university research and training in the relevant scientific areas. German universities in the 19th century were leaders in organic chemistry and Basel, the center of the Swiss pharmaceutical industry, was the home of the country's oldest university, long a center for medicinal and chemical study. In the United States the government's massive wartime investment in the development of penicillin, as we discussed, profoundly altered the evolution of American industry. In the postwar era, the institutional arrangements surrounding the public support of basic research, intellectual property protection, procedures for product testing and approval, and pricing and reimbursement policies have all strongly influenced both the process of innovation directly and the economic returns (and thus incentives) for undertaking such innovation. We now turn to a brief review of these four key areas.

Public Support for Health Related Research. Nearly every government in the developed world supports publicly funded health related research, but there are very significant differences across countries in both the level of support offered and the ways in which it is spent. In the United States, public spending on health related research took off after the Second World War and is now the second largest item in the federal research budget after defense. (Table 7.1 compares public spending on health care and other fields across a selection of OECD countries.) Despite the fact that the rate of increase in spending has slowed in recent years, federal spending is still roughly equivalent to the research budget of the entire U.S. pharmaceutical industry. Most of this funding is administered through the National Institutes of Health (NIH), although a significant fraction goes to universities. Detailed breakdowns of the nature of this funding are difficult to obtain; one can find divisions by therapeutic class, or disease target, but it is difficult to know how much of this research is oriented toward basic science and how much toward more applied work. The NIH certainly supports both kinds of research. Nevertheless there is consensus that a significant fraction of the support does go toward basic or fundamental science that is widely disseminated through publication in the refereed literature. Both qualitative and quantitative evidence suggests that this spending has had a significant effect on the productivity of those large U.S. firms that were able to take advantage of it (Maxwell and Eckhardt, 1990; Ward and Dranove, 1995;

The Revolution in Molecular Biology

Table 7.1. *Breakdown of National Expenditures on Academic and Related Research by Main Field, 1987*[a]

	Expenditure (1987 M$)						
	UK	FRG	France	Neth.	US	Japan	Average[b]
Engineering	436	505	359	112	1,966	809	14.3%
	15.6%	12.5%	11.2%	11.7%	13.2%	21.6%	
Physical	565	1,015	955	208	2,325	543	21.2%
sciences	20.2%	25.1%	29.7%	21.7%	15.6%	14.5%	
Life sciences	**864**	**1,483**	**1,116**	**313**	**7,285**	**1,261**	**36.3%**
	30.9%	**36.7%**	**34.7%**	**32.7%**	**48.9%**	**33.7%**	
Social	187	210	146	99	754	145	6.0%
sciences	6.7%	5.2%	4.6%	10.4%	5.1%	3.9%	
Arts &	184	251	218	83	411	358	6.8%
humanities	6.6%	6.2%	6.8%	8.6%	2.8%	9.6%	
Other	562	573	418	143	2,163	620	15.6%
	20.1%	14.2%	13.0%	14.9%	14.5%	16.6%	
Total	2,798	4,037	3,212	958	14,904	3,736	

[a] Expenditure data are based on OECD "purchasing power parities" for 1987, calculated in early 1989.
[b] This represents an unweighted average for the six countries (i.e., national figures have not been weighted to take into account the differing size of countries).
Source: Irvine et al. (1990, p. 219).

Cockburn and Henderson, 1996) and that it has played a major role in the emergence of the new biotechnology based firms in the United States (Zucker et al., 1998).

Public funding of biomedical research also increased dramatically in Europe in the postwar period, although the United Kingdom spent considerably less than Germany or France, and total spending did not approach American levels (Table 7.1). Moreover, the institutional structure of biomedical research evolved quite differently in continental Europe than in the United States and the United Kingdom.

In the United Kingdom biomedical research is conducted mainly in the medical schools. The Department of Health and the Department for

R. HENDERSON, L. ORSENIGO, AND G.P. PISANO

Education and Science – particularly through the Medical Research Council (MRC) – have been the main funding agencies. Over the last decade private foundations such as the Wellcome Trust have also emerged as major sources. The MRC funds intramural and especially (approximately two-thirds of the total) extramural research at universities, a much larger proportion than in France.

In France, in contrast, biomedical research is largely performed by CNRS and especially INSERM, which was founded in 1964 to strengthen basic research in the field. In Germany the main actors in biomedical research are the Deutsche Forschungsgemeinschaft (DFG) and the Max Planck Gesellschaft (MP). DFG funds extramural research; MPG receives funds from the federal and state governments for conducting essentially intramural research. After 1972 the newly founded Ministry of Science and Technology (BMFT) emerged as a major actor, sparking sometimes bitter conflict with the other agencies and with universities, particularly with the so-called big science centers which carry out independent research in a limited number of fields.

In addition to these differences in funding, institutional factors in continental Europe also came into play. These not only constrained scientific output, but generally led to science which was far less integrated with medical practice. First, in continental Europe within the medical profession, science does not confer the same status that it does within the Anglo-Saxon countries. Traditionally the medical profession in continental Europe has had less scientific preparation than is typical in either the United Kingdom or the United States. Medical training and practice have focused less on scientific methods per se than on the ability to use the results of research. Moreover, Ph.D.'s in the relevant scientific disciplines have been far less professionally oriented than in the United States or England (Ben-David, 1977; Braun, 1994). Partly as a consequence, within universities medically oriented research has tended to have a marginal role as compared to patient care, especially as compared to that in the United States. Historically the incentives to engage in patient care at the expense of research have been very high: France or Germany has only recently implemented a full-time system designed to free clinicians from their financial ties to patient-related activities. The organizational structure of medical schools has been such as to reinforce this effect. In continental Europe medical schools and hospitals are part of a single organizational entity, whereas in the United States and the United Kingdom they are autonomous actors, which periodically negotiate as to the character of their association. In principle, the European system should have a number of advantages with respect to research and teaching. In practice, the European system has tended to have negative

The Revolution in Molecular Biology

consequences for research as patient care has tended to absorb the largest fraction of time and financial resources. In these systems, resources are not always targeted to specific activities, and given the difficulty of quantifying their cost, even when a fraction of the subsidies provided by the government are supposed to be used for purposes of research and teaching, patient care easily makes inroads into these supposedly "protected" resources (Braun, 1994).

In the United States and the United Kingdom, in contrast, medical schools are usually independent of hospital administrations. This status allows them to give clear priority to their intrinsic goals of research and teaching and often they are able to establish agreements with several different hospitals at the same time.

The weakness of the research function within hospitals in continental Europe is one of the reasons that the decision was made to concentrate biomedical research in national laboratories rather than in medical schools as happened in the United States and the United Kingdom. However, it has often been suggested that the separation of the research from daily medical practice had a negative effect on its quality and especially on the rate at which it diffused into the medical community.

Intellectual Property Protection. In many industries, successful new products quickly attract imitators. But rapid imitation of new drugs is difficult in pharmaceuticals for a number of reasons. One of these is that pharmaceuticals has historically been one of the few industries where patents provide solid protection against imitation. Because small variants in a molecule's structure can drastically alter its pharmacological properties, potential imitators often find it hard to work around the patent. Although another firm might undertake research in the same therapeutic class as an innovator, the probability of its finding another compound with the same therapeutic properties that did not infringe on the original patent could be quite small.[4] However, the scope and efficacy of patent protection has varied significantly across countries.

Both the United States and the majority of the European countries have provided relatively strong patent protection in pharmaceuticals. In contrast, in Japan and in Italy, until (respectively) 1976 and 1978, patent law did not offer protection for pharmaceutical *products*: only *process* technologies could be patented. As a result, Japanese and Italian firms tended to avoid product R&D and to concentrate instead on finding novel processes for making existing molecules.

---

[4] This is not always the case. The history of the discovery of the angiotensin converting enzyme (ACE) inhibitors provides some notable exceptions.

R. HENDERSON, L. ORSENIGO, AND G.P. PISANO

Procedures for Product Approval. Pharmaceuticals are regulated products. Procedures for approval have a profound impact on both the cost of innovating and firms' ability to sustain market positions once their products have been approved. As in the case of patents, there are substantial differences in product approval processes across countries.

Since the early 1960s most countries have steadily increased the stringency of their approval processes. However, it was the United States, with the Kefauver–Harris Amendment Act in 1962, and the United Kingdom, with the Medicine Act in 1971, that took by far the most stringent stance among industrialized countries, followed by the Netherlands, Switzerland, and the Scandinavian countries. Germany and especially France, Japan, and Italy have historically been much less demanding.

In the United States, the 1962 amendments were passed after the thalidomide disaster. They introduced a proof-of-efficacy requirement for approval of new drugs and established regulatory controls over the clinical (human) testing of new drug candidates. Specifically, the amendments required firms to provide substantial evidence of a new drug's efficacy based on "adequate and well controlled trials." As a result, after 1962 the the U.S. Food and Drug Administration (FDA) shifted from a role as essentially an evaluator of evidence and research findings at the end of the R&D process to an active participant in the process itself (Grabowski and Vernon, 1983).

The effects of the amendments on innovative activities and market structure have been the subject of considerable debate (see, for instance, Chien, 1979; Peltzman, 1974). They certainly led to large increases in the resources necessary to obtain approval of a new drug application (NDA), and they probably caused sharp increases in both R&D costs and the gestation times for new chemical entities (NCEs), along with large declines in the annual rate of NCE introduction for the industry and a lag in the introduction of significant new drug therapies in the United States when compared to Germany and the United Kingdom. However, the creation of a stringent drug approval process in the United States may have also helped create an isolating mechanism for innovative rents. Although the process of development and approval increased costs, it significantly increased barriers to imitation, even after patents expired. Until the Waxman–Hatch Act was passed in the United States in 1984, generic versions of drugs that had gone off patent still had to undergo extensive human clinical trials before they could be sold in the U.S. market, so that it might be years before a generic version appeared even once a key patent had expired. In 1980, generics held only 2% of the U.S. drug market.

The institutional environment surrounding drug approval in the

The Revolution in Molecular Biology

United Kingdom was quite similar to that in the United States. Regulation of product safety began in 1964 and from the very beginning relied heavily on formal academic medicine, in particular on well-controlled clinical trials, to demonstrate the safety and efficacy of new drugs. Extensive documentation and high academic standards were required of all submissions. The Committee on Safety of Drugs (CSD) (since 1971 the Committee on Safety of Medicines [CSM]) comprised independent academic experts, voluntarily organized and supported by the industry. The system was based on a strong cooperative attitude between the CSD/CSM, industry and academe, and effectively imposed very high standards on the industry (Davies, 1967; Wardell, 1978; Hancher, 1990; Thomas, 1994). As in the United States after 1962, the introduction of a tougher regulatory environment in the United Kingdom was followed by a sharp fall in the number of new drugs launched into Britain and a shakeout of the industry. A number of smaller, weaker firms exited the market and the proportion of minor local products launched into the British market shrank significantly. The strongest British firms gradually reoriented their R&D activities toward the development of more ambitious, global products (Thomas, 1994).

Japan represented a very different case from either the United States or the United Kingdom. In Japan, prior to 1967 any drug approved for use in another country could be sold without going through additional clinical trials or regulatory approval. As soon as the drug was listed in an accepted official pharmacopoeia, it could be sold in Japan (Reich, 1990). At the same time non-Japanese firms were prohibited from applying for drug approval. Thus Japanese firms were simultaneously protected from foreign competition and given strong incentives to license products that had been approved overseas. Under this regime the primary technology strategy for Japanese pharmaceutical companies became the identification of promising foreign products to license-in (Reich, 1990).

The Structure of the Health Care System and Systems of Reimbursement. Perhaps the biggest difference in institutional environments across countries was in the structure of the various health care systems. In the United States, pharmaceutical companies' rents from product innovation were further protected by the fragmented structure of health care markets and by the consequent low bargaining power of buyers. Moreover, unlike in most European countries (with the exception of Germany and the Netherlands) and Japan, drug prices in the United States are unregulated by government intervention. Until the mid-1980s the overwhelming majority of drugs were marketed directly to physicians,

R. HENDERSON, L. ORSENIGO, AND G.P. PISANO

who largely made the key purchasing decisions by deciding which drug to prescribe.

The ultimate customers – patients – had little bargaining power, even in those instances in which multiple drugs were available for the same condition. Because insurance companies generally did not cover prescription drugs (in 1960, only 4% of prescription drug expenditures were funded by third-party payers), they did not provide a major source of pricing leverage. Pharmaceutical companies were afforded a relatively high degree of pricing flexibility. This pricing flexibility, in turn, contributed to the profitability of investments in drug R&D.

Drug prices were also relatively high in other countries that did not have strong government intervention in prices, such as Germany and the Netherlands. In the United Kingdom, price regulation was framed as voluntary cooperation between the pharmaceutical industry and the Ministry of Health. This scheme left companies to set their own prices, but a global profit margin with each firm was negotiated which was designed to assure each of them an appropriate return on capital investment, including research. The allowed rate of return was negotiated directly and was set higher for export oriented firms. In general, this scheme tended to favor both British and foreign R&D intensive companies which operated directly in the United Kingdom. Conversely, it tended to penalize weak, imitative firms as well as those foreign competitors (primarily the Germans) trying to enter the British market without direct innovative effort in loco (Burstall, 1985; Thomas, 1994).

In Japan, the Ministry of Health and Welfare set the prices of all drugs, using suggestions from the manufacturer based on the drug's efficacy and the prices of comparable products. Once fixed, however, the price was not allowed to change over the life of the drug (Mitchell et al., 1995). Thus, whereas in many competitive contexts prices began to fall as a product matured, this was not the case in Japan. Given that manufacturing costs often fall with cumulative experience, old drugs thus probably offered the highest profit margins to many Japanese companies, further curtailing the incentive to introduce new drugs. Moreover, generally high prices in the domestic market provided Japanese pharmaceutical companies with ample profits and little incentive to expand overseas.

III. The Revolution in the Biological Sciences and Changing Competence in Pharmaceutical Research and Development

The revolution in genetics and molecular biology that began more than 40 years ago with Watson and Crick's discovery of the double helix struc-

The Revolution in Molecular Biology

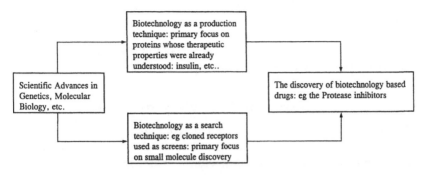

Figure 7.1

ture of deoxyribonucleic acid (DNA) and that continued with Cohen and Boyer's discovery of the techniques of genetic engineering has had an enormous impact on the nature of pharmaceutical research and development and on the organizational capabilities required to introduce new drugs.[5] As we discuss later, application of these advances initially followed two relatively distinct technical trajectories (see Fig. 7.1). One trajectory was rooted in the use of genetic engineering as a process technology to manufacture proteins *whose existing therapeutic qualities were already quite well understood* in large enough quantities to permit their development as therapeutic agents. The second trajectory used advances in genetics and molecular biology as tools to enhance the productivity of the discovery of conventional "small molecule" synthetic chemical drugs.

More recently, as the industry has gained experience with the new technologies, these two trajectories have converged. Contemporary efforts in biotechnology are largely focused on the search for large molecular weight drugs that must be produced using the tools of genetic engineering but whose therapeutic properties are not, as yet, fully understood. Understanding the distinction between these two trajectories is of critical importance to building an understanding of the history of the industry since the two require quite different organizational competencies and had quite different implications for industry structure and for the nature of competition across the world. Whereas in some regions – particularly the United States – the ability to manufacture proteins in quantity triggered an explosion of entry into the industry and a proliferation of new firms, the use of genetics as a tool for small molecule discovery appears to have reinforced the dominance of the large, global

---

[5] Biotechnology has also had far ranging impacts on a number of other fields, including diagnostics and agriculture. For the purposes of this chapter we limit ourselves to considering its impact on human therapeutics.

R. HENDERSON, L. ORSENIGO, AND G.P. PISANO

pharmaceutical firms at the expense of smaller regional players. As a first step toward understanding these outcomes and the ways in which they were shaped by both local competition and institutional structures, we focus on the organizational capabilities required to exploit biotechnology as a process technology and then turn to its use as a research tool.

A. Biotechnology as a Process Technology

Historically, most drugs have been derived from natural sources or synthesized through organic chemical methods. Although traditional production methods (including chemical synthesis and fermentation) allowed the development of a wide range of new chemical entities and many antibiotics, they were not suitable for the production of most proteins. Proteins, or molecules composed of long interlocking chains of amino acids, are simply too large and complex to synthesize feasibly through traditional synthetic chemical methods. Those proteins that were used as therapeutic agents – notably insulin – were extracted from natural sources or produced through traditional fermentation methods. However, since traditional fermentation processes – which were used to produce many antibiotics – could only utilize naturally occurring strains of bacteria, yeast, or fungi, they were not capable of producing the vast majority of proteins. Cohen and Boyer's key contribution was the invention of a method for manipulating the genetic characteristics of a cell so that it could be induced to produce a specific protein. This invention made it possible for the first time to produce a wide range of proteins synthetically and thus opened up an entirely new domain of search for new drugs – the vast store of proteins that the body uses to carry out a wide range of biological functions.

Since the human body produces approximately 500,000 different proteins, the great majority of whose functions are not well understood, in principle Cohen and Boyer's discovery thus opened up an enormous new arena for research. However, the first firms to exploit the new technology chose to focus on proteins such as insulin, human growth hormone, tPA, and factor VIII – proteins for which scientists had a relatively clear understanding of the biological processes in which they were involved and of their probable therapeutic effect. This knowledge greatly simplified both the process of research for the first biotechnology-based drugs and the process of gaining regulatory approval. It also made it much easier to market the drugs since their effects were well known and a preliminary patient population was already in place.

Thus for those firms choosing to exploit this route, the organizational capabilities that have been most critical to success have been those of

The Revolution in Molecular Biology

manufacturing and process development: learning to use the new rDNA techniques as a process to produce natural or modified human proteins. The development of this competence created significant challenges for nearly all of the established pharmaceutical firms since it required both the creation of an enormous body of new knowledge and a fundamental shift in the ways in which manufacturing process development was managed inside the firm.

The manufacture of small molecular weight drugs is essentially a problem in *chemical* process R&D. It draws primarily on chemistry and chemical engineering, disciplines which have existed in academia and industry since the 18th century and in which there is a long history of basic scientific research. As a result much of the relevant theoretical knowledge has been codified in scientific journals and textbooks, and in searching for and selecting alternative chemical processes for the development of small molecular weight drugs, the pharmaceutical firm has at its disposal a wealth of scientific laws, principles, and models which describe the structure of relationships among different variables (e.g., pressure, volume, temperature). Thus process research chemists approaching the manufacture of a small molecular weight drug can often begin their work by deriving alternative feasible synthetic routes from theory.

The characteristics of the knowledge base underlying successful biotechnology process development are quite different. Recall that the major discovery underlying the field was only made in 1973, so that in comparison to small molecular weight drugs, biotechnology is in its infancy. Moreover, although there has been extensive basic scientific research in molecular biology, cell biology, biochemistry, protein chemistry, and other relevant scientific disciplines, most of this work has been geared toward the problems of product "discovery" or to the identification of potentially important proteins, rather than to their manufacture. There has been very little basic research conducted on the problems of engineering larger-scale biotechnology processes. Thus process developers in biotechnology have little theory to guide them in the development of new manufacturing processes.

Perhaps just as importantly, there is a long history of practical experience with chemical processes whereas process developers in biotechnology have almost no practical experience to draw on. The chemical industry emerged in the 18th century, and chemical synthesis has been used to produce pharmaceuticals since the late 1800s. Through this experience a large body of heuristics has evolved, which is widely used to guide process selection, scaleup, and plant design. Most pharmaceutical firms have also developed standard operating procedures for production activities such as quality assurance, process control, production scheduling, changeovers, and maintenance. Experience with these routines provides concrete starting points for development and guidance about what types of process techniques are feasible within an actual production environment.

In contrast, some observers were initially skeptical that recombinantly engineered processes could be scaled-up at all. The first biotechnology-based pharmaceutical to be manufactured at commercial scale, recombinant insulin, was approved by regulatory authorities in 1982. Since that time, only about 25 biotechnology-based therapeutics have been approved for marketing. When a company develops and scales up a specific new biotechnology process, it is not only likely that it represents the company's first attempt – it is probably also the first time *anyone* has attempted the process.

These differences imply that an organization developing a process for a protein molecule needs not only new technological or scientific capabilities, but also different organizational capabilities than those required for the development of a manufacturing process for a new small molecular weight compound. As described in Pisano (1996), biotechnology process development requires the capability to "learn by doing" in the actual production environment since it is virtually impossible to "learn before doing" in the laboratory. In contrast, small molecule pharmaceutical process development requires the capability to exploit the rich theoretical and empirical knowledge base of chemistry through laboratory research or what Pisano (1996) refers to as "learning before doing."

B. Biotechnology as a Research Tool

The new techniques of genetic engineering have also had a significant impact on the organizational competencies required to be a successful player in the pharmaceutical industry through their impact on the competencies required to discover "conventional," small molecular weight drugs. However, although the adoption of biotechnology as a process technology was unambiguously competence destroying for incumbent pharmaceutical firms, adoption of biotechnology as a search tool was only competence destroying for those firms who had not made the transition from "random" to "guided" drug discovery. For those firms that had made the transition, the tools of genetic engineering were initially employed as another source of "screens" with which to search for new drugs. For example, the techniques of genetic engineering allow researchers to clone target receptors, so that firms can screen against a "pure" target rather than against, for example, a pulverized solution of rat's brains that probably contains the receptor. They can also allow for the breeding of rats or

The Revolution in Molecular Biology

mice that have been genetically altered to make them particularly sensitive to interference with a particular enzymatic pathway. Firms choosing to use biotechnology-based research tools thus had to strengthen their scientific capabilities. Nevertheless, the use of biotechnology as a research tool proved to be less destructive of existing research competences than the use of biotechnology for drug production, provided that the firm had already made the transition to guided or "science-driven" drug discovery. Firms that had not adopted this new approach to drug discovery found the biotechnology challenge to be forbidding.

The transition from random to guided drug discovery required the development of both a large body of new knowledge and substantially new organizational capabilities in drug research. So-called random drug discovery drew on two core disciplines: medicinal chemistry and pharmacology. Successful firms employed battalions of skilled synthetic chemists and pharmacologists, who managed smoothly running large-scale screening operations. Although a working knowledge of current biomedical research might prove useful as a source of ideas as to possible compounds to test or as a source of suggestions for alternative screens, by and large it was not critical to employ researchers at the leading edge of their field or to sustain a tight connection to the publicly funded research community, and firms differed greatly in the degree to which they invested in advanced biomedical research.

The ability to take advantage of the techniques of guided search, in contrast, required a very substantial extension of the range of scientific skills employed by the firm; a scientific work force that was tightly connected to the larger scientific community and an organizational structure that supported a rich and rapid exchange of scientific knowledge across the firm (Gambardella, 1995; Henderson and Cockburn, 1994). The new techniques also significantly increased returns to the scope of the research effort (Henderson and Cockburn, 1996).

Managing the transition from random to guided drug discovery was thus not a straightforward matter. In general, the larger organizations who had indulged a "taste" for science under the old regime were at a considerable advantage in adopting the new techniques and smaller firms, firms that had been particularly successful with the older regime, and firms that were much less connected to the publicly funded research community were much slower to follow their lead (Gambardella, 1995; Cockburn et al., 1998).

These differences were critical in shaping responses to the use of biotechnology as a research tool since for those firms that had made the transition to guided drug discovery, the adoption of the tools of genetic engineering to provide an additional resource in the search for small

R. HENDERSON, L. ORSENIGO, AND G.P. PISANO

molecule drugs was a fairly natural extension of the existing competence base. Molecular geneticists could be hired to provide an additional scientific discipline among many, and the screens that they provided could be easily accommodated within the existing procedures by which research was conducted. Thus the larger, more scientifically sophisticated firms were at an enormous advantage in employing biotechnology as a research tool in the search for small molecule drugs (Zucker and Darby, 1996; 1997), and this advantage shaped national responses to the biotechnology revolution. It continues to shape responses as the two trajectories have begun to converge.

C. The Discovery of Biotechnology-based Drugs

More recently the pursuit of biotechnology has come to require new competencies in drug research because it has fundamentally shifted both the domain and the methods of search for new therapeutic agents. Whereas in the traditional synthetic chemical world researchers search among the entire universe of small molecules, biotechnology research focuses on the more than 500,000 proteins present in the human body. This search requires quite different technical and organizational capabilities since it calls for firms to develop deep understanding of the role of particular proteins in causing disease.

As we outlined, early entrants into biotechnology sidestepped the need to develop these new competencies through a focus on proteins that either were already in use as drugs (insulin, human growth hormone, factor VIII) or whose functions were relatively well understood but that had not historically been available in large enough quantities to support commercial development (e.g., tissue plasminogen activase [tPA], erythropoetin [EPO]). The majority of the biotechnology-based proteins that have been approved for marketing to date were all developed this way, and they all emerged without the firms' concerned needing to develop new research competencies.

The larger firms who adopted biotechnology as a search tool sidestepped the need to develop significantly new competencies through their use of the technology as a source of screens. Their initial focus remained on the discovery of small molecular weight drugs, so that their historic competencies in small molecule search and synthesis were not initially challenged.[6]

---

[6] In our view the new techniques of combinatorial chemistry and rapid throughput screening had a similar impact in that they were largely competence enhancing rather than competence destroying. For a slightly different view of the impact of biotechnology on the competence of established pharmaceutical firms see Zucker and Darby (1997).

The Revolution in Molecular Biology

As these avenues have become exhausted, two new strategies have emerged. The first explores the therapeutic properties of a known protein. The case of beta interferon is an example of this type of "protein in search of a use" strategy in action. Almost every organization involved in the early days of the biotechnology industry pursued the development of beta interferon. Initially there was optimism that it would be an effective cancer treatment, but preliminary development along these lines proved frustrating and it was subsequently explored as a treatment for a wide range of diseases. To date its only formally approved use is as a treatment for multiple sclerosis, a therapeutic indication that was never mentioned in the early discussions of the drug. This search strategy thus mirrors in an intriguing way the "random" approach to conventional drug research discussed earlier, although it explores the effects of large molecules in humans rather than the effects of small molecules in animals or in chemical screens.

A second strategy has been to focus on a specific disease or condition and to attempt to find a protein that might have therapeutic effects. Here detailed knowledge of the biological characteristics of specific diseases is an essential foundation for an effective search. For example, researchers working in cancer, Acquired Immunodeficiency Syndrome (AIDS), and autoimmune diseases have focused on trying to discover the proteins responsible for modulating the human immune system. Since this type of knowledge is fundamental to the guided or science based approach to drug discovery that has been adopted by the leading pharmaceutical companies in pursuit of new small molecular weight drugs, it is a strategy that draws heavily on the competencies of established firms. However, since its focus is on large molecules as therapeutic agents, it also requires the process development capabilities that have typically been developed by the new entrants to the industry.

Both strategies also draw on organizational competencies in clinical development, marketing, distribution, and the process of gaining regulatory approval more typically associated with the larger established pharmaceutical companies since they both result in the discovery of "novel" agents, or agents whose therapeutic properties are not completely understood.

IV. Patterns of Industry Evolution

Thus the techniques of molecular biology had dramatic implications both for the discovery of new drugs, on the one hand, and for the ways in which they were manufactured, on the other. "Biotechnology," in the

R. HENDERSON, L. ORSENIGO, AND G.P. PISANO

Table 7.2. *Patent Applications at the European Patent Office, 1978–1993*

	World Patent Shares (%) 1978-1993	No. Firms 1978-1986	No. Firms 1987-1993
USA	36.5	213	303
Japan	19.5	108	185
UK	5.9	39	64
Germany	12.0	45	58
France	6.0	37	52
CH	4.2	11	19

popular sense, has provided an important additional source of new drugs, but, as discussed, it is by no means the only way in which these techniques have changed the industry. Each trajectory – biotechnology-based proteins and the use of genetics as a tool in the search for conventional drugs – has been associated with different organizational regimes and patterns of industry evolution across countries.

Tables 7.2 and 7.3 present some summary data that provide a preliminary view of some of these differences. Table 7.2 shows the number of firms active in biotechnology across the world for the periods 1978–1986 and 1987–1993, as defined by their actively patenting at the European patenting office. Although the United States clearly hosts the majority of firms, notice that the Japanese are also very highly represented. Table 7.3 illustrates the dramatic differences in institutional forms visible across the world that we have discussed. Newly founded firms are far more impor-

The Revolution in Molecular Biology

Table 7.3. *Activity in Genetic Engineering by Type of Institution*

	% Patents by institution, European patent office data		
	NBFs	Established corporations	Universities & other research insitutions
1978-1986			
USA	43.2	34.5	22.3
Japan	0.00	87.7	12.3
Germany	0.01	81.8	17.7
UK	27.3	49.1	23.6
France	18.7	21.5	59.8
Switzerland	0.00	92.9	7.1
Netherlands	12.7	56.4	30.9
Denmark	0.00	93.5	6.5
Italy	0.00	95.7	4.3
1987-1993			
USA	40.4	38.1	20.7
Japan	3.1	86.9	10.0
Germany	3.0	80.0	17.0
UK	23.7	44.7	31.6
France	16.7	35.0	48.3
Switzerland	4.7	89.0	6.3
Netherlands	20.0	62.5	17.5
Denmark	5.7	92.5	1.9

tant in the United States and the United Kingdom than they are elsewhere, whereas the public sector plays a disproportionately important role in France. New firms (NBFs) play a negligible role in the industry in Japan, Switzerland, and Germany. We do not have the comprehensive data that would allow us to match trajectory to institution type, but we believe

R. HENDERSON, L. ORSENIGO, AND G.P. PISANO

that the vast majority of the NBFs initially pursued the first trajectory, or a focus on biotechnology as a process technology, and the established firms – with the important exception of the Japanese firms entering the industry from fermentation related fields – largely pursued the second trajectory, or a focus on the use of biotechnology as a research tool in the search for small molecule drugs. Newly founded firms were initially far more successful than the established firms in bringing new biological entities to market, for example. Zucker and Darby present an analysis of 21 new biological entities approved for the U.S. market by 1994: 7 were discovered by small independent firms, 12 by small firms that were subsequently acquired, and only 2 by established pharmaceutical firms acting "in their own right" (Zucker and Darby, 1996).

More recently, as the two trajectories have merged, there has been an explosion of intracompany agreements, the majority between the NBFs and the larger, established firms. Many companies that were initially slow to respond to the opportunities offered by the new science have attempted to "catch up" through joint research agreements or the outright purchase of promising new firms. For example, out of 95 biotechnology drugs that entered clinical trials in the United States between 1980 and 1988, 15 were developed solely by pharmaceutical firms, 36 were developed solely by biotechnology firms, and 44 were developed jointly by pharmaceutical and biotechnology firms (Bienz-Tadmore et al., 1992).

We next explore these geographical differences in more detail as prelude to our discussion in Section V of the degree to which they can be explained by differences in the institutional structure surrounding the industry across the different regions of the world.

A. The United States

In the United States, the use of biotechnology as a process technology was the motive force behind the first large-scale entry into the pharmaceutical industry since the early post–World War II period. The first new biotechnology start-up, Genentech, was founded in 1976 by Herbert Boyer (one of the scientists who developed the recombinant DNA technique) and Robert Swanson, a venture capitalist. Genentech constituted the model for most of the new firms. They were primarily university spin-offs and they were usually formed through collaboration between scientists and professional managers, backed by venture capital. Their specific skills resided in the knowledge of the new techniques and in the research capabilities in that area. Their aim consisted in applying the new

The Revolution in Molecular Biology

scientific discoveries to commercial drug development.[7] Genentech was quickly followed by a large number of new entrants (Ryan et al., 1994). Entry rates soared in 1980 and remained at a very high level at least until 1985. By the beginning of 1992, there were 48 publicly traded biotechnology companies specialized in pharmaceuticals and health care and several times this number still privately held.

The first biotechnology product, human insulin, was approved in 1982, and between 1982 and 1992, 16 biotechnology drugs were approved for the U.S. market. As is the case for small molecular weight drugs, the distribution of sales of biotechnology products is highly skewed. Three products were major commercial successes: insulin (Genentech and Eli Lilly), tPA (Genentech in 1987), and erythropoietin (Amgen and Ortho in 1989). By 1991 there were over 100 biotechnology drugs in clinical development and 21 biotechnology drugs with applications submitted to the FDA (Grabowski and Vernon, 1994): this was roughly one-third of all drugs in clinical trials (Bienz-Tadmore et al., 1992). Sales of biotechnology-derived therapeutic drugs and vaccines had reached $2 billion, and two new biotechnology firms, Genentech and Amgen, have entered the club of the top eight major pharmaceutical innovators (Grabowski and Vernon, 1994).

Established pharmaceuticals initially played a less direct role in this application of biotechnology, at least in the United States. Zucker and Darby show that of all the firms in their sample (U.S. firms with "affiliated" or "linked" "stars") taking out worldwide genetic-sequence patents between 1980 and 1990, 81% were dedicated biotechnology firms (Zucker and Darby, 1996). Most of the major companies invested in biotechnology R&D through collaborative arrangements, R&D contracts, and joint ventures with the new biotechnology start-ups (Arora and Gambardella, 1990; Pisano, 1991; Barbanfi et al., 1998). As we outlined, the application of molecular biology to the development of protein-based drugs required a completely different set of competencies in both drug discovery and process development. Incumbents were thus poorly positioned to exploit the technical opportunities afforded by the new trajectory through in-house research or manufacturing. However, because the competencies required for clinical development, regulatory approval, and marketing were essentially the same for biotechnology and traditional synthetic drugs, incumbents were sought out as partners who

---

[7] In principle, these techniques could have a wide range of applications, including in agriculture, chemicals, or food. However, the majority of the new firms focused on pharmaceuticals, particularly on diagnostics, and the use of monoclonal antibodies and therapeutics.

R. HENDERSON, L. ORSENIGO, AND G.P. PISANO

could help commercialize the fruits of the new firms' R&D. Thus, during
the 1970s and 1980s, a market for know-how emerged in biotechnology
with the start-up firms positioned as upstream suppliers of technology
and R&D services and established firms positioned as downstream
buyers who could provide capital as well as access to complementary
assets (Pisano and Mang, 1993).

With the notable exception of Eli Lilly, in the United States the more
established firms moved slowly to exploit the techniques of genetic
engineering as a *production* tool. In general they began to acquire the
technology through processes of collaboration – both with small biotechnology firms and directly with university laboratories – and then moved
gradually to a process of outright acquisition either through the acquisition of small firms or through a process of internal group building. As of
this writing a few of the larger pharmaceutical firms – including, most
prominently, Merck, Hoffman La Roche, and Eli Lilly – have quite
sophisticated "biotechnology" capabilities in-house, but the large firms
have yet to emerge as major players in the large molecular weight drug
market in their own right (Zucker and Darby, 1996).

In retrospect it is clear that in the early days of the new technology
the new firms filled two crucial roles. On the one hand, they acted as
"middlemen" in the transfer of technology between universities – which
lacked the capability to develop or market the new technology – and
established pharmaceutical firms that lacked technical expertise in the
new realm of genetic engineering but that had the downstream capabilities needed for commercialization (Orsenigo, 1995). On the other, the
new firms also made a major contribution in responding to the *systemic*
or *architectural* implications of biotechnology. The pursuit of large molecular weight compounds as drugs not only required new competencies
in both research and process development, but also altered the relationship between these. Historically, process development and research had
been managed as highly separate activities. Since genetic engineering is,
at its roots, a process technology, it inherently involves a far higher
degree of *integration* between research and process development, on the
one hand, and manufacturing activities on the other. Thus one of the critical institutional roles played by the small U.S. start-ups was to develop
an entirely new set of "architectural" competencies that enabled them
to act as effective integrators across research, manufacturing, and process
development.

Whereas the use of genetics as a source of large molecular weight
drugs was pioneered, at least in the United States, by newly founded
firms, this was not the case with the use of genetic technology as a tool
for the discovery of traditional, or small molecular weight drugs. This

The Revolution in Molecular Biology

technology was pioneered largely by established firms – although there was enormous variation across firms in the speed with which the new techniques were adopted.

For those firms that were already heavily investing in fundamental research and in which participation in the broader scientific community was already recognized to be of value, the new knowledge presented itself as a natural extension of existing work. They might have been exploring the mechanisms of hypertension, for example: knowledge of the genetic bases of this mechanism was a fairly easily accommodated "competence," and in general these firms moved quite quickly to adopt the new techniques (Gambardella, 1995; Zucker and Darby, 1997). Within the United States, for example, firms such as Merck, Pfizer, and SmithKline-Beecham made the transition relatively straightforwardly.

Those firms that had been more firmly oriented toward the techniques of random drug design, however, found the transition much more difficult. It was often hard to recruit scientists of adequate caliber if the firm had no history of publication or of investment in basic science, and once they were recruited it proved to be difficult to create the communication patterns that the new techniques required. The new techniques probably also significantly increased returns to scope. As drug research came to rely increasingly on the insights of modern molecular biology, discoveries in one field often had implications for work in other areas, and firms that had the size and scope to capitalize on these opportunities for cross-fertilization – and the organizational mechanisms in place to take advantage of these opportunities – reaped significant rewards. Thus one of the major impacts of the revolution in molecular biology has been to drive a wedge between those firms that have been able to absorb the new science into their research efforts and those that are still struggling to make the transition (Zucker and Darby, 1997).

B. Europe and Japan

The exploitation of genetics as a tool to produce proteins as drugs in Europe and Japan lagged considerably behind that in the United States and proceeded along different lines. The most striking difference, of course, is the virtual absence of the phenomenon of the specialized biotechnology start-ups in Europe and Japan, with the partial exception of the United Kingdom and other isolated cases, notably in France (Genetica and afterward Transgene), at least until the second half of the 1980s.

This difference is particularly striking given that in most European countries, and in Japan, governments (at various levels: the European

295

R. HENDERSON, L. ORSENIGO, AND G.P. PISANO

Commission [EC], national, and local governments) have devised a variety of measures to foster industry–university collaboration and the development of venture capital to favor the birth of new biotechnology ventures. To date the results of these policies have not been particularly impressive, although the increase in the rate of formation of new bio-technology based firms in the 1990s may reflect the fact that these policies are now beginning to have an impact. Ernst and Young (1994) suggests that there are now approximately 380 biotechnology companies in Europe. Britain is the European country with the largest number of NBFs, followed by France, Germany, and the Netherlands (Escourrou, 1992; SERD, 1996). Recent data, moreover, suggest a dramatic increase of NBFs in Germany, with different sources estimating their number in the 400–500 range or as more than 600 (Coombs, 1995).

However, very few of these companies resemble the American proto-type. Many of the new European firms are not involved in drug research or development but are instead either intermediaries commercializing products developed elsewhere or active in diagnostics, in the agricultural sector (especially in the Netherlands), or in the provision of instrumen-tation and/or reagents (Merit, 1996). Moreover, some of these compa-nies (especially the most significant ones, like Celltech and Transgene) have been founded through the direct support and involvement of both governments and large pharmaceutical companies rather than through the venture capital market.

The contribution of this new breed of companies to the development of European biotechnology remains to be seen. They already seem to be suffering from the disadvantages of entering the market relatively late. Only the earliest entrants are significant innovators, and some of the most successful have already been acquired (or predict that they will be acquired shortly) by U.S. companies, like many of their American counterparts.

In the absence of extensive new firm founding, most of the innovation in biotechnology in mainland Europe has occurred within established firms. In France there has been significant entry, largely from firms diver-sifying into biotechnology and from other research institutions, while in Germany there has been almost no entry at all. Thus in mainland Europe a few firms account for a large proportion of biotechnology patents, and innovation in biotechnology rests essentially on the activities of a rela-tively small and stable group of large established companies. However, whereas the majority of the established American firms adopted the techniques of genetic engineering as a manufacturing tool primarily through acquisition and collaboration with the small American start-ups,

The Revolution in Molecular Biology

the European firms showed considerable variation in the methods through which they acquired the technology.

The British (Glaxo, Wellcome, and to a lesser extent ICI) and the Swiss companies (particularly Hoffman La Roche, Ciba Geigy, and Sandoz) moved earlier and above all more decisively in the direction pioneered by the large U.S. firms in collaborating with or acquiring American start-ups. Firms in the rest of Europe tended to focus primarily on the establishment of a network of alliances with local research institutes, although it is worth emphasizing that German companies lagged somewhat behind. Hoechst signed a 10-year agreement with Massachusetts General Hospital as early as 1981, but Bayer did not enter seriously until 1985. In general the Germans made little progress in the field and they are not now considered to be among the leaders in European biotechnology. In some countries (e.g., Italy), the scientific community took the lead in the attempt to promote the commercial development of genetic engineering, through the establishment of linkages and collaboration with the pharmaceutical industry. The biggest European innovators are a research institution, Institut Pasteur, and two companies that have not been traditional players in the pharmaceutical industry: Gist-Brocades and Novo Nordisk.

In Japan, entry in biotechnology was pioneered by the large food and chemical companies with strong capabilities in process technologies (e.g., fermentation), like Takeda, Kyowa Hakko, Ajinomoto, and Suntory. Although these firms have strong competencies in process development, they generally lack capabilities in basic drug research. During the 1980s there was concern among some U.S. observers that biotechnology would be the next industry in which Japanese firms achieved dominance, but to date this has not occurred and there has been only limited entry into the pharmaceutical industry via biotechnology.

Within Europe the large British and Swiss firms – Glaxo, Wellcome, and Hoffman La Roche, for example – were also able to adopt the technology. Those firms that had smaller research organizations, that were more local in scope, or that were more oriented toward the exploitation of well-established research trajectories – in short, those firms that had not adopted the techniques of "rational" or "guided" drug discovery – have found the transition more difficult (Henderson and Cockburn, 1994; Gambardella, 1995). Although data speaking directly to this issue are difficult to obtain, it appears that many of the smaller American companies, and almost all of the established French, Italian, German, and Japanese companies have been slow to adopt the tools of biotechnology as an integral part of their drug research efforts.

R. HENDERSON, L. ORSENIGO, AND G.P. PISANO

V. National Systems of Innovation and the Evolution of the Pharmaceutical Industry: How Did Institutions Matter?

This brief description of the impact of the revolution in molecular biology on the pharmaceutical industry highlights the diversity of responses across the world and suggests that there are a number of "stylized facts" to be explored in examining the relationship between "national systems of innovation" and the evolution of the industry. *First, why was it the case that the use of molecular biology as a production tool was pioneered in the United States by small, newly founded firms; in Japan by firms diversifying into the industry from other fields; and in Europe largely by established pharmaceutical firms?* Why did new entrants play a much smaller role in the European context? *Second, did national systems of innovation play a role in shaping the diffusion of the use of molecular biology as a research tool?* This technology was in almost every case pioneered by established pharmaceutical firms, yet its rate of adoption varied widely across the world. *Last,* what can the answers to these questions tell us about the central question addressed by this book: that of the degree to which competitive advantage in this industry was shaped by firm versus national characteristics?

A. The Evolution of "Biotechnology"

The question of why the phenomenon of the small, independently funded biotechnology "start-up" was initially an American one is an old and much discussed issue. One of the reasons that it cannot be answered definitively is that the answer is to a large degree overdetermined. As the discussion of Section III suggested, the use of molecular biology as a production technology was a competence destroying innovation for the vast majority of the established pharmaceutical firms. In the United States a combination of factors made it possible for small, newly founded firms to take advantage of the opportunity this created. These factors included a favorable financial climate; strong intellectual property protection; a scientific and medical establishment that could supplement the necessarily limited competencies of small, newly founded firms; a regulatory climate that did not restrict genetic experimentation; and, perhaps most importantly, a combination of a very strong local scientific base and academic norms that permitted the rapid translation of academic results into competitive enterprises. In Europe (although to a lesser extent in the United Kingdom) and in Japan many of these factors were not in place, and it was left to larger firms to exploit the new technology.

298

The Revolution in Molecular Biology

A Strong Local Scientific Base and Academic Norms That Permitted the Rapid Translation of Academic Results into Competitive Enterprises. The majority of the American biotechnology start-ups were tightly linked to university departments, and the very strong state of American academic molecular biology clearly played an important role in facilitating the wave of start-ups that characterized the eighties (Zucker et al., 1998). The strength of the local science base may also be responsible, within Europe, for the relative British advantage and the relative German and French delay. Similarly the weakness of Japanese industry may partially reflect the weakness of Japanese science. There seems to be little question as to the superiority of the American and British scientific systems in the field of molecular biology, and it is tempting to suggest that the strength of the local science base provides an easy explanation for regional differences in the speed with which molecular biology was exploited as a tool for the production of large molecular weight drugs.

Although this explanation might seem unsatisfying to the degree that academic science is rapidly published and thus, in principle, rapidly available across the world, the American lead appears to have been particularly important because in the early years of the industry the exploitation of "biotechnology" required the mastery of a considerable body of tacit knowledge that could not be easily acquired from the literature (Zucker et al., 1998; Pisano, 1996).

The transmission of this kind of tacit knowledge was probably facilitated by geographic proximity (Jaffe et al., 1993). In the case of biotechnology, however, several authors have suggested that the U.S. start-ups were not simply the result of geographic proximity (Zucker et al., 1998). These authors have suggested that the flexibility of the American academic system, the high mobility characteristic of the scientific labor market, and, in general, the social, institutional, and legal context that made it relatively straightforward for leading academic scientists to become deeply involved with commercial firms were also major factors in the health of the new industry.

The willingness to exploit the results of academic research commercially also distinguishes the U.S. environment from that of either Europe or Japan. This willingness has been strengthened since the late 1970s and the passage of the Bayh–Dole Act, and the resulting role of universities as seedbeds of entrepreneurship has probably also been extremely important in the takeoff of the biotechnology industry.

In contrast, links between the academy and industry – particularly the ability freely to exchange personnel – appear to have been much weaker in Europe and Japan. Indeed, the efforts of several European govern-

R. HENDERSON, L. ORSENIGO, AND G.P. PISANO

ments were targeted precisely toward the strengthening of industry–university collaboration, and it has been argued that the rigidities of the research system of continental Europe and the large role played in France and Germany by the public, nonacademic institutions have significantly hindered the development of biotechnology in those countries.

The importance of these kinds of factors, as distinct from the strength of the science base per se, as being absolutely critical to the wave of new entry in biotechnology that occurred in America in the early eighties is given further credibility by the rate at which the use of molecular biology diffused across the world.

Access to Capital. It is commonly believed that lack of venture capital restricted the start-up activity of biotechnology firms outside the United States. Clearly, venture capital – which is to some extent a largely American institution – played an enormous role in fueling the growth of the new biotechnology based firms. However, at least in Europe, there appear to have been many other sources of funds (usually through government programs) available to prospective start-ups. In addition, the results of several surveys suggest that financial constraints did not constitute a significant obstacle to the founding of new biotechnology firms in Europe (Ernst and Young, 1994; Merit, 1996). In addition, although venture capital played a critical role in the founding of U.S. biotechnology firms, collaborations between the new firms and the larger, more established firms provided a potentially even more important source of capital. This raises the question, Why couldn't prospective European or Japanese biotechnology start-ups turn to established pharmaceutical firms as a source of capital? Though we can only speculate, a plausible answer revolves around the evolution of the market for know-how. The way the market for know-how in biotechnology evolved created many opportunities for European and Japanese companies to collaborate with U.S. biotechnology firms. Although some United States–based NBFs such as Amgen, Biogen, Chiron, Genentech, and Genzyme pursued a strategy of vertical integration from research through marketing in the U.S. market, most firms' strategies emphasized licensing product rights outside the United States to foreign partners. Thus to an even greater extent than many established U.S. pharmaceutical firms, European and Japanese firms were well positioned as partners for U.S. NBFs. Given the plethora of U.S. NBFs in search of capital, European and Japanese firms interested in commercializing biotechnology had little incentive to invest in local biotechnology firms. Even in the absence of other

The Revolution in Molecular Biology

institutional barriers to entrepreneurial ventures, start-ups in Europe or Japan might have been crowded out by the large number of United States–based firms eagers to trade non-U.S. marketing rights for capital.

Intellectual Property Rights. In Section II we discussed the degree to which strong patent protection (or the lack of it, in the case of Italy and Japan) has shaped the industry's history. The establishment of clearly defined property rights also played a major role in making possible the explosion of new firm foundings in the United States, since the new firms, by definition, had few complementary assets that would have enabled them to appropriate returns from the new science in the absence of strong patent rights (Teece, 1986).

In the early years of "biotechnology" considerable confusion surrounded the conditions under which patents could be obtained. In the first place, research in genetic engineering was on the borderline between basic and applied science. Much of it was conducted in universities or was otherwise publicly funded, and the degree to which it was appropriate to patent the results of such research became almost immediately the subject of bitter debate. Millstein and Kohler's groundbreaking discovery – hybridoma technology – was never patented, whereas Stanford University filed a patent for Boyer and Cohen's process in 1974. Boyer and Cohen renounced their own rights to the patent but nevertheless were strongly criticized for having being instrumental in patenting what was considered to be a basic technology. Similarly a growing tension emerged between publishing research results versus patenting them. Whereas the norms of the scientific community and the search for professional recognition had long stressed rapid publication, patent laws prohibited the granting of a patent to an already published discovery (Merton, 1973; Kenney, 1986; Etzkowitz, 1996). In the second place the law surrounding the possibility of patenting life formats and procedures relating to the modification of life-forms was not defined. This issue involved a variety of problems, but it essentially boiled down first to the question of whether living things could be patented at all and second to the scope of the claims that could be granted to such a patent (Merges and Nelson, 1994).

These hurdles were gradually overcome. In 1980 Congress passed the Patent and Trademark Amendments of 1980 (Public Law 96-517), also known as the Bayh–Dole Act, which gave universities (and other nonprofit institutions, as well as small businesses) the right to retain the property rights to inventions deriving from federally funded research.

301

R. HENDERSON, L. ORSENIGO, AND G.P. PISANO

The 1984 passage of Public Law 98-620 expanded the rights of universities further, by removing certain restrictions contained in Bayh–Dole regarding the kinds of inventions that universities could own, and the right of universities to assign their property rights to other parties. In 1980, the U.S. Supreme Court ruled in favor of granting patent protection to living things (*Diamond* v. *Chakrabarty*), by granting a patent to a scientist working for General Electric who had induced genetic modifications on a *Pseudomonas* bacterium that enhanced its ability to break down oil, and in the same year the second reformulation of the Cohen and Boyer patent for the rDNA process was approved. In the subsequent years, a number of patents were granted establishing the right for very broad claims (Merges and Nelson, 1994). Finally, a one-year grace period was introduced for filing a patent after the publication of the invention.

It is often stressed (see, for instance, Ernst and Young, 1994) that the lack of adequate patent protection was a major obstacle to the development of the biotechnology industry in Europe. First, the grace period introduced in the United States is not available: any discovery that has been published is not patentable. Second, the interpretation has prevailed that naturally occurring entities, whether cloned or uncloned, cannot be patented. As a consequence, the scope for broad claims on patents is greatly reduced and usually process rather than product patents are granted. In 1994 the European Parliament *rejected* a draft directive that attempted to strengthen the protection offered to biotechnology.

Although it is clear that stronger intellectual property protection is not unambiguously advantageous, as the controversy surrounding the NIH's decision to seek patents for human gene sequences clearly illustrated, our suspicion is that at least in the early days of the industry the United States reaped an advantage from its relatively stronger regime. (For more on this difficult and complex subject see Merges and Nelson, 1994.)

A Regulatory Climate That Did Not Restrict Genetic Experimentation. Although public opposition to genetic engineering was a significant phenomenon in the United States in the earliest years of the industry (Kenney, 1986) it has quickly became less important and in general the regulatory climate in the United States has been a favorable one. In contrast, opposition to genetic engineering research by the "Green" parties is often quoted as an important factor hindering the development of biotechnology, especially in Germany and in other Northern European countries, and public opposition to biotechnology is said to have been a

The Revolution in Molecular Biology

factor behind the decision of some companies to establish research laboratories in the United States.

B. The Use of Molecular Biology as a Research Tool

Explaining variations in the rate of adoption of molecular biology as a research tool across the regions of the world is, in contrast, rather more difficult. In general the techniques were adopted first by the large, globally oriented U.S., British, and Swiss firms. Adoption by the other European firms, and by the Japanese, appears to have been a much slower process. At first glance the relative strength of the local science base and the degree to which university research was connected to the industrial community appear to be as important an explanation here as it was in understanding the case of the diffusion of "biotechnology." American and U.K. science was arguably rather more advanced than either Japanese or mainland European: hence the slow diffusion of the new techniques to Japanese and mainland European pharmaceutical firms. Unfortunately this explanation is made much less plausible by the Swiss case. The Swiss companies established strong connections with the U.S. scientific system, suggesting that geographic proximity played a much less important role in the diffusion of molecular biology as a research tool.

A second possible explanation is that diffusion was shaped by the relative size and structure of the various national pharmaceutical industries. Henderson and Cockburn (1996) have shown that between 1960 and 1990 there were significant returns to size in pharmaceutical research, and that post 1975 these returns have come primarily from the exploitation of economies of scope. They interpret this as suggesting that the effective adoption of the techniques of guided search and rational drug design placed a premium on the ability to integrate knowledge within the firm and thus that the larger, more experienced firms may have been at a significant advantage in the exploitation of the new techniques. To the degree that those firms that had already adopted the techniques of "rational" drug discovery were at a significant advantage in adopting molecular biology as a research tool, as we have argued, the preexistence of a strong pharmaceutical national industry, with some large internationalized companies, may thus have been a fundamental prerequisite for the rapid adoption of molecular biology as a tool for product screening and design. The U.S. pharmaceutical industry has traditionally been internationally oriented and – at least from the early 1980s – open to international competition in the domestic market. But in many European countries (particularly in Italy and to a lesser extent in

R. HENDERSON, L. ORSENIGO, AND G.P. PISANO

France), the pharmaceutical industry was highly fragmented into relatively small companies engaged essentially in the marketing of licensed
products and – as far as R&D is concerned – in the development of minor
products for the domestic markets.

Notice, however, that although size or global reach may have been a
necessary condition, the failure of the largest German and Japanese firms
to adopt these techniques suggests that it was not sufficient. The largest
Japanese and German firms were arguably quite as international and
quite as large as the Swiss.

The most plausible explanation is that institutional variables – particularly the stringency of the regulatory environment and the nature of
the patent regime – were also important. As mentioned in Section II,
there is now widespread recognition that the introduction of the
Kefauver–Harris Amendments had a significant impact in inducing a
deep transformation of the U.S. pharmaceutical industry, particularly
through raising the cost and complexity of R&D. Partly as a result many
U.S. firms were forced to upgrade their scientific capability.

Similarly, as discussed in Section II, the two European countries whose
leading firms did move more rapidly to adopt the new techniques –
Britain and Switzerland – appear to have actively encouraged a
"harsher" competitive environment. The British system encouraged
the entry of highly skilled foreign pharmaceutical firms, especially the
American and the Swiss, and a stringent regulatory environment also
facilitated a more rapid trend toward the adoption by British companies
of institutional practices typical of the American and Swiss companies:
in particular, product strategies based on high-priced patented molecules, strong linkages with universities, and aggressive marketing strategies focused on local doctors. The resulting change in the competitive
environment in the home market induced British firms to pursue strategies aiming less to the fragmentation of innovative efforts into numerous minor products than to the concentration on a few important
products that could diffuse widely into the global market. By the 1970s,
the ensuing transformations of British firms had led to their increasing
expansion into the world markets.

Lacy Glenn Thomas has suggested that the slowness with which the
vast majority of the European firms (British and Swiss excluded)
adopted the techniques of guided drug discovery reflected much weaker
competitive pressures in their domestic markets (Thomas, 1994). The
Japanese experience also looks in many respects like that pursued in
Europe outside Switzerland and the United Kingdom. In Japan legal and
regulatory policies combined to frame a very "soft" competitive environment that appears to have seriously slowed the adoption of modern

The Revolution in Molecular Biology

techniques by the Japanese pharmaceutical industry. As a result of the combination of patent laws, the policies surrounding drug licensing, and the drug reimbursement regime, Japanese pharmaceutical firms had little incentive to develop world-class product development capabilities, and in general they concentrated on finding novel processes for making existing foreign or domestically originated molecules (Mitchell et al., 1995; Reich, 1990). Moreover, Japanese firms were protected from foreign competition and simultaneously had strong incentives to license products that had been approved overseas. Under this regime the predominant technology strategy for Japanese pharmaceutical companies became the identification of promising foreign products to license-in (Reich, 1990).

Mitchell, Roehl, and Slattery (1995) have noted that some of these institutional factors are beginning to change, and that these changes are starting to have effects on the R&D strategies and capabilities of some (but not all) firms participating in the Japanese pharmaceutical sector. After 1967, foreign-originated products required clinical testing in Japan in order to be approved for sale. After 1976, drug products could be patented. After 1981, the ministry changed its pricing policy such that prices for established drugs are reviewed periodically and compared against those of newer drugs. Together these factors have combined to increase the incentives for original research. Recent evidence suggests that the share of new chemical entities approved in the United States that originate from Japan has increased substantially in recent years – from 4% in the 1970s to around 25% in 1988 (Mitchell et al., 1995) – despite the fact that perhaps because they lack a history of strong internal R&D, it is taking time for Japanese pharmaceutical companies to match world-class research capabilities.

Strong domestic competition, the existence of appropriate incentive mechanisms toward aggressive R&D strategies, and integration into the world markets thus certainly appear to be important explanatory variables in an analysis of variations in the diffusion of the new technologies in drug screening and design across *regions*. Notice, however, that they appear to tell us little about variations in diffusion across *firms*.

Most of the firms that rapidly adopted the new techniques were large multinational or global companies, with a strong presence, at least as research is concerned, in the United States and generally on the international markets. Zucker and Darby present some evidence that size alone is a reasonable predictor of adoption, at least in the United States (Zucker and Darby, 1996). As we have suggested, we suspect that this correlation reflects the fact that adoption is highly correlated with the degree to which firms have made the transition to guided drug discov-

R. HENDERSON, L. ORSENIGO, AND G.P. PISANO

ery. By and large these were larger firms that had early developed a "taste" for science and that were able to build and sustain tight links to the public research community (Gambardella, 1995; Cockburn and Henderson, 1996; Zucker and Darby, 1997). Here institutional factors appear to have been a necessary but not sufficient condition: to the extent that the adoption of the new techniques involved also the successful adoption of particular, academiclike, forms of organization of research within companies (Henderson, 1994), and this process was in turn influenced by the proximity and availability of first-rate scientific research in universities, it was much easier for American (and to a lesser extent British) firms to adopt them.

From this perspective, it is tempting to suggest that the origin of the American advantage in the use of biotechnology as a research tool, as well as a process technique, lies in the comparatively closer integration between industry and the academic community, as compared to that in other countries. One might also speculate that this was – at least to some extent – the result of the strongly scientific base of the American medical culture and – relatedly – of the adoption of tight scientific procedures in clinical trials. Through this mechanism, American companies might have to develop earlier and stronger relationships with the biomedical community and with molecular biologists in particular. Segregation of the research system from both medical practice and close contact with commercial firms (as in France and possibly in Germany) has been highlighted as a major factor hindering the transition to molecular biology in these two countries (see, for instance, Thomas, 1994).

VI. Conclusion

The processes by which advances in basic science influence commercial R&D play a critical role in the prosperity of modern economies. The case of the revolution in molecular biology and its impact on the pharmaceutical industry provides an intriguing window into how these processes work, and how they both shape and are shaped by institutional forces. In the case of "biotechnology," or the use of molecular biology as a production technique, advances in basic science made several of the core competencies of existing firms – particularly those related to process development and manufacturing – obsolete. In the United States, institutional flexibility on a wide range of dimensions led to the formation of specialized biotechnology firms which could provide these competencies and bridge the gap between basic university research, on the one hand, and clinical development of drugs, on

The Revolution in Molecular Biology

the other. Thus the NBFs were, in many ways, an institutional response to the technical opportunities created by new scientific know-how.

The case of biotechnology as a research tool presents a different but complementary picture. This trajectory was born within the confines of established pharmaceutical firms, and institutional factors appear to have played a "necessary" rather than "sufficient" role in its diffusion. The use of biotechnology as a research tool was adopted by pharmaceutical firms as a way to use molecular biology to enhance the value and productivity of their existing assets and competencies and was in this sense "competence enhancing." But it was only competence enhancing for *some* pharmaceutical firms – those that were already oriented toward "high science" research and that were already firmly embedded in the global scientific community. Thus this case is one of existing institutional arrangements and structures shaping, rather than creating, the path of technical change. Forces facilitating institutional flexibility and responsiveness played a less prominent role in this domain. This may help to explain why U.S. firms have been joined by both Swiss and British firms as leaders in the application of molecular biology to small molecule discovery.

What conclusions can be drawn as to the role of national systems of innovation? Different systems of innovation may be better suited to promoting different types of innovation. Since competence destroying technical change generally requires the emergence of new organizations, new organizational forms, and new institutional arrangements, it may tend to flourish in locations which support institutional flexibility and variety. This appears to have been the case in biotechnology. In contrast, relative performance across countries in the adoption of rational drug design, a competence enhancing trajectory developed by incumbents, varied more with the strength of existing institutional arrangements such as the strength of links between universities and other factors favoring incumbents' access to the new technology. Clearly, a better understanding of the processes by which basic scientific advances influence commercial R&D activities, and how these processes work in different environments, would seem to be a potentially fruitful research trajectory in its own right.

References

Abernathy, W.J., and J. Utterback (1978) "Patterns of Industrial Innovation," *Technology Review*, June–July, pp. 40–47.

Arora, A., and A. Gambardella (1990) "Complementarity and External Linkage:

R. HENDERSON, L. ORSENIGO, AND G.P. PISANO

The Strategies of the Large Firms in Biotechnology," *Journal of Industrial Economics*, Vol. 37 (4), pp. 361–379.

Barbanti, P., A. Gambardella, and L. Orsenigo (1998) "The Evolution of the Forms of Organization of Innovative Activities in Biotechnology," *Biotechnology/International Journal of Technology Management*.

Ben-David, J. (1977) *Centers of Learning: Britain, France, Germany and the United States*, McGraw-Hill, New York.

Bienz-Tadmore, Brigitta, Patricia A. Decerbo, Gilead Tadmore, and Louis Lasagna (1992, 1992) "Biopharmaceuticals and Conventional Drugs: Clinical Success Rates, "*Bio/Technology*, Vol. 10, pp. 521–525.

Braun, D. (1994) *Structure and Dynamics of Health Research and Public Funding: An International Institutional Comparison*, Kluwer Academic Publishers, Amsterdam, The Netherlands.

Burstall, M.L. (1985) *The Community's Pharmaceutical Industry*, Commission of the European Communities, Brussels.

Chien, R.I. (1979) *Issues in Pharmaceutical Economics*, Lexington Books, Lexington, Mass.

Clymer, H.A. (1975) "The Economic and Regulatory Climate: U.S. and Overseas Trends," in R.B. Helms (ed.), *Drug Development and Marketing*, American Enterprise Institute, Washington, D.C.

Cockburn, Iain, and Rebecca Henderson (1996) "Public-Private Interaction in Pharmaceutical Research," in *Proceedings of the National Academy of Sciences* 93/23 (November 12, 1996), pp. 12725–12730.

Cockburn, Iain, Rebecca Henderson, and Scott Stern (1997) "Fixed Effects and the Diffusion of Organizational Practice in Pharmaceutical Research," MIT Mimeo.

Cockburn, Iain, Rebecca Henderson, and Scott Stern (1998) "Balancing Research and Production: Internal Capital Markets and Promotion Policies as Incentive Instruments," MIT Mimeo.

Coombs, Aston (1996) *The European Biotechnology Yearbook*, EBUS, Netherlands.

Davies, W. (1967) *The Pharmaceutical Industry: A Personal Study*, Pergamon Press, Oxford.

Ernst and Young (1994) "Biotechnology in Europe," *Ernst & Young Annual Report*, London.

Ernst and Young (1995) "European Biotech 95: Gathering Momentum, *Ernst & Young Annual Report*, London.

Escourrou, N. (1992) "Les Sociétés de Biotechnologie Européennes: Un Reseau Très Imbriqué," *Biofutur*, July–August, pp. 40–42.

Etzkowitz, Henry (1996) "Conflict of Interest and Commitment in Academic Science in the United States," *Minerva*, Vol. 34(3), pp. 326–360.

Gambardella, A. (1995) *Science and Innovation in the US Pharmaceutical Industry*, Cambridge University Press, Cambridge.

Grabowski, H. and J. Vernon (1994) "Innovation and Structural Change in

The Revolution in Molecular Biology

Pharmaceuticals and Biotechnology, "*Industrial and Corporate Change*, Vol. 3(2), pp. 435–450.

Grabowski, H. and J. Vernon (1983) *The Regulation of Pharmaceuticals*, American Enterprise Institute for Public Policy Research, Washington, D.C., and London.

Hancher, L. (1990) *Regulating for Competition: Government, Law and the Pharmaceutical Industry in the United Kingdom and France*, Oxford University Press, Oxford.

Henderson, Rebecca (1994) "The Evolution of Integrative Competence: Innovation in Cardiovascular Drug Discovery," *Industrial and Corporate Change*, Vol. 3(3), pp. 607–630.

Henderson, Rebecca and Iain Cockburn (1994) "Measuring Competence? Exploring Firm Effects in Pharmaceutical Research," *Strategic Management Journal*, Vol. 15, Winter special issue, pp. 63–84.

Henderson, Rebecca and Iain Cockburn (1996) "Scale, Scope and Spillovers: The Determinants of Research Productivity in Drug Discovery," *Rand Journal of Economics*, Vol. 27(1), pp. 32–59.

Irvine, J., B. Martin, and P. Isard (1990) "Investing in the Future: An International Comparison of Government Funding of Academic and Related Research," *Edward Elgar*, p. 219.

Jaffe, Adam B., Manuel Trajtenberg, and Rebecca Henderson (1993) "Geographic Localization of Knowledge Spillovers as Evidenced by Patent Citations," *Quarterly Journal of Economics*, August 1993, pp. 578–598.

Kenney, M. (1986) *Biotechnology: The Industry-University Complex*, Cornell University Press, Ithaca, N.Y.

Laragh, J.H. et al. (1972) "Renin, Angiotensin and Aldosterone System in Pathogenesis and Management of Hypertensive Vascular Disease," *American Journal of Medicine*, Vol. 52, pp. 644–652.

Lippman, S.A. and R.P. Rumelt (1982) "Uncertain Imitability: An Analysis of Interfirm Differences in Efficiency Under Competition," *The Bell Journal of Economics*, Vol. 13(2), pp. 418–438.

Maxwell, Robert A. and Shohreh B. Eckhardt (1990) *Drug Discovery: A Case Book and Analysis*, Humana Press, Clifton, N.J.

Merges, R. and R.R. Nelson (1994) "On Limiting or Encouraging Rivalry in Technical Progress: The Effect of Patent Scope Decisions," *Journal of Economic Behavior and Organization*, Vol. 25, pp. 1–24.

MERIT (1996) "The Organization of Innovative Activities in the European Biotechnology Industry and Its Implications for Future Competitiveness, Report for the European Commission," Maastricht.

Merton, D. (1973) in N.W. Starer (ed.), *The Sociology of Science: Theoretical and Empirical Investigation*, University of Chicago Press, Chicago.

Mitchell, W., T. Roehl, and R.J. Slattery (1995) "Influences on R&D Growth Among Japanese Pharmaceutical Firms, 1975–1990," *Journal of High Technology Management Research*, Vol. 6(1), pp. 17–31.

R. HENDERSON, L. ORSENIGO, AND G.P. PISANO

Myers, Stewart and Christopher Howe (1997) "A Life Cycle Financial Model of Pharmaceutical R&D," MIT Program on the Pharmaceutical Industry, Working Paper #41-97.

Nelson, R.R. (ed). (1992) *National Systems of Innovation*, Oxford University Press, Oxford.

Orsenigo, Luigi (1995) *The Emergence of Biotechnology*, Pinter Publishers, London.

Peltzman, Sam (1974) *Regulation of Pharmaceutical Innovation: The 1962 Amendments*. Washington American Enterprise Institute for Pubic Policy, Washington, D.C.

Pisano, G. (1991) "The Governance of Innovation: Vertical Integration and Collaborative Arrangements in the Biotechnology Industry," *Research Policy*, Vol. 20, pp. 237–249.

Pisano, G. (1996) *The Development Factory: Unlocking the Potential of Process Innovation*, Harvard Business School Press, Boston.

Pisano, Gary and Paul Y. Mang (1993) "Collaborative Product Development and the Market for Know-How: Strategies and Structures in the Biotechnology Industry," in R. Rosenbloom and R. Burgelmon (eds.), *Research on Technological Innovation, Management and Policy*, Vol. 5, JAI Press, Greenwich, Conn.

Reich, Michael (1990) "Why Japanese Don't Export More Pharmaceuticals: Health Policy as Industrial Policy," *California Management Review*, Winter, pp. 124–150.

Ryan, A., J. Freenan, and R. Hybels (1994) "Biotechnology Firms," in G. Carroll and M. Hannan (eds.), *Organizations in Industry Strategy, Structure, and Selection*, Oxford University Press, New York.

SERD (1996) "The Role of SMEs in Technology Creation and Diffusion: Implications for European Competitiveness in Biotechnology, Report for the European Commission," Maastricht.

Statman, Meir (1983) *Competition in the Pharmaceutical Industry: The Declining Profitability of Drug Innovation*, American Enterprise Institute, Washington, D.C.

Teece, D.J. (1986) "Profiting from Technological Innovation: Implications for Integration, Collaboration, Licensing and Public Policy," *Research Policy*, Vol. 15(6), pp. 185–219.

Thomas, L.G., III (1994) "Implicit Industrial Policy: The Triumph of Britain and the Failure of France in Global Pharmaceuticals," *Industrial and Corporate Change*, Vol. 3(2), pp. 451–489.

Tushman, M.L. and Anderson, P. (1986) "Technological Discontinuities and Organizational Environments," *Administrative Science Quarterly*, Vol. 31, pp. 439–465.

Ward, Michael and David Dranove (1995) "The Vertical Chain of R&D in the Pharmaceutical Industry," *Economic Inquiry*, Vol. 33, pp. 1–18.

Wardell, W. (1978) *Controlling the Use of Therapeutic Drugs: An International Comparison*, American Enterprise Institute, Washington, D.C.

The Revolution in Molecular Biology

Zucker, G. Lynne, and Michael R. Darby (1996) "Costly Information in Firm Transformation, Exit, or Persistent Failure," *American Behavioral Scientist*, Vol. 39, pp. 959–974.

Zucker, G. Lynne, and Michael R. Darby (1997) "Present at the Revolution: Transformation of Technical Identity for a Large Incumbent Pharmaceutical Firm After the Biotechnological Breakthrough," *Research Policy*, Vol. 26(4, 5), pp. 429–447.

Zucker, Lynne, Michael Darby, and Marilynn Brewer (1997) "Intellectual Human Capital and the Birth of U.S. Biotechnology Enterprises," *American Economic Review*, June, Vol. 87(3).

ELSEVIER

Research Policy 32 (2003) 971–990

www.elsevier.com/locate/econbase

Mapping technological trajectories of the Green Revolution and the Gene Revolution from modernization to globalization

Govindan Parayil*

Information and Communications Management Programme, Faculty of Arts and Social Sciences, AS3 #04-16, National University of Singapore, Singapore 117570, Singapore

Received 22 August 2001; accepted 1 August 2002

Abstract

The dynamics of technology development along the technological trajectories of the Green Revolution and the Gene Revolution could be explicated by the social morphologies of modernization and globalization. The Green Revolution was shaped by the exigencies of modernization, while the Gene Revolution is being shaped by the imperatives of neo-liberal economic globalization. Innovation, development, and diffusion of technologies followed different trajectories in these two realms because of being part of different innovation systems. Considerations of private gain and profit in the form of high returns to shareholders of agro-biotech corporations of global reach, largely, determine the dynamics of technological innovation in the Gene Revolution. Technology transfer and local adaptive work in the Green Revolution was carried out in the international public domain with the objective of developing research capacity in post-colonial Third World agriculture to increase food production to avert hunger-led political insurrection during the Cold War. Differentiating these two trajectories is important not only due to the normative implications inherent in comparing the impacts of these two "revolutions", but also due to the important lessons we learn about how different contexts of innovation in the same technology cluster could evolve into contrasting research policy regimes.
© 2002 Elsevier Science B.V. All rights reserved.

Keywords: Technological trajectory; Innovation; Biotechnology; Green revolution; Gene revolution; Modernization; Globalization

1. Introduction

The objective of this paper is to compare the "Green Revolution" and the "Gene Revolution"—two different technological trajectories of modern biotechnology—at a macro level. Many analysts tend to assume or confuse that the Gene Revolution is a continuation of the Green Revolution by other means, such as by infusing advanced genetic technology into agricultural production. Conway (1998) exhorts the urgency of incorporating advances in genetic technol-

ogy to make the Green Revolution environmentally more sustainable. He sees the recent advances in molecular genetics in agricultural biotechnology as a continuation of the innovation process in the Green Revolution. "The Green Revolution strikes gold" is Guerinot's (2000) characterization of the invention of the so-called "Golden Rice." Guerinot makes this claim without realizing that the invention of the "Golden Rice" was not part of the Green Revolution, which was essentially an international technology transfer/assistance programme in the public domain to improve agricultural productivity of Third World farmers. The so-called "Golden Rice" is owned by Syngenta, world's largest biotech multinational,

* Tel.: +65-6874-6320; fax: +65-6779-4911.
E-mail address: icmgp@nus.edu.sg (G. Parayil).

972 *G. Parayil/Research Policy 32 (2003) 971–990*

whose mission is not to feed β-carotene enriched rice to vitamin A deficient children in the developing world, but to enhance the profit of its shareholders through this biotech innovation. In a critical commentary on agro-biotech corporations that promote GM crops for profits without any regard to their potential harms, the Thistle (2001) argues that the recent biotech revolution is "a natural outgrowth of the Green Revolution" without paying attention to the different history and politics of these two socio-technological changes. Serageldin (1999a) also sees the difference between the Green Revolution and the Gene Revolution in terms of only the scientific advance of the latter over the former. He argues that the Gene Revolution should be turned into a "doubly green revolution" such that "increased productivity and natural resource management are in balance."

The major thrust of this paper, therefore, is to show that the Green Revolution and the Gene Revolution are entirely different socio-technological systems in that these two "revolutions" involved different technological trajectories that were moulded under different social, political, and economic contexts. I will show that the contexts of these two revolutions have been progressively constructed. While the Green Revolution ended in the 1980s, the Gene Revolution took off in the 1990s under a different socio-technological regime and it was not a technological reincarnation of the Green Revolution under a different name. Differentiating these two trajectories is important not only due to the normative implications inherent in comparing the impacts of these two "revolutions", but also due to the important lessons we learn about how different contexts of innovation in the same technology cluster could evolve into contrasting research policy regimes.

The differential dynamics of these two trajectories are apparent from the fact that despite efforts to revive the Green Revolution under changed international political economic contexts during the late 1980s and 1990s, the crucial international technology transfer and research capacity development regime collapsed due to the transformation and reconfiguration of the milieu of biotech innovation in the wake of globalization. Efforts to revive the Green Revolution under the rubric of a "Doubly Green Revolution" (Conway, 1998) did not take off due to privatization of the research infrastructure and technology transfer regime

after the end of the Cold War, and also due to the influence of neo-liberal doctrines curtailing international aid and capital and technology flow gravitating to new players on the international arena.

In this paper, we will look at the processual dynamics of technological change in biotechnology to trace its trajectories since the mid-1950s. Biotechnologies of various forms and styles have been part of almost all societies and civilizations since antiquity. By biotechnology I mean all human endeavours to alter living things to produce food, fodder, and fiber. More specifically, it means "[A]ny technological application that uses biological systems, living organisms, or derivatives thereof, to make or modify products or processes for specific purposes".[1] Conventional methods of tissue culture, cell fusion, selection and cross-breeding of plants involved in the Green Revolution to the recombinant DNA and genome shotgun sequencing techniques employed in the Gene Revolution would fall under the spectrum of modern biotechnology. Biotechnology is conceptualized in an expansive sense, as a package of artifacts, practices, and knowledge, to produce agricultural crops. The temporal focus will be from around the middle of the 20th century to the present, while the spatial focus will be on the Third World as well as Western industrialized nations. We will look at what global political economy and societal factors shaped the Green Revolution (1950–1980) and how a different set of such factors are shaping the Gene Revolution (1980 to the present) and their policy implications.

2. Technological trajectories of biotechnology

The clustering of innovations and the dynamics of technology development in modern biotechnology shows a clear pattern of technological trajectories. Nelson and Winter (1977) and Dosi (1982) posited the concept of technological trajectory as the path blazed by the normal developmental dynamics of a technology. Technological trajectory is defined by a technological paradigm that emerges from specific selection environment as the solution to the particular

---

[1] This is the definition given to biotechnology in the 1992 Rio Convention on Biological Diversity or CBD. Quoted in Adler (2000, p. 175).

G. Parayil/Research Policy 32 (2003) 971–990   973

cluster of technological problems in question (Dosi, 1982). Dosi showed how the trajectory is circumscribed by such factors as raw materials, scientific laws, industrial conflicts, functional constraints, market forces, laws and regulations. Technological trajectories and paradigms are useful heuristics to explain ex-post facto the developmental dynamics of new technologies within neo-Schumpeterian evolutionary framework of firm-based innovation. In this scheme, the process of technological innovation remains an uncertain activity of selection, "niche" finding, adaptation and other learning and problem solving activities within the milieu of firms and associated actors.[2] Devoid of a clear account of the extant political, economic, historical, and cultural factors in shaping the innovation process, technology dynamics remains a partially useful concept, but fails to provide a clear and comprehensive account of technological change.

In order to present a comprehensive account of the processes and factors that shape technology, one needs to discard all seemingly internalist–externalist or content–context dichotomy and, instead, should follow a contextualist approach to unpack the complexity of how technologies evolve as a contingent and self-organizing process.[3] The self-organizing notion is within the ambit of evolutionary change of technological systems where the selection mechanism is shaped by specific societal forces. The agency question falls within the confines of the society–technology dialectic, and mutual production and reproduction becomes a self-organizing process that is non-deterministic. The implication is that technological change is not an autonomous process, but rather the shaping of technology involves diverse societal agents with differential power relationships derived from their strategic

positions in the social matrix. Thus the heuristic of contextualism is to recognize both the contingent and emergent nature of technology development, such that the particular political, economic, historical and epistemic factors shaping technology could be discerned through careful analysis (Staudenmaier, 1985; Latour, 1987; Bijker and Law, 1992; Callon, 1987; Pinch, 1996; Misa, 1992).[4] In order to explicate the complexity of technological change, a useful approach is to look at the social shaping of technology in a contextualist spectrum of macro-level processes within the epistemic context of technology as knowledge.[5] Contextualism enjoins both constructivist and evolutionary modes of theorizing on technological change by being reflexive of the agential milieu of the actor-network mobilized along these two technological trajectories.[6]

We could explicate the dynamics of technology development in modern agro-biotech innovation systems by following the contextualist frameworks of modernization and globalization. In this article, I argue that the technological trajectory of the Green Revolution was shaped by the exigencies of modernization in post-colonial societies, while the technological trajectory of the Gene Revolution is being shaped by the imperatives of neo-liberal economic globalization. It will be shown that in addition to different set of external factors acting as focusing devices for the selection environment, technological trajectories in these two realms are also shaped by various techno-scientific factors and micro-level processes. It is important to note at the outset that these two major socio-technological changes are very different in nature—the Green Revolution refers to

---

[2] The fact that uncertainty always pervades technological innovation is well known (Rosenberg, 1998). However, it is possible to identify and analyze the factors behind many uncertainties and possibly steer the course of technological change, to a large extent, through appropriate technology policy measures (Rosenberg, 1998; Parayil, 1999).

[3] As Latour (1987, 1988) has clearly shown, any claim of a divide between internal (knowledge) and external (society) of techno-science is a non-starter in conceptualizing technological change. Technology and society are inextricably bound up in a "seamless web" (Hughes, 1986) of human and non-human actors involved in mutual shaping and reshaping of technology and society (Callon, 1987).

[4] If we substitute the word "science" for "technology" in Kuhn's (1970) classic opening sentence in *The structure of scientific revolutions*, one could find the historicity of technology's shaping similar to that of science's.

[5] Intellectual roots of this argument could be traced to the systems model of Hughes (1983, 1987), the actor-network model of Latour (1987), Callon (1987), and Law (1987), social construction of technology (SCOT) model of Pinch and Bijker (1987), contextualist spectrum of Merrit Roe Smith and Reber (1989), and technology as knowledge argument of Layton (1974, 1989), Vincenti (1984, 1990), and Rosenberg and Vincenti (1978). Thomas Misa's (1992) argument that a meso-level analysis captures the complexity of technological change is also a precursor to the contextualism argument presented here.

[6] The contingent and emergent nature of technology dynamics is eminently captured in most of the pioneering works cited above.

974   *G. Parayil/Research Policy 32 (2003) 971–990*

a major project of international technology transfer in the post-colonial era, while the Gene Revolution refers to a process of knowledge intensive innovation which takes place in the industrialized world whose impact will have tremendous implications on agricultural technology and food safety and security of both the developed and developing worlds.

The Green Revolution is very much a product of technological innovation in the international public domain where Western and Third World governments, public supported non-profit national and international agricultural research institutions, universities, multilateral aid agencies, and Western charitable organizations collectively worked together to increase agricultural productivity. The technology transfer and local adaptive work was carried out by public agencies. The objective was to transfer and develop knowledge and research capacity in agricultural innovation and transfer new agriculture practices and necessary technological inputs to farmers in strategically important Third World countries from Western countries during the Cold War.

Unlike the Green Revolution, the Gene Revolution is being shaped by dominant forces in the international private domain where the technological innovation process is determined, largely, by private capital and its quest for profit, market share and shareholder value. The technological trajectory is being shaped by the tension between the public and private domains because of the exigencies of globalization. Most of the knowledge that is being mobilized and utilized by the private agro-biotech corporations to develop proprietary technologies comes from local and global public knowledge domains (government, academe, and global/local intellectual commons). The trend is to privatize the means and sources of knowledge production and to deploy strategies to enclose knowledge commons through intellectual property right regimes.[7]

Both the Green Revolution and the Gene Revolution are different versions of global and national innovation systems in agricultural biotechnology. We could reconfigure several macro models of innovation to capture the dynamics of innovation in these two

realms of biotechnology.[8] However, my objective is not to demonstrate the effectiveness of these models to capture the dynamics of innovation in biotechnology. Without going through an exhaustive analysis of the applicability of any of these models, it could be argued that the trilateral approach of looking only at government, industry, and academe acting as the backbones of the innovation process needs to be bolstered by expanding the agential milieu to include other significant actors (such as NGOs) involved in this complex process.[9] But before proceeding further let us look at the technological change in modern agriculture in its two phases mentioned earlier.

3. The Green Revolution and its dynamics

3.1. What is the Green Revolution?

The beginning of the so-called "Green Revolution" could be traced to the year 1954 when Norman Borlaug invented strains of "miracle" dwarf wheat in Mexico at what is now known as the International Maize and Wheat Improvement Center or CIMMYT. The precursors to the Green Revolution could be traced to the spectacular increase in productivity of hybrid maize (corn) experienced by US farmers in the 1930s and 1940s and to the Rockefeller Foundation-led effort in the 1940s to increase the productivity of wheat and maize in Mexico (Kloppenburg, 1988). Also the agricultural research system that evolved during the post-colonial period in many Third World countries that helped to create the Green Revolution had benefited from the colonial agricultural research system

---

[7] For an excellent exposition of the concept of intellectual commons and a rationale for public regulation of intellectual commons to prevent its ruin through uncontrolled exploitation, see Dawson (1998).

[8] Leading exponents of these models of innovation are: Freeman (1987), Lundvall (1992), and Nelson (1993, 2000) on national innovation system (NIS); Gibbons et al. (1994) on "Mode 2"; Saxenian (1994, 2000) and Acs et al. (2000) on regional economies and innovation networks; and, Etzkowitz and Leydesdorff (1997, 2000) and Leydesdorff (2000) on the "Triple Helix" of university–industry–government relations. In the case of agricultural innovation, many scholars use the term "National Agricultural Innovation System" or NAIS (Hall et al., 2000).

[9] However, as will be argued later, a government–university–industry interactive model is appropriate, but with limitations (as discussed in the concluding section), for capturing the innovation process in biotechnology. We could argue that the Green Revolution had followed some aspect of the NIS or NAIS mode, while the Gene Revolution may be following the Triple Helix mode.

G. Parayil/Research Policy 32 (2003) 971–990   975

established by imperial governments to support the production of commercial export crops such as cotton, rubber, tea, coffee and spices (Hall et al., 2000).

The term "Green Revolution" refers to the changes in agriculture technology and mode of practice of agriculture experienced by some Third World countries, mostly in Asia and Latin America. As a result of the changes in their technological capacity and agrarian relations, these countries experienced considerable increase in the production and productivity of basic cereals like wheat and rice (Parayil. 1992, 1999; Binswanger and Ruttan, 1978).[10] It involved a set of technology policy measures and practices to force nature to be bountiful through the transfer and adaptation of simple technological fixes. In terms of the rate of change of technological change, the Green Revolution was the first major radical innovation in agriculture for several centuries since the introduction of the plough.[11] It is an instance of a relatively "successful" technology transfer, in terms of increases in per capita cereal production.[12] It presents a circumscribed way of looking at a narrowly defined technology policy objective of increasing agricultural productivity.

Generally, the Green Revolution involved the use of seeds of high-yielding varieties (HYVs), primarily of wheat and rice, and the adoption of a "modern" package of agricultural tools and practices involving chemical fertilizers, tractors, pesticides, controlled water, mechanical threshers, electric and diesel pumps, and so forth. These changes were instituted in place of the "traditional" agricultural practice involving the use of seeds whose genetic composition went back thousands of years. "Traditional" technologies also include

wooden ploughs, waterwheels, and bullock carts. In the traditional mode, animals and humans provided the energy required for all agricultural activities. Finally, traditional agriculture is dependent on the vagaries of natural irrigation provided by rains. While the traditional mode involved subsistence production, the modern practice was based on market production. More details of the technical and institutional aspects of the Green Revolution will be presented with occasional reference to India's Green Revolution to flesh out the details of the technology transfer, innovation, and diffusion of this technological event in a concrete setting. India is considered an example of the success of the Green Revolution that made the country self sufficient in food grain production, and even an exporter of grains.

Subsistence farming is often characterized by an "exclusion effect,"[13] which is a tendency on the part of peasant farmers to resist modernization and technological innovation, specifically, to resist radical innovations. This tendency to maintain status quo, and consequently not to undertake innovations, culminated in depressed agricultural productivity, which in turn forced many Third World governments to formulate and implement a new agricultural policy to break out of the stasis in production. The situation was compounded by droughts that culminated in a near-famine situation in the 1960s in many Asian countries. The prevailing policy was for the agricultural sector to fend for itself, and it was expected to provide the surplus to sustain the emerging industrial sector. The decline in food production due to negligence and the lack of cooperation from nature (droughts) forced the governments to depend on imported food to avert hunger and famines. This frightening food situation forced Third World governments, under heavy external pressure from aid donors and the World Bank, to introduce a new agricultural policy for increasing food production (details of the external political economy factors to be discussed anon).

The agricultural policy was aimed at increasing land productivity by introducing a technological solution

---

[10] The term itself was coined by William Gaud in a speech entitled "The Green Revolution: Accomplishments and Apprehensions" given at the meeting of the Society for International Development in 1968. For more details on this naming episode, see Dalrymple (1979) and Binswanger and Ruttan (1978).

[11] This claim must be qualified because agriculture involved numerous innovations through the ages, such as the introduction of water wheels, selective breeding, crop rotation and so on. However, these changes occurred gradually, as opposed to the innovations involved in the Green Revolution and the Gene Revolution, which occurred in a short time span of half century in human history.

[12] In countries such as India, Pakistan, Philippines, Mexico, Turkey and Indonesia, rice and wheat yields more than doubled in two decades. See Gordon Conway (1998) for productivity statistics, especially Chapter 7.

[13] Dosi (1982) defines "exclusion effect" in a different context involving technological paradigms as a tendency on the part of engineers and R&D personnel to be "blind" to other notional possibilities of technological innovation besides the one they "select" as the solution to a given technological problem.

G. Parayil / Research Policy 32 (2003) 971–990

in the form of the Green Revolution package of technology. This was partly indigenously developed, but mostly developed and transferred from the West to India and other countries. There was no intention of introducing land reforms or changing agrarian relations such as introducing serious land reforms. The policy objective was to target medium to large farmers who were encouraged to adopt the new agricultural practices. The Green Revolution has made countries like India self-sufficient in food grains *production* even though its spread was uneven because of inequalities in land tenure, education, infrastructural facilities and so on. The Green Revolution also caused enormous environmental problems, an important topic beyond this paper.[14] The Green Revolution may be characterized as the new technological paradigm that replaced the old paradigm characterized by subsistence farming.

Succinctly put, the technological change in agriculture may be seen as the transformation of the newly derived knowledge in agricultural technology, both local and foreign, into food. The high-yielding varieties (HYVs) of rice suited to tropical conditions in Southeast and South Asia were developed in the 1960s at the International Rice Research Institute (IRRI) in Manila. The earlier varieties developed at the IRRI were based on genetic materials (germplasm) drawn from China, Taiwan, and Indonesia (Binswanger and Ruttan, 1978). These high yielding rice varieties were extensively and successfully introduced in several Third World countries in the 1960s. Semi-dwarf wheat originated in Japan in the late 19th century, and the two most important varieties used for international breeding programmes were Akakomugi and Daruma (Dalrymple, 1988). The Japanese crossed Daruma with several American varieties. The most productive variety that arose from these experiments was known as Norin 10. Norin 10 was introduced into the US

in 1946 and was crossed with several native varieties by the US Department of Agriculture scientists (Dalrymple, 1986). In 1948, scientists in Washington State crossed Norin 10 with Brevor, a native variety. In 1954, the Norin–Brevor cross was taken to Mexico where Norman Borlaug and his colleagues developed several varieties of the HYVs of wheat seeds that were later transferred to India and other Third World countries (Parayil, 1992).

A reconstruction of the history of the Green Revolution shows that four protagonists played crucial roles in its implementation. They are the local and national governments of recipient Third World countries, multilateral and bilateral donor agencies, international agricultural research institutions, as well as the farmers and peasants of these countries. The institutions under the government of India, for example, which planned and coordinated the transfer and diffusion of the new technology were the Ministry of Food and Agriculture, and the Indian Council of Agricultural Research (ICAR), along with the various agricultural research institutes and agricultural universities (Parayil, 1992).

The multilateral and bilateral donor agencies were the Ford Foundation, the Rockefeller Foundation, the World Bank, and the US Agency for International Development (USAID). The two key international agricultural research institutions were the IRRI in Manila and the International Maize and Wheat Improvement Center (CIMMYT) in Mexico.[15] Several more international agricultural research institutes were established after these. In 1971, all the international agricultural research institutes were brought under the umbrella of the Consultative Group on International Agricultural Research (CGIAR).[16]

The farmers and peasants by adopting and adapting the new agricultural technology to their particular situ-

---

[14] It is beyond the scope of this article to give a complete account of the social, economic and environmental impact of the Green Revolution in India and other countries. The Bhopal gas explosion in India, which killed over 15,000 people during and after the aftermath, is connected to the Green Revolution. The Union Carbide factory was set up to produce synthetic pesticides. As a result of the changes in agricultural practices, hundreds of thousands of agricultural workers are injured and killed by agricultural machines, chemical fertilizers and pesticides. These "mini Bhopals" occur because the workers operate and handle these dangerous substances and tools without proper guidance and protective gears. For details of Bhopal, see Parayil (1998).

[15] The IRRI, instituted in 1960, was the joint effort of the Ford Foundation and the Rockefeller Foundation. The Philippines government provided the land for the institute. The CIMMYT, instituted in 1959, was also the joint effort of the Ford and Rockefeller Foundations, and the Mexican government provided the land.

[16] Instituted in 1971, the CGIAR is jointly sponsored and supported by the Rockefeller and Ford Foundations, the World Bank, the Food and Agricultural Organization (FAO), the United Nations Development Programme (UNDP), and recently, the United Nations Environment Programme (UNEP). See Anderson (1998) and Kloppenburg (1988) for details. IRRI and CIMMYT are two of the core research institutions under CGIAR.

G. Parayil/Research Policy 32 (2003) 971–990   977

ation made the Green Revolution a "successful" technology transfer event. The development and spread of the Green Revolution involved different learning processes. The establishment of agricultural universities, patterned after the land-grant universities of the US, is an important event in the history of the Green Revolution that helped to transfer "modern" agricultural knowledge from the US to several Third World countries. USAID helped with investments and logistical support to start up the land-grant type universities, the Rockefeller Foundation helped with the development of a national agricultural research system, and the Ford Foundation helped with farm extension work (Read, 1974; Lele and Goldsmith, 1989).[17]

3.2. Modernization paradigm and the Green Revolution

When the process of modernization began more than two centuries ago in Europe, a new productive relationship emerged during the ensuing Industrial Revolution as a result of the rapid proliferation of new industrial and military technologies.[18] The conflictive practices of various social actors during this period of rapid social change centered on such categories as production, consumption, power, and experience saw the rise of industrial capitalism. The rise of industrial capitalism concomitantly caused a radical shift in the modes of scientific and technological knowledge production and utilization. The transformation in the means and modes of production set off a long wave of innovation clusters around such key technologies as steam engine, electricity, chemicals, electronics, computers, and, most recently, the Internet and genetics. The coming of mass industrial society with the on-

set of modernization was also characterized by the proliferation of nation states as the unit of political economy. The essential features of this stage of social transformation, or what Beck (1992) calls the period of "first modernity," were controllability, certainty, security, linear progress, and convergence. Technological innovation during this stage was notable for energy intensive production, efficiency of operation, systematization, communication based on bottom–up dynamics, and vertical integration of firms and businesses.

Although most of Asia, Africa and Latin America were under colonial rule, European modernity did not make any serious impact on the social and economic structures of the colonies until after World War II when many former colonies began to gain independence and started to formulate their own development agendas. The modernization project with the infusion of new technologies implemented through different economic development models began in earnest in the former colonies during the 1950s and onwards. The common denominator of the economic development models was modern technology, the rapid infusion of which was expected to materialize through its transfer from industrialized nations. The modernization project also considered foreign aid in capital and technology as vital for achieving development. The basic assumption of modernization theory is convergence, an important ontological premise of the project that first appeared in Europe during the Enlightenment. That is, the world is on a particular Eurocentric path of economic and social change engendered by the ideals of Enlightenment; the West arrived there first, and the rest is expected to reach there eventually through a catching up process.[19]

It is axiomatic in modernization theory that Third World ("traditional") societies could be transformed through a concerted project of economic development, which can be achieved by changing the means of production (technology) and by transforming and remoulding archaic social structures that lack the wherewithal for technological innovation. It is assumed that changes in the means of production would entail a

---

[17] According to Read (1974), hundreds of researchers and agricultural experts from American land-grant universities went to India and other destinations to help with the establishment of agricultural universities and their research facilities. Also, thousands of Third World scientists trained in the US and UK returned to India and other nations of their origin to teach and conduct research.

[18] It is instructive to note here that before the onset of the Industrial Revolution in Europe several critical technologies and scientific ideas reached Europe before the 16th century from China, India, and Arabia. These included the magnetic compass, printing and paper making, the water mill, cast iron, iron-chain suspension bridge, piston bellows, metallurgy, the loom, the lathe, gun powder, paper, chemistry and mechanics (Needham, 1954, 1969; Hall, 1957; Adas, 1989; Law, 1987).

[19] This meta-model of modernization and the ensuing universalist narrative of progress is under attack from post-colonial and postmodern theorists of development and change. For a discursive excursus of the modernization project of "development" as an excuse for normalizing the Third World, see Escobar (1995).

change in the relations of production. Modernization can, thus, be achieved by adopting the "right" policies by the government. By formulating and implementing the "right" package of policies, the state and other agents of economic power could induce technological change, where technological change could be equated to a problem solving activity. This minimalist, though profoundly effective, model can be a useful heuristic to understand technological innovation. However, the economic development and modernization drive of Third World countries stalled and became victim to the contingencies of geo-strategic power struggle between the super powers during the Cold War.

The Green Revolution is an unlikely spill over, or perhaps spin-off, of Cold War geopolitics. It is an ironic and unexpected outcome of the campaigns by the US and its allies to check the expansion of the "Red Revolution" in the Third World. The creation of the international system of innovation that caused the increase in agricultural productivity was "closely associated with an American foreign policy that saw that food security problems, particularly in Asia, could lead to political instability and the spread of communism" (Hall et al., 2000, p. 74). The prevailing thinking was influenced by the belief that centralized scientific research institutes of international scope "could solve the generic problem of increasing the biological potential of important food crops and that this would lead to increased food production" (Hall et al., 2000, p. 72). Perkins (1997) persuasively reconstructs the political ecology of yield transformation in agriculture between 1945 and 1970 to the vagaries of geopolitics during the Cold War in which the solutions to American "national security" concerns could be found by alleviating the food deficit of strategically important Third World countries such as India, Indonesia, Philippines, and Mexico through the Green Revolution technologies. The post-colonial political and economic situation in countries of Asia, Africa, and Latin America was rather dismal. Food production was struggling to keep up with the population growth. Fighting hunger and reversing near-famine conditions in countries that still had not fallen to the "communist domino" became an important strategic policy consideration for the US and its allies like UK (Perkins, 1997).

American-led effort to transfer and locally adapt agricultural innovation converged with state-led efforts in Third World countries to modernize their economies. The inventions and innovations leading to the Green Revolution package of technologies came from the efforts of government and non-profit organizations. The public sector and private non-profit organizations played important part in the transfer and establishment of research capacity for Third World agricultural technology development. The establishment of agricultural universities, research and extension stations, irrigation facilities, seed and fertilizer distribution systems were all built by governments, in most Third World countries, to spread the Green Revolution. Markets played only a secondary role in the diffusion of the technology, while the primary role of diffusing the technology came from governmental efforts in tandem with non-profit organizations. Unlike modern biotechnology innovations, profits and private property rights were not key concerns of the developers of the technology. Political, social and, to some extent, humanitarian considerations were emphasized over proprietary social relations in the development and diffusion of the new agricultural technologies.

3.3. Dynamics of the technological trajectory in the Green Revolution

Although the exact nature of the trajectory and outcome of technological change cannot be predicted in advance, the temporal dynamics of this process shows one important feature, that technological change follows an evolutionary course and that it may be possible to steer the course of development of technology in a "desirable" way by influencing the selection environment.[20] Two key theoretical insights about technological change that need to be highlighted here are the cumulative and evolutionary nature of this process. Ideas, practices, theories, and laws from the past do pass on to the development of newer technologies (Layton, 1974). The reason why this aspect tends to be neglected, or does not seem apparent, is due to the tacit nature of technological knowledge (Vincenti, 1990).

In shaping technological trajectory, the "selection" process is characterized by instruction, understanding, experience of learning by doing, and, finally, cognitive

---

[20] Schot (1992) presents an innovative approach to influencing and shaping the selection environment for the development of "cleaner" and "safer" technologies by using the methodology of "constructive technology assessment".

change. The term "evolution" is used in this context as an explanatory metaphor or meta-model. Technological change is construed as a selective-retention process that is adapted to a sequential process of variation and selection.[21] Within the milieu of socio-cultural evolution, adaptive learning and perception lead to the accumulation and change of technological knowledge. The evolutionary concept is important in explaining technological change because it captures the temporal nature of this phenomenon.

The cumulative nature of technological change, on the other hand, implies that technological change is irreversible. By cumulative, it is not meant a theory of accretion of everything past from that particular technological trajectory. The major idea here is that the functional attributes and basic design principles and operational guidelines of a technological system, to a large extent, remain invariant. In the physical and biological sciences, paradigmatic changes may occur due to new experimental discoveries and revolutionary theoretical advances. However, in the case of technology its fundamental functional attributes remain more or less invariant. The functional attributes of the technology remain invariant until a radically different technological paradigm inaugurates a new technological trajectory (Dosi, 1982; Constant, 1973).[22] Old and new technological paradigms do co-exist, and it is rarely a case that the old trajectory becomes totally extinct when the new trajectory is formed. Technological change is fundamentally influenced and moulded by its antecedents in a path-dependent way (Rosenberg, 1994). The idea of technological improvement as a means for efficient action remains the same in all societies. That is, more out of less. Improved efficiency, increased productivity, less cost, less human intervention for avoiding hazardous conditions, and so forth are achieved by improving on the existing technologies. Therefore, it is important that all models of technological change must account for both the cumulative and evolutionary nature of this process.

From the above vantage point, technological change can be characterized as a problem-solving activity. Although it is a problem solving activity conducted in need-based or need-induced circumstances, the solutions do not simply appear on a technology shelf, as often postulated in the production function model of technological change in neo-classical economics (Sahal, 1981). Extant political, economic, social, and institutional factors, including government policies, organize the problem-solving activity, but by themselves, do not provide the solutions. The technological problems vary from one situation to another and their solutions vary according to the degree of complexity of the technological system. Some of the technological problems are low efficiency, adverse environmental conditions, simple functional failures, imbalances between artifacts of different vintages, and inadequate organizational structures. These problems can be the direct or indirect result of climatic and geographic constraints, natural disasters, social and cultural demands for change, simple economic wants, military demands, varying resource positions, and other contingent factors.

The selection environment was moulded by the combined efforts of the international donor agencies, Third World governments, and the international and national research institutes. The transfer, diffusion, location specific adaptation, and the indigenous development of the new technological knowledge, ultimately, transformed the existing "traditional" knowledge system, in areas where the Green Revolution made a lasting impact. The case clearly delineates an active selective-retention process at different stages of learning by doing. There evolved a technological algorithm of how new knowledge could be utilized in the transformative process of increasing productivity of the material output, in this case into food.

The technological algorithm evolved as a means for simplifying the complex knowledge of a new agricultural system into a simplified form that farmers and peasants could easily understand in order to produce the desired results. In simple terms, it can be equated to a decision-rule-making process that includes all the protagonists associated with the Green Revolution. The role of the government and international bilateral and multilateral agencies was to create a proper selection environment for the technology users and developers, in this case, peasants-farmers, local

---

[21] The intellectual foundation for this evolutionary epistemology was first presented in Campbell (1974), who followed the pioneering works of Popper on the evolutionary nature of knowledge creation and its change.

[22] This, however, does not preclude the interpretative flexibility of old and new technologies because of the enrollment of new actors into the network (Bijker, 1995; Bijker and Law, 1992).

extension workers, and research and development personnel. Unlike natural selection, where chance occurrences are the norm, in technological selection, persuasion and the perception of the need for the new knowledge, decide the outcome. The possibilities for variation are limited and fixed, ex ante, to a large extent. As a result, the public agencies can guide the path of selection in most instances. The technological change involved new ways of doing things. However, the fundamental nature of agricultural practice did not change. The changes related to using new seeds and seed preparation, weeding and using chemical pesticides, watering, fertilizer application, and other such activities. Droughts and other "natural" disasters, inadequate organizational structures, pre-industrial technologies, social and cultural demands for change, simple economic wants, and the varying resource positions of regions and states were the problems that influenced the variation processes for the selection of the Green Revolution package of technology within the larger political economic context of the Cold War.

4. The Gene Revolution and its dynamics

4.1. The Gene Revolution in agriculture

The increase in productivity associated with the Green Revolution began to taper off in the 1980s (Conway, 1998; Strauss, 2000). Advances in cellular and molecular biology have opened up new vistas in agricultural technology since the mid-1970s, with particular impetus being felt during the 1990s because of advances in genomics spurred on by computers and information technology. The landmark scientific events in modern biology in general, and biotechnology in particular, were the rediscovery in 1900 of Gregor Mendel's revolutionary work on the genetic basis of inheritance, the identification of DNA as the physical carrier of genetic information by Max Delbruck in 1938, the discovery of the double helical structural model of the DNA molecule by James Watson and Francis Crick in 1953, and the recombinant DNA experiments pioneered by Stanley Cohen and Herbert Boyer in 1973 (Ruttan, 2001, pp. 370–374). The landmark Cohen–Boyer genetic experiments became the basis for recombinant DNA (r-DNA)

technique, which enabled the splicing of genes by transferring genes (genetic information) from one organism into another.[23] McMillan et al. (2000) claim that the "biotech revolution" started in 1973 when the Cohen–Boyer r-DNA technique was invented. According to Kenney (1986, p. 23), the Cohen–Boyer gene splicing invention "was the single pivotal event in the transformation of the 'basic' science of molecular biology into an industry." However, the commercial significance of these new biotech innovations did not materialize until 1980 when the US Supreme Court extended patent protection to genetically modified organisms with its landmark ruling on Diamond versus Chakrabarty.

Although the claim that the "Gene Revolution"[24] in agriculture began in 1973 is understandable given the invention of the r-DNA technique in that year, the actual impact of the invention was felt only in the late 1970s with the unprecedented flow of scientific knowledge from the academy to the biotech industry. The strong knowledge network forged between major research universities and biotech corporations paved the way for the genetic revolution in modern biotechnology. The biotech revolution gained momentum in the early 1980s when large corporations began investing huge amounts of R&D capital for developing transgenic crops with the above scientific and technological advances acting as powerful knowledge base and selection environment for innovation and technology development. The selection environment was shaped by the huge influx of private capital and the unprecedented technology transfer arrangements between industry and university. Paarlberg (2000b) argues that the green signal for corporate investment in genetic technology came only after governments like the US extended intellectual property protection to transgenic organisms. Some experts claim that consequently a "Second Green Revolution" or a "Doubly

---

[23] See Cohen et al. (1973). The Cohen–Boyer r-DNA technique was described in the US Patent no. 4,237,244 issued to them in 1980 as "Process for Producing Biologically Functional Molecular Chimeras" (Kloppenburg, 1988, pp. 193–194).

[24] Unlike an authoritative source for the naming of the "Green Revolution" (see footnote 10), it is difficult to ascertain who came up with the term "Gene Revolution". The term appears in Serageldin (1999b), among other sources. Buttel et al. (1985) used the term "biorevolution" to signify the political economy of this staged transition from the Green Revolution.

G. Parayil/Research Policy 32 (2003) 971–990   981

Green Revolution" (Conway, 1998) began with the marriage of genetic engineering and modern agricultural practices. However, I will show that the Gene Revolution is not a continuation of the Green Revolution. They follow different systems of innovation and trajectories of technology development and diffusion.

Breeding steps using modern plant engineering techniques is a straightforward and clinical process than the trial and error process involved in the Green Revolution. While traditional selective breeding techniques improve the quality and yields of crops, genetic engineering techniques enable direct manipulation of plants through inserting, altering, or removing genes for specific purposes. Genetic engineering is a controlled mechanism involving precision and rigour, while traditional plant breeding involves trial and error that may involve the transfer of unwanted genes to the host organism (Nielson et al., 2001). Genetic technologies share a qualitative superiority over conventional methods of plant breeding because they can bypass the conventional method's reliance on sexual means to transfer genetic information (Kloppenburg, 1988). Genetic technologies "permit the modification of living organisms with an unprecedented specificity and allow a qualitatively different degree of genetic transformation" (Kloppenburg, 1988, p. 3).

There are essentially two techniques in modern biotechnology for plant breeding—molecular markers and genetically engineering transgenic crops. Molecular markers involve identifying specific genes from the DNA sequences in plant genomes with specific traits. Molecular marker tools are then used to screen varieties of plants for genes that confer resistance to specific diseases. Using this technique, plant breeders speed up the development of new varieties with the desirable traits (Arends-Kuenning and Makundi, 2000).

In genetic engineering, first the required objectives are determined in advance, such as resistance to pests, salinity, drought and so on, or improving nutritional levels, or delivering vaccines. The next step is to identify the sources of a gene in plants, animals or fungi that would offer solution to the particular problem being identified. The gene is isolated and its sequence is decoded to understand the structure of the gene and, if necessary, to redesign the gene to suit the new environment. The isolated gene is inserted into a single cell nucleus of the target plant and develops

the cell into complete plants. The new transgenic plants are tested in isolated environment for resistance and other characteristics before commercial application. This technique is controversial because genes that confer the specific traits may come from different plant species or even from animals or virus. Well-known transgenic crops are *Bt*-corn, roundup ready soybean, and "Golden Rice".[25] Plant breeding in this new biotechnology is aimed at addressing specific needs, such as increasing tolerance to herbicides and pesticides, creating tolerance to drought or salinity or pests, and delivering vaccines and nutrients, unlike increasing crop yield as the primary concern in the Green Revolution.

The US is by far the biggest adopter of genetically engineered crops, almost three-fourths of the total crop area devoted to GM crops, while the other major producers are Argentina, Canada, and China (Nielson et al., 2001). Global area of transgenic crops planted in 1999 was 39.9 million ha, of which the US (72%), Argentina (17%), and Canada (10%) are the principal global GM crop producers (James, 2001). Developing countries that conducted transgenic crop field trials included Argentina, Belize, Bolivia, Chile, China, Costa Rica, Cuba, Egypt, Guatemala, India, Malaysia, Mexico, South Africa, Thailand, Turkey, and Zimbabwe

---

[25] The bacterium *Bacillus thuringiensis* (*Bt*) produces crystalline proteins that kill pests and hence the transfer of the specific genes from this bacterium confers insecticide property to corn, cotton and other crops (Conway, 1998, p. 152). The *Bt* gene induced toxin is claimed to be lethal to certain pests (such as stem borers), but not to humans and other organisms in the ecosystem. Roundup Ready soybeans developed by Monsanto have a gene added that confers tolerance to Monsanto herbicide Roundup (glyphosate) that kills several types of weeds (Paarlberg, 2000b). The so-called "Golden Rice" variety was developed by Ingo Potrykus and colleagues at the Swiss Federal Institute of Technology in Zürich by engineering the provitamin A (β-carotene) through a biosynthetic pathway into carotenoid-free rice endosperm. The golden hue of the rice is due to the β-carotene in the endosperm. For details of the engineering process, see Ye et al. (2000). Although developed by scientists working at publicly supported universities with partial funding from the Rockefeller Foundation, ironically, some seventy or so patents on various gene segments, sequencing processes and other developmental techniques used for creating this rice variety are held by private agro-biotech companies. At the final count Zeneca Agrichemicals, now known as Syngenta (largest biotechnology company in the world after merging with the agricultural division of Novartis), owns "Golden Rice". For a detailed exposition of the controversy surrounding "Golden Rice" see Christensen (2000).

(James, 1997). Among developing countries, China and Argentina are by far the largest producers of GM crops. Experimental planting of GM plants and crops are increasingly being adopted in India, Mexico, Philippines, Thailand, and Brazil. Experimental trials and local adaptive research are currently going on "Golden Rice" to transfer the added nutritional capacity from this japonica variety into indica variety of rices (which are consumed by most Asians) at rice research stations in India, China, and the International Rice Research Institute in Manila (IRRI, 2000).

4.2. Globalization and the Gene Revolution

The modernization project characterized variously by such structural and functional categories as industrial capitalism (Marx), rationalization and bureaucratization (Weber), and systemic and functional differentiation (Parsons) underwent a qualitative change towards the end of 20th century. Industrial societies began to show structural transformation in their modes of interactive and intersubjective practices based on the notional possibilities of new technologies and modes of production centered on information and networks (Castells, 1996). This structural transformation of industrial societies is variously characterized, but most commonly, as "post-industrial society" (Bell, 1973), or as "informational capitalism" (Castells, 1996). In this new phase of capitalism knowledge became the most important factor of production as opposed to machines, labour, and natural resources, the predominant factors of production of the industrial society (Drucker, 1993).[26] The coming of the post-industrial society[27] coincided with the newest phase of economic globalization as well. The so-called "new economy" is characterized by flexible production and free movement of capital on a global scale.[28] Replacement of industrial capitalism by financial

capitalism and the intensification of "free" trade on a global scale are the other features of globalization.

The new phase of globalization is in certain definite ways different from its previous forms of the 16th and 19th centuries (UNDP, 1999). The new era of globalization is marked by: (i) the emergence of *new markets* which link foreign exchange and capital markets on a global scale operating in real time or selected time, thus eliminating spatial differentials; (ii) strengthening of *new and old actors* like WTO, IMF and other supra-national entities that exercise authority over national governments of Third World countries; (iii) the rise of MNCs which enjoy greater economic and political power over economically weaker governments where they operate plants but often disregard environmental and labour norms that would often be deemed violations in their home bases; (iv) enactment of *new rules* such as multilateral agreements on trade, services and intellectual property backed by strong enforcement mechanisms; (v) proliferation of new technological tools, such as biotechnologies, information and communication technologies that co-ordinate market operations; (vi) the rise of a *network society* marked by structural changes in social morphology of regions connected by the new tools and markets; (vii) the rise of resistance movements, such as NGOs through better articulation of civil society norms; and (viii) the exclusion of large areas of the world (such as Africa) from the benefits of trade and technological innovation.[29] The emergence of genetic agriculture coincides with the newest phase of globalization while the Green Revolution was part of industrial modernity.

Public sector (governments and government supported institutions) and non-profit international research institutions, international donor agencies like the World Bank, and philanthropic institutions like the Ford and Rockefeller Foundations played the key roles in the invention, innovation and diffusion of Green Revolution technologies. Private sector actors, which are predominantly multinational corporations, play the leading role in the innovation and diffusion of agricultural biotechnology related to the Genetic Revolution. The technological trajectory is shaped by the imperatives of private property institutions,

---

[26] Of course capital, labour, and natural resource were the predominant factors in classical economics. Neo-classical economics dismissed natural resource (land) because of its perceived lack of marginal value in the production function.

[27] Other analysts characterize the age as post-Marxist, or post-capitalist, or post-modern, or post-Fordist.

[28] The neo-liberal position on globalization professes the free movement of all factors of production. But the free movement of labor is, ironically, not a reality yet for obvious reasons.

---

[29] More detailed exposition of these characteristics of contemporary globalization can be found in UNDP (1999), Castells (1996), and Held et al. (1999).

G. Parayil/Research Policy 32 (2003) 971–990   983

market forces, global finance, and transnational (and in certain cases national) regulatory institutions. The contingencies and imperatives of economic globalization shape the technological trajectory. New plants and crops are being developed not to solve problems of hunger and deprivation, but mostly to increase shareholder values of companies that have invested heavily in R&D efforts in the biotechnology sector. Consumer preferences are more important than farmer's rights and interests in the development and diffusion of genetic agricultural technology, and the trend is to develop technology suited for the interests of large biotech firms. Very little feedback and input is derived from public agencies and farmers in technological innovation in the Gene Revolution.

4.3. Dynamics of the technological trajectory in the Gene Revolution

The selection environment of genetic-based agro-biotechnology is being shaped by the investment decision of private biotech corporations, the tension between the public and private domains in matters related to property rights in life forms, regulatory battles, consumer rights activism, and resistance to technological development from sources within civil society. While "donor fatigue" (Anderson, 1998) and tightening of funds for public benefit scientific research (Alston et al., 1998) have reduced financial support for agriculture research in the public domain, private sector investment in agriculture R&D has been increasing at impressive rates in recent years. Six large agriculture biotechnology companies spent more than 75% of global research and development and other investment money in biotechnology in 1998 (Krattiger, 1999). Private corporations registered almost all transgenic crops approved for planting as proprietorial technology (James, 1998). The total annual budget for 1998 of 16 international agricultural research centers under the Consultative Group on International Agricultural Research (CGIAR) was only US$ 345 million, while a single biotechnology corporation increased its R&D expenditure to US$ one billion per year (James, 1999). Public sector investment in agricultural technology declined considerably since the 1980s as some Third World countries improved their food production and maintained "a long period of sustained growth" in production (Alston et al., 1998).

The end of the Cold War and the rise of neo-liberal politics saw Western aid, particularly American aid, to Third World countries to decline in both real and absolute terms. The general consensus in the West is that markets are the best arbiters of technology development and selection, and that the government better get out of the way of the private sector. "Inappropriability of benefits," otherwise known as the market failure argument of public goods, is claimed as a major reason given for the decline in investment in agriculture R&D in the public domain (Alston et al., 1998).

The progressive removal of trade barriers in agricultural commodities sector with the arrival of a global trade regime under the World Trade Organization (WTO) since 1995 favoured the private sector players to invest in biotechnology. The liberalization of intellectual property rights of patents, copyrights and trade secrets in plants, animals, genetic knowledge, and traditional knowledge systems under the TRIPS[30] agreements favoured the rights of private investors in the biotechnology sector than small farmers and producers. This allowed large biotechnology multinationals with unlimited resources for R&D and new product development to reap monopoly benefits obtained through patenting engineered life forms and new protein molecules from hitherto unavailable intellectual property rights.[31] Padrón and Uranga (2001, p. 315) rightly argue that intellectual property protection to biotechnological inventions is "a burden too heavy for the patent system" to adjudicate and resolve.

Some factors that led to the increased role of the private sector in agricultural research systems in recent

---

[30] The GATT negotiations that concluded in the establishment of the WTO contained an Annex on Trade-Related Aspects of Intellectual Property Rights (the "TRIPS Agreement"). The TRIPS Agreement gives wide coverage to intellectual property protection to patents, copyright and neighboring rights, trademarks, geographical indications, industrial designs, layout-designs (topographies) of integrated circuits, and undisclosed information. For more details on the TRIPS Agreement, see World Intellectual Property Organization (1997), Chapter 28.

[31] The trend started when the US Supreme Court in the landmark Diamond versus Chakrabarty case upheld, by a margin of one vote, the patent on a genetically engineered oil-eating bacterium. The US Supreme Court ruled that "Everything under the sun made by man is patentable" (quoted in the Economist, 2001). The US statute on patent rights essentially became the norm for the rest of the world. As David (1993) argued, the US managed to impose its position on patents and trade rights at international forums because of its economic and political might.

time are: (i) the declining funding levels in most developing countries due to various internal and external factors (Alston et al., 1998); (ii) weak management and bureaucratic inefficiencies in the public research domain leading to budgetary cuts of research funds (Echeverria et al., 1996); (iii) the intensification and institutionalization of intellectual property rights at global levels, especially after the TRIPS Agreement, opening up potentially greater opportunities for private research investments over public investments in agricultural technology; (iv) the rapid pace of privatization and encouragement given to private competition with parastatals in agriculture input supply in many countries, and the growing commercialization of agriculture as a world-wide phenomenon and the great demand for purchased inputs as a result of increasing competition in domestic and international markets (Pray and Umali-Deininger. 1998); and, (v) the commodification of traditional knowledge in agriculture and the legal loopholes to treat traditional knowledge as intellectual commons unprotected by IPR statutes (Dutfield, 2000). And the spectacular advances in molecular biology overcame the final biological barrier to the entry of private capital in agro-biotech business, although the process had already set in motion with the advent of hybridization in plant breeding to commercialize seed production during the Green Revolution.

The technological trajectory of agro-biotechnology is highly influenced by the tension between the public and private domains in biotech innovation and knowledge creation. The concept of property and the easiness of appropriating publicly available knowledge, including the gold-rush mentality to patent and commoditize traditional knowledge of plants, farming systems, and living matter derived from the global intellectual commons, has become a powerful shaper of agro-biotechnology. The tendency to mediate the conflict between the public and private domains in biotechnological research through property rights at the expense of other important concerns like bio-diversity, distributive justice, cultural rights, and ecological stability will have tremendous impact on molding the future trajectory of agro-biotechnology. In understanding innovation and technology development one must pay adequate attention to the public and private aspects of technology (Nelson. 1989). Padrón and Uranga

(2001) see the problem straddling both the production and consumption side of modern biotechnology. On the production side they argue that the biotechnology industry depends on public science more than other industries and the privatization of such public knowledge not only fosters free riding but also deters innovation in socially beneficial ways.[32] On the consumption side, they argue, regulatory issues become serious because biotechnology is not "just another step on the continuum of technological development," but rather "a radical change with profound ontological implications" (Padrón and Uranga, 2001, p. 318).

While globalization and the rise of the knowledge economy and MNCs are key agents in shaping agricultural biotechnology, in order to understand the dynamics of innovation and commercialization of the technology one needs to see the process through the theoretical lens of "reflexive modernity" (Giddens. 1991; Beck, 1997) and "world risk society" (Beck, 1999). Various "knowledgeable agents" reflexively shaping genetic technology through self-confrontation is a key facet of technology dynamics here, albeit the power of the agents is differentially distributed according to the norms of the political economy of global financial capitalism. An array of life-style choices available as a result of spectacular technological innovations fomenting doubts about the "natural order," and concurrently spawning "new ethical spaces" (Giddens. 1994) for individuals to engage in the debates concerning the very shaping of such innovations is at the heart of reflexive modernity (Kerr and Cunningham-Burley. 2000).

The very fact that there is more profit to be made in the trade of inputs in agriculture than the output is also an important consideration in the value chain of agro-biotechnology commoditization. That various industries catering to the input side (seeds, fertilizers, pesticides, finance, mechanical tools, and educational institutions) of the production function emerged as significant players is crucial to this commodification

---

[32] McMillan et al. (2000) present empirical evidence to show that public science, that is, knowledge originated from universities, research institutions, government laboratories, and so on played a critical role in the success of US biotechnology firms. For a detailed and systematic examination of the contribution of public science (knowledge authored at institutions conducting scientific research funded by the government) to industrial technology, see Narin et al. (1997).

G. Parayil / Research Policy 32 (2003) 971–990   985

process. The proprietary nature of genetic knowledge and agricultural processes at different stages of input–output matrix under the new trade regime in a growing "weightless economy" (Quah, 1997) of "informational capitalism" (Castells, 1996) makes it ever more attractive for private capital to venture in agricultural research systems for new product development.[33]

Another key agent that plays a vital role in shaping agro-biotechnology is social movement NGOs that oppose globalization and genetic technology. Environmental NGOs that oppose the release of genetically modified organisms (GMOs) into the environment and the production and distribution of genetically modified (GM) foods are important agents in regulating research and trade in agricultural biotechnology. Indigenous rights activists that oppose the commodification and piracy of traditional plants and agricultural knowledge, and farmers' movements that oppose infringement on their right to owning and propagating seeds and farming systems are also influential, to some extent, in influencing the trajectory of biotechnology. For example, biotech giant Monsanto's plan to commercialize its so-called "terminator" seeds using genetic use restriction technology (GURT) whereby farmers would be unable to save and use harvested seeds was shelved due to intense pressure from farmers worldwide (Wright, 2001). Consumers in developed nations, particularly in Western Europe are also important agents in shaping the policy agenda for technological innovation and the regulatory regime pertaining to GMOs and GM food (Paarlberg, 2000a; Grossman and Endres, 2000). The politically sensitive nature of genetic technology in Europe has a crucial bearing on shaping both the regulatory regime and technological innovation in agro-biotechnology in large Western production centers such as the US, Canada and Australia. In fact, Paarlberg (2000b) argues that the future of GM food and agro-biotech will be shaped by the looming confrontation between the US-based industry groups and European consumers.

The rather advanced and complex regulatory regime in developed countries is in stark contrast to the near absence of any meaningful regulatory system in developing countries. This is particularly serious because large developing countries such as China, India, Brazil, Mexico, Argentina, and Indonesia are seriously engaged in the production of transgenic crops and the consequences to public health and possible ecological concerns become secondary issues to these countries when contrasted to their concerns for economic growth through increased agricultural productivity. The successful decoding of the rice genome by the giant Swiss MNC Syngenta has raised alarms about the future course of plant breeding in this important food source for more than half of the humans in the world (Wade, 2002).[34] Plant breeding in the new agro-genetics innovation system appears to be directed for the commercial gains of large MNCs than for improving the food security of poor consumers in the Third World.

Finally, technological change in agro-biotechnology must be analyzed within the framework of technological risks in both local as well as global contexts. The trade in technology-related products have become global in nature and the transfer of technology is now increasingly within the province of MNCs as part of their global investment strategies. Because of the competition for foreign direct investment from Third World countries, MNCs are trying to find destinations where regulatory regimes (labour, environmental and consumer) and intellectual property rights are more to their favour. Therefore, the management of technological risks associated with agro-biotech must be formulated as a global risk management problem to be formulated and implemented and coordinated with international, regional, and local agencies. Ultimately, it boils down to issues of the governance of science and technology and constructing effective technology foresight tools and mechanisms, such as constructive technology assessment (CTA), environmental impact assessment (EIA) and sustainable development indicators (SDI). A proper way to getting started on these line of future actions would be to follow an effective precautionary principle that is not too lax and broad as private

---

[33] However, the material foundations of the "weightless economy" and "informational capitalism" cannot be denied or wished away as Law and Hetherington (2000) have eloquently argued.

[34] The important scientific journal *Science* (5 April 2002) published two separate works on sequencing the rice genome by scientific teams from China and Syngenta. Syngenta's earlier controversial role in its acquisition of the "Golden Rice" may be the reason for its foray into rice genomics being watched with alarm by some concerned scientists. See, Wade (2002) and Christensen (2000) for details.

investors and MNCs prefer and not too narrow and restrictive as deep ecologists and anti-biotechnology activists want.[35] How to find that constructive middle ground where the promises of agricultural biotechnology could benefit all stakeholders (particularly poor farmers and consumers in the Third World) is an important agenda for technology policy development to meet the challenges and exigencies of economic globalization.

5. Discussion and conclusion

The Gene Revolution in agricultural biotechnology was not a natural follower of the technological trajectory of the Green Revolution. The innovation, development, and diffusion of agricultural technologies followed different paths because of being part of different global and national innovation systems. The technological paradigms of these two phases of biotechnology were shaped within different societal, political, economic, and epistemic contexts. Existing social, political, economic, and ecological conditions shape the manner in which new technologies are developed and disseminated. As shown earlier, the Green Revolution and the Gene Revolution followed different technological trajectories. However, as Kloppenburg (1988, p. 4) persuasively argued, "an understanding of the 'old' biotechnologies is a prerequisite to understanding of the 'new' biotechnologies". While the forces of industrial modernity shaped the Green Revolution within the context of post-colonial politics during the Cold War, the Gene Revolution is being molded by the contingencies of globalization and "intellectual capitalism" (Granstrand, 2000) which gives impetus to private ownership of agricultural knowledge and artifacts. State and other public sector agencies, international donor and research institutions, and other non-profit organizations played a decisive

role in shaping the Green Revolution. Geo-strategic considerations emanating from the politics of hunger during the Cold War, the national aspirations of Third World governments to attain food self-sufficiency, and the goodwill of scientists and technologists were the catalysts in the technology development and transfer regime of the Green Revolution. The Green Revolution was shaped by the convergence principle embedded in the social morphology of modernity, where the idea of economic development and improved standard of living spreading to the Third World through state-led efforts to channel technology and development aid from the West reigned supreme. Although there was no Marshall Plan to modernize Third World economies, the contingencies of the Cold War prompted the West to find a quick technological fix to avert hunger-led insurrection and possible communist takeover of key Third World nations without demanding drastic changes in the social relations of production and distribution of their agrarian sector—putatively the crucial economic sector in which most people sought their sustenance. The technology transfer programme under the Green Revolution was a "success" in a circumscribed way of increasing food production. However, serious distributional, environmental, and occupational hazards were not addressed because these were not the priority of the developers and promoters of the technology.

The model of innovation system in the Green Revolution follows a non-traditional global–local innovation system involving governments and public agencies, universities, and private international charitable agencies. Instead of industry acting as the third element of the usual innovation structure, we have non-profit international charitable organizations acting as agents of technology development and diffusion. Although there was some feedback received from farmers and local extension agents in the innovation process at a later stage, the innovation system may still be reconfigured in a trilateral mode.[36] Despite the active role of certain important societal agents (farmers, extension agents and consumers) in the innovation network of the Green Revolution, they did not, however, become an autonomous agent to

---

[35] The precautionary principle included in the Biosafety Protocol adopted by the United Nations Convention on Biological Diversity (CBD) in Cartagena in 1999 and Montreal in 2000 should be the basis for biotechnology development. CBD adopted the same text of the precautionary principle included in the 1992 Rio Declaration: "Where there are threats to serious or irreversible damage, lack of full scientific certainty shall not be used as a reason for postponing cost-effective measures to prevent environmental degradation," quoted in Adler (2000, p. 194).

[36] This trilateral reconfiguration may be undertaken within the "NIS" or "NAIS" mode of agricultural innovation, as described in footnotes 8 and 9.

G. Parayil / Research Policy 32 (2003) 971–990   987

be categorized as a fourth element of the innovation system.[37]

Considerations of private gain and profit in the form of high returns to shareholders of agro-biotech companies of global reach determine the dynamics of technological innovation in the Gene Revolution. As Arends-Kuenning and Makundi (2000, p. 333) argue, the crops and plant traits developed under such regime "will be the ones expected to make the largest profits". The imperatives of global financial architecture, unrestricted trade, and unhindered rights of intellectual property in life forms and organic molecules, and global risk concerns under the juggernaut of economic globalization are the catalysts of technology development in the Gene Revolution. Technological innovation, development, and transfer in the agro-biotech sector are being influenced by the tension between the public and private domains where the asymmetry in power relations in favour of the former until recently is now tilting towards the latter.

The technological trajectory will be affected significantly by the success in mediating between the public and private domains by de-centring the primacy of IPR regime in biotechnology research. Ways must be found to adapt the IPR regime such that innovations in agro-biotechnology could be stimulated through new institutional frameworks that encompass broader social goals than patents and private gain. As Peter Drahos (1999, p. 443) puts it eminently, "such adaptation must be governed by the public purpose that is embedded in patent law and the broader public ethic, rather than by private purposes". Instead of the present trend of proprietarianism in IPR, an instrumental attitude should prevail because there is nothing essentialist about private property (Drahos, 1996), other than it being a socially constructed entity that is mediated and moulded by prevailing social relations. As one observer puts it imaginatively, "Focusing on the problems of justifying intellectual property is important not because these institutions lack any sort of justification, but because they are not so obviously or easily justified as many people think. We must begin

to think more openly and imaginatively about the alternative choices available to us for stimulating and rewarding intellectual labor" (Hettinger, 1989: 52).

Biotechnology should be treated as a primary good containing a basket of information and artifacts crucial for human well being, such that public or private efforts to create artificial scarcity in the domain through unjustified regulation or unconditional enclosing of this intellectual commons could be thwarted. Setting the research agenda in biotechnology sector should not be left to the marketplace or the Leviathan alone. Using the presently construed "Triple-Helix" of government–industry–academe model of global and national system of innovation to self-organize and self-govern itself for the larger public good is inadequate. To understand the complexity of the technological trajectory in agricultural biotechnology, we should look at the innovation system made up of government, academe, industry, *and* NGOs. Because of the important role NGOs (both formally organized social groups as well as unorganized general public concerned about biotechnology) play as a key stakeholder in (re)shaping biotechnology in recent times through resistance and regulatory pressure as shown in this article, the innovation system should be re-conceptualized to include this key societal agent. How such an innovation model should be configured is an important research problem for another occasion.[38]

Acknowledgements

An earlier version of this article was presented in a colloquium at the Institut für Geschiste, Departement für Geistes-, Sozial- und Staatswissenaschaften, Swiss Federal Institute of Technology Zürich on 2 February 2001. Comments and suggestions for improvement of the manuscript from the editor, Michel Callon, and two anonymous referees of *Research Policy* are gratefully acknowledged. Thanks are also extended to Greg Felker, Robert Ferguson, and T.T. Sreekumar for constructive comments on an early draft of the article.

---

[37] This is in contrast to the fourth significant agent made up of NGOs which played a key role in shaping the Gene Revolution. This significant player should be added to the existing trilateral models to form a quadruple helix. Mehta (2002) argues that this fourth helix should be called the "public".

---

[38] A preliminary thought on this point is to develop a quadruple helix of state–university–industry–NGOs relationship, bearing in mind the extremely complex nature of the interactive relationships in such a structural configuration.

References

Acs, Z., de la Mothe, J., Paquet, G., 2000. Regional innovation: in search of an enabling strategy. In: Acs, Z. (Ed.), Regional Innovation, Knowledge and Global Change. Pinter, London, pp. 37–52.

Adas, M., 1989. Machines as the Measure of Men: Science, Technology, and Ideologies of Western Dominance. Cornell University Press, Ithaca, NY.

Adler, J.H., 2000. More sorry than safe: assessing the precautionary principle and the proposed international biosafety protocol. Texas International Law Journal 35, 173–205.

Alston, J.M., Pardey, P.G., Roseboom, J., 1998. Financing agricultural research: international investment patterns and policy perspectives. World Development 26 (6), 1057–1071.

Anderson, J.R., 1998. Selected policy issues in international agricultural research: on striving for international public goods in an era of donor fatigue. World Development 26 (6), 1149–1162.

Arends-Kuenning, M., Makundi, F., 2000. Agricultural biotechnology for developing countries: prospects and policies. American Behavioral Scientist 44 (3), 318–349.

Beck, U., 1992. Risk Society: Towards a New Modernity. Sage, London.

Beck, U., 1997. The Reinvention of Politics: Rethinking Modernity in the Global Social Order. Polity Press, Cambridge.

Beck, U., 1999. World Risk Society. Polity Press, Cambridge, UK.

Bell, D., 1973. The Coming of Post-Industrial Society: A Venture in Social Forecasting. Basic Books, New York.

Bijker, W.E., 1995. Of Bicycles, Bakelites, and Bulbs: Towards a Theory of Sociotechnical Change. MIT Press, Cambridge, MA.

Bijker, W.E., Law J., (Eds.), 1992. Shaping Technology/Building Society: Studies in Sociotechnical Change. MIT Press, Cambridge, MA.

Binswanger, H., Ruttan, V.W., 1978. Induced Innovation: Technology, Institutions and Development. Johns Hopkins University Press, Baltimore.

Buttel, F.H., Kenney, M., Kloppenburg, J.R., 1985. From green revolution to biorevolution: some observations on the changing technological bases of economic transformation in the Third World. Economic Development and Cultural Change 34 (1), 31–55.

Callon, M., 1987. Society in the making: the study of technology as a tool for sociological analysis. In: Bijker, W.E., Hughes, T.P., Pinch, T., (Eds.), 1987. The Social Construction of Technological Systems: New Directions in the Sociology and History of Technology. MIT Press, Cambridge, MA, pp. 83–106.

Campbell, D.T., 1974. Evolutionary epistemology. In: Schlipp, P.A. (Ed.), The Philosophy of Karl Popper. Open Court, La Salle, IL, pp. 413–463.

Castells, M., 1996. The Rise of the Network Society. Blackwell Scientific Publishers, Malden, MA.

Christensen, J., 2000. Golden Rice in a Grenade-Proof Greenhouse. The New York Times, 21 November, pp. D1 and D5 (Science Times).

Cohen, S., Chang, A., Boyer, H., Helling, R., 1973. Construction of biologically functional bacterial plasmids in vitro. Proceedings National Academy of Science USA 70, 3240–3244.

Constant, E., 1973. A model of technological change applied to the Turbojet Revolution. Technology and Culture 14, 553–572.

Conway, G., 1998. The Doubly Green Revolution: Food for All in the Twenty-First Century. Cornell University Press, Ithaca, NY.

Dalrymple, D.G., 1979. The adoption of high-yielding grain varieties in developing nations. Agricultural History 53 (4), 704–726.

Dalrymple, D.G., 1986. Development and Spread of High-Yielding Wheat Varieties in Developing Countries. US Agency for International Development, Washington, DC.

Dalrymple, D.G., 1988. Changes in wheat varieties and yields in the United States, 1919–1984. Agricultural History 64 (4), 20–36.

David, P.A., 1993. Intellectual property institutions and the Panda's thumb: patents, copyrights, and trade secrets in economic theory and history. In: Wallerstein, M.B., Mogee, M.E., Schoen, R.A. (Eds.), Global Dimensions of Intellectual Property Rights in Science and Technology. National Academy Press, Washington, DC, pp. 19–61.

Dawson, A.C., 1998. The intellectual commons: a rationale for regulation. Prometheus 16 (3), 275–289.

Dosi, G., 1982. Technological paradigms and technological trajectories. Research Policy 11 (3), 147–162.

Drahos, P., 1996. A Philosophy of Intellectual Property. Dartmouth, Aldershot, UK.

Drahos, P., 1999. Biotechnology patents, markets, and morality. European Intellectual Property Review 9, 441–449.

Drucker, P.F., 1993. Post-Capitalist Society. HarperBusiness, New York.

Dutfield, G., 2000. The public and private domains: intellectual property rights in traditional knowledge. Science Communication 21 (3), 274–295.

Echeverria, R.G., Trigo, E.J., Byerlee, D., 1996. Institutional change and effective financing of agricultural research in Latin America. World Bank Technical Paper No. 330. World Bank, Washington, DC.

Economist. 2001. Patently absurd. The Economist Quarterly (5) 36–40.

Escobar, A., 1995. Encountering Development: The Making and Unmaking of the Third World. Princeton University Press, Princeton.

Etzkowitz, H., Leydesdorff, L., 1997. A Triple Helix of University–Industry–Government Relations. Cassell Academic, London.

Etzkowitz, H., Leydesdorff, L., 2000. The dynamics of innovation: from National Systems and "Mode 2" to a Triple Helix of university–industry–government relations. Research Policy 29, 109–123.

Freeman, C., 1987. Technology Policy and Economic Performance: Lessons from Japan. Pinter, London.

Giddens, A., 1991. Modernity and Self-Identity. Polity Press, Cambridge.

Giddens, A., 1994. Living in a Post-industrial Society. In: Beck, U., Giddens, A., Lash, S. (Eds.), Reflexive Modernization: Politics, Tradition and Aesthetics in the Social Order. Polity Press, Cambridge, pp. 56–109.

Granstrand, O., 2000. The shift towards intellectual capitalism—the role of infocom technologies. Research Policy 29, 1061–1080.

Grossman, M.R., Endres, A.B., 2000. Regulation of genetically modified organisms in the European Union. American Behavioral Scientist 44 (3), 378–434.

Guerinot, M.L., 2000. The Green Revolution strikes gold. Science 287 (1), 241–243.

Hall, A., Clark, N., Rasheed, S.V., Sivamohan, M.V.K., Yoganand, B., 2000. New agendas for agriculture research in developing countries: policy analysis and institutional implications. Knowledge, Technology and Policy 13 (1), 70–79.

Hall, R., 1957. Epilogue: the rise of the west. In: Singer, C., et al. (Eds.), A History of Technology, vol. III. Oxford University Press, Oxford.

Held, D., McGrew, A., Goldblatt, D., Perraton, J., 1999. Global Transformations: Politics, Economics and Culture. Stanford University Press, Stanford.

Hettinger, E.C., 1989. Justifying intellectual property. Philosophy and Public Affairs 18 (1), 31–52.

Hughes, T.P., 1983. Networks of Power: Electrification in Western Society, 1890–1930. Johns Hopkins University Press, Baltimore.

Hughes, T.P., 1986. The seamless web: technology, science, etcetera, etcetera. Social Studies of Science 16, 281–292.

Hughes, T.P., 1987. The evolution of large technological systems. In: The Social Construction of Technological Systems: New Directions in the Sociology and History of Technology. MIT Press, Cambridge, MA, pp. 51–82.

IRRI, 2000. The Reward of Rice Research, Annual Report 1999-2000. International Rice Research Institute, Manila.

James, C., 1997. Global Status of Transgenic Crops in 1997. ISAAA Briefs No. 5, International Service for Acquisition of Agri-biotech Applications, Ithaca, NY.

James, C., 1998. Global Review of Commercialized Transgenic Crops: 1998. ISAAA Briefs No. 8, International Service for Acquisition of Agri-biotech Applications, Ithaca, NY.

James, C., 1999. Global Status of Commercialized Transgenic Crops: 1999. ISAAA Briefs No. 12, International Service for Acquisition of Agri-biotech Applications, Ithaca, NY.

James, C., 2001. Transgenic crops worldwide: current situation and future outlook. In: Qaim, M., Krattiger, A.F., von Braun, J. (Eds.), Agriculture Biotechnology in Developing Countries: Towards Optimizing the Benefits for the Poor. Kluwer Academic Publishers, Boston, pp. 11–24.

Kenney, M., 1986. Biotechnology: The University–Industrial Complex. Yale University Press, New Haven.

Kerr, A., Cunningham-Burley, S., 2000. On ambivalence and risk: reflexive modernity and the new human genetics. Sociology 34 (2), 283–304.

Kloppenburg, J.R., 1988. First the Seed: The Political Economy of Plant Biotechnology, 1492–2000. Cambridge University Press, Cambridge.

Krattiger, A.F., 1999. Networking biotechnology solution with developing countries: the mission and strategy of the International Service for the Acquisition of Agri-Biotech Application. In: Hohn, T., Leisinger, K.M. (Eds.), Biotechnology of Food Crops in Developing Countries. Springer-Verlag, Vienna, pp. 25–33.

Kuhn, T.S., 1970. The Structure of Scientific Revolutions. University of Chicago Press, Chicago.

Latour, B., 1987. Science in Action: How to Follow Scientists and Engineers Through Society. Harvard University Press, Cambridge, MA.

Latour, B., 1988. The Prince for machines as well as for machinations. In: Elliott, B. (Ed.), Technology and Social Processes. Edinburgh University Press, Edinburgh, pp. 20–43.

Law, J., 1987. On the social explanation of technical change: the case of the Portuguese maritime expansion. Technology and Culture 28 (2), 227–252.

Law, J., Hetherington, K., 2000. Materialities, spatialities, globalites. In: Bryson, J.R., Daniels, P.W., Henry, N., Pollard, J., (Eds.), Knowledge, Space, Economy. Routledge, London, pp. 34–49.

Layton, E.T., 1974. Technology as knowledge. Technology and Culture 15 (1), 31–41.

Layton, E.T., 1989. Through the looking glass or, news from lake mirror image. In: Cutcliffe, S.H., Post, R.C. (Eds.), In Context: History and the History of Technology. Lehigh University Press, Bethlehem, PA, pp. 29–41.

Lele, U., Goldsmith, A.A., 1989. The development of national agricultural research capacity: India's experience with the Rockefeller Foundation and its significance for Africa. Economic Development and Cultural Change 37 (2), 305–344.

Leydesdorff, L., 2000. The triple helix: an evolutionary model of innovations. Research Policy 29, 243–255.

Lundvall, B.-Å., (Ed.), 1992. National Systems of Innovation: Towards a Theory of Innovation and Interactive Learning. Pinter, London.

McMillan, G.S., Narin, F., Deeds, D.L., 2000. An analysis of the critical role of public science in innovation: the case of biotechnology. Research Policy 29, 1–8.

Mehta, M., 2002. Regulating biotechnology and nanotechnology in Canada: a post-normal science approach for inclusion of the Fourth Helix. In: Proceedings of the Paper Presentation at the International Workshop on Science, Technology and Society: Lessons and Challenges, National University of Singapore, 18–19 April 2002 (Mimeo).

Misa, T.J., 1992. Theories and models of technological change: parameters and purposes. Science Technology and Human Values 17, 3–12.

Narin, F., Hamilton, K.S., Olivastro, D., 1997. The increasing linkage between U.S. technology and public science. Research Policy 26, 317–330.

Needham, J., 1969. The Grand Titration: Science and Society in East and West. Allen & Unwin, London.

Needham, J., 1954-1988. Science and Civilization in China, vols. 1–7. Cambridge University Press, Cambridge.

Nelson, R.R., 1989. What is private and what is public about technology? Science, Technology and Human Values 14 (3), 229–241.

Nelson, R.R. (Ed.), 1993. National Innovation Systems: A Comparative Study. Oxford University Press, New York.

Nelson, R.R., 2000. National innovation systems. In: Acs, Z. (Ed.), Regional Innovation, Knowledge and Global Change. Pinter, London, pp. 11–26.

Nelson, R.R., Winter, S.G., 1977. In search of useful theory of innovation. Research Policy 6, 36–76.

Nielson, C.P., Robinson, S., Thierfelder, K., 2001. Genetic engineering and trade: panacea or dilemma for developing countries. World Development 29 (8), 1307–1324.

Padrón, M.S., Uranga, M.G., 2001. Protection of biotechnological inventions: a burden too heavy for the patent system. Journal of Economic Issues 35 (2), 315–322.

Paarlberg, R., 2000a. Promise or peril? Genetically modified crops in developing countries. Environment 42 (1), 19–27.

Paarlberg, R., 2000b. The global food fight. Foreign Affairs 79 (3), 24–38.

Parayil, G., 1992. The green revolution in India: a case study of technological change. Technology and Culture 33 (4), 737–756.

Parayil, G., 1998. The 'revealing' and 'concealing' of technology. Asian Journal of Social Science 26 (1), 17–28.

Parayil, G., 1999. Conceptualizing Technological Change: Theoretical and Empirical Explorations. Rowman & Littlefield, Lanham, MD.

Perkins, J.H., 1997. Geopolitics and the Green Revolution: Wheat, Genes, and the Cold War. Oxford University Press, New York.

Pinch, T., Bijker, W.E., 1987. The social construction of facts and artifacts: or how the sociology of science and the sociology of technology might benefit each other. In: The Social Construction of Technological Systems: New Directions in the Sociology and History of Technology. MIT Press, Cambridge, MA, pp. 17–50.

Pray, C.E., Umali-Deininger, D., 1998. The private sector in agricultural research systems: will it fill the gap? World Development 26 (6), 1127–1148.

Quah, D.T., 1997. Increasingly weightless economies. Bank of England Quarterly Bulletin (2) 49–56.

Read, H., 1974. Partners With India: Building Agricultural Universities. University of Illinois Press, Urbana-Champaign.

Rosenberg, N., 1994. Exploring the Blackbox: Technology, Economics and History. Cambridge University Press, Cambridge.

Rosenberg, N., 1998. Uncertainty and technological change. In: Neef, D., Siesfeld, G.A., Cefola, J. (Eds.), The Economic Impact of Knowledge. Butterworth-Heinemann, Boston, pp. 17–34.

Rosenberg, N., Vincenti, W.G., 1978. The Britannia Bridge: The Generation and Diffusion of Technological Knowledge. MIT Press, Cambridge, MA.

Ruttan, V.W., 2001. Technology, Growth, and Development: An Induced Innovation Perspective. Oxford University Press, New York.

Sahal, D., 1981. Alternative conceptions of technology. Research Policy 10, 2–24.

Saxenian, A., 1994. Regional Advantage: Culture and Competition in Silicon Valley and Route 128. Harvard University Press, Cambridge, MA.

Saxenian, A., 2000. Regional Networks and Innovation in Silicon Valley and Route 128. In: Acs, Z. (Ed.), Regional Innovation, Knowledge and Global Change. Pinter, London, pp. 123–138.

Schot, J.W., 1992. Constructive technology assessment and technology dynamics: the case of clean technologies. Science, Technology and Human Values 17 (1), 36–56.

Serageldin, I., 1999a. From Green Revolution to Gene Revolution. Economic Perspectives 4 (4), October 1999, IIP E-Journals, US Department of State. At: http://usinfo.state.gov/journals/ites/1099/ijee/bio-serageldin.htm (accessed: 24 May 2002).

Serageldin, I., 1999b. Biotechnology and food security in the 21st century. Science 285 (16 July), 387–389.

Smith, M.E., Reber, S.C., 1989. Contextual contrasts: recent trends in the history of technology. In: Cutcliffe, S.H., Post, R.C. (Eds.), In Context: History and the History of Technology. Lehigh University Press, Bethlehem, PA, pp. 133–149.

Staudenmaier, J.M., 1985. Technology's Storyteller: Reweaving the Human Fabric. MIT Press, Cambridge, MA.

Strauss, M., 2000. When Malthus Meets Mendel. Foreign Policy No. 119, Summer, pp. 105–112.

Thistle, 2001. Continuing the Green Revolution: The corporate assault on the security of the global food supply. The Thistle 13(4), June/July. At: http://web.mit.edu/thistle/www/v13/4/food.html (accessed: 24 May 2002)

Vincenti, W.G., 1984. Technological knowledge without science: the invention of flush riveting in American airplanes, ca. 1930–ca. 1950. Technology and Culture 25, 540–576.

Vincenti, W.G., 1990. What Engineers Know and How They Know It. John Hopkins University Press, Baltimore.

UNDP (United Nations Development Programme), 1999. Human Development Report 1999. Oxford University Press, New York.

Wade, N. 2002. Experts Say They Have Key to Rice Genes. The New York Times, April 5, 2002 (http://www.nytimes.com/2002/04/05/science/05RICE.html)

World Intellectual Property Organization (WIPO), 1997. Introduction to Intellectual Property Theory and Practice. Kluwer Law International, London.

Wright, B.D., 2001. Intellectual property rights challenges and international research collaboration in agricultural biotechnology. In: Qaim, M., Krattiger, A.F., von Braun, J. (Eds.), Agricultural Biotechnology in Developing Countries: Towards Optimizing the Benefits for the Poor. Kluwer Academic Publishers, Boston, pp. 289–314.

Ye, X., Al-Babili, S., Klöti, A., Zhang, J., Lucca, P., Beyer, P., Potrykus, I., 2000. Engineering the provitamin A (β-carotene) biosynthetic pathway into (carotenoid-free) rice endosperm. Science 287 (14), 303–305.

[4]

PERSPECTIVES

Engaging biotechnology companies in the development of innovative solutions for diseases of poverty

Hannah E. Kettler and Sonja Marjanovic

Global policy interventions are required to ensure that the scientific and technological innovations that biotechnology companies are advancing are applied to the health needs of people in the developing world. We present an overview of the present role of biotechs in 'diseases of poverty' R&D, and outline ways to facilitate higher levels of involvement in the future.

Infectious parasitic, viral and bacterial diseases, such as malaria, human immunodeficiency virus (HIV), tuberculosis, African trypanosomiasis and leishmaniasis, account for one-half to two-thirds of all healthy years lost in the developing world. (Healthy years lost are measured in disability-adjusted life years. This measure takes into account both the number of healthy years lost due to premature death, and the disability caused by debilitating but non-fatal diseases and injuries.) More than 90% of the 14 million people who succumb to these diseases each year live in the developing world[1]. FIGURE 1 uses the example of malaria to demonstrate the significant overlap between regions with the highest burden for a certain disease and regions with the highest poverty levels — the basis for the phrase 'diseases of poverty'.

The development of new drugs, vaccines and diagnostics is a crucial component of the multifaceted approach that is needed to prevent, treat and effectively control diseases of poverty. However, the urgent need for new products continues to outstrip supply.

Predicted market size is a crucial driver of R&D investment in the private, for-profit, pharmaceutical and biotechnology industries, which are the main repositories of the skills and resources that are required to turn ideas into marketable pharmaceutical products. Despite the large number of affected people, there is little or no viable market for new products for diseases of poverty; the majority of patients in the developing world struggle to pay for off-patent, essential medicines that are relatively cheap, let alone new products.

R&D and biotechnology

Biotechnology companies operate within a complex network of universities and research institutes, venture capitalists, specialized service providers, regulatory agencies and large pharmaceutical companies that assist in the commercialization of products and technologies[2,3]. With a few notable exceptions — such as Genentech and Amgen, which have evolved into fully integrated enterprises — the majority of biotechnology companies operate within the discovery and preclinical stages of R&D of new drugs and vaccines. Biotechs advance diverse sets of technologies, new drug targets and optimized leads for transferral into the clinic, and downstream development by pharmaceutical companies. The growing interdependence of the success of large and small companies is demonstrated by data covering the sources of product discovery and market introduction.

According to a 2001 survey of the pharmaceutical and biotechnology sectors[4], biotechnology companies discovered 45% of the biomedical products on the market in that year (universities accounted for 10%; pharmaceutical companies accounted for 45%), up from 5% in 1993. However, biotechnology companies brought only 20% of these products through the development process to market. From the perspective of large pharmaceutical companies, more than 30% of the products in their pipelines were in-licensed or developed elsewhere with company sponsorship. The same study estimates that by 2005 more than 50% of pharmaceutical companies' revenues will be derived from products that are discovered, researched and developed by an outside organization.

The technological and product outputs of these R&D-intensive biotechnology companies could prove invaluable in the development of new weapons to fight diseases of poverty specifically, and to advance global health more generally. A study lead by the Joint Centre for Bioethics at the University of Toronto listed the top ten biotechnologies identified by biotechnology and global health experts as having the potential to improve health in developing countries[5] (BOX 1).

Consider, for example, what has happened since a consortium of scientists from the Wellcome Trust Sanger Institute (Cambridge, UK), The Institute for Genomic Research (Maryland, USA) and Stanford University (California, USA) completed sequencing the malaria genome — a scientific accomplishment that is likely to advance our understanding of the pathogenesis of malaria infection, the molecular mechanisms by which resistance to antimalarial drugs develops and the molecular complexity of the parasite[6]. Biotechs were among the first to transform these scientific advances into business opportunities. Several companies have discovered new malaria vaccines, including Aquila Biopharmaceuticals, CEL-SCI, Epimmune and Novartis. Associated product development projects have yielded

PERSPECTIVES

Estimate of world malaria burden

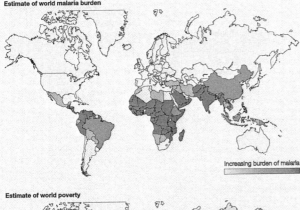

Increasing burden of malaria

Estimate of world poverty

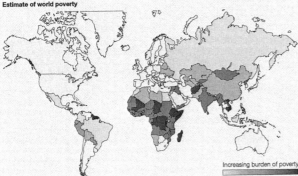

Increasing burden of poverty

Figure 1 | **The relationship between malaria and poverty.** Source of data: Roll Back Malaria InfoSheet (http://www.who.int/inf-fs/en/InformationSheet10.pdf).

needle-free injection devices for vaccine injection (Bioject Medical Technologies) and kits for parasite detection (AMRAD Corp.).

Despite the potential for biotechnology companies to significantly contribute to the management of diseases of poverty, the most effective means of harnessing the capabilities and resources of this sector are yet to be elucidated. There are several impediments that must be overcome.

Obstacles and disincentives

Funding and markets. Most biotechnology companies are wholly dependent on external sources of finance, and have yet to generate a profit[7,8]. Few are in a position to donate technologies, and raising funds to work on neglected-disease projects in-house is difficult and time consuming. The expectation of poor returns on investments in neglected diseases, due to small or non-existent viable markets for products, discourages all of the primary sources of financing for biotechnology companies (private venture capitalists, public investors, shareholders and pharmaceutical companies) from supporting such projects. So, creating 'viable markets' for neglected-disease products is an urgent policy priority if we are to motivate greater engagement of private companies and investors with the problems of neglected diseases. At the moment, the opportunity costs of pursuing a malaria project, rather than, say, a cancer or arthritis project (which have large, well-understood markets), will almost always make the former unaffordable, even if some of the costs of research could be covered by public grants. There are few public or philanthropic funds

that help for-profit companies commercialize and develop projects for diseases of poverty. The Bill and Melinda Gates Foundation is one of few organizations that explicitly places a priority on research into diseases of poverty, but it has made few direct grants to companies. In the United States, companies can apply for Small Business Innovative Research grants for specific pieces of work, but these would not cover the costs of the entire development process.

Public relations. Creating positive public relations is an often-cited incentive for embarking on diseases-of-poverty projects. However, gains in reputation from media exposure can by no means be guaranteed. Large companies have been inundated with negative press about their contributions, or lack thereof, to global health. This further discourages the participation of small companies, who cannot afford to defend their level of contribution in response to criticism from the press and non-government organizations. Even when press coverage is complimentary, some industry representatives argue that this is insufficient justification for investment in activities that do not add to shareholder value[9].

Human resources and capabilities. Small companies often lack the personnel to investigate the potential value of project opportunities for neglected diseases. Competing for funding from the myriad of new global health initiative partnerships and other public sources is time consuming and labour intensive. In addition, many biotechs lack the skills required to design and manage a global health strategy — the global health network involves players which most biotechs have had little experience in working with. In addition, developing countries have unique and unfamiliar institutional socioeconomic healthcare environments, which makes designing and conducting clinical trials of new therapies, as well as the distribution of products, all the more difficult.

Poorly developed access and delivery systems. Biotech and pharmaceutical companies are discouraged from undertaking new R&D programmes for neglected diseases by the fact that patients in the developing world are presently not able to access the existing, often affordable, cost-effective and essential health technologies, drugs and vaccines. In interviews, some large pharmaceutical firms proposed that global health funds would be better spent in the creation of effective distribution and treatment administration systems, rather than on (or in addition to)

R&D investments. The rationale behind this opinion is that once companies see marketed products being used effectively they will be more likely to consider investing in new products[9].

Inexperienced partners. From the perspective of potential collaborators (see later discussion of public–private partnerships (PPPs)), working with small, young biotechnology companies can be highly management intensive, expensive and risky. This is due to the untested nature of much of the technology and science of biotechs, and the high attrition rates associated with every drug or vaccine project.

Benefits and opportunities
Despite the hurdles outlined in the previous section, surveys of biotechnology enterprises indicate that many are genuinely interested in contributing to the management of diseases of poverty[10]. Some of the reasons for this interest are discussed here.

Access to the finances for neglected disease research is particularly attractive to biotechnology companies that lack revenue streams from licensed-out products and technologies. Neglected-disease projects might allow biotechnology companies to establish proof of their technology concept, to explore new applications for therapeutic molecules and to gain valuable experience in product development. Such opportunities might provide invaluable spillovers in terms of locating new industry partners, as well as new financiers for other, potentially more profitable, diseases areas.

To finance other projects in their pipeline, most small biotechnology companies need to license-out proprietary innovations at relatively early stages in the development cycle, conceding at least partial control over their innovations and reducing the value of accruable rents. In view of the opportunity costs and trade-offs that licensing-out entails, biotechnology companies might find the prospect of participating in a PPP attractive, particularly if they are ensured retention of control over valuable assets. (Depending on how the contract is structured, this might mean control over 'non-neglected markets or non-neglected-disease applications'.)

Biotechnology companies place significant importance on networking, reputation building and gaining industrial exposure. Young biotechnology companies in particular might consider participation in 'global-scale' initiatives to be an attractive means of showcasing their technologies and expertise, of establishing

their profile, and of expanding their contact base in industry and the public sector.

Biotechnology companies have demonstrated their ability to successfully advance a niche market strategy by the active pursuit of orphan drug candidates in the United States following the introduction of the Orphan Drug Act in 1983 (TABLE 1). As of 2000, biotechnology companies had sponsored 70% of the more than 900 orphan-designated projects in the United States, and 50% of all approved biotechnology products had orphan status. If policies were put in place to boost the expected return on the investment in presently neglected diseases, companies might choose to pursue diseases of poverty as a means of demonstrating the value of their technologies in a less competitive niche area[11].

Biotechnology and PPPs
PPPs, such as the Malaria Vaccine Initiative (MVI), the Medicines for Malaria Venture (MMV), the Global Alliance for Tuberculosis Drug Development (GATB) and the International AIDS Vaccine Initiative (IAVI), provide an important vehicle for engaging biotechnology companies in global health research. Under contract agreements, biotechnology and pharmaceutical companies commit to applying their technological and scientific expertise to a neglected disease in exchange for financial, human and networking resources from the PPP staff. These resources might include assistance in designing the product to meet the targeted populations' specific needs; finding suitable research, contract and manufacturing partners; and helping to manage the project.

Almost twenty product development PPPs have been established during the past five years (TABLE 2). Although the organizational and business models of individual PPPs differ, they all share the common mission of advancing the development of drugs, vaccines and diagnostics for diseases of poverty to make new, effective and affordable products available to patients who require them. PPPs leverage the skills and resources present in both the public and private sectors, and use venture capitalist-like investment strategies (driven by a social, not-for-profit mission) and virtual R&D management tools (such as outsourcing, contract research and portfolio management) to achieve their objectives. As a group, PPPs have raised more than half a billion US dollars of finance, as well as substantial in-kind resources from public research centres, such as the National Institute of Allergy and Infectious Diseases (NIAID), the UK Medical Research Council and Walter Reed, and from

pharmaceutical companies, such as Glaxo-SmithKline, Pfizer and Bayer[12,13].

As illustrated by the following examples, biotechnology companies have already contributed to PPPs in a number of different ways.

Celera. Celera has donated the development and commercialization rights for their CRA-3316 compound to the Institute for One World Health (IOWH) for all parasitic diseases in all markets. CRA-3316 is a cysteine proteinase inhibitor discovered, screened and optimized by Khepri (acquired by Axys, in turn acquired by Celera) that IOWH is initially applying to Chagas disease in the Latin American region where the disease is most prevalent. IOWH is managing the synthesis of the new compound, and will work with NIAID to complete the necessary preclinical animal testing of the drug. With successful results, IOWH will file an Investigational New Drug application with the US FDA.

Chiron. Chiron has licensed a chemical compound to GATB for an undisclosed amount, and GATB is developing it further through contracts with other partners. Chiron has waived any royalties from sales in developing countries, but retains a 'grant-back' option to make and sell the medicine in countries within the Organisation for Economic Cooperation and Development. If Chiron exercises its grant-back option, it will have to reimburse GATB for all of the development costs. However, it has a financial commitment in the case of failure, significantly reducing its risks. Should Chiron choose not to take the project back in the late clinical trial, manufacturing or distribution stages, GATB has the right to seek another partner.

> **Box 1 | Top ten biotechnologies***
>
> • Molecular diagnostics
> • Recombinant vaccines
> • Vaccine delivery systems
> • Bioremediation
> • Sequencing pathogen genomes
> • Female-controlled protection against sexually transmitted diseases
> • Bioinformatics
> • Nutritionally enhanced genetically modified crops
> • Recombinant therapeutic proteins
> • Combinatorial chemistry
>
> *These are the top ten global health biotechnologies according to REF. 5.

PERSPECTIVES

Table 1 | **Orphan drug legislation: product qualifying criteria and incentives**[19]

Legislation details	United States	Japan	European Union
Year initiated	1983	1993	1999
Qualification criteria	Condition affecting fewer than 200,000 in the US, around 1 case per 1,000 people, or if a vaccine, preventive drug, or diagnostic drug, fewer than 200,000 per year; or proof of no expectation to recover R&D costs from US sales even if the disease affects more than 200,000.	Condition affecting fewer than 50,000 Japanese, around 1 case per 2,500.	Condition affecting less than 1 per 2,000 people in the EU with no 'satisfactory method authorized'; or for selected serious conditions, proof that sales in the EU will not recover investment costs.
Market exclusivity	Seven years of exclusive approval rights.	Post-approval monitoring period is extended to between six and ten years before re-examination.	Ten years of exclusive approval rights from centralized approval process.
Tax credits	Tax credit equal to 50% of qualified clinical research expenses for the taxable year.	A 6% tax credit for orphan related R&D expenses incurred in Japan and a maximum 10% reduction in the corporate tax rate.	Orphan drugs are eligible for member state incentives to support R&D and, when applicable, small and medium-sized enterprises.
Development grants	Annual budget approximately US $12 million. Clinical trials awarded $150,000–$300,000 per year in direct costs for up to three years.	Grants available to subsidize orphan drug development.	Limited EU level funds are available. It is hoped that member states will earmark grant money to support orphan drug research.
Regulatory review	Approval requirement assistance, including access to fast-track approval, and waived FDA user fees.	Expedited regulatory review.	Fast-track approval process considered. Fund to waive part or all of user fees.

OxxonPharmaccines, Therion and Targeted Genetics. These three companies are working with IAVI on a contract basis to apply and develop their specific platform technologies (PrimeBoost technology, pox virus vectors and adeno-associated viral vectors, respectively) to an AIDS vaccine. In exchange for funding to cover the costs of the project, these companies have agreed to make any successful product available at 'an affordable price'. Although the research takes place within the individual companies, IAVI staff are actively involved in managing the project, and have a role in pulling together the academic, clinical and manufacturing partners needed to advance the product.

Maxygen. Maxygen is using its Molecular Breeding technology to help IAVI identify and produce antigens to block gp120, a glycoprotein on the envelope of HIV virions that facilitates binding of the virus to its target cells. This project, specifically designed to combat Indian and African clades of HIV, is being financially supported by the Rockefeller Foundation.

Distinct but complementary to the contributions of the biotechnology sector, large pharmaceutical companies have, for the most part, contributed in-kind resources to PPPs. Such resources include compound libraries, facilities, staff time and experience with drug development.

A call for action
PPPs are one vehicle through which biotechnology companies can contribute to the development of new products for diseases of poverty. Progress to date has demonstrated that mutually beneficial deals between for-profit companies and public global health groups are possible. However, the specific mission of PPPs is to accelerate a portfolio of products for neglected diseases, not to facilitate participation of the biotechnology industry *per se*. With limited budgets and

ambitious timelines, PPPs (supported by scientific advisory boards drawn from academia and industry) select projects that best meet their respective mandates. As a consequence, there are many potentially valuable products, ideas and technologies in the public and private sectors that are not being funded at present.

If more biotechnology companies are to be encouraged to take a strategic business decision to pursue technical solutions for global health priorities, a package of incentives and provisions that cater specifically to their needs and capabilities must be established. Such a package might consist of government-backed incentives — using the orphan drug legislation as a model — that reduce the costs and risks of, and enhance the expected returns from, investing in diseases of poverty; innovative funding mechanisms to cover project and overhead costs, such as Advanced Purchase Contracts; and the establishment of an initiative for biotechnology and global health, the function of which is to broker the interface between biotechnology companies and the opportunities and initiatives in the global health R&D arena.

An incentives policy to encourage R&D. Reports, company surveys and evidence from the orphan drug legislation in the United States indicate that biotechnology companies would respond positively to a package of incentives that encompasses both 'push' (cost and risk reducing) and 'pull' (expected return-enhancing) components[14–17]. Technically, diseases of poverty already qualify for orphan status according to both the US and EU definitions, but the guarantee of seven to ten years of market exclusivity — the crucial 'pull' component of the orphan drug legislation — does not apply in cases in which there are no paying markets.

Given the interdependence of small and large companies in the R&D process, any new policy must target both segments of the industry. There must be a balance between up-front finances and a viable market as a reward should any company successfully develop a new product. An incentive package for global neglected diseases might include fast-track product approval; funding to support clinical trials of products in whatever regions are relevant (the Orphan Drug Act only covers trials within the United States); tax credits for qualifying preclinical and clinical research; a choice between market exclusivity for the product in development or extending patent protection to another product; and the establishment of a public office (equivalent to the Office for Orphan Product Development at the US FDA) that provides information about the incentives package and assists companies in their applications.

The implementation of 'transferable patent exclusivity' could face considerable political resistance in light of escalating healthcare and pharmaceutical costs in developed countries. However, it is the 'pull' component that many pharmaceutical and biotechnology companies have advocated. In December 2003, the DG Research Unit of the European Union commissioned a report that explores the costs and benefits of such an idea; the results of this report are expected in 2004.

As an alternative to 'transference of market exclusivity' as a 'pull' incentive, many have advocated the establishment of a guaranteed purchase fund that would supplement any market that exists for neglected diseases[18]. The Bill and Melinda Gates Foundation supports a policy working group that is exploring the feasibility of establishing a legal contract between the company and some payer (be that a government or a pool of for-profit and not-for-profit investors) to guarantee purchase of any product at some pre-established price. To be successful, such arrangements would need to overcome logistical, political and credibility hurdles, including what price to set, how to deal with any second- or third-generation products for any one targeted disease, and how to ensure their use. However, given the crucial role that an enhanced market might have as an incentive, an imperfect purchase commitment might arguably be better than none.

When thinking about the potential impact of such incentives policies, it is important to acknowledge that for some diseases the scientific obstacles are such that even the prospect of a larger market will not be sufficient to attract new private-sector contributions.

Innovative funding vehicles. Biotechnology companies need access to sustainable sources of funding to build a business model that includes neglected diseases as a priority. However, commercial investors and PPPs cannot and/or are unlikely to ensure such finance.

The Rockefeller Foundation, the Program for Appropriate Technology in Health, the Bill and Melinda Gates Foundation and others have explored the feasibility of establishing new philanthropic investment funds to pursue the dual goals of public health and financial sustainability. Such funds would pursue an investment strategy clearly focused on concrete, philanthropically driven social returns. At a minimum, a return of invested capital plus a modest financial profit would be

Table 2 | Product development PPPs*

Name	Founded	Product/disease	Industry partners	Key funders
AREAS	2003	Vaccines/TB	Merck, Corixa	Gates
Drugs for Neglected Disease Initiative	2003	Drugs/Chagas disease, African trypanosomiasis, leishmaniasis	Strategy in development	MSF
European Malaria Vaccine Initiative	1998	Vaccines/malaria	Not available	European Commission; Irish, Dutch, Swiss and Danish governments
Foundation for Innovative New Diagnostics	2003	Diagnostics/TB	Strategy in development	Gates
Global Alliance for TB Drug Development	2000	Drugs/TB	Chiron	Gates
Global Microbicide Project	2000	Microbicides	Not available	Rockefeller, Gates, Hewlett Packard, Mellon, UN Population Fund
Hookworm Initiative	2000	Vaccines/hookworm	None yet	Gates
Infectious Disease Research Institute	1993	Vaccines/leishmaniasis	Corixa	Gates, NIH
Institute for One World Health	2001	Drugs/Chagas disease, leishmaniasis, diarrhoeal diseases, malaria	Celera	Gates
International AIDS Vaccine Initiative	1996	Vaccines/HIV	Targeted Genetics, Oxxon, Alphavax, Maxygen, Therion, Berna Biotech AG, Cobra Therapeutics Ltd, IDT GmbH	Gates, Rockefeller, UNAID; Danish, Swedish, Canadian, Irish, Dutch, UK and US governments; Levi Strauss Foundation, World Bank
International Partnership for Microbicides	2002	Microbicides	Strategy in development	Gates
Japanese Pharmaceutical Malaria Drug Partnership	1999	Drugs/malaria	14 major Japanese pharma companies	Japanese government
LAPDAP	1998	Drugs/malaria	GSK	DFID
Malaria Vaccine Initiative	1999	Vaccines/malaria	Adprotech, GSK, Apovia, Progen Industry, Bharat Biotech, Oxxon, Biotech Australia	Gates
Medicines for Malaria Venture	1999	Drugs/malaria	GSK, Roche, Ranbaxy, Bayer, Immtech, Schering Plough, Bristol-Myers Squibb	Gates, Rockefeller, Wellcome Trust, World Bank, DFID, IFPMA, Dutch and Swiss governments, ExxonMobile Corp.
Meningitis Vaccine Project	2001	Vaccines/meningitis	Synco, BiosYnth, Serum Institute of India	Gates
Pediatric Dengue Vaccine Initiative	2001	Vaccines/dengue	Strategy in development	Gates, Rockefeller

*Source: http://www.ippph.org. DFID, Department For International Development; GSK, GlaxoSmithKline; IFPMA, International Federation of Pharmaceutical Manufacturers Association; LAPDAP, Lapdap Antimalarial Product Development Partnership; MSF, Médecins Sans Frontières; NIH, National Institutes of Health; PPP, public–private partnership; TB, tuberculosis; UN, United Nations; UNAID, Joint United Nations Programme on HIV/AIDS.

PERSPECTIVES

expected, because one of the clearest indications of financial sustainability is a return on invested capital. Experienced venture capitalists, working in collaboration with philanthropic investors and PPPs, would manage the investments, and put together a feasible, strategic funding package of sufficient magnitude to support the companies' proposed project and/or business plans.

A broker for biotechnology companies in global health. A broker for global health is one of the recommendations that came out of a project conducted in 2001–2002 by a team at the Institute for Global Health at the University of California San Francisco, funded by the Rockefeller Foundation.

This strategy is based on the premise that many biotechnology companies have technologies that are relevant to global health, but lack crucial information about how to get involved in neglected-diseases projects, and about the growing network of funds, foundations and initiatives working on R&D issues. The broker's goal would be to work with companies to cultivate global health applications for technologies and products that they have in their portfolios.

For companies committed to pursuing these applications, the broker's team would help identify key partners in developed and developing countries, and facilitate access to the resources that are needed to successfully complete a project. Existing PPPs and other initiatives would benefit from the added information about technologies and companies that the 'brokering' institution could provide.

Final thoughts

The proactive steps that the US Biotechnology Industry Organization (BIO) have recently taken to promote the importance of diseases of poverty are a fundamental indicator of the biotechnology sector's commitment to finding viable opportunities to contribute to neglected-disease initiatives. In December 2002, BIO and the Bill and Melinda Gates Foundation co-hosted the first 'Biotechnology for Global Health' conference, a partnering forum at which, through roundtable discussions, panels, one-on-one meetings and company presentations, biotechnology companies and global health PPP initiatives met and exchanged information about themselves and their areas of expertise. BIO is now exploring the feasibility and potential value of establishing a new initiative, BioVentures for Global Health, to support and advance the involvement of the biotechnology industry in global health research.

The efforts of BIO can only succeed if other key global players, including public funders, governments of developed and developing countries, non-government organizations and pharmaceutical companies in the global health R&D space, match its energy and commitment. In addition to focused neglected-disease R&D initiatives and policies, a global regulatory and legal environment that encourages innovation must continue to be built and supported. Global players must debate and make transparent decisions on complex issues, including intellectual property rights for research and regulatory standards for essential global health products.

Although finding new private participants to devote time and resources to the challenge of neglected diseases is important, it is also necessary to realistically assess the scope and potential value of biotech contributions and keep this in perspective within the context of global health needs. Biotechnology companies must not be regarded as an alternative to pharmaceutical companies in the pursuit of new products. Rather, their role must well be to complement pharmaceutical contributions and to plug the gap at early stages of the R&D process (for example, commercializing academic research) in which it has proven difficult to engage pharmaceutical firms. By engaging biotechnology companies in earlier stages of drug and vaccine R&D for neglected diseases, pharmaceutical companies might be encouraged to contribute to the later, less speculative but more expensive stages of clinical testing, manufacturing and marketing.

The ultimate goal of the global health network of policy makers, funders and 'doers' (non-government organizations, companies, health clinics and so on) is not to increase industry participation in global health or even to develop new products. Positive impacts on health only eventuate when effective and affordable tools, and knowledge for treatment and prevention, are delivered in a sustainable manner to patients in the developing world. Looking ahead, private industry, including biotechnology companies, has an essential role to play in designing and implementing appropriate healthcare solutions for diseases of poverty.

Hannah E. Kettler is at Global Health Initiatives, Bill and Melinda Gates Foundation, PO Box 23350, Seattle, Washington 98102, USA.
e-mail: hannahk@gatesfoundation.org

Sonja Marjanovic is at the Judge Institute of Management Studies, University of Cambridge, Trumpington Street, Cambridge CB2 1AG, UK.
e-mail: sm423@cam.ac.uk

doi:10.1038/nrd1308

1. World Health Organization. *The World Health Report 2002. Reducing Risk, Promoting Healthy Life* (WHO, Geneva, 2002).
2. Barley, S. R., Freeman, J. & Hybels, R. C. in *Networks and Organisations: Structure, Form and Action* Ch. 12 (eds Nohria, N. & Eccles, R.) 311–347 (Harvard Business School Press, Watertown, 1992).
3. Powell, W. W., Koput, K. W. & Smith-Doerr, L. Interorganisational collaboration and the locus of innovation: networks of learning in biotechnology. *Adm. Sci. Q.* **41**, 116–145 (1996).
4. Ashton, G., Kettler, H., Saunders, E. & McAuslane, J. *External Collaboration and Licensing in Pharmaceutical R&D* (CMR International, London, 2001).
5. Daar, A. S. *et al.* Top ten biotechnologies for improving health in developing countries. *Nature Genet.* **32**, 229–232 (2002).
6. Gardner, M. J. *et al.* Genome sequence of the human malaria parasite *Plasmodium falciparum. Nature* **419**, 498–511 (2002).
7. Casper, S. & Kettler, H. *The Road to Sustainability in the UK and German Biotechnology Industries* (Office of Health Economics, London, 2000).
8. West, J. The mystery of innovation: aligning the triangle of technology, institutions and organisation. *Aust. J. Manage.* **26**, 21–45 (2001).
9. Kettler, H. & White, K. *Valuing Industry Contributions to Public Private Partnerships for Health Product Development* (Initiative on Public Private Partnerships, 2003).
10. Marjanovic, S. *The Role of Biotechnology Firms in the 'Global Health Challenge': Investigating the Potential for Smaller Biotech Contributions to 'Remedying' Neglected Diseases of the Developing World.* Unpublished thesis, Cambridge Univ. (2002).
11. Milne, C., Kaitin, K. & Ronchi, E. *Orphan Drug Laws in Europe and the US: Incentives for the Research and Development of Medicines for the Diseases of Poverty* (CMH Working Paper Series, Paper No. WG2: 9, 2001).
12. Kettler, H. & Towse, A. *Public–Private Partnerships for Research and Development: Medicines and Vaccines for Diseases of Poverty* (Office of Health Economics, London, 2002).
13. Nwaka, S. & Ridley, R. Virtual drug discovery and development for neglected diseases through public–private partnerships. *Nature Rev. Drug Discov.* **2**, 919–928 (2003).
14. Kettler, H. *Biotechnology for Global Health: An Options and Feasibility Study.* Final version prepared for the Rockefeller Foundation (2002).
15. Kremer, M. in *Innovation Policy and the Economy* (eds Jaffe, A. B., Lerner, J. & Stein, S.) (2001).
16. WHO/IFPMA Roundtable. *Drug Development Working Group Priority Infectious Diseases Requiring Additional R&D* Draft (2001).
17. Grabowski, H. *Increasing R&D Incentives for Neglected Diseases: Lessons from the Orphan Drug Act* (Duke Univ., Durham, 2003).
18. Glennerster, R. & Kremer, M. The purchase precommitment as a supplement to patents and government-funded research: a better way to spur medical research and development. *Regulation* **23**, 34–39 (2001).
19. Kettler, H. E. *Narrowing the Gap Between the Provision and Need for Medicines in Developing Countries* (Office of Health Economics, London, 2000).

Competing interests statement
The authors declare that they have no competing financial interests.

(i) Online links

FURTHER INFORMATION
Bill and Melinda Gates Foundation:
http://www.gatesfoundation.org/default.htm
Biotechnology Industry Organization: http://www.bio.org/
Drugs for Neglected Diseases Initiative:
http://www.dndi.org/
Global Alliance for TB Drug Development:
http://www.tballiance.org/
Institute for One World Health:
http://www.oneworldhealth.org/
International AIDS Vaccine Initiative: http://www.iavi.org/
Malaria Vaccine Initiative: http://www.malariavaccine.org/
Medicines for Malaria Venture:
http://www.mmv.org/pages/page_main.htm
Office of Health Economics: http://www.ohe.org
Access to this interactive links box is free online.

Part II
Science and Innovation

[5]

Journal of Economic Behavior and Organization 24 (1994) 91-114. North-Holland

Evaluating technolological information and utilizing it*
Scientific knowledge, technological capability, and external linkages in biotechnology

Ashish Arora

Heinz School of Public Policy and Management, Carnegie Mellon University, Pittsburgh, PA, USA

Alfonso Gambardella

Bocconi University, Milan, Italy and University of Urbino, Urbino, Italy

Received February 1992, final version received February 1993

Recent research in the theory of innovation has suggested that the role of collaborative alliances in the innovation process has become very important. In this paper we argue that firms differ significantly in their ability to benefit from these collaborative relationships. We relate the internal knowledge base of the firm to the number of collaborative agreements (external linkages) that it enters into with other agents (small start-up biotechnology firms, and universities) endowed with complementary assets for innovation. We distinguish between scientific and technological capabilities, the former relating to the evaluation of information, and the latter to the utilization. We test the implications of the model using data on a sample of 26 large U.S. chemical and pharmaceutical companies active in the biotechnology sector in the 1980s using econometric techniques appropriate for count events.

Key words: Innovation; Technological collaboration; Information

JEL classification O31; L22; D81

1. Introduction

One important part of the intellectual legacy of Schumpter must surely be

Correspondence to: A. Arora, The H. John Heinz III School of Public Policy and Management, Carnegie Mellon University, Pittsburgh, PA 15213-3890, USA.

*Financial support from the TEG program at the Center for Economic Policy Research is acknowledged. A.G is grateful to the Lynde and Bradely foundation for financial support. We wish to thank Tim Bresnahan, Paul David, Nathan Rosenberg, and Ed Steinmueller, and an anonymous referee for helpful comments, and Gary King for the software used for some of the regressions in this paper. Our thanks also to members of seminars at CMU, UC San Diego, and the NBER, for their comments. The usual caveats apply and all remaining errors are our responsibility alone.

the distinction between innovation and invention. The distinction is important because it emphasizes that there is a great deal more to the process of innovation than an 'idea'. One fruitful way of conceptualizing the distinction is as a division of labor in the innovation process [cf. Grandstrand and Sjölander (1990)]. Innovation encompasses several functions, and the skills and competencies required for one function may not be the same as those required for the other, and may be better suited for different types of firms or institutions.[1] In their survey article, Cohen and Levin (1989) note that there is a pressing need to investigate the circumstances under which a division of labor takes place between agents generating new knowledge and those engaged in its commercialization. The increasing cost of innovation, and the growing complexity and multidisciplinarity of knowledge has given a particular impetus to a better articulated division of labor in science based industries.

However, a division of labor in innovation is rather more difficult than in, say, pin making, because the intermediate input, knowledge and information, is difficult and costly to transfer via arms length market relationships. As a result one sees a variety of different institutional forms. In the past decade even large established firms have entered into systematic collaborative linkages with universities, and with small research intensive firms.[2] These linkages have taken different forms such as contractual and non-contractual agreements, equity swaps, and acquisitions. Our earlier research showed that the different strategies of external linkages are mutually complementary, which suggests that firms systematically differ in the extent to which they form external collaborative linkages in general [Arora and Gambardella (1990)].

But are all firms equally well placed to enter into such collaborative linkages? Previous research suggests that there are firm specific attributes that affect the value that a firm derives from such relationships [e.g. Teece (1986), Nelson (1990)]. Cohen and Levinthal (1989) argue that firms invest in R&D for two purposes. On the one hand, they invest in R&D to generate innovations; on the other hand, R&D serves as a device to exploit external research. Rosenberg (1990) argues that in-house basic research is necessary to

---

[1]Indeed, Jewkes, Sawyer and Stillerman (1958) have argued that innovation is realized through the interaction of firms that are distinguished by size, expertise, and other attributes. Along similar lines, Williamson (1975) has argued that an efficient innovation system might be characterized as one where smaller firms dominate the early, invention stage, and are subsequently acquired by larger firms which are responsible for development and marketing. Arrow (1983) argues that small firms have a comparative advantage in the early stages of innovation. Stigler (1956) predicted the growth of specialized firms which would provide research services for other firms.

[2]A few authors have even argued that the locus of innovation has shifted from the large firm to a *network* of inter-organizational relations, in which it is difficult to identify the innovator with a specific type of organization. See for instance Piore and Sabel (1984), *Business Week* (1986), and Imai (1988).

monitor the flow of scientific information in the outside world.[3] Both of these studies emphasise the potential synergies between external and internal knowledge. However, their discussion does not deal with the multi-dimensionality of knowledge, and therefore does not fully unpackage the source of the synergies.

In this paper we distinguish between two different types of knowledge based capabilities: The *ability to evaluate* information, and the *ability to utilize* information. The question that we address in this paper is 'What is the role of the scientific and technological capabilities in the ability and willingness of a firm to enter into collaborative linkages in the innovation process?'

We develop our model in the context of the newly emerging biotechnology sector. Biotechnology is a major scientific and technological breakthrough. In the 1980s, it has exerted a profound influence on the scientific and technological basis of the chemical, and especially, of the pharmaceutical industry.[4]

There are two main reasons why we focus on biotechnology. First, due partly to the complex knowledge base underlying biotechnology, the development of biotechnology has involved an intricate network of alliances amongst large established chemical and pharmaceutical firms, small research-intensive companies (the so-called New Biotechnology Firms or NBFs) and universities [Arora and Gambardella (1990), Pisano et al. (1988)]. Second, biotechnology has had a strong science-push origin, and is possibly the sector par excellence where science plays a major role. This is important because we seek to distinguish between the impact of scientific and technological capabilities.

Although biotechnology is particularly apt a context in which to make our point, we believe that our analysis has more general applicability. We provide a conceptual framework for analyzing how in-house scientific and technological capabilities affect the ability and the willingness of a firm to derive value from external linkages. Also, while we carry out our discussion in terms of collaborative alliances, the model applies mutatis mutandis to the decision on in-house research projects as well. In section 2 we develop a model of the relations between large firms and external parties in biotechnology to analyze the impact of different dimensions of the internal knowledge base on the decision to establish linkages with other agents. We than test these implications. The econometric specification and the data are discussed

[3] Mowery (1981) shows that, during 1921–1945, large U.S. firms, when starting new innovation projects, also contracted out part of the research to specialized institutes. The empirical results obtained in Arora and Gambardella (1990) indicate that the number of patents in biotechnology held by large firms is positively related to the number of external linkages.

[4] By biotechnology we mean the *r*DNA technology, the monoclonal antibody technology, the PCR technology, as well as molecular biology and molecular genetics. For a 'technical' introduction to biotechnology, see OTAF (1984), Daly (1985) or Olson (1986).

in section 3. Section 4 presents the empirical results. Section 5 concludes the paper.

2. The model

In this section we characterize the optimal expected number of collaborative venture (henceforth external links) of the firms. Consider therefore a large, established corporation, at the point of deciding whether or not to enter into a collaborative agreement with an NBF. The latter typically offers an 'idea'. For instance, it may have synthesized in an E. Coli culture a protein which is closely related to some physiological disorder; or it may have discovered a way of inducing a gene to express itself. In some instances, a useful technique relating to the way of separating a particular protein from other products produced by a cell cultures or a delivery system may be on offer. The large firm is expected to fund the additional research that may be needed and then, possibly, to license in the product.[5]

One can conceive of each such link as the purchase of an *option* on a 'project' rather than an investment in the project. The initial investment made in a typical external agreement is not very large relative to the total R&D budgets for drug research. Typical sums involved in these linkages are of the order of about $1–5 million. The total expenditure involved in introducing a drug, including marketing, could easily be of the order of a $100 million. Looking upon the choice problem of the firm as that of the purchase of an option is particularly apt when the NBF has not completed its research and the agreement is made before the results of the research are known. Even when the product is relatively well developed – for instance, when the NBF has successfully produced a protein with good *potential* therapeutic properties – it is still true that large investments of financial and scientific and technological resources are required to determine optimal dosage and indications, and to carry out the long clinical trials that are needed to satisfy the regulatory requirements of non-toxicity and efficacy. The bulk of expenditures are incurred during this period of development, which may take on average from three to six years.[6]

In other words, even if the scientific uncertainties are substantially resolved, there are still technological and commercial uncertainties to be resolved.[7] The resolution of these uncertainties requires time, and, more

---

[5]For a more detailed discussion of what the two parties to the agreement bring to the table, see Arora and Gambardella (1990), and Pisano et al. (1988), and the references cited therein.

[6]The report submitted by Glaxo on its anti-emetic drug, Zofran, is reported to have run to 80 volumes, of 500 pages each.

[7]The case of *TPA* (tissue plasminogen activator), a genetically engineered drug produced by Genentech illustrates the kind of issues involved. *TPA* passed the toxicity test of the FDA; it also proved capable of dissolving blood clots. And although it was found to be no more effective than Streptokinase, a much older drug, costing only a tenth as much, *Activase*, the name under which Genentech markets *TPA*, still managed to record sales of over $200 million in 1989.

importantly, requires that the firm's researchers be able to get first hand experience with the product or the technique. There are instances of a large firm funding research or entering into joint ventures and then not renewing their commitments after the initial contract expires or selling its stake in the venture.[8]

To model the complex process of how these large firms seek out partners for their external links we need to make some simplifying assumptions. The most heroic of these is the neglect of the motivations of the partner – the NBFs and the universities. In fact, in the model, they exist only in the form of potential projects which the (large) firm may select. In a formal sense, we assume that the firm faces a horizontal supply curve of projects, but one which slopes sharply upwards after a certain point. The reader may think of each NBF as offering a certain fixed number of projects, and similarly for each relevant university department. We shall assume that this population of projects is large enough so that any strategic or congestion type effects can be neglected. It should also be emphasized that the object of analysis is not the decision on the acquisition of an NBF. Rather, it is the decision to form some sort of a strategic alliance which is far more narrowly circumscribed. We shall also use the same model to describe the links with the NBFs as well as with the universities. The reasons for such agreements may differ according to whether the agreement is with an NBF or with a university. This issue is further discussed in section 3.

Transaction cost perspectives

Our conceptualization differs from traditional analyses of the so called 'make-buy' decision which have been framed largely in terms of transaction cost economics. This perspective would suggest that firms would attempt to internalize the knowledge assets required for innovation, and hence, that firms with strong in-house research capabilities would be less likely to enter into external linkages such as strategic alliances.

Pisano (1990) has provided the strongest case for this perspective in the context of biotechnology. Based upon a sample of some 92 biotech projects at the pre-clinical and earlier stage, for 30 large pharmaceutical firms, he found that firms were more likely to do the R&D exclusively in-house, the smaller the number of NBFs active in the area, the greater the number of previous projects the firm had accomplished (through in-house efforts alone) in the particular therapeutic area, and possibly, the greater the fraction of their sales accounted for by pharmaceuticals. Aside from certain technical

---

[8]In a conversation with the authors, the vice president for research of a major pharmaceutical corporation pointed out that these projects varied greatly (ex post) in quality, and that many turned out to be 'mistakes'. However, and this is crucial to our conceptualization, he noted that the 'mistakes' were financially not very serious for they 'did not show up on the bottom line'.

problems (e.g. failure to control for firm effects), the point is that the empirical evidence is amenable to alternative interpretations. For instance, the smaller the number of NBFs in a particular area, the less likely it is that the large firm would be able to find a suitable partner with the appropriate research capabilities. But a more important problem with Pisano's analysis is that it implicitly assumes that the choice problem facing the large firm can be adequately represented by assuming that the project is initiated by the large firm which then decides whether to use external sub-contractors or not. Our understanding of the industry is that the bulk of the new projects in the sector are initiated by NBFs, some of which may be offered to large firms for their participation. Large firms thus have two distinct (but not independent) choices: Of the projects offered to them by NBFs, to choose the ones in which to participate, and to choose projects for in-house R&D. Of the latter, they may choose to involve external agents as well. Hence, the number of exclusively in-house projects in an area would reflect a strategic decision to focus in-house research efforts on that area. In other words, the measure of in-house capability used by Pisano could very easily represent a strategic decision, and the failure to control for firm effect could cause serious bias.

Assumptions and notation

The rest of the story is as follows. The firm faces a two step choice. In the first stage, the firm selects m projects out of a population of N. For each of these m projects, the firm has to incur a cost of K. At the point that the firm selects these m projects, it does not observe the true payoff from the project, X, Instead it observes a signal that is correlated with X. This signal could include the scientific reports of the experiments results and techniques associated with the protein in question, and quality of the scientific and technical personnel of the NBF (or the university researchers). After having selected a project (and incurred the cost K) the firm then observes X. It then decides whether to develop the project or not. If it chooses to develop the project, its payoff is $X - D - K$, where D is the development cost. If firm decides not to go ahead with the project its payoff is $-K$.[9]

Let Y represent the real value signal that the firm observes. How much information the signal conveys to the firm depends, inter alia, on the existing capabilities of the firm that are related to its scientific capabilities, denoted here by θ. Similarly, the value of the project depends upon the ability of the

[9]We conjecture that a more gradual resolution of uncertainty would not greatly change the conclusions. After we wrote this paper it was pointed out to us that our formal model was similar to one that Nelson (1961) had used to discuss the value of parallel research projects. A little thought will reveal that an additional parallel project can be thought of as an additional option. The difference is that in the case of parallel projects, the options are not independent. In fact, in Nelson's model only one option is exercised. In Appendix B we discuss a general model where an arbitrary number of interdependent options can be bought and exercised.

firm to utilize effectively the know-how it is purchasing. This ability we label technological capability and represent by γ.

Both X and Y are assumed to have compact support. In particular, suppose that the support of X is given by $(0, B)$. Let $F(x \mid y; \theta, \gamma)$ represent the cumulative distribution function of X, conditional on Y, and let $G(\cdot)$ and $g(\cdot)$ represent the distribution and the density function, respectively, of Y. We assume, without great loss of generality, that all functions are smooth and differentiable in their arguments.

We also make the following assumptions:

Assumption 1.

$$F(x \mid y'; \theta, \gamma) \succeq_{F.O.S.D.} F(x \mid y; \theta, \gamma) \; \forall y' \geq y,$$

where $\succeq_{F.O.S.D.}$ stands for *first order stochastically dominate*

Assumption 2.

$$F(x \mid y; \theta^1, \gamma) \succeq_{S.O.S.D.} F(x \mid y; \theta^2, \gamma) \; \forall \theta^1 > \theta^2,$$

where $\succeq_{S.O.S.D.}$ stands for *second order stochastically domin̦ es.*

Assumption 3.

$$F(x \mid y; \theta, \gamma^1) \succeq_{F.O.S.D.} F(x \mid y; \theta, \gamma^2) \; \forall \gamma^1 \geq \gamma^2.$$

Definition. A distribution $F(x \mid y)$ *first order stochastically dominates* another distribution $F(x \mid y')$ if $F(x \mid y) \geq F(x \mid y') \; \forall x \in [0, B]$.

A distribution $F(x \mid \theta)$ *second order stochastically dominates* another $F(x \mid \theta')$ *if*

(i) $E(X \mid \theta) = E(X \mid \theta')$,
(ii) $\int_0^\tau F(x \mid \theta) \, dx \leq \int_0^\tau F(x \mid \theta') \, dx \; \forall \tau \in [0, B]$.

Assumption 1 makes precise the sense in which Y is a signal for X. Intuitively, one would expect that a high value of Y would lead the decision maker to expect that the project is more attractive. [Milgrom (1981)]. Assumption 2 can be thought of as saying that firms which have superior evaluation capabilities receive signals which are less 'noisy' – they have a conditional distribution of X that is more heavily concentrated around the mean. Assumption 3 is quite straight forward. A firm with higher in-house technological capability has a higher expected valuation for any project, for a given value of the signal. Put differently, for any given value of the signal, a firm with superior technological capability expects to do better with the project.

We would like to highlight two assumptions that we have made that considerably simplify the analysis. The first is that the payoffs are independent across projects. In turn this assumption implies that the cost of implementing projects does not rise with the number of projects implemented. In Appendix B we show that one can allow for inter-dependence of payoffs without affecting our results. The second assumption we make is that the firm can choose from, and implement, a maximum of N projects (where N can be different for each firm). As an empirical matter, firms tend to specialize in a few therapeutic areas. In these therapeutic ares, firms acquire facilities and expertise to manufacture the pharmaceutical, as well as the complementary equipment that may be required.[10]

The optimal number of options

The firm's maximization problem can be shown to be equivalent to the choice of an optimal threshold value of the signal, y^*. To see this, note that for every potential project, the payoffs and costs incurred by the firm are independent of all other projects that the firm may have selected. Hence, for each project, the decision on whether to select it or not depends only on the value of the signal. Furthermore, since all projects have the same distribution (ex ante) of returns, the optimal threshold value of the signal will be the same for all projects. The firm's objective function can be written as:

$$\max_{y^*} \Pi \equiv \left[\int_D^B (x-D) \, dF(x \mid \omega; \theta, \gamma) - K \right] (1 - G(y^*))N, \tag{1}$$

where ω is the event $\{Y \geq y^*\}$.

The maximand consists of two parts. The second term is simply the expected number of projects that the firm will select. The first term, enclosed in the square brackets, represents the expected net payoff from any one project for which the signal is at least as high as y^*.

Consider the first term. Integrating by parts it can easily be seen to be equal to

$$\left[(B-D) - \int_D^B F(x \mid \omega; \theta, \gamma) \, dx - K \right].$$

---

[10]For instance, the suppliers of diagnostic kits such as Abbot, also supply hospitals with the equipment used along with the diagnostic kits. As can be imagined, a firm with no prior experience in the diagnostic markets would not be likely to undertake a project for the production of a diagnostic kit because it would lack the required complementary knowledge (and physical) assets.

The term inside the integral sign is a conditional probability. This can be simplified as follows.

$$F(x\,|\,\omega;\theta,\gamma)=\Pr\,(X\leq x\,|\,Y\geq y^*)$$

$$=\frac{1}{1-G(y^*)}\int_{y^*}F(x\,|\,y;\theta,\gamma)g(y)\,\mathrm{d}y. \tag{2}$$

We can therefore write (1) as

$$\max_{y^*}\Pi\equiv\left[(B-D-K)-\int_D^B\frac{1}{1-G(y^*)}\left(\int_{y^*}F(x\,|\,y;\theta,\gamma)g(y)\,\mathrm{d}y\right)\mathrm{d}x\right](1-G(y^*)). \tag{3}$$

Assuming an interior solution, the first order condition is given by

$$\frac{\partial\pi}{\partial y^*}=-[B-D-K](g(y^*))+\int_D^B F(x\,|\,y^*;\theta,\gamma)g(y^*)\,\mathrm{d}x=0. \tag{4}$$

One can also see that the second derivative is unambiguously negative at any stationary point.

$$\frac{\partial^2\pi}{\partial y^{*2}}=-[B-D-K](g'(y^*))+\int_D^B F_y(x\,|\,y^*;\theta,\gamma)g(y^*)\,\mathrm{d}x$$

$$+\int_D^B F(x\,|\,y^*;\theta,\gamma)g'(y^*)\,\mathrm{d}x. \tag{5a}$$

Upon substituting (4), we get

$$\mathrm{RHS}=\int_D^B F_y(x\,|\,y^*;\theta,\gamma)g(y^*)\,\mathrm{d}x.$$

Using Assumption 1 we get

$$F_y(x\,|\,\cdot)\leqq 0. \tag{5}$$

Notice that the actual number of projects selected by the firm, m, has a binomial distribution with parameters $(N,1-G(y^*))$. The average number of projects selected is of course, $N(1-G(y^*))$. It is straight forward to see that the latter is decreasing in y^* and we need only look at the effect of various parameters on y^* in order to get the qualitative effects.

Proposition 1. $\partial y^*/\partial\gamma\leqq 0$.

Proof. By totally differentiating (4), we get

$$\left(\frac{\partial^2 \Pi}{\partial y^{*2}}\right) \frac{\partial y^*}{\partial \gamma} = -\left[\int_D^B F_\gamma(x\,|\,y^*;\theta,\gamma)g(y^*)\,\mathrm{d}x\right]. \tag{6}$$

We have already established that the first term on the LHS is negative at any interior stationary point. By Assumption 3, we have $F_\gamma(x\,|\,\cdot)\leqq 0$. Hence we must have $\partial y^*/\partial \gamma \leqq 0$.   Q.E.D.

This result states that, all else held constant, firms with higher in-house technological capabilities, which allow them to derive greater expected payoffs from any project, will select a greater number of projects.

Proposition 2. y^* is non decreasing in θ.

Proof. By Assumption 2, we know that

$$E(X\,|\,y;\theta^1,\gamma) = E(X\,|\,y;\theta^2,\gamma), \tag{7a}$$

$$\int_0^\tau F(x\,|\,y;\theta^1,\gamma)\,\mathrm{d}x < \int_0^\tau F(x\,|\,y;\theta^2,\gamma)\,\mathrm{d}x, \tag{7b}$$

$$\forall y, \forall \tau, \forall \theta^1 > \theta^2.$$

Integrating by parts on both sides of eq. (7a) we get

$$\int_0^B F(x\,|\,y;\theta^1,\gamma)\,\mathrm{d}x = \int_0^B F(x\,|\,y;\theta^2,\gamma)\,\mathrm{d}x. \tag{8}$$

Rewriting one gets

$$\int_D^B (F(x\,|\,y;\theta^2,\gamma) - F(x\,|\,y;\theta^1,\gamma))\,\mathrm{d}x = \int_0^D (F(x\,|\,y;\theta^1,\gamma) - F(x\,|\,y;\theta^2,\gamma))\,\mathrm{d}x \tag{9}$$

By (7b) we know that the RHS of (9) is non positive. Hence the LHS of (9) must be non positive as well. In turn, this implies that $\int_D^B F(x\,|\,y;\theta,\gamma)\,\mathrm{d}x$ is increasing in θ. The rest of the proof is straightforward and is similar to that for Proposition 1.

Proposition 2 states that firms which are better able to judge the true value of the project from the signal (in the sense defined above) will be more focussed in their search. They will be more discerning in the projects they select and will have a higher threshold value. This would result in fewer projects selected on average.

Table 1

Definition of variables.

AWF_{it}	= Number of agreements of the ith firm with other firms in biotechnology during 1983–1989.
AWU_{it}	= Number of agreements of the ith firm with universities and other non-profit institutions in biotechnology during 1983–89.
PAT_i	= Number of U.S. patents in biotechnology of the ith firm applied for by 1984 and granted by 1987.
RDS_{it-1}	= Ratio of R&D to total sales of the ith firm in the year $t-1$.
SPS_i	= The average number of scientific publications by the employees of the ith firm during the period 1975–1982 deflated by the total sales in millions of dollars in 1986.
$LOGSAL_i$	= Logarithm of total sales in millions of dollars of the ith firm in 1986.
PHS_i	= Ratio of pharmaceutical sales to total sales of the ith firm in 1986.

As pointed out earlier, these results have been generated by a model which can best be described as a partial equilibrium model and can be thought of as characterizing the demand function of the large firm. In the empirical analysis we explicitly invoke the assumption that the firms face a perfectly elastic supply of projects (up to N). Large firms are likely to have far greater bargaining power than NBFs. Also, given that we lack data on the small NBFs and their motivations, this assumption seems to us to be the most sensible way to allow us to empirically test the model and it is to this that we now turn.

3. The specification

The dependent variables

We estimate two equations. The dependent variables are: (i) the number of agreements of the ith firm with other firms in biotechnology during 1983–1989, AWF_{it}; (ii) the number of agreements of the ith firm with universities and other non-profit research centers (government laboratories, foundations, etc.) in biotechnology during 1983–1989, AWU_{it}. Our sample consists of 26 large U.S.-based firms. Most of them are pharmaceutical companies. A few of them are large chemical firms with a special interest in biotechnology. Most AWFs are agreements with NBFs. A few agreements of the firms in our sample involve large firms as well, mostly cross border marketing agreements, or relating to the supply of equipment or materials. In AWU we have also included collaborative links with non-profit research institutions such as Scripps. The data, including the list of firms, and the sources for data are described in Appendix A.

The explanatory variables

In the model developed in section 2, we distinguished between the ability of the firm to evaluate external knowledge, and the ability to utilize it. The more problematic issue is one of deciding upon measures for these capabilities. We suggest that the in-house scientific capabilities of the large firms is a good measure of their skills in selecting 'good' projects. Scientific capability enables the firm to reduce the uncertainty about the outcome of individual projects [Nelson (1982), David, Mowery and Steinmueller (1988)]. It has been argued that science provides information that helps restrict the search for successful innovations at the downstream applied research and development stages. As noted earlier, Rosenberg (1990) has suggested that firms perform basic research as a part of maintaining an in-house research capability which is useful primarily for being 'plugged in' to external information flows. Since a great deal of useful information in biotechnology is science based, an in-house scientific capability is crucial for evaluating and assessing information originating outside of the firm's boundaries.

As a measure of in-house scientific capabilities, we use the ratio of the average number of scientific papers published by the personnel of the firms between 1975–1983 to the total sales. Since we wanted a measure of the stock of scientific capability, we wanted a reasonably long period over which to measure scientific publications. These data were conveniently available and provide a good measure of the stock of scientific capability that existed in the firms prior to the start of our sample period. Nelson (1990) notes that pharmaceutical companies allow their scientific personnel to publish quite freely. Halperin and Chakrabarti (1987) have found that publications of scientific papers are very highly correlated with the number of elite researchers employed by the firm (more so than patenting by the firms).

The ability to utilize knowledge depends upon a host of factors that we have chosen to label technological capability. In the empirical analysis, we use two measures of the technological capabilities of the large firms. The first variable is the total number of biotechnology patents of the firms in our sample applied for in the U.S. by 1984. It is intended to capture the extent to which the firm has invested in biotechnology related research in the past years its technological skills.[11] The use of patents is supported by the results in Nairn et al. (1987) who find that in the pharmaceutical industry, patents are highly correlated with other measures of corporate technological strength such as expert options, highly cited publications in the area, and the number of new drugs registered.

The second variable is the ratio of R&D expenditures to sales, between 1982–1988 lagged once. We expect this variable to measure the extent to which the firm is a 'research-oriented' firm. Since these are large corporations, one would expect a research intensive firm to also have in place other

---

[11]As we are dealing with patents applied for by 1984, we have a measure of the predetermined technological skills of the large firms in this field before the 1983–1989 flows of agreements in our sample.

assets like a marketing and distribution network. In other words, while R&D is not the only activity important for successful commercialization, one could reasonably expect that a (large) firm which relies upon research based products would have invested in other complementary assets needed for the successful commercialization of the innovation. Since our other measures are stock measures which do not extend beyond the start of the sample period, we used one period lags in R&D intensity to capture any possible changes in the technology orientedness of the firm in the sample period. However, as might be expected, R&D intensity tends to be stable over time, and the choice of the length of the lag does not materially affect the results, and neither does the use of a stock measure for R&D intensity.[12]

We would like to emphasize that the nomenclature adopted here for the two types of capabilities is no more than suggestive. We use these terms in a broader sense than perhaps is commonly understood. For instance, we do not mean to suggest that the ability to utilize technology depends solely upon a narrow range of laboratory skills. In the same vein, one needs more than purely academic scientific knowledge to evaluate the commercial value of scientific projects. Furthermore, the distinction is a conceptual one: It is likely that both of these abilities may depend upon the presence of similar (or even the same) resources and competencies. As an empirical matter, in most industries one may not always be able to tell them apart. However, as pointed out earlier, pharmaceuticals (and particularly since the advent of biotechnology) is the most science based of all industries, and one can legitimately distinguish between technological capability and scientific capability.

The set of explanatory variables includes time-dummies to account for factors common to all firms in any given year. We also use the logarithm of the total sales of the firm in 1986, and the ratio of the pharmaceutical sales to the total sales of the firm in 1986. The latter two variables are intended to control for firm differences in the *scale* of search. Recall from section 2 that the expected number of projects undertaken by the firm depended, inter alia, upon N, the population of projects from which the firm chooses. The total number of potential partners may be increasing in the size of the firm, as well as in the extent to which the firm is involved in the health-care sector.

The same specification is used for both the AWF and the AWU equations. This needs elaboration and qualification for there are important ways in which the two differ. First and foremost, one would imagine that the agreements with universities are driven by considerations which relate to longer term objectives than is the case with the linkages with other firms. In

---

[12]In order to allay any concerns that may arise from our practice of deflating these variables by average sales, we would like to point out that we also ran the regression without deflating R&D and scientific publications by total sales. The results are remarkably similar, and are available on request.

addition, the firm may be able to recruit promising young researchers, and may even be able to attract senior researchers as consultants. Furthermore, our data pertain only to formal agreements and collaborations. We know that there are a great many informal agreements between individual researchers and departments on the one hand, and large corporations on the other, which are not included in our data. To some extent, these relationships can substitute for more formal ties with a university.

This implies that universities are more likely than the NBFs to provide to the large firms ways of augmenting its stock of knowledge and techniques. The firm gains access to top researchers who can guide its in-house research efforts. It becomes especially important to have this access when new bodies of knowledge become critical for the innovation process, when the rate of advance is rapid, and when the firm lacks sufficient familiarity with the new knowledge.[13] This line of reasoning underscores the notion that collaborative agreements with universities are different from those with other firms. Nonetheless, there is an important sense in which our specification could be justified. In biotechnology, the distance between science and commercialization is relatively short. Therefore, the research carried on by universities can and does directly lead to new products. One needs only to point to the hybridoma technology, which till recently, was the basis for the overwhelming majority of marketed biotechnology products, and which came directly out of the Medical Research Laboratory at Cambridge University. To some extent, therefore, an agreement with a universities is like an option on an asset with uncertain value: It is also a way of augmenting the scientific capabilities of the firm, and helping the firm enter a new technological area.

Estimation issues

From the model in section 2, the expected number of projects is a function of the threshold y^*, and of the scale of search, N. As is customary, we assume that the log of the expected number of agreements can be written as

$$\log(E\, AWF_{it}) = \alpha_t + \alpha_p PAT_i + \alpha_{rd} RDS_{it-1} + \alpha_{sp} SPS_t$$

$$+ \alpha_s LOGSAL_i + \alpha_{phs} PHS_i, \tag{10}$$

where E is the expectation operator, and the α's are parameters to be estimated (α_t are time dummies). A similar equation is estimated with AWU_{it} as the dependent variable.

The dependent variables, AWF and AWU, are count events, i.e. non-

---

[13]In fact our impression is that the rapid pace of scientific and technological advance caught most of the major firms unprepared and without adequate in-house expertise in areas like molecular biology and molecular genetics.

negative integers. Moreover, they have a fairly short range (see table 2). In such cases, Ordinary Least Squares produces inefficient estimators. A novel feature of our analysis is that we provide a formal model which incorporates the uncertainty in the decision making process and shows that the dependent variable will have a binomial distribution taking non-negative integer values. In contrast, previous uses of the Poisson and the negative binomial model have not offered an economic justification, appealing instead solely to the count event nature of the dependent variable [see for instance, Hausman et al. (1984), Hall et al. (1986), King (1989c)].

We first use the Poisson approximation to the Binomial. We then generalize to a Negative Binomial (NB) distribution to allow for the parameter of the Poisson distribution to be a random variable with a *gamma* distribution instead of being deterministic (cf. Hausman et al. (1984)]. For greater efficiency, we also estimate (10) and (11) jointly, using Seemingly Unrelated Poisson Regressions (SUPR), which is analogous to the well known SURE.[14]

We would like to point out one important aspect of our empirical specification. We utilize observations over time on the same set of firms. We treat all the observations on a particular firm as representing draws from the same distribution (except for the obvious time effect common to all firms). We interpret measures of technological capability as proxies for the ability to utilize, and those of scientific capability as proxies for the ability to judge. Since we are not imputing any direct structural role to our empirical measures, we are justified in not explicitly modelling unobserved firm characteristics. In our judgement, a more serious issue is the treatment of lagged dependent variables. A given agreement with an NBF, or a university would be expect to raise the scientific or the technological capabilities of the firm with some lag, and therefore, would deserve to be included as an explanatory variable in explaining collaborative alliances suitably far off in the future. We did not use lagged depend variables because all our explanatory variables (apart from the time dummies and RDS) had only firm specific dimensions and for any given firm, were constant over time. It is also the reason that we could not model the process as a dynamic search process.

4. The empirical results

This section presents the results from estimating eqs. (10) and (11). Table 2 presents the descriptive statistics of the variables used in the estimations. The positive correlation (0.6) between RDS and PHS in 2B reflects the fact that pharmaceuticals is a more research intensive business than other lines of

---

[14]Unlike SURE which pertains to continuous variables, SUPR increases the efficiency of estimation even when the set of explanatory variables is the same for both the equations provided the cross correlation across equations is non-zero. See King (1989a, 1989b).

Table 2

(a) Descriptive statistics

Variable	Mean	Std. Dev.	Minimum	Maximum
AWF	1.615	1.878	0.000	7.000
AWU	0.423	0.758	0.000	3.000
PAT	3.961	7.124	0.000	28.000
RDS	0.078	0.032	0.034	0.144
SPS	0.042	0.048	0.006	0.213
PHS	0.472	0.338	0.003	1.000
LOGSAL	8.555	8.661	4.631	10.209

(b) Correlation matrix of explanatory variables

	PAT	RDS	SPS	LOGSAL	PHS
PAT	1.00				
RDS	0.34	1.00			
SPS	0.21	0.41	1.00		
LOGSAL	0.15	−0.38	−0.61	1.00	
PHS	0.08	0.60	0.22	−0.49	1.00

Notes: For definition of variables see table 1. All dated variables (*AWF, AWU* and *RDS*) are measured at their 1986 values. Number of firms = 26.

business such as cosmetics. As anticipated earlier, our measure of scientific capability is correlated with our measures of technological capability. The correlations between PAT and RDS is somewhat lower than that between RDS and SPS, largely because both of the latter are deflated by sales. Moreover, the correlation between PAT and RDS is higher than that between PAT and SPS. Nonetheless, it does suggest that one must be circumspect in generalising our results. Table 3 reports the results for the AWF and AWU equations using the Poisson, NB, and SPUR specifications.[15]

Consider the AWF-equation first. From table 3a, the signs of the coefficients are the same in all three estimation procedures, and even the point estimates are remarkably similar across the three specifications. Table 4 shows that this pattern is displayed even under the Ordered Probit specification. From table 3a, one can see that PAT and RDS have a positive impact on AWF. Together, we expect these variables to measure the ability of the firm to successfully utilize outside knowledge for generating and commercializing innovations. Following our model in section 2, we expect

[15]Both as a methodological exercise, and to check the robustness of our specification, we also estimated both equations using the Ordered Probit specification We also estimated both equations using the absolute values of R&D and Scientific Publications (without deflating by sales). The changes in specification did not materially affect our results, and hence the results for these specifications are not reported here. They are available on request

Table 3

Variable	Poisson	Negative binomial	SUPR
(a) *AWF*			
PAT	0.016	0.016	0.016
	(0.008)	(0.007)	(0.008)
RDS	11.862	11.667	11.384
	(3.406)	(3.441)	(3.723)
SPS	−6.100	−5.401	−5.503
	(2.084)	(2.112)	(2.195)
LOGSAL	0.425	0 411	0.403
	(0.108)	(0.109)	(0.110)
PHS	0.230	0.193	0.212
	(0 318)	(0.325)	(0.320)
CONSTANT	−4.576	−4.394	−4.375
	(1.066)	(1.070)	(1.131)
Delta (δ)	–	−1.137	–
		(0.542)	
Theta (θ)	–	–	0 061
			(0 071)
Log Likelihood	−66.484	−63.462	−457.558
(b) *AWU*			
PAT	0.017	0.018	0.018
	(0 011)	(0.011)	(0.008)
RDS	−5.467	−5.086	−7.779
	(5.772)	(5.958)	(6.706)
SPS	3.850	4.808	5 125
	(3.636)	(3.558)	(3 861)
LOGSAL	0.350	0.329	0.343
	(0.181)	(0.178)	(0.178)
PHS	1.583	1.518	1.691
	(0.725)	(0.738)	(0.754)
CONSTANT	−6.086	−5.747	−5.473
	(1.884)	(1.858)	(1.947)
Delta (δ)	–	−0.749	–
		(0.453)	
Theta (θ)	–	–	0.061
			(0.071)
Log Likelihood	119.54	−115.49	−457 558

Notes: The total number of observations is 178. Delta is the over-dispersion parameter (δ) in the Negative Binomial specification. SUPR involves jointly estimating the *AWF* and *AWU* equations. See text for details. Log Likelihood for SUPR is the joint likelihood for both equations. Standard Errors (Robust), calculated from analytic first and second derivatives (Eicker-White) are given in parentheses. All equations include time dummies for three years 1983, and 1985–1989. *CONSTANT* thus refers to the year 1984. Theta (θ) is the cross equation correlation coefficient in the SUPR specification.

that these variables lower the threshold value (of the signal) for the projects that the firm is willing to undertake. This in turn increases the number of projects carried out by the firms, and therefore AWF. SPS has a negative impact on AWF. As discussed in section 3, we believe that the in-house scientific skills of the firms, which we measure by the number of scientific papers, are correlated with their ability to evaluate alternative projects. Our model in section 2 implies that a greater capability for selecting 'good' projects raise the threshold value (of the signal) for the projects that the firm plans to undertake. Ceteris paribus, this lowers the number of projects that they undertake, and therefore reduces AWF.

Consider now the AWU equation. Again, the estimates are fairly similar whether one considers Poisson, NB, or SUPR. It is obvious that the AWU results are nót as robust as the AWF-results. In addition, the signs of RDS and SPS are the opposite of those in the AWF equation. The AWU results are not completely unexpected. Our theory treats each collaborative linkage as the purchase of an option. A priori considerations suggest, and the results appear to confirm, that AWU is more likely to be a more long-term decision, and driven by considerations that are not quite captured by our model. We present our results here to emphasize that these are important topics for further research.

5. Conclusions

The knowledge base needed to generate innovations increasingly has become complex and multi-disciplinary. In many science based industries, potential innovations originate with small research intensive firms which usually lack the downstream competencies and assets necessary to commercialize those innovations. The acquisition of these downstream competencies is a slow and complex evolutionary process. As a result, established firms are more likely to possess these competencies. This is the basis of a division of labor in innovation, where innovation involves systematic exchange of technological information between firms.

Our analysis has been carried out from the perspective of the established firms in the industry. Starting from the premise that even though the scientific and technological opportunities present in the form of the university research and small research intensive firms are 'available' to all firms in the industry, the latter differ in their ability to benefit from these opportunities. Their ability to take part in the division of labor in science based industries depends upon their in-house knowledge assets. We distinguish two separate components of the latter, the ability to evaluate and the ability to utilize. Even though both allow a firm to extract greater value from participating in the division of innovative labor, they have different implications for the behavior of the firm. According to our model, the ability to utilize raises the number of innovation ventures. Firms with better ability to evaluate are more selective and focus on fewer but more valuable linkages. Although it is difficult to find suitable measures of scientific and technological capabilities, our empirical measures performed well as far as external linkages with other firms are concerned. However, linkages with universities do not appear to be adequately described by our model. It appears that universities play a somewhat different role in the innovation process. They appear to be more important as sources of scientific information and capabilities, rather than as sources of new innovations.

Our findings suggest two promising avenues for further research. First, our findings suggest that scientific capability, broadly defined to mean the application of general and abstract principles to industrial problems, is becoming a key factor determining competitiveness in a number of technologically sophisticated and important sectors. The growing proximity between science and technology, the former typically the preserve of academics, and the latter of firms, also implies that we can expect a change in the old division of labor between industry and university. The growing perception that universities can, and ought to, play a more important role in promoting national competitiveness places the linkages between private firms and universities squarely on the agenda for future research.

Second, collaborative linkages of various sorts amongst firms themselves are growing in number and importance. Our research has highlighted the importance of technological and scientific capabilities that affect the value that a firm derives from participating in such a division of innovative labor. However, the dynamic processes through which these capabilities are acquired, and in turn, how the development of such capabilities itself changes the patterns of specialization and collaborative linkages, are not yet understood and remain important challenges for future research.

Appendix A

Our sample consists of all the large U.S. pharmaceutical corporations and those large U.S. chemical corporations that are active in biotechnology. For all these firms, *Bioscan* reports information on the partner, the type of agreement (research contracts, licensing agreements, joint-ventures, etc.), the product category, and the date. The date is not available for a few agreements in *Bioscan*. These were included as 1983 agreements, as were all agreements in 1983 and earlier years. Thus, AWF and AWU in 1983 is a 'residual' category which contains all agreements before the big upsurge of external linkages of the large firms in biotechnology during the mid-1980s [see *New York Times*: also from *Bioscan*]. AWF and AWU include all types of agreements, viz. research contracts, licenses, joint-ventures, etc. We did not distinguish betweem them partly because their precise nature was not always obvious from the descriptions and because it appeared quite likely that there was some flow of scientific and technological know-how in all such linkages.

Data on scientific publications were obtained from the *Science Citation Index* [various years]. Some of these data are also reported in Halperin (1986). Nominal R&D expenditures are from the Security & Exchange Commission (SEC) 10K forms of the companies. The deflator is the industry cost of R&D index developed by the Battelle Institute [*Battelle*, (1989)]. Data on total sales were obtained from *Moody's Industrial Manual*, (various years), and pharmaceutical sales data were taken from *Medical Advertising News*, 'Top 50 pharmaceutical companies', Special report, September 1, 1987. The number of patents in biotechnology in the U.S. of each individual firm are taken from the OTAF (1987) data base which lists the institutions and the number of patents per institutions granted in the U.S. up to Dec. 1987, and applied for by 1984. Since we are using the number of patents applied for by 1984, this is likely to reflect the stock of internal knowledge of the firm up to 1984, i.e. before most agreements in biotechnology took place.

The list of companies is given below.

Abbot Laboratories	American Cyanamid
American Home Products	Baxter International
Becton Dickenson	Bristol-Myers
Corning Glass	Dow Chemicals
Du Pont	Eastman Kodak
ICN Pharmaceuticals	Johnson & Johnson
Eli Lilly	Marion Laboratories
Merck	Monsanto
Pfizer	Proctor and Gamble
Rorer	Schering Plough
Smithkline	Squibb
Sterling Drugs	Syntex
Upjohn	Warner Lambert

Appendix B

Extensions of the basic model: Increasing costs and interdependent payoffs

In the text, we had assumed that the cost of development D, the gross payoffs $F(x|y)$, and the cost of selecting a project, K, to be independent of the number of projects selected. However, it also true that the theoretical results obtained in the text are more general and carry over when one allows for increasing costs of taking on projects.

Suppose one were to write $K(m)$, $K' > 0$, where m is the number of projects that the firm selects. (In the language of the paper, m is the number of options that the firm purchases, K is the cost of the option, and D is the cost of exercising the option.) This formulation implies that the net payoffs from a project is no longer independent of the number of projects that are selected. To keep the modelling of the search process simple, we will assume that the firm receives all the signals and then chooses from amongst those. As before, let the signals be denoted by $Y_1 \ldots Y_N$. Let z_i^N represent the ith order statistics of the sample so that, for instance, z_1^N represents the maximum value of all the signals from a sample of size N, where N is understood to be the total number of potential projects from which the firm may choose. Since the projects are no longer independent in the net payoffs, there will be a threshold value y_m^* implicitly defined by the following equation

$$\int_D^B (x-D)\, dF(x \mid y_m^*; \theta, \gamma) = K(n).$$

The following properties of the threshold values are immediate.

R1. $z_m > y_m^*$ implies that the total profits are increased if the corresponding mth project is undertaken.

Proof. The proof follows directly from Assumption 1 in the text and the definition of the threshold value.

R2. $y_m^* > y_j^*$ for all $j < m$.

Proof. Since $K(m) > K(j) \,\forall\, j < m$,

$$\int_D^B (x-D)\, dF(x \mid y_m^*; \theta, \gamma) > \int_D^B (x-D)\, dF(x \mid y_j^*; \theta, \gamma).$$

By integrating by parts on both sides we get, after rearranging,

$$\int_D^B \{F(x \mid y_m^*; \theta, \gamma) - F(x \mid y_j^*; \theta, \gamma)\}\, dx > 0.$$

Hence $F_y < 0 \Rightarrow y_m^* > y_j^* \,\forall\, j < m$.

This says that the marginal project has a higher threshold value than the infra-marginal projects.

R3. $z_m > y_m^*$ implies $z_j > y_j^*$ for all $j < m$.

$z_m < y_m^*$ implies $z_j < y_j^*$ for all $j > m$.

Proof. This result follows from the definition of z_m and the previous result.

From the preceeding, it is quite obvious that the way the form could proceed would be to order all the signals, and then choose the projects it wishes to undertake. Clearly, if it chooses a project with a signal y_i, then it must also choose all projects with signals y_j, where $y_j > y_i$. In other words we can focus upon the order statistics z_i from now on with the understanding a project corresponding to a higher order statistic is undertaken logically prior to one corresponding to a lower order statistic.

One also easily show that y_m^* has the same qualitative comparative static properties as y^* in the text. Formally,

R4. y_m^* is decreasing in γ.

Proof. $(B-D) - \int_D^B F(x \mid y_m^*; \theta, \gamma) \, dx = K(m)$

$$\Rightarrow \left[\int_D^B F_y(x \mid y_m^*; \theta, \gamma) \, dx \right] \frac{\partial y_m^*}{\partial \gamma} = - \int_D^B F_y(x \mid y_m^*; \theta, \gamma) \, dx$$

$$\Rightarrow \frac{\partial y_m^*}{\partial \gamma} < 0.$$

R5. y_m^* is increasing in θ.

Proof.   By reductio ad absurdum.

Let $\bar{\theta} > \theta$. Let $\overline{y^*}$, y^* be the corresponding threshold values. Suppose $\overline{y^*} < y^*$. By definition

$$\int_D^B F(x \mid \overline{y^*}; \bar{\theta}, \gamma) \, dx = \int_D^B F(x \mid y^*; \theta, \gamma) \, dx = B - D - K(m),$$

$$\overline{y^*} < y^* \Rightarrow \int_D^B F(x \mid y^*; \bar{\theta}, \gamma) \, dx < \int_D^B F(x \mid \overline{y^*}; \bar{\theta}, \gamma) \, dx,$$

$$\Rightarrow \int_D^B F(x \mid y^*; \theta, \gamma) \, dx > \int_D^B F(x \mid y^*; \bar{\theta}, \gamma) \, dx.$$

But from the proof of Proposition 2 in the text it follows that the last inequality is false. Hence the result asserted is true.

Proposition 1'.   *The expected number of projects undertaken is increasing in γ.*

Proof. Rather than operating directly on the expected value of m, we will in fact prove the stronger result that an increase in γ implies that the distribution of m first order stochastically dominates. Let $H_m(\ )$ represent the distribution function for m, the number of projects that the firm undertakes.

Then $H(k) = \Pr\{m \leq k\} \equiv \Pr\{z_k \leq y_k^*\} \equiv G_k(y_k^*)$, where $G_k(y_k^*)$ represents the distribution function of z_k.

$$\left\{\frac{\partial G_k(y_k^*)}{\partial \gamma}\right\} = g_k(y_k^*)\frac{\partial y_k^*}{\partial \gamma} < 0. \text{ (by R4)}$$

$$\Rightarrow H_\gamma(k) < 0.$$

$$\Rightarrow \frac{\partial \mathscr{E}(m)}{\partial \gamma} > 0, \text{ where } \mathscr{E} \text{ is the expectation operator.} \qquad Q.E.D.$$

Proposition 2′. The expected number of projects undertaken is decreasing in θ.

Proof. The proof of the proposition follows very closely the line of proof of the previous proposition, using R5.

References

Arora, A. and A. Gambardella, 1990, Complementarities and external linkages: The strategies of the large firms in biotechnology, The Journal of Industrial Economics 37, no. 4, 361–379.

Cohen, W. and D. Levinthal, 1989, Innovation and learning: The two faces of R&D, Economic Journal 99, Sept., 569–596.

Cohen, W. and D. Levinthal, 1990, Absorptive capacity: A new perspective on learning and innovation, Administrative Science Quarterly 35, 153–176.

Daly, P., 1985, The biotechnology business (Francis Pinter, London).

David P. and J. Stiglitz, 1979, Analysis of factors affecting the R&D choices of firms, CEPR memorandum no. 232 (Stanford University, Stanford, California).

David P., D. Mowery and E. Steinmueller, 1988, The economic analysis of payoffs from basic research: An examination of the case of particle physics research, CEPR Working paper no. 122 (Stanford University, Stanford, California).

Granstrand, O. and S. Sjölander, 1990, The acquisition of technology and small firms by large firms, Journal of Economic Behavior and Organization 13, no. 3, 367–386.

Halperin, M., 1986, The publications of U.S. industrial scientists: A company and industry analysis, Ph.D. thesis (Drexel University, Philadelphia).

Halperin, M. and A. Chakravorti, 1987, Firm and industry characteristics influencing publications of scientists in large American companies, R&D Management 17, no. 3, 169–173.

Hausman J., B. Hall and Z. Grilliches, 1984, Econometric models for count data with an application to the patents–R&D relationships, Econometrica 52, no. 4, 909–938.

Hall, B., Z. Grilliches and J. Hausman, 1986, Patents and R&D: Is there a lag?, International Economic Review 27, no. 2, 265–283.

Imai, K., 1988, Japan's corporate networks, paper prepared for the Minit-JPERC Conference, Stanford University, Aug., 22–24 (Stanford, California).

King, G., 1989a, Unifying political methodology: The likelihood theory of statistical inference (Cambridge University Press, Cambridge, UK).

King, G., 1989b, A seemingly unrelated Poisson regression model, Sociological Methods and Research, 17, no. 3, 235–255.
King, G., 1989c, Event count models for international relations: Generalizations and applications, International Studies Quarterly 33, no. 2, 123–147.
Milgrom, P., 1981, Good news and bad news: Representation theorems and applications, The Bell Journal of Economics, Winter, 380–391.
Mowery, D., 1981, The emergence and growth of industrial research in American manufacturing 1899–1945 (Ph.D. Thesis, Stanford University).
Nairn, F. et al., 1987, Patents as indicators of corporate technological strength, Research Policy 16, 143–155.
Nelson, R., 1960, The economics of parallel R&D efforts, Review of Economics and Statistics.
Nelson, R., 1982, The role of knowledge in R&D efficiency, Quarterly Journal of Economics 97, 453–470.
Nelson, R., 1990, Capitalism as an engine of progress, Research Policy 19, 193–214.
Olson, S., 1986, Biotechnology: An industry comes of age (National Academy Press, Washington DC).
Orsenigo, L., 1989, The emergence of biotechnology: Institutions and markets in industrial innovation (Francis Pinter, London).
Pisano, G.P., W. Shan and D. Teece, 1988, Joint ventures and collaborations in the biotechnology industry, in: David Mowery, ed., International collaborative ventures in US manufacturing (Ballinger Publishing Co., Cambridge MA).
Pisano, G.P., 1990, The R&D boundaries of the firm: An empirical analysis, Administrative Science Quarterly 35, 153–176.
Rosenberg, N., 1990, Why do firms do basic research?, Research Policy 19, 165–174.
Stigler, G., 1956, Industrial organization and economic progress, in: L. White, ed., The state of social sciences (University of Chicago Press, Chicago).
Teece, D., 1986, Profiting from technological innovation, Research Policy, 15, 285–305.
Vaccà, S., 1986, L'Economia delle relazioni tra Imprese: Dall'espansione dimensionale allo sviluppo per reti esterne, Economia & Politica Industriale 51, 3–41.
Williamson, O.E., 1975, Markets and hierarchies: Analysis and anti-trust implications, A study in the economics in international organization (The Free Press, London)

Bibliography of periodicals and trade publications cited.
Arthur Young Inc., 1988, Biotech 89: Commercialization (Mary Ann Liebert Inc., New York).
Arthur Young Inc., 1989, Biotech 90: Into the new decade (Mary Ann Liebert Inc., New York).
Bioscan, 1990, The biotechnology corporate directory service (Oryx Press, Phoenix, Arizona).
New York Times, 1988, Drug makers try biotech partners, Sept., 30, pg. C1.
Office of Technology Assesment (OTA), 1984, Commercial biotechnology: An international analysis (US Congress, OTA-BA-218, Washington, DC).
Office of Technology Assessment and Forecasting (OTAF), 1987, Technology profile report: Genetic engineering, December (U.S. Patents and Trademarks Office, Washington, DC).
Science Citation Index, various years (Institute for Scientific Information (ISI), Philadelphia).

[6]

Strategic Management Journal, Vol. 15, 85–100 (1994)

KNOWLEDGE, INTEGRATION, AND THE LOCUS OF LEARNING: AN EMPIRICAL ANALYSIS OF PROCESS DEVELOPMENT

GARY P. PISANO

Graduate School of Business Administration, Harvard University, Boston, Massachusetts, U.S.A.

This paper uses data on 23 process development projects in pharmaceuticals to explore the broader issue of how organizations create, implement, and replicate new routines. A framework is presented which links approaches to experimentation and the structure of underlying knowledge. Although the concept of learning-by-doing is well accepted in the literature, the framework here suggests that where underlying scientific knowledge is sufficiently strong, effective learning may take place outside the final use environment in laboratories (i.e., 'learning-before-doing'). This proposition is tested by comparing how an emphasis on laboratory experimentation impacts process development lead times in two different technological environments: traditional chemical-based pharmaceuticals and new biotechnology-based pharmaceuticals. The data indicate that in chemical-based pharmaceuticals—an environment characterized by deep theoretical and practical knowledge of the process technology—more emphasis on laboratory experimentation (learning-before-doing) is associated with more rapid development. In contrast, in biotechnology-based pharmaceuticals—an environment in which process technology is often characterized as being more of an 'art' than a science—a greater emphasis on laboratory experimentation does not seem to shorten process development lead times. These results suggest that there is no one best way to learn, but that different approaches may be required in different knowledge environments.

INTRODUCTION

The past decade has witnessed a renewed interest by scholars in the role that organizational capabilities, resources, and other firm-specific assets play in competitive performance (e.g., Teece, 1982; Nelson and Winter, 1982; Wernerfelt, 1984; Hayes, Wheelwright, and Clark, 1988; Prahalad and Hamel, 1990; Chandler, 1990). Terms such as 'core competence' and 'organizational capabilities' have joined entry barriers, strategic groups, and kindred terms in the lexicon of strategic management. While the concept

of learning has long fascinated organizational theorists, the proposition that competitive advantage stems from firm-specific skills and capabilities has made learning a focal point of concern in fields such as competitive strategy, organizational behavior,[1] and technology and operations management (Hayes *et al.*, 1988; Hayes and Pisano, 1994).

A growing body of empirical evidence indicates that firms in the same industry often possess significantly different levels of capabilities along such performance dimensions as quality (Garvin, 1988), product development speed (Clark and Fujimoto, 1991; Iansiti, 1994), research pro-

---

Key words: Organizational learning, process R & D, pharmaceuticals, experimentation

CCC 0143–2095/94/100085–16
© 1994 by John Wiley & Sons, Ltd.

---

[1] See the February 1991 special issue of *Organization Science*.

ductivity (Henderson and Cockburn, 1994), and manufacturing productivity (Hayes and Clark, 1986; Bailey, Bartelsman, and Haltiwanger, 1994). If proficiency at a particular activity (such as manufacturing) is critical to competitive advantage, and such proficiency can be improved over time, then learning must play a central role in the competitive advantage of firms. Without learning, it is difficult to imagine from where a firm's unique skills and competencies would come.

While learning can take many forms and occur in many different organizational settings, empirical research on the topic has focused almost exclusively on the learning curve (Wright, 1936; Hirsch, 1952; Rapping, 1965; Hirschmann, 1964; Alchian, 1959; Arrow, 1962; Stobaugh and Townsend, 1975; Lieberman, 1984). This research has documented the tendency for manufacturing performance to improve with cumulative production experience and has provided an empirical foundation for the concept of 'learning-by-doing' (Arrow, 1962). The learning curve, however, reflects only a narrow slice of the broader phenomenon of organizational learning. Firms routinely create and implement new organizational and technical processes through purposeful planning and R&D prior to the start of production. Such planning and R&D are by no means limited to technological innovation. Firms can also 're-engineer' a wide range of business processes such as customer services, order fulfilment, and distribution. One of the chief challenges of innovation lies not only in designing the process, but also implementing and replicating it within the firm's operating environment. If organizational capabilities are embedded in routines, as many scholars are now suggesting (see, for example, Nelson and Winter, 1982), then how firms go about designing, implementing, and replicating such routines must be a central facet of organizational learning.

This paper attempts to shed light on organizational learning by reporting empirical evidence from a study of the development of new production processes in the pharmaceutical industry. The strategy behind this paper is to use the development of production processes as a window into the broader phenomenon of the creation of new organizational capabilities. Production processes are but one of a broader class of organizational routines that can be found through-

out an enterprise (Garvin, 1994). Like other types of routines, production processes have an organizational dimension as well as a technical dimension. However, because production processes are the output of formal and reasonably well-documented R&D projects, their development is more amenable to empirical research than other types of organizational processes.

Based on detailed observation of 23 process development projects, this paper identifies two strategies for learning: learning-by-doing and learning-*before*-doing. The basic thesis explored is that each of these approaches is appropriate in different knowledge environments. The paper is organized as follows: the following section provides a conceptual framework for analyzing development as a learning process. This framework posits that the chief challenge of process development is to learn about and predict how different technical choices will influence performance in the actual future operating environment. The state of prior knowledge about the process technology determines the appropriate strategies for acquiring the requisite feedback. In environments where prior knowledge is weak, high-fidelity feedback requires experiments in the actual production environment ('learning-by-doing'). In contrast, when reliable theoretical models and heuristics exist, laboratory experiments, simulation, and other forms of 'learning-*before*-doing' can be productively harnessed. The process development cycle is described in pharmaceuticals and highlights critical differences between biotechnology and chemical pharmaceutical process development. Data from 23 pharmaceutical process development projects are used to examine the impact of different learning strategies on development lead times. The paper concludes with a discussion of implications for further research on capability-based approaches to strategy.

DEVELOPMENT AS A LEARNING PROCESS

Although there are many ways an organization can acquire knowledge, there is broad consensus in the literature that organizational learning is a problem-solving process triggered by gaps between actual and potential performance (Von Hippel and Tyre, 1993; Dosi and Marengo,

1993; Iansiti and Clark, 1994). As a result of a stream of studies demonstrating an empirical link between cumulative production experience and manufacturing performance (Wright, 1936; Hirsch, 1952; Rapping, 1965; Hirschmann, 1964; Alchian, 1959; Arrow, 1962; Stobaugh and Townsend, 1975; Lieberman, 1984), the concept of learning-by-doing has figured prominently in discussions of organizational learning. The basic premise behind the concept of learning-by-doing is that only through actual production experience can an organization discover process problems that cause a gap between actual and potential performance. Thus, as Von Hippel and Tyre (1993: 25) argue: 'The need for learning-by-doing indicates that the innovation process will often be iterative and that developers typically *can't* "get it right the first time"'.

Indeed, anyone who has tried to learn a new skill (such as driving a car) would appreciate that practice (driving around in the high school parking lot with an instructor) is no substitute for actually performing the skill repetitively in the actual use enviornment (a real road with real Boston drivers). However, although it may be impossible to get everything right the first time, organizations also routinely attempt to anticipate and correct as many problems as possible *before* starting production. That is, not all problem-solving associated with learning is a reaction to on-line feedback. For example, when developing a new product, organizations attempt to anticipate the needs and preferences of future customers (Clark and Fujimoto, 1991). When an organization undertakes process R&D to proactively identify potential problems and to design solutions to those problems before production starts, they are engaging in what might be referred to as 'learning-*before*-doing'. The choice between learning-by-doing and learning-before-doing is clearly a matter of degree. The framework below is used to explore the conditions that might lead an organization to emphasize one of these learning strategies over the other.

Process development: Learning and integration

The fundamental challenge of process development is quite similar across industries despite differences in specific activities. The starting point for process development is a description of the product, or a product design. In chemicals,

this might be a written description of the molecule, a formula for the required set of reactions, and other data characterizing the molecule. Product designs normally also include a set of functional specifications as targets. At the time process development starts, of course, the description may be incomplete or in a state of flux. While a well-specified product design might allow a sufficiently skilled person to build a replica of the product, it does not include a specific set of instructions for economically making large quantities. This is the role of the process development. The output of process development is an organizational routine for production. In pharmaceuticals, the organizational routines for manufacturing processes include technical specifications (such as equipment designs, reaction conditions, raw materials) and a complete set of standard operating procedures and instructions used by operators and computers to monitor and control the process. Process development creates the organizational routines needed to replicate knowledge embedded in a product design.

Process developers start with a set of targets for process performance. These might be framed in terms of unit cost, capacity, yields, quality levels, critical tolerances, or other operating characteristics. To simplify the exposition, let C represent the set of performance characteristics of the process when operated under expected commercial conditions. The performance of the process (C) is determined by choices over a set of process parameters (p) which define the technical and organizational characteristics of the routine under development (e.g., raw materials, sequence of reactions, reaction temperatures and pressures, control procedures, etc.). The goal of the process developer is to find a set of process parameters, p, which either optimizes $C^*(p^*)$ or at least achieves minimum target levels of it.

There is ample evidence from research on product development that integration across functional boundaries (Clark and Fujimoto, 1991), system–component interfaces (Iansiti, 1994), different scientific knowledge bases (Henderson and Cockburn, 1994), and sequences of projects (Iansiti and Clark, 1994) plays a critical role in development performance. Integrated problem-solving is also critical for successful process development. For example, in developing a new chemical process, choices about which chemical solvents to use must be integrated with equipment design decisions since some solvents are too corrosive for certain types of vessels. Cross-functional integration between R&D and manufacturing is a particularly salient issue in process development. Since process performance is affected by interactions between technical choices (e.g., the duration of a drying cycle) and the actual operating conditions and capabilities of the future manufacturing site (e.g., how the plant's drying equipment is operated and maintained), technical choices must be tightly integrated with operating choices and conditions. Indeed, given these interaction effects, it makes little sense to describe a process technology in isolation from the actual operating environment. A process technology *is* ultimately a set of technical choices embodied in a set of operating routines.

In searching for $C^*(p^*)$ the process developer needs feedback about how the process will ultimately perform in the future factory environment. Gaining insights about critical interactions between process and operating variables is an important element of this search. Integrative capabilities rest on having mechanisms in place to generate and facilitate the requisite feedback loops. There are a variety of approaches for generating and facilitating feedback, but one of the most important is experimentation (Ulrich and Eppinger, 1995). Experiments lie at the heart of iterative search processes.

Experiments can be conducted in different ways and under different conditions. *Laboratory tests* of the process during development are one way to generate feedback. However, when a process is tested in the laboratory, the researcher does not actually get to observer $C(p)$; instead they observe laboratory performance ($L(p)$). Whether such tests are a good simulation of actual production conditions, and thus provide a good basis for predicting performance, will depend on the differences between laboratory test conditions and actual operating conditions. If tests are performed under conditions which differ significantly from actual operating conditions, then test results may not be an accurate predictor of future performance. There are many reasons for the lack of fidelity in these tests. In chemical processes, differences in scale can impact process performance; factory workers may perform certain operating tasks differently

from Ph.D. chemists in the laboratory; equipment may be different; there may be subtle differences in the raw materials available for research purposes vs. those available in commercial quantities. In many situations intervening variables are not known. A major challenge of process development is to make predictions about $C(p)$ based on observations of $L(p)$.[2]

There are two approaches to minimizing the error between test results and actual operating results. One approach is to make test conditions as close to actual operating conditions as possible. For example, one might run test batches in a pilot production facility or even a commercial manufacturing plant, rather than in the laboratory. While this provides much higher-fidelity feedback about future process performance, it can also have the added benefit of allowing developers to gain a deeper understanding of the factory environment. In-factory tests create a direct feedback loop between the developers and the production environment.[3] The idea that some things can only be learned by running the process in the factory is consistent with the idea of learning-by-doing. This suggests that production plants, rather than laboratories, should become venues for experimentation as early as possible in the development cycle.

Despite the learning value of development in the plant, one must keep in mind that factory experiments are relatively costly. They use capacity which might be deployed to make saleable products. In addition, the experiments themselves might be more costly because of minimum efficient batch sizes or require investments in specialized equipment. For example, one company in the current study calculated that test batches produced in the commercial factory cost 45 times more than batches produced in smaller-scale development facilities. Also, due to the availability of sophisticated instrumentation, laboratory experiments can generally be conducted with a much greater degree of control and precision than factory tests.

---

[2] For simplicity, the added complication that one can have observational errors even in the laboratory, due to instrument calibration, or other factors is ignored. Thus, the true $L(p)$ might differ from the observed $L(p)$.

[3] Tyre and Von Hippel (1993) find that given the 'situated nature' of learning, the physical location of where problems get solved can have an important impact on how they get solved.

An alternative to doing experimentation in the factory is to have knowledge that allows one accurately to predict performance under actual operating conditions (C(p)) from test results observed in the laboratory (L(p)). Such knowledge might be embedded in formal or informal models containing the relevant underlying variables, their interactions, and their impact on outcomes. The model might be based on theory (e.g., the laws of thermodynamics) or an accumulated body of experience. It is not uncommon for developers to use 'rules of thumb' or heuristics. In chemical processes, for example, developers will sometimes refer to a process as having a 'linear scale-up'. This means that results at small scale can be extrapolated to larger scale with a high degree of predictability. Thus, efficient search not only requires researchers to have good knowledge about L(p) but also how L(p) maps into C(p). On the role of scientific knowledge in supporting efficient R&D, see Nelson (1982).

This sample framework suggests that the appropriate learning strategy depends on the state of knowledge characterizing the technology.[4] Where underlying basic theoretical knowledge is strong, one may know enough about the critical variables and their behavior to design laboratory experiments that provide a reasonably accurate prediction of expected commercial performance. If the researcher knows and therefore can control for enough of the critical variables in the laboratory, there should be fewer surprises when the process is transferred into the commercial setting. Under these conditions, it may be efficient to carry out process development under laboratory conditions, and transfer the process to a pilot or commercial plant only after the process design is largely completed. In these cases, prior scientific and practical knowledge provides predictive models (e.g., 'if we observe L'(p') in the laboratory, we can expect C'(p') in the plant'). Such models, even if they are quite informal and even tacit, help problem-solvers to predict how different pieces of the puzzle might fit under different conditions. These models and the basic knowledge underlying them represent mechanisms of integration.

In contrast, where theoretical knowledge is weak and experience limited, too many of the critical variables may be unknown, making it virtually impossible to predict how the process tested in the laboratory will perform when run in the factory. Much of what is learned in the laboratory could be irrelevant or misleading. Unless developers have been lucky, a rework of the process may be required to get the process to operate as planned. For such situations, the feedback necessary for integration can only be generated by experiments conducted under conditions which are as close to actual operating conditions as possible. Without good predictive models, integration requires learning by doing.

PROCESS DEVELOPMENT IN PHARMACEUTICALS: BACKGROUND

This study focuses on the development of production processes for therapeutically active chemical or biochemical compounds used in drugs.[5] Since process development activities are part of larger product development projects, a brief overview of product development in pharmaceuticals would be helpful. *Product development* in pharmaceuticals begins with the discovery and synthesis of a molecule which scientists believe to have desirable therapeutic effects. The development of a compound into a drug product involves a sequence of tests to determine its safety, efficacy, and proper dosage strength and form. The compound is first tested on laboratory animals to determine if it has any toxic side effects. If it appears safe, the drug is then tested on human patients to further determine safety (Phase I clinical trials), efficacy at different dosage strengths (Phase II trials), and overall efficacy (compared with existing treatments or a placebo) in a large patient sample (Phase III trials). Data collected from these clinical trials are then submitted to regulatory

---

[4] Most process technologies lie somewhere along the continuum between having very strong theoretical or experiential knowledge bases and very weak ones. Bohn and Jaikumar (1992) have developed a useful framework known for characterizing different 'stages of knowledge'.

[5] Thus, the sample excludes the process of formulating the final drug form (e.g., capsule, tablet, cream, liquid) taken by patients. For clarification, the term 'chemical' is used to describe small molecules synthesized through traditional organic chemical methods. 'Biochemicals' is used to describe large protein molecules produced from genetically engineered cells (biotechnology).

authorities (e.g., the Food and Drug Administration—FDA—in the U.S.A.) for review. The drug can only be sold commercially after the FDA (or its equivalent outside the U.S.A.) formally approves it. The entire drug development cycle can take anywhere from 3 to 12 years from the time a compound is discovered until it is approved for sale.

Process development occurs somewhat in parallel with product development. While the specifics of process development for chemical and biotechnology-based drugs are quite different, the basic challenge is the same. Initially, when product research scientists (e.g., chemists or molecular biologists) first discover or synthesize a new molecule, they have a technique for producing it in very small quantities. The techniques used by discovery scientists are not commercially viable production processes. They are generally capable of producing extremely small quantities of the compound, at very high cost, and at very low purity levels. A commercial process must be capable of producing the compound in relatively large quantities (metric tons vs. grams), in extremely pure form, at economically feasible cost levels, and within relevant regulatory constraints.

Processes go through three (often iterative) development stages: process research, pilot development, and commercial plant scale-up. *Process research* involves defining the basic structure of the process. For chemical processes, this usually involves searching for and selecting among alternative 'synthetic routes' or the sequences of reactions used to synthesize the molecule. Process research for biotechnological processes typically involves deciding which type of cell (e.g., bacterial or mammalian) will be used to produce the protein, what genetic manipulations will be required, and what type of purification processes will be required. The goal of process research is to define the basic architecture of the process, rather than all the details. In this sense, it is akin to the 'concept development' phase in most product development projects. As an organizational process, this stage of the development cycle has some distinctive characteristics. It is generally performed by Ph.D. chemists or molecular biologists in laboratory settings using very small-scale equipment (such as shake flasks and test tubes). Although small-scale experiments generate important data and validate knowledge,

important aspects of the problem-solving process are conceptual. For example, process research chemists often start their search for potential synthetic routes by examining the literature and deriving possibilities from theory. In one project in the sample, the process research group initially identified 27 theoretically possible ways to synthesize the desired molecule. Before any physical experiments were run on these, further modeling and analysis were done to weed out the processes with serious problems (e.g., toxic by-products or excess complexity). Eventually, these 'thought experiments' narrowed the field down to four possible synthetic processes. Attempts were then made to run these processes in small-scale laboratory equipment. Two of the processes did not work at all. Of the two other viable processes, one appeared to have characteristics that would make it attractive for commercial development. This one was advanced to the pilot development phase.

Pilot development involves scaling up the process to some intermediate scale and selecting reaction parameters (such as timing, temperature, pressure) which optimize the efficiency of the process. Because scale can affect the behavior of both chemical and biochemical processes, pilot scale production serves to uncover process problems and typically triggers additional refinements of the process. In some cases, attempts to scale up run into insurmountable problems and a new round of process research is required to find a new basic process. In contrast to process research, the character of pilot development is much more empirical, as it relies heavily on data generated from actual pilot production runs (or analysis of the output of those runs), rather than on theory. In many companies, a different organization is responsible for pilot production and often the people in this organization have different backgrounds than in process research (e.g., chemical engineers instead of chemists). As a result, technical problems are often framed and solved in very different ways between process research and process development. In process research, problems and their solutions are framed in terms of the basic chemistry or biochemistry of the process. A problem with low yield, for example, might trigger a search for different synthetic routes (in chemicals) or a change in the host production cell (in biotechnology). In contrast, pilot development tends to focus on physical or mechanical solutions (e.g., changing the flow rate of the process or altering the equipment design). These differences are partly due to differences in personnel background (e.g., scientist vs. engineers), but they also reflect the impact of different physical environments. In a good research laboratory, equipment is never supposed to be a constraint on experimentation; equipment is transparent. In contrast, equipment and other aspects of the operating environment are precisely what define a pilot development facility.[6]

Finally, *commercial start-up* involves not only scaling up the process to commercial scale, but also transferring it and adapting it to the plant where the product will be produced for commercial scale. Transfer procedures typically include documenting the process in detail and transferring these documents, along with development scientists, to manufacturing sites. This is also a phase where unanticipated problems often can and do arise. The transfer process is complete once the plant can make a set number of batches of materials which meet quality specifications. The commercial start-up phase can be challenging because it is where the world of process R&D meets (and often clashes with) the realities of the plant. Whereas process research emphasizes conceptual exploration, deepens fundamental knowledge, generates plausible alternatives to technical problems, and lays foundations for further development, commercial start-up revolves around the immediate and pragmatic problem of getting a process up-and-running within rigid timelines. How smoothly this phase goes depends on how well problem-solving during research and pilot development have integrated knowledge about the factory environment.

Organizations have choices about allocating effort (e.g. resources) across these three stages of process R&D. For example, some organizations tend to emphasize process research and invest a significant share of process R&D resources before they ever test the process in the pilot plant. The philosophy behind this strategy is that identifying and solving as many technical problems as possible up front leads to smoother and more rapid development in subsequent stages. Other organizations prefer to do a larger share of their

development in the factory. These organizations generally subscribe to the philosophy that it is impossible to 'get it right the first time'. Only by going into the actual production environment can you discover critical process interactions. Each of these approaches represents different strategies for learning. As discussed earlier, each of these different strategies should be appropriate in different technological environments.

Although the development cycle can be described generically, there are critical differences in the nature of the technological environment between chemical development and biotechnology development. These differences stem from the respective maturities of the underlying scientific fields, the development of relevant scientific theory, and the availability of process engineering heuristics. Chemical process R&D utilizes chemistry and chemical engineering—disciplines which have existed in academia and industry since the eighteenth century (Haber, 1958). There is a long history of basic scientific research in both chemistry and chemical engineering conducted by universities and chemical-producing companies. Much of the relevant theoretical knowledge has been codified in scientific journals and textbooks. In searching for and selecting alternative chemical processes, the developer has at their disposal a wealth of scientific laws, principles, and models which describe the structure of relationships between different variables (e.g., pressure, volume, temperature). As noted earlier, process research chemists in pharmaceuticals often begin their work by deriving alternative feasible synthetic routes from theory. Perhaps just as importantly, there is a long history of practical experience with chemical processes. The chemical industry emerged in the eighteenth century, and chemical synthesis has been used to produce pharmaceuticals since the late 1800s. Through this experience, a large body of engineering heuristics have evolved which are widely used to guide process selection, scale-up, and plant design. In addition, computer-aided modeling is performed to simulate the impact of different process variables on yields, cost, throughput, and capacity. The knowledge base is not just technical, but also extends to organizational issues. Through cumulative experience, pharmaceutical companies have developed organizational routines and standard operating procedures. Well-established routines for quality assurance

---

[6] For a discussion of how the physical environment can impact problem-solving, see Tyre and Von Hippel (1993).

and process control, production scheduling, changeovers, maintenance, and other production activities define clear constraints about the feasibility of different process technologies within an actual production environment.

The characteristics of the knowledge base underlying biotechnology process development is quite different from those described for chemical synthesis. In comparison to chemical-based drugs, biotechnology is in its infancy. The major discovery triggering commercial R&D on therapeutic recombinant proteins was only made in 1973. The first commercial biotechnology enterprises were founded in the mid-1970s. Although there is extensive basic scientific research in molecular biology, cell biology, biochemistry, protein chemistry, and other relevant scientific disciplines, most of this work has been geared toward the problems of product discovery. Compared with the chemical world, very little basic research has been done on the problems associated with engineering larger-scale biotechnology processes. Not only do process developers in biotechnology have little theory to guide them in searching for and selecting alternatives, they also have very little practical experience. The first biotechnology-based pharmaceutical to be manufactured at commercial scale—recombinant insulin—was approved by regulatory authorities in 1982; and since that time, only a total of about 25 biotechnology-based therapeutics have been approved for marketing. Indeed, there was initially skepticism by some observers that recombinantly engineering processes could even be scaled up. Researchers interviewed during the study generally described biotechnology process development as involving 'more art than science'.

Compared with chemical synthesis, biotechnology process technology is a regime characterized by relatively immature theory and thin practical experience. Using Bohn and Jaikumar's (1992) terminology, bioprocessing technology can be considered at a lower 'stage of knowledge' than synthetic chemical process technology. The weaker knowledge base underlying biotechnology production means that the ability of laboratory research to bring together, integrate, and generate the relevant knowledge will be limited. It is difficult to characterize processes in the laboratory. Feedback from laboratory experiments is likely to be noisy. Process development

performance is likely to hinge on the experiments conducted under conditions more closely resembling the final production environment. In contrast, the chemical process environment, with its rich base of theoretical and practical knowledge, provides better opportunities to explore options, characterize the process and make predictions about process performance in laboratory settings. In this environment, high process development performance is likely to hinge on exploiting opportunities for learning during process research.

EMPIRICAL ANALYSIS

Data and sample

The data used in the analysis are drawn from a larger study on process development performance in the pharmaceutical industry. Since the type of information required for the analysis is not publicly available, it was necessary to gain the cooperation from pharmaceutical companies willing to participate in the study. Because these data are highly proprietary, the names of the participating firms and details of specific projects, other than aggregate statistics, cannot be disclosed. Data for the present analysis were collected from 23 process development projects; 13 projects involved the development of traditional chemical processes and 10 involved new biotechnology-based processes. In total, 11 organizations participated in the study (five established drug companies, five relatively young biotechnology firms, and a biotechnology division of a major pharmaceutical firm). For each project, data were collected on the history and timing of critical project events, resources expended, and the details of approaches used to identify and solve problems. These data were obtained through a combination of in-depth interviews with project participants, questionnaires, and proprietary company documents. In total, the data collection process spanned 2 years, and involved close to 200 interviews with personnel from participating R&D sites and plants in the U.S.A. and Europe.

The nature of the data collection process is one reason the sample size is relatively small. A second factor limiting the sample size was the population of potential projects. Each process development project in the sample was associated with the development of a new molecular entity—a relatively rare event in the pharmaceutical industry. The largest and most productive pharmaceutical firms rarely launch more than one new molecular entity in any given year, and many companies have gone several years without launching any. The situation for biotechnology-based drugs—an emerging area in pharmaceuticals—is even more constraining. Since 1982, only 25 biotechnology-based drugs have been developed and approved. Thus, although the sample of 23 is relatively small, it actually represents a significant share of the total number of projects by all companies completed during the time frame of the study. The small sample size obviously involves trade-offs. On one hand, it severely constrains the statistical analysis. On the other hand, it permits a very deep examination of individual projects which in turn provided insights into the development processes, the nature of the problem-solving, the types of variables to include in the statistical analyses, and the appropriate measures for such variables.

Dependent variable

In turbulent environments there is strategic value in being able to develop new capabilities rapidly. Given the strategy of using process development as a window into the broader phenomenon of organizational learning, the dependent variable is the elapsed *lead time* (in months) between the start of the process development project and its successful completion.[7] A process development project was considered to have started when the organization first began to explore ways of producing the molecule that would be feasible at a larger scale (i.e., the start of process research). A project was considered completed only after the process technology was successfully transferred to the commercial plant and could be operated consistently, within desired performance specifications. That is, the project was viewed as completed only when a fully operational production routine had been established.

---

[7] Occasionally, process development projects are temporarily halted or are idle because of exogenous factors such as a delay in clinical trials. These idle periods were subtracted out of our measure of process development lead time.

Independent variables

Learning strategy

Learning strategies in development are characterized by the allocation of effort to different phases of the project. As discussed earlier, process development projects go through three phases: (1) process research; (2) pilot development; and (3) commercial start-up. Using data on the number of person-hours invested in the project over different phases, two variables were constructed:

RESEARCH % = percentage of total project person-hours expended prior to the first pilot batch of production.

PILOT DEV % = percentage of total project person-hours expended between the first pilot batch of production and the start of technology transfer to the commercial plant.

A high percentage of project resources expended during the process research phase indicates that the organization is focusing its efforts on laboratory-based learning and small-scale experiments. Because the time between the start of the project and the first pilot batch can vary significantly across projects, and because this may have an influence on the resources expended, a third variable was created to control for the lead time before the first pilot batch:

PILOT-1 LEAD = the number of months elapsed between the beginning of the project and the first pilot batch of production.

Organizational structure

Prior research on development suggests that organizational structure will have an important influence on development performance in general, and lead times in particular. Specifically, more integrated structures have been shown to be associated with rapid development (Clark and Fujimoto, 1991; Iansiti, 1994). In the sample, two types of organizational structures for process development were observed. In one set of projects, the 'upstream' research activities (such as defining the basic chemistry of the process) were performed in a different organizational subgroup from the 'downstream' development activities (such as process optimization and scale-up in the plant). The other set of projects were characterized by an integrated structure in which a single group was responsible for all phases, from initiating process research through transfer to and scale-up in the commercial manufacturing site. Based on the previous research, the integrated structure would be expected to have shorter development lead times:

INTEGR = 1 if the project used an integrated organizational structure

= 0 otherwise.

Technical content

A dummy variable, CHEM, was coded as 1 if the project involved the development of a traditional synthetic chemical process, 0 if it was a biotechnology project. In preliminary analyses, the effects of additional content control variables, such as number of chemical steps in the process, the scale of the output, and the therapeutic class of the drug, were examined. These other variables did not improve the statistical quality or insight of the models. Since their inclusion also did not impact the other effects examined in the model, they were dropped from further analysis and are not reported here.

Descriptive statistics

Table 1 presents means and standard deviations of the continuous variables, and a frequency distribution for the dummy variable INTEGR, for both the full sample and the chemical and biotechnology subsamples. These descriptive statistics provide a picture of some of the areas where differences exist between the chemical and biotechnology projects. With respect to overall process development lead times, there is a relatively large difference between the chemical projects and the biotechnology projects (80.15 months on average for the chemical projects and 41.40 months on average for the biotechnology

projects). This difference of nearly 40 months is interesting in light of the fact that biotechnology is the newer process technology. The higher variance relative to the mean process development lead time for biotechnology projects indicates that biotechnology projects may be associated with greater uncertainty. To the extent that novelty is also associated with difficulty, biotechnology process development projects might have been expected to take longer than traditional chemical projects. The fact that this is not the case in the sample might be due to 'an entrepreneurial firm' effect. All but one of the biotechnology projects were undertaken by relatively smaller and younger entrants into the pharmaceutical business. By virtue of their smaller size and entrepreneurial structures and systems, biotechnology firms may have communication and integrated problem-solving capabilities supporting fast development. Unfortunately, the sample does not include enough variance between firm type and project type to test this hypothesis directly. Interestingly, however, in the one case in the sample where an established pharmaceutical firm undertook a biotechnology project, its lead time performance was actually superior to that of the biotechnology firms developing similar processes. The statistical analysis will provide further insights about the extent to which these differences can be attributed to particular development strategies and approaches, and whether those strategies and approaches tend to be associated with either chemical or biotechnology projects.

With respect to the percentage of resources expended during the research phase, the differences between the average chemical and biotechnology projects appears to be much smaller than the variance within each class. That is, while the biotechnology projects, on average, show a slightly more aggressive strategy for early investments in process research, this does not appear to be a biotechnology-specific attribute. There are some chemical projects that also show very aggressive 'front-loading' of resources early in the project. Likewise, there are a number of biotechnology projects where relatively few resources were invested in process research prior to the start of pilot production. In contrast, there appear to be some very significant differences during the pilot development phase of projects. In the chemical projects, a much greater share

Table 1. Descriptive statistics: Means and frequencies (standard deviations in parentheses)

	Full sample $(n = 23)$	Chemical projects $(n = 13)$	Biotech. projects $(n = 10)$
Percentage of total project hours invested in process research phase (RESEARCH %)	13.70 (13.41)	12.69 (15.60)	15.00 (10.59)
Percentage of total project hours invested in pilot development phase (PILOT DEV %)	42.26 (28.96)	55.15 (29.18)	25.50 (19.04)
No. of months between start of project and first pilot batch (PILOT-1 LEAD)	15.39 (11.28)	18.92 (12.58)	10.80 (7.64)
PROCESS DEVELOPMENT LEAD TIME	63.30 (27.07)	80.15 (20.88)	41.40 (16.34)
Frequency			
Integrated organizational structure	14	4	10

of total development resources were expended during pilot production than in the biotechnology projects (55% vs. 25%). This suggests that a much greater share of the process development in biotechnology projects is going on *after* the process is transferred into the plant. This strategy of doing development in the plant is consistent with the earlier discussion that the weaker knowledge structure characterizing biotechnology should require a greater emphasis on learning-by-doing. Finally, one of the most striking differences between the two subsamples is the distribution of organizational structures. All 10 of the biotechnology projects utilized a single integrated development group responsible from the start of process research through the final scale-up and validation in the plant. In contrast, only four out of the 13 chemical projects used such a structure.

The statistical analysis is done in two stages. Ordinary least squares (OLS) analysis is used to estimate the overall impact of the above variables on development lead times for the entire sample. These results provide a picture of the overall impact of different development strategies, but they do not indicate differential impacts across the chemical and biotechnology classes. A second set of models was therefore estimated separately for the two subsamples to test the hypothesis that process research will be more productive (in terms of reducing lead times) in chemical projects than in biotechnology projects.

OLS results for full sample

For the full sample of projects, the following model was estimated using OLS:

$$\text{LEAD TIME}_i = \beta_0 + \beta_1 \text{CHEM}_i + \beta_2 \text{RESEARCH \%}_i + \beta_3 \text{PILOT-1 LEAD}_i + \beta_4 \text{PILOT DEV \%}_i + \beta_5 \text{INTEGR}_i + e_i$$

where all variables are defined as before. Results of this analysis are shown in Table 2.

Three versions of the model were estimated. Model 1 represents the base case in which only technical class differences are controlled. The results confirm the earlier discussion about the

96   *G. P. Pisano*

Table 2.   Regression results, full sample: Process development lead time (standard deviations in parentheses)

	Model 1	Model 2	Model 3	Model 4
Constant	41.40***	42.57***	36.96***	46.72***
	(6.03)***	(6.25)	(7.39)	(11.48)
CHEM	38.75	27.60***	22.98***	15.91
	(8.02)	(7.07)	(7.71)	(9.98)
Percentage of total project hours invested in process research phase (RESEARCH %)		−0.886***	−0.776***	−0.776***
		(0.270)	(0.277)	(0.275)
No. of months between start of project and first pilot batch (PILOT-1 LEAD)		1.121***	1.064***	1.034***
		(0.344)	(0.339)	(0.338)
Percentage of total project hours invested in pilot development phase (PILOT DEV %)			0.180*	0.197*
			(0.133)	(0.133)
Integrated organizational structure (INTEGR)				−9.858
				(8.91)
Adj. R^2	0.504	0.69	0.70	0.707
F	23.34	17.35	14.04	11.62
p	< 0.01	< 0.01	< 0.01	< 0.01

*** $p < 0.01$; ** $p < 0.05$; * $p < 0.1$

lead time advantage of biotechnology-based projects. Model 2 examines the impact of concentrating resources on the research phase of the projects (e.g., the period before the process is first tested at pilot scale). After controlling for the duration of this period (PILOT-1 LEAD), the model shows that a greater concentration of resources during the research phase is associated with a shorter overall development lead time. The negative coefficient on RESEARCH % is significant at $p < 0.01$. Model 3 adds the variable on the percentage of resources expended *during* the pilot phase of development (PILOT DEV %). In this model, the estimated impact of RESEARCH % remains negative and statistically significant but the coefficient on PILOT DEV % is positive, although relatively weak.

Model 4 includes the effect of organizational structure. Consistent with the previous research on product development, the estimated sign of the coefficient on INTEGR is negative, suggesting that integrated process R&D organizations tended to be able to develop process more quickly. However, the standard error on the coefficient is relatively large and thus our confidence in this effect is limited. Part of the problem may be related to the level of detail captured in the variable. Previous studies of the

product development have probed organizational structure at a much greater level of detail and have included metrics of team structures and types of managers associated with the project (see, for example, Clark and Fujimoto, 1991; Iansiti, 1994). By contrast, the metric used here is admittedly quite crude and does not capture many of the important underlying aspects of organizations affecting integration. Perhaps an even greater problem in interpreting this result is the high correlation between organizational structure and the technology class.

Another interesting aspect of the results is the impact of different development strategy variables on the dummy variable, CHEM. Model 1, which makes no adjustments for development strategy or organizational structure, indicates that the average chemical project took 38.75 months longer than the average biotechnology project. However, as shown in Table 2, the coefficient on the technology class dummy receded with the addition of each development strategy variable. It is difficult to interpret the coefficient on CHEM in Model 4 because of the high correlation with INTEGR. However, in Model 3, which does not include INTEGR, the coefficient on CHEM is 22.98 (vs. 38.75 in the completely unadjusted model). This suggests that some share of the difference between the lead times in chemical and biotechnology projects can be explained by the specific approaches used on individual projects, rather than anything inherent in the technical requirements of the projects themselves. While the technical environment may create constraints and opportunities, what individual firms do and how individual projects are managed seems to matter a great deal.

The earlier discussion suggested that the technical environment may matter in other ways. The appropriateness of different practices and approaches may vary depending on characteristics of the knowledge environment. The analysis below examines the hypothesis that the leverage of research should be greater in the chemical segment due to the stronger knowledge base than in the biotechnology segment.

Analysis of the impact of research effort in chemical technology vs. biotechnology

From the earlier discussion, research is expected to have a greater pay-off (in terms of reduced overall lead times) in chemical projects than in biotechnology projects. To test this hypothesis, a version of the model was estimated with all variables, except RESEARCH %. The residuals from this model provide a measure of lead time adjusted for these other factors. They represent the variance in development lead times that cannot be explained by the length of the pilot period (PILOT-1 LEAD), the concentration of resources in the pilot development phase (PILOT DEV %), organizational structure (INTEGR), or technology class (CHEM). Adjusted lead time was then regressed against RESEARCH % separately for each subsample using a simple OLS model:

$$\text{Adjusted Lead Time}_i = \beta_0 + \beta_1 \text{RESEARCH \%}_i + e_i$$

Results from these regressions are shown in Table 3 and regression plots for each subsample are provided in Figures 1 and 2. The results are strongly consistent with the hypothesis. The coefficient on RESEARCH % for the chemical subsample is negative and highly significant. A greater share of resources expended during the research phase of chemical process development projects is associated with shorter lead times.

Table 3. Regression results, analysis of residual effects: Dependent variable = adjusted lead time[a] (standard errors shown in parentheses)

	Chemicals ($n = 13$)	Biotech. ($n = 10$)
Constant	10.238* (5.082)	−2.198 (7.221)
Percentage of total project hours invested in process research phase (RESEARCH %)	−0.807*** (0.259)	0.147 (0.400)
Adj. R^2	0.42	> 0.01
F	9.716	0.134
p	< 0.01	0.72

*** $p < 0.01$; ** $p < 0.05$; * $p < 0.10$
[a] Lead Time has been adjusted to control for the effects of differences in CHEM, PILOT-1 LEAD, PILOT DEV %, and INTEGR.

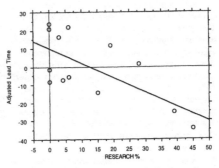

Figure 1.   Regression for chemical subsample

Organizations undertaking chemical process development projects appear to be able to learn-before-doing. For the biotechnology subsample, in contrast, the regression is insignificant and the estimated coefficient is slightly positive. In biotechnology, additional focus on research does not appear to provide leverage for shortening lead times.

Further analysis of the biotechnology plot suggested an interesting pattern. Although the overall relationship is not statistically significant, three outliers (marked on the graph with shaded points), appear to be masking a relatively strong *positive* relationship between Adjusted Lead

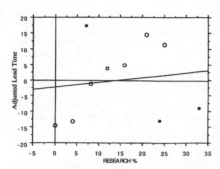

Figure 2. Regression for biotechnology subsample. Shaded points are 'outlier' points referred to in the text. Note: the estimated regression line shown in the plot is for all biotechnology cases, and does not exclude marked outliers

Time and RESEARCH % for most of the biotechnology cases. This leads to further investigation of the three outliers to identify any factors that might differentiate them from the other biotechnology projects. One factor appeared to stand out very strongly: all three outlier projects were undertaken by organizations with relatively more biotechnology process development experience than the others. Clearly, a few outliers do not constitute a trend and conclusions cannot be drawn at this time. However, this investigation suggests that the relationships between experience, firm-specific knowledge, and learning strategies may be worthy of further analysis. One plausible hypothesis is that experienced firms have accumulated deeper technical knowledge that can be tapped through research. A firm with little experience may be forced to 'learn-by-doing' until it accumulates enough understanding of the underlying technical parameters and interactions. It should be stressed that this is offered here as a plausible hypothesis for further investigation, rather than as a conclusion. Subsequent papers from this study will focus on these issues.

CONCLUSIONS AND IMPLICATIONS

The strategy in this paper has been to use process development as a vehicle to explore the broader phenomenon of organizational learning. While process development is but one of many possible activities that leads to the creation of new organizational knowledge, it has two characteristics that make it useful for this purpose. First, to the extent organizational knowledge is embodied in routines, the study of process development provides some insight into how such routines are created, implemented, and replicated. Second, going back to Schumpeter (1934), it has been well understood that organizational learning requires integration of new and existing knowledge. The integration required during development projects (Clark and Fujimoto, 1991; Iansiti, 1994; Henderson and Cockburn, 1994) is a microcosm of the learning processes within organizations (Iansiti and Clark, 1994).

The results of the analysis indicate that there is no one best approach to learning (learning-by-doing vs. learning-before-doing), but that it depends on the nature of the firm's knowledge environment. Deep knowledge of the effect of specific variables and their interactions increases the leverage of research and other forms of learning-before-doing. Learning-by-doing is required when organizations lack the underlying knowledge needed to simulate and predict effects 'off-line'.

Two caveats to the findings should be stressed. First, like most other studies of development, the small sample size has placed severe constraints on the power of the statistical analysis. The results presented here suggest some interesting patterns that can hopefully be further validated in other studies in other industry settings. Second, while process development in pharmaceuticals may be a useful window into broader issues, much more empirical analysis is required in other industry contexts and for other types of organizational activities to get a more complete picture. It might be useful to test the hypothesis as it relates to organizational innovations. Extrapolating the results of this paper, the most rapid approach to implementing an organizational innovation (such as a new product development process, a new way of handling customer complaints, a new incentive plan for the sales force, etc.) may depend on the structure of knowledge characterizing the specific organizational technology. For example, if the contemplated change lies in an area where there is well-developed and empirically validated theory and where the firm

Knowledge, Integration, and the Locus of Learning   99

has experience making similar changes in the past, then more effort in planning and organizational design might be valuable in accelerating implementation. In contrast, where organizational theory and practice are not well developed, detailed up-front planning may accomplish little. Instead, the organizations may need to experiment by implementing a specific change and observing how it works in practice.

The general conclusion that different approaches to learning may be required in different types of environments has some potentially interesting implications for strategy. Resource-based views of strategy emphasize the value of knowledge and organizational competencies as competitive assets (e.g. Winter, 1987). The framework and data presented here draw attention to the interaction between the knowledge base of the firm and its competencies. Qualitatively different types of organizational competencies are required to exploit different types of knowledge bases. The locus of strategically valuable resources and competencies may vary accordingly. For example, in environments characterized by rich scientific knowledge bases and detailed understanding of underlying causes and effects (higher stages of knowledge to use Bohn and Jaikumar's terminology), resources supporting research may be critical to competitive advantage. In contrast, in environments where technology is more art than science, resources that support learning-by-doing capabilities are likely to be very valuable. The discussion suggests that the appropriate characterization of the technical environment should go beyond the usual delineation by R&D intensity. Both of the environments included in this study—chemical process technology and bioprocessing—are highly R&D intensive. Yet they differ fundamentally in the degree to which process R&D is driven by theory and prior experience. Further research mapping characteristics of the environment (technical as well as competitive) into requirements for organizational processes is a fruitful area for further research.

Finally, some aspects of an organization's knowledge environment may be idiosyncratic. Thus, even within the same industry or same technology area, different firms may need to utilize different approaches to learning. Similarly, as firms gain experience in a technology through learning-by-doing, their knowledge base becomes

deeper and they may have opportunities to be more proactive in their learning. Since technical environments are rarely static, and knowledge bases of individual organizations evolve as a matter of course, learning processes within a firm may need to change over time. Whether and why some organizations can adapt their internal processes more successfully than others are critical issues in understanding organizational learning and the ability of firms to sustain high performance over long periods of time.

ACKNOWLEDGEMENTS

This paper is part of a larger study on process development in pharmaceuticals. I am grateful to Steven Wheelwright for his collaboration on this project and to the companies who participated in the study. I would like to express my appreciation to the editors of this special issue for helpful comments on an earlier draft. I am also thankful to Kim Clark, Robert Hayes, Marco Iansiti, Richard Nelson, Richard Rosenbloom, and Eric von Hippel for helpful comments, and to Sharon Rossi for excellent research assistance. All errors and omissions remain entirely my responsibility. The financial support of the Harvard Business School Division of Research is gratefully acknowledged.

REFERENCES

Alchian, A. (1959). 'Costs and output'. In M. Abramowitz (ed.), *The Allocation of Economic Resources: Essays in Honor of B. F. Haley*. Stanford University Press, Stanford, CA, pp. 23–40.

Arrow, K. (April 1962). 'The economic implications of learning by doing', *Review of Economic Studies*, 29, pp. 166–170.

Baily, M. N., E. Bartelsman and J. Haltiwanger (May 1994). 'Downsizing and productivity growth: Myth or reality?', Center for Economic Studies, Discussion Papers, U.S. Department of Commerce, Washington, DC.

R. Bohn and R. Jaikumar (1992). 'The structure of technological knowledge in manufacturing', Harvard Business School working paper.

Chandler, A. D., Jr. (1990). *Scale and Scope: The Dynamics of Industrial Competition*. Harvard University Press, Cambridge, MA.

Clark, K. and T. Fujimoto (1991). *Product Development Performance: Strategy, Organization, and Management in the World Auto Industry*. Harvard Business School Press, Boston, MA.

Dosi, G. and L. Marengo (1993). 'Some elements of an evolutionary theory of organizational competencies'. In R. W. England (ed.), *Evolutionary Concepts in Contemporary Economics*. University of Michigan Press, Ann Arbor, MI, pp. 234–274.

Garvin, D. (1988). *Managing Quality*. Free Press, New York.

Garvin, D. (1994). 'The processes of organization and management', Harvard Business School working paper #94–084.

Haber, L. (1958). *The Chemical Industry During the Nineteenth Century*. Oxford University Press, Oxford.

Hayes, R. H. and K. B. Clark (1986). 'Exploring the sources of productivity differences at the factory level'. In K. Clark, R. Hayes and C. Lorenz (eds.), *The Uneasy Alliance: Managing the Productivity–Technology Dilemma*. Harvard Business School Press, Boston, MA, pp. 151–188.

Hayes, R. H. and G. Pisano (January–February 1994). 'Beyond world-class: The new manufacturing strategy', *Harvard Business Review*, pp. 77–86.

Hayes, R., S. Wheelwright and K. Clark (1988). *Dynamic Manufacturing: Creating the Learning Organization*. Free Press, New York.

Henderson, R. M. and I. Cockburn (1994). 'Measuring competence? Exploring firm effects in pharmaceutical research', *Strategic Management Journal*, Winter Special Issue, **15**, pp. 63–84.

Hirsch, W. (1952). 'Manufacturing progress function', *Review of Economics and Statistics*, **34**, pp. 143–155.

Hirschmann, W. (1964). 'Profiting from the learning curve', *Harvard Business Review*, **42** (1), pp. 125–139.

Iansiti, M. (1994). 'Real-world R&D: Jumping the product generation gap', *Harvard Business Review*, **71** (3), pp. 138–149.

Iansiti, M. and K. B. Clark (1994). 'Integration and dynamic capability: Evidence from product development in automobiles and mainframe computers', *Industrial and Corporate Changes*, Special Issue (forthcoming).

Lieberman, M. (1984). 'The learning curve and pricing in the chemical processing industries', *Rand Journal of Economics*, **15**(2), pp. 213–228.

Nelson, R. (August 1982). 'The role of knowledge in R&D efficiency', *Quarterly Journal of Economics*, pp. 453–470.

Nelson, R. and S. Winter (1982). *An Evolutionary Theory of Economic Change*. Harvard University Press, Cambridge, MA.

Pralahad, C. K. and G. Hamel (May–June 1990). 'The core competence of the corporation', *Harvard Business Review*, pp. 79–91.

Rapping, L. (1965). 'Learning and the World War II production functions', *Review of Economics and Statistics*, **48**, pp. 98–112.

Schumpeter, J. (1934). *Theory of Economic Development*. Harvard University Press, Cambridge, MA.

Stobaugh, R. and P. Townsend (1975). 'Price forecasting and strategic planning: The case of petrochemicals', *Journal of Marketing Research*, **12**, pp. 19–29.

Teece, D. J. (1982). 'Towards an economic theory of the multi-product firm', *Journal of Economic Behavior and Organization*, **3**, pp. 39–63.

Tyre, M. J. and E. Von Hippel (1993). 'The situated nature of adaptive learning in organizations', Sloan School of Management Working Paper #BPS-3568-93.

Ulrich, K. T. and S. Eppinger (1995). *Product Design and Development*. McGraw-Hill, New York.

Von Hippel, E. and M. Tyre (1993). 'How the "learning by doing" is done: Problem identification in novel process equipment', Sloan School of Management Working Paper #BPS 3521-93.

Wernerfelt, B. (1984). 'A resource-based view of the firm', *Strategic Management Journal*, **5** (2), pp. 171–180.

Winter, S. G. (1987). 'Knowledge and competence as strategic assets'. In D. J. Teece (ed.), *The Competitive Challenge*. Ballinger, Cambridge, MA, pp. 159–184.

Wright, T. P. (February 1936). 'Factors affecting the cost of airplanes', *Journal of Aeronautical Science*, **3**, pp. 122–128.

[7]

Economies of Scale in Experimentation: Knowledge and Technology in Pharmaceutical R&D

PAUL NIGHTINGALE

(CoPS Innovation Centre, SPRU, University of Sussex, Brighton BN1 9RF, UK. Email: p.nightingale@susx.ac.uk)

This paper explores how changes in genetics, database, high-throughput screening and bioinformatics technologies have allowed pharmaceutical firms to exploit economies of scale in experimentation. Traditional craft-based, sequential experimentation in chemistry and biology has been complemented by firstly, the automated, mass-production analysis of populations and secondly, by 'in silico' experimentation using simulations and databases. The changes are analysed within a Chandlerian framework that highlights how increases in the 'throughput' of R&D are dependent on organizational and managerial responses to systemic uncertainty.

1. Purpose

This paper aims to show how pharmaceutical companies have attempted to exploit economies of scale in both chemical and biological experimentation in order to improve the throughput of their R&D processes. In doing so, it aims to extend the Chandlerian framework into R&D and explores the organizational and managerial implications of a shift towards the automated analysis of large populations of samples.

2. Introduction

Do economies of scale exist in pharmaceuticals experimentation? And if so, do they provide a competitive advantage for large firms and therefore favour a particular organizational form, in a way that is consistent with Chandler's framework? At first glance, one might be sceptical. Economies of scale are traditionally found in production systems where the physical nature of the

Industrial and Corporate Change   Volume 9   Number 2   2000

——————————— *Economies of Scale in Experimentation* ———————————

technology allows dramatic increases in the size of the production processes. The outputs of these processes are tangible goods rather than the intangible knowledge produced by R&D.

Similarly, Chandler's theory of the modern business enterprise focuses on how large manufacturing firms supplanted the 'invisible hand' of the market to coordinate the production of goods (Chandler, 1990). His internalist explanation stresses how substantial economies of speed, scale and scope were achieved by large firms that combined investments in high fixed-cost capital goods with the managerial hierarchies required to coordinate the throughput of materials through high volume production systems. These high cost investments in technology and organization then paved the way for lower unit-cost production.

The modern pharmaceutical industry does not fit this pattern. It is R&D intensive rather than production intensive (Pavitt, 1984) and consequently has not made major changes to its capital:labour ratios as the costs of production are dwarfed by the costs of development and marketing (OTA, 1993; Scherer, 1993; but see Pisano, 1998). Pharmaceutical firms' R&D processes tend to produce only two or three new drugs a year, which are sold to healthcare organizations under heavily regulated conditions rather than to mass markets. Moreover, an internalist perspective fails to account for the way that pharmaceutical firms are embedded in scientific, technical and regulatory networks (see e.g. Gambardella, 1992; Stankiewicz, 1993; Gabrowski and Vernon, 1994; McKelvey, 1994; Thomas, 1994; Galambos and Sewell, 1995; Zucker and Darby, 1997; Galambos and Sturchio, 1998; Henderson *et al.*, 1998; Martin, 1998). Lastly, the industry relies on patents rather than volume production to secure its profits (Pavitt, 1984; Levin *et al.*, 1987).

Having highlighted the substantial differences, it is important to note that economies of scale *do* exist within pharmaceutical experimentation. Jansen (1996), for example, reports how one pharmaceutical firm increased its screening capacity 100-fold over 4 years. This paper will argue that these changes can be explained using a Chandlerian framework that has been modified in two ways. First, since the output of R&D is knowledge and documentation for regulators, rather than tangible products, the framework needs to explore the impact of technology on the generation of intangible capital, in particular distributed, technology-specific knowledge. Secondly, since economic rents in the pharmaceutical industry come from obtaining patent protection after solving the complex technical problems involved in drug development, the framework needs to recognize the economic importance of increasing the throughput and capacity of the R&D process. This throughput is dependent on avoiding failures and experimental 'dead ends,' and so

—————————— *Economies of Scale in Experimentation* ——————————

depends on the capacity to solve problems, rather than the physical characteristics of production processes.

These two modifications have been made by exploring the interaction between the technologies and cognitive processes involved in innovation; in particular, by examining the interaction between the economic incentives to improve the throughput of R&D and the organizational, cognitive and technical constraints on technical change. Rosenberg, in a series of important articles on the technologies involved in innovation, has pointed out that, unlike production, the capital goods of R&D are not high fixed cost machinery but the instrumentation that has the potential to reduce the costs and increase the productivity of research. He suggests that advances in scientific knowledge (including methods and instrumentation) do not produce technology directly, but instead gradually reduce the cost of solving complex technical problems (Rosenberg, 1974, 1992, p. 389; cf. Mowery and Rosenberg, 1979, 1989, pp. 214–217).[1]

If Rosenberg's insights into instrumentation and improved scientific knowledge are correct, and the paper will argue that they are, then there seems no reason why pharmaceutical firms should not invest in technology if they can ensure that the costs are covered by an increased throughput of drugs *à la* Chandler. Since the tacit knowledge required to solve technical problems is embodied in people and embedded in organizations, it is, to use a Chandlerian term, 'highly inter-dependent'. As a consequence, economic advantages can be obtained by organizing the division of labour to ensure that high cost activities are exploited to the full (Chandler, 1990, Lazonick, 1991, Babbage 1836, cf. Rosenberg, 1994:24–46). This in turn requires the coordination of flows between interdependent activity cells.

Moreover, as Rosenberg (1974) notes, investments in research and instrumentation open up new markets that had previously been inaccessible because of the complexity of the technical problems involved. In the pharmaceutical industry solving technical problems first is important because it generates the patent protection required to exclude competitors from markets (Levin *et al.*, 1987; OTA, 1993). This increases the value of R&D and improves profitability.

This paper will show that large pharmaceutical firms have invested in new experimental technologies and organizational changes to reduce the cost of finding new drugs (and to find new drugs that were previously too technically complex to develop), and moreover they have done this by exploiting

---

[1] This view is increasingly received wisdom in innovation studies (Vincenti, 1990; Pavitt, 1996). A number of authors have highlighted the importance of person-embodied tacit problem solving skills coming from basic research (Rosenberg, 1990; Pavitt, 1991; Salter and Martin, 2000).

————————————— *Economies of Scale in Experimentation* —————————————

experimental economies of scale within the R&D process. The question then remains 'how is this possible?' The empirical evidence will show that both chemistry and biology have shifted from craft-based, sequential processes of experimentation on single compounds to automated mass-production processes of parallel experimentation on populations complemented by computer simulations, or what are referred to in this paper as *in silico* experiments [cf. Rothwell's (1992) fifth-generation innovation process].

This change has been highlighted in the technical literature. Gelbert and Gregg (1997, p. 669) note that

> the process of drug discovery has been rapidly evolving over the last two to three decades. Prior to that time it focused primarily on empiricism . . . this has radically changed with the rapid introduction of cutting edge scientific technologies . . . The changes in the drug discovery process, which have allowed orders of magnitude improvements in the efficiencies in some steps of the process, have been due to the complex interplay between a number of rapidly evolving new technologies . . .

which Andrade and Sander highlight as the '. . . "high-throughput", "massively parallel", robotised and miniaturised methods of biological experimentation' (1997, p. 675).

The paper will argue that, unlike scale economies in production that generally increase the scale of production processes, these economies of scale in experimentation have dramatically reduced the size of the experimental unit. As part of this change, traditional, craft-based scientific approaches have been complemented by automated processes whereby:

- The nature of scientific understanding becomes more fundamental.
- Experimentation shifts from a single unit to the population.
- Experimentation shifts from a craft process to an automated mass-production process.
- The scale of experimentation undergoes fundamental changes.
- The cycles of 'trial and error' experimentation are complemented by computer simulations.
- Complementary screening is performed *in silico* by computer simulations.

This process has happened in both chemistry and biology, and has required the introduction of information and visualization technologies that could control and collect the vast quantities of data generated by these processes and present it in such a way that scientists could tacitly understand its implications.

The paper is structured as follows. The following section positions the paper within the literature and explores how knowledge is used in pharmaceutical firms. Sections 4 and 5 illustrate the theory by showing how the small-molecule pharmaceutical innovation process has changed over time to illustrate economies of scale in both chemical and biological experimentation. The last section discusses the empirical evidence in light of the theory and draws conclusions about future directions of research.

2.1 Limitations of the Methodology

The paper uses a case study methodology to illustrate the changing nature of experimental technology. Case study research is limited in its generalizability and creates dangers of producing results that are time-, sector-, country- and technology-specific. The changes in experimentation have been abstracted from empirical work in seven large US and European firms, and a number of weaknesses are present in the data. Firstly, it overemphasizes technical change in one particular drug discovery methodology without placing it in the context of organizational, cultural and regulatory change or any alternative methodologies. Secondly, it overemphasizes discovery at the expense of marketing. Thirdly, it necessarily oversimplifies the nature of technical implementation, and care must be taken not to extrapolate from the very clean, rational stylized process to the nitty-gritty of the real world. Fourthly, it downplays the inherent uncertainty involved in technical change and many of the innovations highlighted here, in particular computer aided molecular discovery and high throughput screening (HTS), have not yet lived up to their initial expectations. Lastly, it cannot be emphasized enough that the changes described here are complements to traditional 'wet' chemistry and biology, both of which continue to play the key role in drug discovery.

The paper alleviates these weaknesses by linking the empirical evidence to an established Chandlerian framework in an attempt to produce a more holistic understanding of the complexities of large pharmaceutical firms by analysing one of their subsystems—R&D. The empirical evidence should be interpreted as supporting Rosenberg's and Chandler's insights, and demonstrating the existence of economies of scale in intangible capital. The next section will explore the nature of innovation before exploring the specifics of pharmaceutical experimentation.

3. *Knowledge and Pharmaceutical Experimentation*

The role of knowledge in technical change has been the subject of an increasing amount of attention in the 1990s. However, the exact role of knowledge in the mechanisms that produce technology is poorly understood and often the established theoretical explanations are directly contradicted by empirical evidence (Pavitt, 1987, 1996, 1997). There have been two major traditions of work in the economics literature on the nature of knowledge. The first neoclassical tradition adopts a logical structure developed from nineteenth century applications of energy models to allocation problems (Mirowski, 1989). In doing so, its primary concern is with appropriability, and following Arrow (1962), it tends to treat knowledge in terms of the allocation of information while assuming that the utilization of that information is unproblematic (Mowery and Rosenberg, 1989, pp. 4–7).

The second Schumpeterian tradition, while recognizing the importance of appropriability, has tended to stress other factors. In particular, the sector-specific variations in knowledge use (Pavitt, 1984), its tacit and inherently uncertain nature (Freeman, 1982; Nelson and Winter, 1982; Freeman and Soete, 1998), its cumulative features (Dosi, 1982) and the considerable difficulty of utilizing it (Mowery and Rosenberg, 1989; Pavitt, 1997). In this second tradition knowledge is treated as a capacity that is embedded in the organizational routines of firms (Nelson and Winter, 1982; Dosi *et al.*, 1999). Learning and experimentation within firms allow them to solve complex technical problems and comprehend external sources of information (Cohen and Levinthal, 1990; Pavitt, 1990). Once cognitive problems are solved, firms must overcome their 'core-rigidities' and reconfigure their internal and external organization to turn technologies into products (Leonard-Barton, 1995; Granstrand *et al.*, 1997; Pavitt, 1997). In pharmaceuticals this process is complicated by the intellectual property issues (not analysed in this paper) associated with different techniques (Heller and Eisenberg, 1998). Rosenberg has argued that the ability to solve technical problems is dependent on a range of tangible technologies, in particular advanced instrumentation, which can reduce the costs of problem solving (Rosenberg, 1974, 1992).

What impact new technologies have on innovation has been the subject of recent debate. Some, such as Dasgupta and David (1994), have argued that information technology will lead to a codification of tacit knowledge, while I have argued that tacit knowledge is an intrinsic feature of technical change (Nightingale, 1998), and therefore is unlikely to be codified in the way argued by Dasgupta and David. While this paper is not about the 'conversion' problems associated with moving from one category (knowledge as a capacity)

----------------- *Economies of Scale in Experimentation* -----------------

into another (information as a state), it does explore how the technologies used in innovation generate information and how this information is understood. To do this, the paper follows in the second Schumpeterian tradition and explores the relationship between cognitive issues and the changing technologies of pharmaceutical experimentation by treating knowledge as an embodied cognitive capacity, quite distinct from organizational forms, technologies and information. The impact of changes in the technologies of experimentation on tacit knowledge requirements will be discussed in the conclusion.

Previous work (Nightingale, 1998), building on the work of Turro (1986), Barrow (1988), Vincenti (1990) and Searle (1995), produced a theory of innovation based on a 'direction argument'. In the theory science is understood as a problem-centred, social practice of exploring and mapping patterns in nature, where 'patterns' refers to the constant relationships and symmetries that may or may not be written down as laws of nature. These patterns exist in the behaviour of the real world, in the scientist's tacit knowledge and as abstract mathematical structures that map out the changing relationships being investigated (Oppenheimer, 1956; Barrow, 1988).[2] Once these patterns are understood, they can be extrapolated from 'known starting conditions' to approximate 'unknown end results'. However, this extrapolation of patterns for prediction is often uncertain, even if the starting conditions are well specified, as firstly, the symmetry between the behaviour of the real world and the patterns codified by scientists is not always conserved, and secondly, because various errors can grow in nonlinear ways.[3]

Innovation processes, on the other hand, often start with 'known end results' and attempt to find the initially 'unknown starting conditions' that will produce the desired behaviour. Symmetry breaking means that these 'unknown starting conditions' cannot be found directly using scientific knowledge, which can only be used to go in the opposite direction—from known starting conditions to unknown end results. In effect, science is a one way

----

[2] Tacit knowledge is defined as a category of neurophysiological causation that provides context to actions (Nightingale, 1998).

[3] Nonlinear error growth occurs when small differences in starting conditions produce large differences in end results. Symmetry breaking occurs when the symmetry between an abstract law of nature and its real world outcome is unstable. For example, a perfect cone balanced on its tip will always fall, as any deviations, even down to quantum fluctuations, will create an increasing downwards force upon it. The direction in which the cone falls breaks the unstable symmetry between the laws of nature and its physical outcome, and creates extra information not contained in the original law of nature—creating extra complexity (Barrow, 1988). For simple systems the mathematical laws of physics are often all that is required to describe the system, while for systems with large amounts of symmetry breaking additional empirical information is required. This is why physicists can explore patterns mathematically, while chemists and biologists who deal with more complex (i.e. more symmetry breaking) phenomena must rely more on experiments.

——————————— *Economies of Scale in Experimentation* ———————————

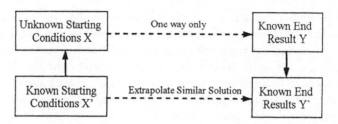

FIGURE 1. Innovation moves from known end result Y to unknown starting conditions X, via Y' and X'.

TABLE 1. Technological Traditions for the Modification Methodology

Desired end result	Q: What causes similar end result?	Extrapolate functional solution
Cure disease	→ modified natural product	→ modify natural product

TABLE 2. Technological Traditions for Structure-based Drug Discovery

Desired end result	Q: What causes similar end result?	Extrapolate functional solution
Cure disease	→ a biochemical mechanism	→ prevent the mechanism
Prevent mechanism	→ preventing protein catalysis	→ prevent protein catalysis
Incapacitate protein	→ block active site or cleft	→ block active site
Block active site	→ bind a molecule to active site	→ find molecule that binds
Molecule that binds	→ specific 3-D chemistry	→ find molecule to match it

street and technical change is going in the opposite direction. Consequently, engineers rely on the assumption that similar problems will have similar solutions. They then use their tacit knowledge to 'see' how the problem they face relates to similar problems they have faced in the past (Fergusson, 1977, 1992; Vincenti, 1990; Nightingale, 1998). If this similar problem has a known solution, the technologist can extrapolate an uncertain, similar solution, making the problem more specific (Vincenti, 1990; see Figure 1).

This knowledge about which solutions are appropriate for which problems is technology-specific and generates ways of solving problems called 'technological traditions'. Two different traditions of drug discovery are illustrated in Tables 1 and 2. Table 1 shows the technological tradition based around modifying a natural product with a known effect, e.g. the development of aspirin from willow bark. Table 2 illustrates the technological traditions associated with an alternative methodology, structure-based drug discovery, developed in the mid-1970s. This methodology is based around a 'lock and key' philosophy of finding a molecule that will bind to a protein and turn it off.[4] This prevents the protein catalysing a disease-causing biochemical mechanism and hopefully stops the disease. While there are a number of other methodologies involved in modern drug discovery, this paper will concentrate on structure-based drug discovery and the impact of various experimental technologies on it. Its technological traditions form a downwards spiralling hierarchy that turns the initial vague problem of 'curing a disease' to a series of more specific problems involving finding a molecule to block a three-dimensional (3-D) site within a disease-causing protein.

Once the technological traditions define how a problem will be solved and research has generated the problem's performance criteria, technologists use previous design experience to suggest a likely 'first cut' solution. This 'first cut' solution is the set of starting conditions that the technologists think is likely to approximate the required end result. Since innovation is inherently uncertain, tests are normally required to fine-tune these initial design choices (Petroski, 1986). The testing process serves two purposes: first, it shows if the product will perform as required, and secondly, it helps generate understanding about the relationship between changes in the parameters of the proposed solution (i.e. starting conditions) and the end result (Vincenti, 1990). This new knowledge is then used to modify the solutions for the next round of testing, as shown in Figure 2.

During this cycle of understanding, modifying and testing uncertain solutions, scientific knowledge is used to understand and predict patterns of behaviour, screen out unlikely alternatives and understand how things function. While scientific understanding may be inaccurate, it does provide a route to test assumptions about behaviour (Vincenti, 1990; Gray, 1995) and provides understanding that can be used to modify the design and reduce the number of experimental dead ends that are explored (Deutsch, 1997).

Figure 2 shows that testing can be done in three ways that correspond to the three 'levels' of pattern—the real world, the scientists' mind's eye and the abstract mathematical level. Thus testing can be done by real world experiments, by tacit pattern extrapolation and by extrapolation of mathematical patterns. The extrapolation of mathematical patterns has recently increased in importance because firstly, increased computing power has improved how well discrete computer-based calculations approximate continuous math-

---

[4] In the body some biochemical reactions are catalysed by proteins (enzymes) that stabilize a transition state during reactions. This lowers the amount of energy needed for the reaction to proceed and speeds it up.

Economies of Scale in Experimentation

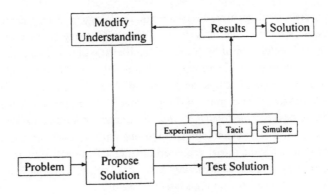

FIGURE 2. The design cycle.

ematical functions, and secondly, because innovations in visualization technologies have allowed engineers and scientists to better understand the results of experiments conducted on mathematical representations in computers. The reliability of these *in silico* experiments depends on the extent of symmetry breaking and nonlinear error growth, which is very high in many areas of medicinal chemistry and biology, making uncertain and often inaccurate *in silico* experiments complements rather than alternatives to real world 'wet' experiments (see footnote 3).

3.1 Technical Traditions to Innovation Process

Because technological traditions define how problems are resolved into more specific subproblems, the initial choices define the performance criteria of later choices. As such, they generate a hierarchy of solutions and subproblems, with initial choices constraining the possible alternatives at lower levels (Vincenti, 1990). For example, the technological traditions for structure-based drug discovery listed in Table 2 generate an ordered sequence of tasks, because some problems can only be attempted after other solutions have been found.

Following down the right hand column on Table 2 produces a sequence of tasks outlined in Figure 3: (i) perform basic biological research to find the disease causing mechanism; (ii) perform research to find an active protein in the disease causing mechanism; (iii) find its 3-D structure; (iv) characterize its active site; (v) select and test molecules to find a lead compound; and (vi) optimize the lead compound and pass onto clinical trials. This in turn forms the basis, and only the basis, for the innovation process outlined in Figure 4.

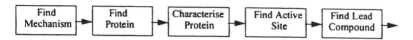

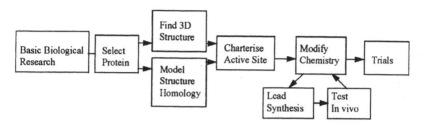

FIGURE 3. The sequence of tasks generated by the technological traditions in Table 2.

FIGURE 4. The sequence of tasks in Figure 3 forms the basis of a stylized innovation process.

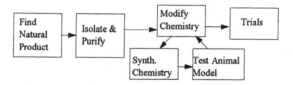

FIGURE 5. The stylized innovation process generated from Table 1.

A similar procedure using the technological traditions in Table 1 has been used to produce Figure 5.

However, the *exact* nature of the innovation process is influenced by technological, economic, institutional and organizational features, which for the pharmaceutical sector include being embedded in heavily regulated clinical trials and regulatory reviews. The next subsection explores the impact of some of the economic constraints within the firm.

3.2 Economic Constraints on Innovation Processes

Firms generally innovate to improve development and production or to produce products that command higher prices in the market. As a result, their innovation processes are constrained by economic and organizational considerations. Within the pharmaceutical industry the cost structure of the innovation process is dominated by the high fixed cost of R&D (~$500m) and the long time scales (typically 10–12 years) involved before revenue is generated (OTA, 1993; Scherer, 1993; Henderson *et al.*, 1998). These high fixed costs have to be spread over a drug development programme that is

——————————— *Economies of Scale in Experimentation* ———————————

effective and efficient, producing large numbers of highly profitable drugs (Drews, 1995). The profit distribution is also highly skewed, with only approximately one-third of drugs covering their development costs, while the throughput of drugs is heavily constrained by very high failure rates, as shown in Table 3 (DiMasi *et al*, 1991; DiMasi, 1995; Sykes, 1997).

In the late 1990s, of 100 molecules that enter development, typically only about 10 [Parket (1998) puts the figure as low as six] will achieve registration. Since the cumulative costs of each stage rises sharply this creates emphasis on reducing failures and in particular late failures. Pharmaceutical firms must therefore ensure that their R&D pipeline produces enough drugs to cover their development costs, with the actual number of drugs developed dependent on the number of development projects and their failure rates. As a consequence the capacity of the R&D pipeline increases with the number of development projects and decreases with increasing failure rates.

This paper will argue in Chandlerian terms that firms invest in R&D to increase the 'throughput' of profitable drugs in order to reduce the corresponding variable cost. If these investments produce a high level of product throughput then pharmaceutical firms can have impressive cost advantages over firms with lower throughput. As Table 3 shows, the major determinant of the capacity of the R&D process is the very high failure rates and it follows that reducing failures, especially costly late failures, is the key to improved profitability. If a company could increase its success rate in Phase II clinical trials from one-in-three to two-in-three, this would have a substantial impact on variable costs and profitability. There are therefore 'throughput' arguments for a minimum efficient size of the type highlighted by Babbage (1836).

Economies of scale are not, however, the only factors influencing minimum efficient size. The pharmaceutical R&D process is characterized by low numbers of new drug introductions—typically between one and three major new medicines a year, per firm. Since the risks of failure are so high, and the cost and time of development so large, a volatile throughput of new products can have a substantial impact on profitability. This volatility affects the cost of capital, and large pharmaceutical firms have developed 'pipelines' of new drugs across a number of therapeutic areas in order to ensure a regular stream of new products. For example, to ensure a throughput of two new drugs a year given a success rate of 20% in the development process, 10 candidates must enter the process each year. Should any one of these projects fail, the implications, while substantial, are not disastrous. The same cannot be said for a small biotechnology start-up with only two potential products. As a consequence, the cost of capital for biotechnology start-ups is substantially higher than for established firms. In Chandlerian terms, a firm specializing on

--------- *Economies of Scale in Experimentation* ---------

TABLE 3. Typical Survival Rates of Drugs Entering Trials

	Preclinical	Phase I[a]	Phase IIb	Phase III[c]	Registra-tion	Market
Survival rate per 1000	1000	480	220	71	61	60
Percentage to market	6	12.5	27	85	97.5	100
Percentage to next phase	48	46	32	87	97.5	

Adapted from Parket (1998).
[a]50–100 healthy volunteers.
[b]200–400 patients.
[c]3000+ patients.

TABLE 4

Dynamic economies of scale and scope in knowledge	Static economies of scale and scope in production
Reduce number of failures	increase capacity
Reduce volatility of throughput	increase scope of production
Reduce time of development	economies of speed
Improve quality	increase value of output
Increase number of products in pipeline	economies of scale

one product line may have the same 'scale' R&D process as other firms and yet still be below the minimum efficient scale because it cannot manage the risks inherent in the process and the resultant volatility in the profit stream in the same way as firms with 10 or so drug candidates entering development each year.[5]

There are other means of improving the economics of R&D that can be directly related to Chandlerian economies of scale and scope (see Table 4). For example, reducing the time of development produces economies of speed as it increases the intensity of (particularly financial) capacity utilization, increasing the number of drugs produced in any given time period. As Scherer has shown, reductions in the time taken for regulatory reviews and clinical trials can also have a considerable impact on cost structures (OTA, 1993; Scherer, 1993). Similarly, economies of scale exist for firms placing drugs through FDA approval porcesses.

Similarly, economies of scale can be found by increasing the number of products in development. This method is limited because many stages, such as clinical trials, do not offer opportunities for economies of scale. Improving the quality of products can be used to increase the value of the R&D

[5] I am grateful to Ed Steinmueller for highlighting the importance of the relationship between economies of scope, uncertainty of 'arrival times' and minimum efficient scale.

—————————— *Economies of Scale in Experimentation* ——————————

process. Quality is determined by *potency* (how tightly a compound binds to a substrate) and *specificity*. High potency allows the dosage to be reduced, reducing the likelihood of side effects. Similarly, a highly specific compound is less likely to interfere with other biological mechanisms. Increasing the quality of products has the potential to increase revenue.

3.3 Increasing Capacity by Reducing Failures: Cost and Quality of Experiments

The previous section argued that ensuring a high throughput of profitable drugs spreads fixed costs further and can increase profitability. Capacity is increased by reducing failures, and failure generally occurs for three reasons: the drug does not affect the target; the drug does affect the target but the target does not affect the disease; and the drug affects the disease target which affects the disease but fails in the market place (Sykes, 1997). The first two reasons are dependent on solving complex technical problems. As Rosenberg (1974, 1990, 1992) has argued, investments in research, training and instrumentation have the potential to allow firms to solve complex problems before their competitors. Pharmaceutical firms therefore invest in research and recoup their investments from an increase in the 'capacity' of their product pipeline.

The economic value of experiments involves a trade-off between their objective cost and the subjective quality of the understanding they produce, as a good understanding of the relationship between chemical structure and biological activity is more likely to produce drugs that will pass clinical trials and go on to be profitable products.[6] The model of problem solving outlined above and described in Figure 2 shows that the subjective quality of the understanding will depend on:

- The starting point of the experiment in terms of the difference between the knowledge the scientist has of the phenomenon and its reality. Thus, complex phenomena are harder to understand, and smarter scientists are better than inexperienced scientists at understanding experimental results.
- The amount of extra understanding that is generated by each cycle, which is a function of the accuracy of the experiment and the 'cognitive

---

[6] Judgements of potential value are based on a tacit understanding of an intrinsically uncertain future. The term 'subjective quality' implies that the ability to judge the 'objective' basis depends on experience. Thus a non-scientist and a scientist would have very different understandings of the potential value of a molecule. But the scientist's training would ensure that his/her perception is closer to the 'truth'.

distance' between the results and the scientists' previous beliefs. An experiment that is testing an irrelevant phenomenon will not generate understanding and a large number of experiments may have to be done to find the correct 'context of similarity'.

- The accuracy of the experiments, which depends on the quality of instrumentation, the presentation of the data and the difficulty of the problem being faced.
- The number of experiments in terms of both the amounts of data generated and the number of experimental cycles that take place.

The objective cost of the experiments on the other hand will be a function of the number of experiments, their fixed cost and their variable cost. As Chandler has shown for mass-production industries, changes in 'scale' can fundamentally affect the relationship between fixed and variable costs. If a scientist does two experiments in parallel using the same equipment, then the variable cost doubles while the fixed cost remains the same. Thus for parallel experiments the cost structure is a function of the fixed cost plus the number of experiments multiplied by the variable cost.

Since the knowledge gained from experiments of a fixed quality is a function of the number that are performed, and the cost structure changes with a shift from serial to parallel experimentation, there is a considerable economic drive towards parallel experimentation if the variable cost can be reduced and the quality maintained. This variable cost will depend on the costs of the reagents, sample and assay, the cost of disposal (potentially very high), the cost of storage and the cost of maintaining a safe environment— all of which are functions of size (i.e. the volume and amounts of reagents). So while Chandlerian production processes got bigger to take advantage of economies of scale, experiments get smaller. An example of this relationship can be seen in Table 5, which highlights the reductions in the volumes of reagent and the increases in the number of experimental samples per test plate.

3.4 Increasing the Value of Experiments

The previous section has argued that there is an economic incentive towards reducing failure in the drug development process as this increases 'capacity'— Chandler's point.[7] Reductions in failures come from improvements in

---

[7] Jansen (1996) highlights the importance of the managerial aspects of change and notes that, of the 100-fold increase in screening capacity over 4 years, the automation provided a 4-fold increase but process improvements provided a 25-fold increase.

—————————— Economies of Scale in Experimentation ——————————

TABLE 5. Summary of Available Microplate Formats Within the 96-well Footprint

Plate density	96	384	864	1536
Array format	8 × 12	16 × 36	24 × 36	32 × 48
Increase over 96-well plate	–	4-fold	9-fold	16-fold
Centre-to-centre spacing (pitch) (mm)	9	4.5	3	2.25
Well diameter (mm)	7	4	2	1.5
Assay volume (range) (μl)	50–300	20–100	5–20	0.5–10
Potential reagent saving	–	4-fold	20-fold	at least 40-fold

Reproduced from Houston and Banks (1997, p. 737).

experimentation which improve the ability of scientists to correctly select therapeutic targets and modify lead compounds—Rosenberg's point. The ability to do this in turn depends on having the tacit knowledge to recognize the 'context of similarity' between changes in starting conditions and changes in end results. In chemistry this is the ability to recognize the relationship between chemical structure and biological activity based on a form of tacit knowledge that medicinal chemists call 'chemical intuition,' and in biology it is the ability to understand biochemical mechanisms and recognize potential sites for therapeutic intervention. There are therefore a number of economic strategies for improving this tacit knowledge and the potential success of experimentation. These are as follows.

Running Experiments in Parallel and Reducing Size The quality and quantity of experiments can be increased by running them in parallel and increasing the number of experimental units so that the results of large numbers of experiments are understood together in relation to one another. This can involve running a population of samples against the same tests or running multiple tests together. For example, in the 1980s molecules were tested sequentially for toxicity, absorption, metabolism and carcinogenicity in a process that might take many years. In the late 1990s the process has moved towards a parallel approach where 'potency, selectivity, metabolism, tissue penetration and carcinogenicity [are] . . . determined simultaneously' (Sykes, 1997, p. 17). Thus, instead of six animals being dosed with a single compound and tested over a week, now a single animal is tested with up to 100 compounds and improved mass-spectroscopy is used to generate equivalent data in a single day (Sykes, 1997).

Improve Quality of Experiments Improvements in instrumentation produce a more accurate understanding of the relationship between cause and effect. Similarly, concentrating on 'more fundamental' features allows experimental scientists to improve their 'context of similarity'. For example, in chemistry this can involve exploring 3-D structure and Quantum Mechanical electrostatic potentials rather than simply 2-D molecules. In biology it can involve a shift from animal models to *in vivo* model systems in tissues, cell lines and individual enzyme substrates. This improved understanding can reduce the number of trial-and-error experiments, improving the cost and time performance.

Perform Experiments in Silico and Use Simulations to Screen Molecules
Experiments in the real world are constrained by cost, the amount of waste they produce, the amount of space that they take up, the amount of time that they take to do, the accuracy with which the results can be detected and interpreted, the physical ability to 'see' results at the microscopic scale, the ability to analyse important phenomena that are disrupted by instrumentation and the extremely fast speeds at which biochemical reactions take place. Since experiments are extrapolating patterns between starting conditions to discover unknown end results, and these patterns can be codified mathematically (in theory), it is theoretically possible to perform simulations instead of experiments. While computer simulations are not accurate enough to replace 'wet' experiments, they are useful as complements. The obvious next step from using simulations to explore biophysical behaviour is to use them to screen samples in order to remove samples that fail to match a given criterion and reduce the number of real world experiments.

The next section will show how pharmaceutical firms adopted these strategies.

4. *The Dynamics of Chemical Experimentation*

This section explores how the technology of experimentation in the pharmaceutical industry has changed over the 1990s. It argues that both biology and chemistry have moved towards the parallel analysis of populations complemented by increased use of *in silico* experimentation on databases.

Within drug discovery, medicinal chemistry experiments are performed to allow scientists to understand the relationship between chemical structure and biological activity. This knowledge is then used to bias where the chemists look in 'chemical space' for compounds to test. Initially experiments were based on sequential analysis of single compounds using animal models, but over the late 1980s and 1990s it has shifted towards the analysis of populations supported by *in silico* computer-aided molecular discovery. Towards the middle of the 1990s, innovation in high throughput screening (HTS)

—————————— *Economies of Scale in Experimentation* ——————————

technologies in chemistry and genetics technologies in biology have allowed scientists to run experiments in parallel and exploit economies of scale.

4.1. Computer-aided Molecular Discovery

One of the main bottlenecks in structure-based drug discovery is finding and modifying the lead compounds that will act as a 'key' and fit into the active site on the protein (the lock) and turn it off. Lead compounds are generally found by screening large numbers of compounds until one is found which binds to the active site on the protein.[8] Medicinal chemists then use their knowledge about the relationship between chemical structure and biological activity to suggest modifications that will improve the performance of the compound. This knowledge is generated by an experimental cycle of proposing and testing potential solutions and then using the results to modify the chemists' understanding of the relationship between chemical structure and biological activity (cf. Figure 2). As Blundell *et al* (1989, p. 448) note, experimental cycles 'can be seen not only as a series of steps leading to an improved . . . product, but also as steps designed to test or falsify the hypothesis generated earlier in the cycle'. The experiments themselves involve moving from the known starting conditions of the proposed solution to the unknown end results of the test. Since this is an extrapolation of patterns of behaviour that can be expressed in mathematical terms, in principle it can be performed by a simulation if sufficient computing power is available (cf. Figure 2). The simulations allow medicinal chemists to model chemical bonds and 3-D structures, and explore how different chemical compounds interact with the protein (Blundell *et al.*, 1988; Gambardella, 1995).

The ability of molecules to fit into the active cleft in disease-causing proteins depends on their 3-D conformation and their electron distribution. The 3-D conformation determines if the molecule will fit, while the electron distribution determines if it will stick. The energy of the molecule can be calculated using the quantum mechanical wave-function Ψ of the molecule and is used to 'quantify interactions inside and outside molecules that reflect molecular properties' (Hann, 1994, p. 132). For example, 3-D structure is determined by minimizing energy calculations, while bonding properties are based on the electron distributions. While algorithms for both types of simulations have been around since the 1960s and supercomputers since the

---

[8] A similar process works with the other main drug targets apart from enzymes: nuclear (hormone receptors), ion channel and seven-transmembrane-domain (7TM) receptors. There are many other methodologies for finding drugs and other ways in which they act. The 'lock and key' methodology is an illustrative rather than representative example.

1970s, simulations were not widely used until high resolution visualization technology was developed in the 1980s that allowed medicinal chemists to understand what the simulations 'showed'. This technology initially came from Evans and Sutherland in 1978, but was quickly followed by Silicon Graphics' faster RASTER system and a range of other products.

Visualization technology is important because it allowed medicinal chemists to explore the biophysics of chemical interactions in 3-D. Beforehand, chemists based their understanding on 2-D pictures of complex molecules. However, because chemistry takes place in a 3-D world, the translation from 3-D chemistry to a 2-D picture and then back to 3-D in the chemist's mind introduces a series of distortions. By having the representation in 3-D, these distortions are removed.[9]

While simulations are very useful, the quantum mechanical nature of the calculations requires large amounts of computing power to produce accurate approximations and simulations are therefore used as complements rather than alternatives to 'wet' chemistry. They do, however, have two very important uses. Firstly, they allow chemists to explore their implicit theories of why certain behaviours take place, and in doing so act as a 'virtual microscope' (Gray, 1995), exploring behaviour that would be impossible to see with instrumentation. Secondly, if the simulations are accurate enough they can be used to screen databases of compounds to select compounds that are likely to be drugs for further empirical testing. While nowhere near as accurate as experiments, this process can be significantly cheaper and faster since 'Ligands can be examined at a rate of 10–100 per minute enabling the examination of a database of 100,000 compounds in less than a week' (Whittle and Blundell, 1994, p. 356). Moreover, the data generated can be compared with experimental data to highlight anomalies. Visualization technology allows this analysis to proceed at a more fundamental level, as it allows medicinal chemists to literally 'see' theoretical features that cannot be picked up by instrumentation—such as different electron densities—in terms of different colours. As such, it allows chemistry to be understood in terms of more abstract and mathematical patterns. For example, molecules can be understood as electron distributions on a 3-D 'backbone' structure or 'pharmacaphor' rather than as atoms and bonds (P. Murray Rust, personal communication). This in turn has the potential to improve the selection of compounds.

While computer simulations were originally introduced to replace experiments, their lack of accuracy and higher than expected costs has limited their

---

[9] As Turro (1986, p. 888) notes 'to many organic chemists, models are the 'pictures' on which the imagination may be exercised and which enhance an intuitive understanding of the object or concept under consideration' (cf. Ugi *et al.*, 1970; Hann, 1994, p. 198).

———————————— *Economies of Scale in Experimentation* ————————————

use and, compared with early expectations, their performance has been disappointing. Moreover, in the early 1990s it became clear that companies following 'rational drug discovery' strategies were in danger of developing very similar drugs with similar performance which automatically restricted their potential market. As a response many pharmaceutical firms have moved away from 'overly rational' approaches and have tried to introduce some 'randomness' into development to distinguish their products from competitors. One of the key technologies in this shift away from rational drug discovery was the development of HTS technologies.

4.2   High-throughput Screening and Population Chemistry

> These new plate formats have arisen as a potential answer to the problematic question being asked at most major pharmaceutical companies: 'How can we screen more targets and more samples cheaply?' (Houston and Banks, 1997, p. 737)

HTS represents a shift from a craft-based experimental cycle whereby scientists test single compounds in series at a human scale to a mass-production process whereby very large numbers of compounds are tested in parallel at the microscale by automated robots. As a result, the yearly throughput of a typical lead discovery group increased from about 75 000 samples tested on about 20 targets to over a million samples tested on over 100 targets (Houston and Banks, 1997, pp. 734–735).[10]

The shift from serial testing to the automated testing of large populations required coordinated changes in all three aspects of the 'chemical' design cycle on the right hand side of Figure 4: i.e. changes in testing, changes in the production of chemicals for testing (proposing and producing uncertain solutions) and changes in how the results of experiments are understood (modifying understanding). These changes are highlighted below.

4.3 High-throughput Screening

HTS is the biological technology that allows large numbers of chemicals to be automatically tested for biological activity. In the 1990s it has 'become the major tool for lead identification, where novel assay formats, assay

---

[10] HTS requires firms to trade off quality, speed and quantity. The automated systems can in principle screen millions of samples a year, but maintaining experimental quality in large mixtures becomes increasingly problematic. However, lead compounds can sometimes be found in months rather than the previous 2 years, improving throughput.

miniaturisation and automation have been utilised to enable cost effective, rapid and successful drug discovery' (Houston and Banks, 1997, p. 734). It comprises a system for data handling, an array of compounds to be tested, a robot to perform the testing and a biological test configured for automation. The test itself is a biological system that has been engineered so that it will produce a detectable signal when it is activated by a compound. This can involve linking it to a fluorescent or radioactive source or the use of bio-sensors.[11] Originally developed as an instrumentation technology, biological assays have gradually decreased in size and improved in sensitivity. As they have become more robust they have been increasingly used for parallel screening.

4.4 Combinatorial Chemistry

The introduction of new high-throughput testing created a bottleneck in the discovery process as the production of compounds did not expand at the same rate. With a shift from being able to test hundreds of compounds to being able to test tens-of-thousands of compounds in the early 1990s, it quickly became obvious that pharmaceutical firms could test all their compounds very quickly. From the perspective of Hughes (1983), the interconnected nature of the innovation process meant that improvements in screening technologies increased demand for compounds and created a 'reverse salient' in synthetic chemistry. This acted as a Rosenbergian 'focusing device,' or Hughesian 'critical problem,' and concentrated innovative activity. Combinatorial chemistry was developed to fill the vacuum—a technology whereby large numbers of compounds are made by the 'systemic and repetitive covalent connection of a set of different "building blocks" of varying structures to each other to yield a large array of diverse molecular entities' (Gordon *et al.*, 1994, p. 1733). Instead of the old 'Woodwardian' serial method of hand-crafting specific compounds, combinatorial chemistry is a mass-production technology that synthesizes large numbers of compounds in parallel, as mixtures using computer-controlled robots [for a review of other methods see Gallop *et al.* (1994), Gordon *et al.* (1994), and and references therein].

---

[11] *In vivo* testing involves trade-offs between the depth and breath of experimental knowledge. As testing has shifted from animal models to the tissue, cell and then biomolecular substrate, the scientist has been able to understand biological phenomena in great depth. But, while *in vitro* tests provide very detailed information about specific substrates, unlike animal tests, they are silent on other biological effects. HTS does not evaluate the drug's absorption, breakdown, toxicity or selectivity. Although some drugs have come directly from HTS, the narrow nature of biological information they provide had generally seen them used in conjunction with traditional medicinal chemistry.

TABLE 6. The Different Domains of Chemical Experimentation

	Real world	*In silico*
Single molecule	serial synthesis and testing	computer-aided molecular discovery—simulations used as 'virtual microscopes'
Populations of molecules	combinatorial chemistry and HTS	QSARs—simulated screens

4.5 Database Innovation

The shift from serial to population experimentation has produced huge amounts of data, requiring changes in how it is understood. Improvements have been seen in data management software, statistical analysis software and the visualization technologies required to represent the large amounts of complex data from experiments on large populations. When experiments were done in a serial fashion experimentation followed the pattern described in Figure 2. With populations, however, it is possible to compare and explore patterns within the population. Thus, a population of compounds that are active against a specific screen can be analysed to see if they contain common features. While this has been a traditional part of medicinal chemistry, in the form of quantitative structure–activity relationships (QSARs), large populations and modern computerized statistical analysis have increased its prominence.

Populations of compounds can also be tested against counterscreens to use statistical techniques to understand why a molecule is selective for a particular substrate. This is part of a shift towards *in silico* analysis of populations whereby data is stored and used to help reanalyse new data. In doing so, anomalies in the patterns between structure and activity in one context can be used to explore old data. This allows a change in the 'sense of similarity,' producing a more detailed understanding of the structure–activity relationship. Since false positives have costly effects, it is important to complement this *in silico* analysis with experimental results (see Table 6).

4.6 Designing Populations *in Silico*

If medicinal chemistry is like fishing for a compound, structural-based drug discovery is 'fishing with a hook' and involves using tacit knowledge to narrow down the volume of the pond where the catch is likely to be found. HTS, however, allows the pond to be divided into small areas that are

—————————— *Economies of Scale in Experimentation* ——————————

searched in parallel and is like 'fishing with a net,' where as much of the pond as possible is trawled (M. Hann, personal communication). Once a trawl has found a 'hit,' established medicinal chemistry techniques based on statistics, intuition and single compound testing are used.

However, the problem of where to trawl is still present. Simply increasing the number of experiments may produce no benefits if they are irrelevant to the problem being solved. The general methodology is to 'trawl' through as diverse a volume of chemical space as possible. However, this raises the question of how to define 'chemical space' and 'chemical diversity' as the structure–activity relationship is context dependent, i.e. a group of compounds that are similar in one context may be very different in another (Gordon, 1995).

Choosing the correct way to describe the compounds before the experiment is difficult because there are a large number of possible 'contexts of similarity,' and the 'correct' one is only known after the experimental results are collected. This is Bradshaw's (1995) paradox: 'We need to know the biological results before we can decide on the appropriate space to represent our compounds'. If it were possible to test all the possible drugs this problem would not exist, but as there are $\sim10^{180}$ possible drugs and $\sim10^{18}$ potential drugs, testing all of them is neither logistically or economically practical. Instead, intelligence and tacit knowledge must be used to bias the test sample towards likely successes and away from irrelevant or repeated compounds.

This rational biasing of the samples proceeds by defining the 'chemical space' where the drug is likely to be found. Various features, like size and toxicology, can be used to remove volumes of the chemical space that are unlikely to produce drugs. The chemical space is then defined by 'descriptors' that correspond to various physical properties that are known to make compounds 'drug like'. The compounds are then 'clustered' together and a representative sample is selected and used to search as much of the chemical space as possible within realistic economic boundaries.[12]

Different firms use different descriptors to analyse chemical space, reflecting an economic trade-off between the quality of the descriptors and the computational cost. Descriptors can also be tested and compared to see how well they pick out known active compounds from previous experimental data. Unsurprisingly, the effectiveness of different descriptors varies in different biological contexts and this knowledge can bias the selection of appropriate descriptors.

For example, 3-D shape is an obvious criterion for 'fitting' into a protein,

---

[12] Clustering several million compounds is often computationally impractical. Compounds are therefore excluded from the statistical analysis based on 'intuition' and knowledge of chemical properties.

──────────────── *Economies of Scale in Experimentation* ────────────────

but the computational cost of measuring, calculating and clustering around the 3-D shape makes it inappropriate for large databases. Work at Chiron (California, USA) in the mid-1990s on 3-D descriptors found very little difference from clusters done on far simpler 2-D 'fingerprints' (Spellmeyer, 1995). In general, diminishing returns to the number and quality of descriptors very quickly set in.[13]

There is also a historical path dependence implicit in the biasing of populations as '[a]ll drug company compound files are biased by the historical programs of that institution, since a disproportionate share of compounds of particular types will have been deposited' (Gordon *et al.*, 1994, 1399). For example, a company may have historical competencies in antiviral drugs and therefore has compounds and experience of making molecules that are biased towards antiviral-'like' drugs.

This bias can be recognized and used to intentionally create libraries. So a medicinal chemist searching for a cure for AIDS could read a technical paper describing how a certain class of molecules bind to an HIV protein (e.g. Rich *et al.*, 1990) and then use this knowledge to bias the library with similar compounds. Libraries are therefore constructed based on knowledge of the relationship between chemical structure and biological activity. As Gordon *et al.* (1994, p. 1399) put it: '*The notion of intentionally biasing a chemical library is a form of drug design*, but not applied to individuals but rather to groups or populations of molecules. [Thus] . . . all libraries are biased in some ways.'

Once the first round of testing has been done, the statistical results can be used to bias a second 'trawl,' as data are now available on the appropriate context for similarity. This rearticulation of understanding can then bias the next round of empirical testing towards more appropriate populations. This involves reusing *in silico* data to design *in vivo* experiments. At Pfizer this computational approach is used to pick up some of the compounds that HTS misses. When applied to a database of compounds, computational searches based on 3-D flexibility found compounds known to be active against HIV-protease that were missed by HTS. The screening of 500 000 precalculated compounds took 1–3 h (Finn, 1995). Thus the simulated *in silico* testing of populations of compounds can be used to complement the real testing of populations of compounds.

───

[13] Interestingly, the chemical diversity of the top 50 and 100 drugs on the market is greater (by various measures) than the chemical diversity of almost all the compound libraries that can be made or bought (in 1995), which has pushed synthesis towards non-standard chemistry (Spellmeyer, 1995). An explanation advanced in the industry is that, because traditional organic chemistry has historically been concerned with copying natural compounds and reactions, it is easier to synthesize compounds that are 'similar' to naturally occurring ones. Pathogens have evolved to be immune to these types of compounds but have no defences against the new non-natural compounds. As K. Pavitt (personal communication) points out, this is an example of a localized search in a complex environment leading to path dependency.

—————————— Economies of Scale in Experimentation ——————————

TABLE 7. The Changing Features of Chemical Experimentation

Changes	Feature		
	Propose solution—(synthesize compounds)	Test solution	Modify understanding
From craft to mass production technologies	From craft-based serial synthesis to combinatorial chemistry.	From testing single compounds in animal models to *in vitro* analysis at biomolecular level using automated, parallel HTS.	From a tacit sense of similarity to using 3-D visualization technologies and statistical analysis to understand populations.
Scale—size	From beaker to micro- and submicro-level; from single to mixtures of compounds.	From 96 to 384, 864, 1536 and 3456 compounds per shelf; increases in speed so that 190 000 compounds can be tested in a few weeks rather than 2 years.	Shift from understanding experiments on single molecule to statistically analysing data on millions of experiments.
Automation	Introduction of robotics to automate and standardize synthesis; automated quality control.	Robots used 24 h a day to produce an approximately 30-fold increase in throughput.	Increased use of visualization and simulations to understand biophysics, and increasing statistical analysis of populations and simulated populations.

As this section has shown through the 1990s, chemical experimentation in the pharmaceutical industry has shifted from single compounds to populations. In doing so, experiments have reduced in size and have become increasingly automated, with computer simulations being used to complement physical experiments. The Chandlerian point about control being the handmaiden of economies of scale is supported by the increased use of computerized databases in the design and interpretation of experiments. Table 7 summaries the evidence and shows how the three stages of the experimental design cycle—(i) proposing solutions, (ii) testing and (iii) modifying understanding in the light of new evidence—have changed. The first row shows how each stage has moved from being craft-based to being part of a mass-production process. The second row shows how increases in the number of tests and decreases in their scale have affected each stage. The last row shows the increased used of technology to perform tasks that were once done by hand.

5. *The Rise of* in Silico *Population Biology*

The previous section outlined changes in the chemistry of drug discovery; this section outlines similar changes in biology. In doing so it necessarily simplifies an enormously complex and research-intensive process as biological experiments shifted from analysing individual proteins to analysing populations of genes.[14] The shift from the protein level to the gene level creates added complications, but the basic process that was seen in chemistry is repeated: experiments that were originally hand-crafted have increasingly become automated processes performed in parallel on populations with complementary analysis of stored and simulated data.[15] The huge amounts of data involved in genetics (a single laboratory can produce more than 100 gigabytes of data per day) has required complementary innovations in information technologies and changes in the way experiments are understood and designed.

These changes have been part of an attempt to increase the throughput of R&D which is limited by the ability to find effective targets for therapeutic intervention. With the 'lock and key' methodology of structure-based drug discovery, starting out with a protein target that plays no role in the disease will doom the project from the start. As a consequence improving understanding of what proteins are involved in which biological mechanisms can reduce failures and improve R&D capacity. However, the biology of disease is hugely complex and a lot of costly, slow and laborious research is needed to unravel it.

Fortunately the possible biological processes involved in diseases are limited by chemical possibilities and evolutionary redundancy (whereby evolution adapts pre-existing features to new functions). Evolutionary redundancy constrains biology because of strict symmetry conditions that govern the relationship between genes and the peptides that make up proteins (i.e. the genetic code). These strict symmetry conditions create strongly conserved patterns within biology that allow knowledge from one biological domain to be extrapolated to another by evolutionary analogy. For example, humans

---

[14] Biological research involves more sophisticated scientific research and experimental cycles that gather data than chemistry, as the mechanisms involved in disease are often unknown.

[15] While genetics has always analysed populations, the 1990s have seen a shift towards the analysis of populations of gene populations, e.g. analysing the differences between the populations of genes of diseased and non-diseased groups, or the analysis of the expression of a given population of genes over time.

——————— *Economies of Scale in Experimentation* ———————

have many metabolic pathways that go back to our early biological ancestors. Since these are shared by creatures with the same ancestors, knowing how they work in a mouse, for example, can be used to understand how they work in humans. Similarly, knowing how one mechanism works in one area of the body can provide understanding about how another similar mechanism works elsewhere. As a consequence, there are relationships between the sequences of similar genes and the function of the corresponding proteins that can be learnt and used to understand the biology of disease and produce targets for therapeutic intervention.

The ability to use this genetic knowledge has developed significantly over the last half century and dramatically through the 1990s. The next subsections will show how initial research on single gene mutation diseases helped develop reusable genetic data and a more fundamental understanding of biology. These data in turn, and the technologies that were developed to use them, allowed biologists to explore the genetics of more complex diseases. Over time, the amount of data has increased, and more and more experiments can be conducted by searching databases and analysing populations of genes *in silico*. These searches have complemented wet biological experimentation, improved the ability of biologists to understand the mechanisms of diseases and hopefully improved the capacity of R&D. As the size and number of databases have increased, more and more research is conducted *in silico* and medicinal biology has become a more theoretical science (Gilbert, 1991; Murray Rust, 1994; Lander, 1996; Gelbert and Gregg, 1997; Weinstein *et al.*, 1997).

This shift has gone through a number of overlapping stages. Firstly, biology was a craft-based 'wet' science, and pharmaceutical research relied on brute empiricism with limited biological understanding. In the 1970s and 1980s, research on genetic mutations generated new technologies and maps of the genome. As sequencing technologies and genetic databases improved, genes could be found within already sequenced data. However, using this information to produce drugs requires understanding the gene's functions. HTS technologies were developed to analyse genetic functions in a number of model organisms and databases were developed that link functional information on genes. The next stage occurred when improved databases and sequencing technologies allowed biologists to analyse populations of genes and build up understanding of mechanisms and models of disease action in order to suggest sites of therapeutic intervention. Diseases are now understood in terms of the differential expression of genes, which in turn cause the production of proteins that catalyse disease-causing biochemical pathways. Since these pathways involve multiple proteins and genes, diseases are understood as polygenetic and multifactorial.

5.1 Science Becomes More Fundamental: the Shift from Classical to Molecular Genetics

Modern genetics has its roots in Gregor Mendel and Weismann's work on inheritance. In the 1940s understanding of the role of genes in biology started to improve after experiments showed that moulds and bacteria with mutated alleles were unable to synthesize essential proteins (enzymes). By the 1950s the connection between genes and proteins was recognized, but it was Watson and Crick's discovery of the structure of DNA that suggested a mechanism to relate genetic material to the rest of biology by explaining how nucleotides relate to peptide sequences. Within a decade the genetic code was broken.[16]

Medical research in the 1970s and 1980s concentrated on finding the single gene mutations that caused inherited genetic disorders such as cystic fibrosis and Huntington's disease, where analysis of the family tree should reveal if the disease is dominant or recessive. Early work on these genetic diseases in the 1950s relied on elucidation of biochemical mechanisms to generate a protein involved in the disease, and from the protein the researchers would work back to find the gene using very slow gel separation of protein fragments. This method involved laborious analysis, whereby biochemical knowledge was needed to suggest a gene 'candidate,' where it us likely to be found and what it is likely to do.

5.2 *In Silico* Experimentation and the Shift Towards Reverse Genetics

The shift towards *in silico* experimentation, where genes are found using computers, happened when research moved from the 'candidate method' towards 'reverse genetics' in the mid-1980s. With 'reverse genetics' the researcher relies on a dense marker map of the genome and good clinical characterization (which is typical of single gene defect diseases) to statistically analyse the correlation between genetic markers and disease states. The

---

[16] Each triplet (codon) of bases on the gene codes for a particular amino acid and additional stop codons code for termination of transcription. There are four possibilities (A, C, G, T) at each base in the triplet, creating 64 codons that code for the 20 amino acids that make up proteins. Within the cell, the DNA in the genome is activated by special enzymes that switch on (express) specific genes and then complementary proteins within the cell use the unzipped gene as a template to produce mRNA, which is then passed to a different part of the cell where it is translated into the amino acid chains that make up proteins. These amino acid chains then fold up into very complex 3-D patterns and the resulting proteins typically have active sites that will catalyse various biochemical reactions.

────────────── *Economies of Scale in Experimentation* ──────────────

higher the correlation between having the disease and having the genome marker, the closer together the marker and the mutated gene are on the genome. Once the gene location is approximated, the next stage is to focus down on that region and pepper it with markers in an attempt to find the exact location of the gene, then find its sequence and produce diagnostics.

These techniques became increasingly possible following the development of automated ways of amplifying and sequencing DNA.[17] Of particular importance were the development of the 'Southern blot,' the polymerase chain reaction (PCR) that allowed DNA to be copied, Saiki's development of thermostable enzymes that allowed the PCR reaction to be automated, automated methods of creating 'primers' that cut the genome at specific points and the discovery of various markers which allowed more detailed genome maps to be developed. Since the data on the location and sequence of genes and markers generated by these technologies could be reused, the scientific community developed a number of publicly available databases.

The development of genome databases followed the sequencing of small (~5000 base pairs) DNA viruses, such as 'Simian virus 40' and 'phage ϕX174' in 1977. As the technology of sequencing improved, larger viruses with important healthcare effects began to be sequenced. Potentially useful biological information coming from the early research groups focused attention on mapping the entire human genome, and in the late 1980s federal funding and money from the Wellcome Foundation was used to set up an international genome sequencing operation. As these databases increased in quantity and quality it became increasingly possible to search for genes within the sequenced data and biology slowly shifted from a predominantly 'wet' experimental science to an increasingly theoretical *in silico* science (Gilbert, 1991).

5.3 Functional Genetics and *in Silico* Screening

While improved sequencing methods and databases have made finding genes and producing diagnostics easier, they have not produced many cures. Finding the gene does not tell you its function, nor does it suggest how that function relates to the mechanism of disease causation, or how that mechanism can be

---

[17] Restriction enzymes were developed in 1966 that allowed DNA to be cut and pasted into bacteria, which could then be used to clone DNA. The development of improved sequencing methods by Sanger in 1975 and Gilber and Maxim in 1977, and the development of automated DNA sequencing by Carruthers and Hood in 1983 made finding the molecular structure of genes easier. These involved making the DNA reproduce various copies of itself of varying lengths using dideoxy bases. If the experiment is run using different radioactively tagged As, Cs, Gs and Ts, and these are then run through a gel, each gel line will separate the DNA out into different lengths. If all four are compared against each other, the resulting separation should show the 'letter' associated with each base in turn.

─────────────── *Economies of Scale in Experimentation* ───────────────

modified. Since 'genetic' diseases account for less than 2% of the disease load, genetic research has not yet had a revolutionary effect on medicine. It has, however, had a profound effect on two areas of medicinal research: firstly, on the development of new technologies associated with rapid gene sequencing, gene engineering and database management, and secondly, it has allowed diseases to be understood at a more fundamental level. Previously, as Table 2 showed, disease biochemistry was typically understood in terms of proteins catalysing biochemical mechanisms. The development of modern medicinal genetics technologies has shifted this thinking, so that rather than starting with the protein, the problem solving process goes back one stage further and asks what causes the production of the proteins that catalyse the biochemical mechanism. The answer provided by the new genetics research is that proteins are caused by the differential expression of genes. Since the linkages between proteins and genes are strongly constrained by the genetic code, this allows medicinal biologists to explore proteins by exploring populations of genes.

With this new methodology, the shift from knowing what genes are involved in disease to finding a cure requires medical biologists to understand how the genes function. Early functional genomics research focused on fruitflies, yeast and nematode worms that had long histories as model organisms as their short life cycles enabled the role of specific genes to be quickly analysed. This information on gene function could be extrapolated to 'similar' genes using evolutionary redundancy (whereby evolution builds on what has gone before and reuses the same materials) within and between species to extend knowledge from one domain to another. Thus genes of medicinal interest in humans can be investigated by analysing 'similar' genes in organisms like yeast, mice, zebra fish and nematode worms. This allows genetic information to link:

- biology to chemistry
- evolution to developmental biology, structural biology, ecology, and medicine
- biological mechanisms between species
- biological mechanisms within the same species
- patterns that contrast diseased and non-diseased states (thus, for genetic diseases differences in the phenotype can be traced to differences in the genome)

As a consequence, the very complex biological investigation of genes' functions has been increasingly complemented by database searches for 'similar' genes with known functions. As with the generation of chemical descriptors,

——————— *Economies of Scale in Experimentation* ———————

the correct 'sense of similarity' is only known after the answer has been found, and considerable tacit knowledge and research experience is needed to find the most useful context (Bork and Bairoch, 1996).

Information from these computational studies is then used to help provide context for the 'wet' biology involved in determining genetic function. This involves ensuring that the gene differentiates between diseased and non-diseased states, finding the mutation and showing its effects. The gene is removed, cloned and inserted into a model organism to examine how it functions. This commonly involves generating transgenic animals where random recombination between the artificial vector and the animal's genome is used to generate either knock-out animals, where the gene is removed, or 'transgenic' animals, where the diseased gene is inserted into the animal's genome. The animals are then analysed to look at the effects on various tissues. This ability to use transgenic animals, to explore the consequences of knocking out genes, and the ability to exploit evolutionary redundancy and use similar sequences to fill in gaps in sequences has enabled medicinal biologists to build up better models of how diseases occur and potentially how they can be cured.

While genetically engineered mouse models have very similar biology to humans, they suffer from high cost and low throughput. As a consequence, high-throughput biological systems have been developed to characterize genes based on yeast, nematode worm or fruitfly systems that have been adapted for robot manipulation and growth in 96-well dishes (Gelbert and Gregg, 1997, p. 670).

5.4 Single Populations to Populations of Populations

As the amount of biological data has increased and the tools for analysing genetic profiles have improved, research has started to explore biology beyond the single gene to populations of genes. New technologies based on expressed sequence tag libraries, 'gene chip' technologies and single nucleotide polymorphisms can now analyse large populations of genes. This allows the exact coding sequence of genes to be found and directly related to proteins, allowing very fast 'shotgun' sequencing. The increase in the speed of sequencing has allowed more genes to be analysed so that scientists can explore different expression levels of thousands of genes in diseased and non-diseased tissues—hopefully showing the genes and proteins involved in disease.[18]

The use of these technologies has shown that large numbers of genes are

---

[18] This has not yet produced increases in R&D throughput as the biological problems are considerably more complex than was initially thought.

———————————— *Economies of Scale in Experimentation* ————————————

involved in many diseases that were traditionally thought not to have a genetic basis (Gelbart, 1998; cf. King *et al.*, 1992). As a consequence, the role of genes in disease has shifted from the concern with single gene mutations that produced diseases in all cases to polygenetic, multifactorial diseases where the genetic variations may dispose a patient towards certain diseases given certain external environmental factors. This shift towards a polygenetic, multifactorial understanding of the role of populations of genes in disease has required innovations in the databases and statistical techniques for analysing genetic data (since many of the established statistical techniques do not work for polygenetic diseases).

The development of these new technologies has in turn required very complex computer systems to control the 'tidal wave of data' coming from these new technologies (Deloukas, 1998). The development of these IT systems has been termed bioinformatics

> a science . . . that uses biological data and knowledge stored in com-
> puter databases, complemented by computational methods, to derive
> new biological knowledge. It is a theoretical biology firmly grounded in
> comprehensive and detailed experimental facts. Currently, bioinformatics is
> making a key contribution to the organisation and analysis of the massive
> amount of biological data from genome sequencing projects and increas-
> ingly from other areas of 'high-throughput,' 'massively parallel,' robotised
> and miniaturised methods of biological experimentation.   (Andrade and
> Sander, 1997, p. 675)

These computer systems are interconnected and constantly updated as new genetic information is released. Programs within the systems automatically explore the new data and inform the relevant research groups that data they are interested in has been made public. These programs also automatically look for new genes within the sequence databases and again automatically inform the relevant parties. This automated analysis of genes and sequence data is done within the databases on 'virtual genes' based on their genetic sequences. While it is a long way from replacing 'wet' biology, it does represent the ultimate in scale reductions as the genes exist only as electronic pulses and their analysis is limited only by the (often substantial) costs of calculation.

As biology has increasingly moved *in silico* the ability to coordinate and control the data to ensure the correct throughput of information into the research programmes has become more important. Large pharmaceutical firms have invested heavily in gaining access to this information, by developing their own systems and by linking to biotech firms (Hopkins, 1998; Martin, 1998; Crowther, 1999; Jones, 1999). From a Hughesian (1983) perspective, the bioinformatics systems and the databases of chemical information form part of the 'control systems' of the overall innovation process, and improvements in R&D throughput are dependent on how the information they contain is used—what Davies (1996) has termed 'economies of system' in his study of telecommunications networks.

As finding the 'drug target' has been the main bottleneck in drug discovery, the development of effective ways of finding 'disease' proteins from their genes has improved the prospects of drug discovery. The development of these new genetics technologies has produced an about turn, so that in the early 1990s the main problem for pharmaceutical R&D was finding targets, while at the end of the 1990s the main problem was deciding which of hundreds of potential targets would be most profitable. The biochemistry of these new targets is better understood and failures are less likely, further increasing the potential capacity of the drug discovery programme. Since the analysis is based on a more fundamental understanding of diseases and their mechanisms, it has the potential (unfulfilled at present) to attack the causes of disease, rather than, as with most modern drugs, the symptoms.

As Table 8 shows, the changes in biology are consistent with the changes in chemistry and with the framework developed earlier.

The next section discusses the implications of these changes and points towards areas where further research is needed.

6. *Discussion and Conclusion*

The empirical evidence has illustrated how pharmaceutical firms have exploited economies of scale in the creation of intangible capital in an attempt to increase the throughput of their R&D processes. The evidence supported Rosenberg's insights that new experimental technologies and research methods can reduce the cost of solving complex technical problems. However, these cost savings are related to the 'throughput' of the R&D process and would not be realized by firms producing small numbers of drugs. Firms require large numbers of drugs in their R&D 'pipeline' to first spread the high fixed costs of these new technologies and secondly to manage the uncertainty and volatility of product success. Pharmaceutical firms have therefore exploited economies of scale and scope across therapeutic areas, suggesting that the organizational and technological features of modern drug discovery may favour a particular organizational form—a large firm with a large R&D department—in a way that is consistent with Chandler's framework.

——————————— *Economies of Scale in Experimentation* ———————————

TABLE 8

Changes	Experimental design and proposing solutions	Testing	Modify understanding
From craft to mass production	From a craft-based wet biology of finding proteins and then genes to an automated high-throughput process of finding genes and their function complemented by *in silico* analysis.	From the use of microscopes and physical instrumentation to the automated generation of data on populations of genes.	Biology has shifted from an empirical science to an increasingly theoretical one based on the computerized statistical analysis of large amounts of empirical and *in silico* information and its representation by advanced visualization technologies.
Changes in scale and size	The experimental unit has shifted from the human body to animal models (1950s), then on to tissues (1960s), the biochemistry of cells and protein mechanism (1970s and 1980s), and onto hundreds of thousands of genes (1990s).	Sequencing has shifted from single genes to populations of thousands of genes. Functional genetics has shifted from single transgenic animals to high-throughput screening systems.	The amount of data produced by genetics laboratories has increased from dealing with 200 or so markers in the early 1990s to 100 gigabytes in 1996 and terabytes of data in 1998.
Automation	Biology has shifted from experiments on laboratory animals to increasingly automated systems where genes and their functions are found and analysed on databases.	Genetic sequencing is increasingly automated, with high-throughput model systems engineered for functional genetics.	Simulations of biological mechanisms based on linkages to databases.

The empirical evidence highlighted the shift from a sequential process of hand-crafted experiments to a parallel process involving the increasingly automated analysis of large populations. Since the costs of experiments are related to the volume of sample, economies of scale in experimentation have produced decreases in size (of the experimental unit)—in contrast to the increases in the size of production technologies. The shift towards performing complementary analysis *in silico* extends this emphasis towards reducing the

──────── *Economies of Scale in Experimentation* ────────

scale of experiments further by making them 'virtual'—whereby costs are constrained by computing power.

The case study evidence, while exploring only a sample of technologies and simplifying a very complex process, has shown that both chemistry and biology have gone through several overlapping and interacting changes:

- Firstly, a shift towards more fundamental science. This involved a shift in chemistry from conceptualizing molecules as static atoms and bonds to seeing them in terms of electron distributions and 3-D dynamic structures. In biology it involved linking biochemical mechanisms to the expression of genes.

- Secondly, a shift towards *complementing* experiments with computer simulations. In chemistry this involved using computers to visualize how the drug molecule bonds to the protein. In biology this involved using databases to analyse genes and their functions.

- Thirdly, a shift towards performing experiments and analysis on populations. In chemistry this involved using HTS and combinatorial chemistry. In biology this involved using various high throughput biological systems to understand biological mechanisms, using genetic information on diseased and non-diseased populations (to find the position of genes), exploring the similarity between families of genes across species (to find similar functions), and exploring populations of genes within an individual (to find how different expression levels relate to different biological features).

- Fourthly, experiments on populations created unprecedented amounts of data that required the introduction of database technologies. In chemistry this involved using statistical analysis to explore similarities between compounds, and databases to cluster various 'descriptors' in the design of experiments. In biology it involved the generation of bioinformatics technologies.

- Lastly, databases were used to conduct complementary *in silico* experiments on old and new data. In chemistry this can be seen in the use of databases to screen populations of compounds for testing and to statistically contrast different screens. In biology the shift has involved searching for genes on databases, and exploring their functions by relating them to 'similar' genes in other environments.

The introduction of these new technologies raises a number of issues about the relationship between scientific knowledge and innovation. The evidence supports the generally accepted view that scientific knowledge does not

——————————— *Economies of Scale in Experimentation* ———————————

produce technology directly. This was explained in terms of the direction argument. An alternative conception of innovation based on the notion of 'technological traditions' was presented where the application of 'similar' solutions to 'similar' problems generated a sequence of tasks that formed the basis for the innovation process.

Since scientific knowledge cannot directly generate a desired outcome, technologists perform a large amount of research to understand the phenomenon they are trying to modify and then rely on their tacit knowledge to suggest an uncertain 'first cut' solution. This is tested and the results of the tests are used to modify how technologists understand the relationship between changing starting conditions and end results, and consequently to 'tune' the technology towards its desired behaviour. The testing process can be done in increasingly accurate ways by relying on tacit understanding, by mathematical pattern extrapolation using simulations and by empirical testing. While high-throughput processes are used at the start of research tasks, the final stages of biological analysis and chemical synthesis are still very craft-based.

The ability of simulations to accurately predict behaviour is limited firstly, by computational cost, secondly, by nonlinear error growth and thirdly, by breakdowns in the symmetry between the laws of nature and their real world outcomes (which generate extra information not contained in the original laws of nature—see footnote 3). Simulations are therefore used differently in chemistry and biology.

Because chemical simulations are often quantum mechanical, the computational cost is very high for large numbers of atoms and more accurate approximations. This limits their accuracy and application, and while simulations are very useful for testing hypotheses, they have not replaced experiments, as was initially hoped. In biology the additional problems of large amounts of symmetry breaking and nonlinear error growth has made the simulation of biochemical reactions very inaccurate and their use is extremely limited. However, evolutionary redundancy and the strict symmetry conditions between genes and proteins create strongly constrained patterns that biologists can exploit.

Biologists use computers to interrelate proteins and their gene sequences, and to explore databases for similar sequences with known functions. The substantial differences within and between chemistry and biology in the use of simulations suggests that thinking of scientific knowledge as an all-encompassing category may be too simplistic for practical applications, as its role in technical change is very problem-specific (and therefore sector-specific).

——————————— *Economies of Scale in Experimentation* ———————————

While this paper has not been specifically about the 'conversion of tacit knowledge,' it does raise issues about its role in technical change. Information technologies have allowed medicinal chemistry and biology to become more fundamental and, while some skills have been replaced—more by the introduction of 'off the shelf' kits than by information technology (see Gilbert, 1991)—the general skill requirements have increased. This is especially true of the additional computational skills required to perform the new techniques. Moreover, since visualization technologies are needed to understand and analyse large populations of data, the notion that tacit knowledge has been converted into 'codified knowledge' seems problematic. Whatever the impact on tacit knowledge, the empirical evidence clearly shows a substantial increase in the amount of electronic information generated during drug discovery, and if 'codification' is taken to mean this, then it is supported by the account described here.

Questions remain about a number of issues. While the effects of these technical changes will be uncertain for a number of years, initial evidence does not suggest radical improvement. While many pharmaceutical companies are stressing that their programmes are of higher quality, there is little evidence that this is translating into improved performance. Care must be taken here as most of the major pharmaceutical firms are concentrating on considerably more complex illnesses than have been attempted before. It may be the case that investment in these new technologies is an 'ante' to remain in a more competitive game. More research needs to be done to see if their attempts to increase efficiency actually improve performance.

This initial failure of 'subcomponent' technologies to contribute towards overall 'systems' performance is partly related to problems of coordination and control of information and knowledge within the innovation process. There are two aspects of this. Firstly, the complexity of the innovation process means that it is very hard to understand the relationship between micro-changes and macro-effects. This problem is not specific to the pharmaceutical industry. The introduction of mass production in manufacturing at the start of the last century required long periods of incremental learning and organizational change before performance improved. As a consequence, the pharmaceutical industry would be expected to take some time to optimize its processes. As Nelson (1991) notes, firms differ substantially in how they adapt their internal organization to new technologies and environments.

Secondly, and more importantly, because the quality of experiments depends on how well they modify technologists' tacit understanding, even if the effects could be calculated they cannot be directly measured. For example, while it is easy to count changes in the number of experiments, the corres-

——————————————— *Economies of Scale in Experimentation* ———————————————

ponding improvements in understanding are far harder to measure. Since
testing the wrong type of compounds is a waste of time, simple quantity
measures are ineffective on their own, and the correct trade-offs between
experimental quality and quantity (highlighted in footnotes 10 and 11) are
extremely uncertain.[19] Furthermore, what effect any changes in one part of
the innovation process have on the rest of the system is even more uncertain
because the correct tacit 'sense of similarity' is only known for sure after the
drug has been found and shown to be a medical and commercial success. For
example, improvements to the trade-off between quantity and quality might
lead to a compound that binds very tightly to a protein; however, it will fail
as a drug if the protein has no effect on the disease. Thus uncertainties in the
system are pervasive, and the key to systems' performance involves controlling and managing these uncertainties.

This inherent systemic uncertainty about the subjective quality of experiments explains why nearly a century separates the introduction of mass
production in manufacturing and R&D. The pharmaceutical industry is not
technologically unsophisticated compared with late-nineteenth-century
small-arms manufacturers; rather, it is attempting a far more complex project.
What is different about the pharmaceutical innovation processes as a
technical system is that it lacks a 'load factor' (Hughes, 1983, pp. 218–221)
that can be used to objectively measure the contribution of each part to the
overall process performance. Production engineers in manufacturing have
well-defined parameters and distributions which they can use to calculate and
optimize throughput, while the complexities of drug discovery mean that
decision making is decentralized to experts, whose choices are often based
on tacit knowledge rather than well-defined parameters. This is because
Bradshaw's paradox applies to the whole innovation process: one cannot be
sure that researchers are looking in the right place until they find what they
are looking for. As a consequence, unlike a physical production system, it is
very hard to relate performance measures of the whole system—i.e. drugs
produced per million dollars of R&D—to performance measures of the component parts.

However, as with Bradshaw's paradox, there are a number of ways of
reducing uncertainty, even if certainty cannot be fully established. The

[19] This uncertainty can be seen in the lack of consensus about the correct strategy. Some firms have
moved towards the 'big hammer' approach and are analysing huge numbers of compounds, while others
are concentrating on much smaller populations. There is no consensus at present as to which approach will
pay off. The difference in strategies in HTS is related to an interesting philosophical issue: given that there
are about 10^{180} possible compounds of the right molecular weight to be drugs (there are approximately
10^{78} particles in the universe), moving from testing 10^4 to testing 10^7 compounds can either be seen as a
1000-fold increase or as an inconsequential waste of time as both 10^4 and 10^7 approximate to zero when
compared with 10^{180}.

─────────── Economies of Scale in Experimentation ───────────

empirical evidence showed how pharmaceutical firms use cycles of experiments to test the implicit hypotheses behind their choices, how they
exploit path-dependent biases in the samples they use for testing, how
they avoid duplicating drugs produced in other firms by 'randomizing' the
search process, how they use more fundamental biology to ensure that the
drugs they are developing will produce the biological effects they desire, how
they develop sophisticated information technology systems to collect and
coordinate the 'tidal waves' of information coming from experiments, and
how they rely on well-trained chemists and biologists to interpret experimental data. Additionally, pharmaceutical firms engage in scientific and
research networks to ensure that their research is world class, internally
peer-review technical choices, attempt to systematize research, achieve
economies of scope by concentrating on specific targets and therapeutic
areas, and reorganize the processes of research based around multi-
disciplinary teams and project-based organizational structures. While these
changes cannot remove the inherent uncertainty associated with drug
discovery, they do create economically important differences in performance.
This paper has only touched on these issues and a more systematic
investigation of the organizational features that turn potential economies
into actual economies of scale is needed.

The relationship between these organizational issues and technological
uncertainty has been the subject of recent research (Clark and Wheelwright,
1992; Brooks, 1995; Pinto and Kharbanda, 1995; Hobday, 1998). While
there is consensus that the traditional functional bureaucracy is efficient at
dealing with well-defined, standardized tasks and more decentralized 'project-
based' forms are better at dealing with high levels of uncertainty, the exact
relationship between organizational architecture and system performance is
badly understood, despite the work of Woodward (1958), Burns and Stalker
(1961), Brooks (1995) and Teece (1996). This is highlighted as an area of
future research.

The effectiveness of different organizational and managerial responses to
systemic uncertainty will have an important effect on firm performance. If
large pharmaceutical firms can 'optimize' the interconnections within the
innovation system and improve overall throughput, then they will increase
their advantages over smaller firms. If, however, the problems are intractable,
then the Chandlerian 'inter-dependencies' between tasks will be reduced and
smaller firms will be in a position to compete on parts of the process. At
present a number of 'platform' biotechnology firms are offering a range of
subcontracted research services, and as many of these new technologies have

——————————— *Economies of Scale in Experimentation* ———————————

reduced the fixed costs of research, the short-term dominance of large firms cannot be assured.[20]

In the medium term the ability of pharmaceutical firms to use genetic information to customize products to specific genetic populations promises to fundamentally influence the throughput of drug discovery, and therefore the boundary between drug discovery and healthcare provision. In general, policy thinking in this area has be guided by rather naïve conceptions of industrial and corporate change, and this is highlighted as an area of future research.

In the short term, however, the ability of large pharmaceutical firms to turn potential economies of scale into actual economies of scale will depend on their ability to organize and manage the flows of information between tasks. Already there is anecdotal evidence that some firms are using bioinformatics systems to successfully link research tasks (Hopkins, 1998; Jones, 1999). However, the literature on innovation in complex systemic technologies would suggest that moving towards large-scale improvements, while possible, will be extremely difficult (Hughes, 1983; Davies, 1996; Hobday, 1998; Nightingale and Poll 2000). As with choosing the appropriate organizational architecture, the use of information technology to link research tasks is too complex to rigidly plan. Instead, firms will have to rely on organizational flexibility, diversity in technical options and a 'muddling through' approach, when faced with technological uncertainty (Stirling, 1998).

That a neo-Chandlerian framework required only a change of emphasis rather than a change of substance to explore the production of intangible capital within pharmaceutical R&D is encouraging. The changes involved incorporating Rosenberg's insights into the role of 'instrumentation as experimental capital-goods,' seeing that economies of scale in experimentation produce reductions in size and recognizing that the systemic uncertainties involved preclude a bureaucratic functional organization. While a narrow technological focus, to the exclusion of issues such as marketing, prevents a holistic analysis of the role of technology and organization in determining the efficiency of particular organizational forms, the changes in the technology of experimentation and the management of intangible capital analysed in this paper suggests that their efficiency is related to increased throughput. While organizational issues require further investigation, as they will affect the minimum efficient size of pharmaceutical firms, the ability to cost effectively exploit the new technologies of experimentation and generate sufficient new drugs to reduce the volatility of product launches means that a particular organizational form—the large firm—may be more efficient

———
[20] I am grateful to an anonymous referee for highlighting this important point.

————————— *Economies of Scale in Experimentation* —————————

than its smaller rivals. If large pharmaceutical firms can overcome the organizational problems associated with high levels of uncertainty, then they should continue to dominate drug development. Considering that Chandler's framework had a deliberately narrow focus (Chandler, 1977, p. 6, 1990, pp. 12–13) and was not intended to explain changes in pharmaceutical technology in the 1990s, the moral of this paper is perhaps that his insights are surprisingly robust.

Acknowledgements

The author would like to thank Ed Steinmueller and an anonymous reviewer for their insightful comments on a previous draft. The paper has benefited from the insights of Pat Beeston, Sarah Crowther, Paul Martin, Andrew Davies, Tim Brady, Howard Rush and Mike Hobday. The usual disclaimers apply. This work was prepared within the Complex Product System Innovation Centre, funded by the UK Economic and Social Science Research Council.

References

Andrade, M. A. and C. Sander (1997), 'Bioinformatics: From Genome Data to Biological Knowledge,' *Current Opinion in Biotechnology*, 8, 675–683.

Arrow, K. (1962), 'Economic Welfare and the Allocation of Resources for Invention,' in R. R. Nelson (ed.), *The Rate and Direction of Inventive Activity*. Princeton University Press: Princeton, NJ, pp. 609–625.

Babbage, C. (1836 repr. 1963), *On the Economy of Machinery and Manufactures*, 4th edn. Frank Cass: London.

Barrow, J. D. (1988), *The World within the World*. Oxford University Press: Oxford.

Blundell, T., D. Carney, S. Gardener, F. Hayes, B. Howlin, T. Hubbard, J. Overington, D. Singh, B. Sibanda and M. Sutcliffe (1988), 'Knowledge Based Protein Modelling and Design,' *European Journal of Biochemistry*, 172, 513–520.

Blundell, T., G. Elliot, S. Gardener, T. Hubbard, S. Islam, M. Johnson, D. Mantafounis, P. Murray Rust, J. Overington, J. E. Pitts, A. Sali, B. L. Sibana, D. Singh, M. Sternberg, M. Sutcliffe, J. M. Thornton and P. Travers (1989), 'Protein Engineering and Design,' *Philosophical Transactions of the Royal Society of London B*, 324, 447–460.

Bork, P. and A. Bairoch (1996), 'Go Hunting in Sequence Databases but Watch Out for the Traps,' *Trends in Genetics*, 12, 425–427.

Bradshaw, J. (1995), 'Pitfalls in Creating a Chemically Diverse Compounds Screening Library,' unpublished document, Glaxo.

Brooks, F. P., Jr (1995), *The Mythical Man Month: Essays on Software Engineering, Anniversary Edition*. Addison Wesley: Reading, MA.

Burns, T. and G. M. Stalker (1961), *The Management of Innovation*. Tavistock: London.

Chandler, A. D., Jr (1977), *The Visible Hand*. Harvard University Press: Cambridge, MA.

――――――――――――― *Economies of Scale in Experimentation* ―――――――――――

Chandler, A. D., Jr (1990), *Scale and Scope: The Dynamics of Industrial Capitalism*. The Belknap Press of Harvard University Press: Cambridge, MA.

Clark, K. B. and S. C. Wheelwright (1992), 'Organising and Leading "Heavyweight" Development Teams,' *Californian Management Review*, 34(3), 9–38.

Cohen W. M. and D. A. Levinthal (1990), 'Absorptive Capacity: A New Perspective on Learning and Innovation,' *Administrative Science Quarterly*, 35, 128–152.

Crowther, S. (1999), 'Patenting Genes: Intellectual Property Rights in Human Genome Research,' unpublished DPhil. Thesis, SPRU, University of Sussex.

Dasgupta, P. and P. David (1994), 'Towards a New Economics of Science,' *Research Policy*, 23, 487–521.

Davies, A. (1996), 'Innovation in Large Technical Systems: The Case of Telecommunications,' *Industrial and Corporate Change*, 5, 1143–1180.

DiMasi, J. A. (1995), 'Trends in Drug Development Costs Times and Risks,' *Drug Information Journal*, 29, 275–384.

DiMasi, J., R. W. Hansen, H. G. Grabowski and L. Lasagne (1991), 'The Cost of Innovation in the Pharmaceutical Industry,' *Journal of Health Economics*, 10, 107–142.

Dosi, G. (1982), 'Technological Paradigms and Technological Trajectories—A Suggested Interpretation of the Determinants and Direction of Technical Change,' *Research Policy*, 11, 147–162.

Dosi, G., L. Marengo, A. Bassanini and M. Valente (1999), 'Norms as Emergent Properties of Adaptive Learning: The Case of Economic Routines,' *Journal of Evolutionary Economics*, 9, 5–26.

Drews, J. (1995), 'The Impact of Cost Containment Initiative on Pharmaceutical R&D,' CMR Lecture, London, 19 June.

Deloukas, P. et al. (1998), 'A Physical Map of 30,000 Human Genes,' *Science*, 282, 744–746.

Deutsch, D. (1997), *The Fabric of Reality*. Allen Lane: London.

Ferguson, E. S. (1977), 'The Mind's Eye: Non-verbal Thought in Technology,' *Science*, 197, 827–836.

Ferguson, E. S. (1992), *Engineering and the Mind's Eye*. MIT Press: Cambridge, MA.

Finn, P. (1995), 'Computational High Throughput Screening,' paper presented at the Conference of the Molecular Graphics and Modelling Society: 'Computational Aspects of Chemical Libraries', April 10–12, 1995, University of Leeds.

Freeman, C. (1982), *The Economics of Industrial Innovation*, 2nd edn. Pinter: London.

Freeman, C. and L. Soete (1998), *The Economics of Industrial Innovation*, 3rd edn. Pinter: London.

Gabrowski, H. and J. Vernon (1994), 'Innovation and Structural Change in Pharmaceutical and Biotechnology,' *Industrial and Corporate Change*, 3, 435–449.

Galambos L. and J. Sewell (1995), *Networks of Innovation: Vaccine Development at Merck, Sharp Dohme and Mulford 1895–1995*. Cambridge University Press.

Galambos, L. and J. L. Sturchio (1998), 'Pharmaceutical Firms and the Transition to Biotechnology: A Study in Strategic Innovation,' *Business History Review*, 72, 250–278.

Gallop, M. A., R. W. Barrett, W. J. Dower, S. P. A. Fodor and E. M. Gordon (1994), 'Applications of Combinatorial Technologies to Drug Discovery, 1. Background and Peptide Combinatorial Libraries,' *Journal of Medicinal Chemistry*, 37, 1233–1250.

Gambardella, A. (1992), 'Competitive Advantages from In-house Scientific Research—The United States Pharmaceutical Industry in the 1980s,' *Research Policy*, 21, 391–407.

Gambardella, A. (1995), *Science and Innovation: The US Pharmaceutical Industry during the, 1980s*. Cambridge University Press: Cambridge.

Gelbart, W. M. (1998), 'Databases in Genomic Research,' *Science*, 282, 659–881.

―――――――――――――――――――――― 356 ――――――――――――――――――――――

─────────────── *Economies of Scale in Experimentation* ───────────────

Gelbert, L. M. and R. E. Gregg (1997), 'Will Genetics Really Revolutionise the Drug Discovery Process?,' *Current Opinion in Biotechnology*, 8, 669–674.

Gilbert, W. (1991), 'Towards a Paradigm Shift in Biology,' *Science*, 349, 99.

Gordon, E. M. (1995), 'Strategies in Combinatorial Chemistry: Practical Applications to Drug Discovery,' paper presented at the Conference of the Molecular Graphics and Modelling Society: 'Computational Aspects of Chemical Libraries,' April 10–12, 1995, University of Leeds.

Gordon, E. M., R. W. Barrett, W. J. Dower, S. P. A. Fodor and M. Gallop (1994), 'Applications of Combinatorial Technologies to Drug Discovery, 2. Combinatorial Organic Synthesis Library Screening Strategies and Future Directions,' *Journal of Medicinal Chemistry*, 37, 1385–1401.

Granstrand, O., K. P. Pavitt and P. Patel (1997), 'Multi Technology Corporations: Why They Have "Distributed" Rather than "Distinctive" Core Competencies,' *California Management Review*, 39(4), 8–25.

Gray, J. (1995), Interview, Unilever, London.

Hann, M. (1994), 'Considerations for the Use of Computational Chemistry Techniques by Medicinal Chemists,' in F. D. King (ed.), *Medicinal Chemistry Principles and Practice*. The Royal Society of Chemistry: London, ch. 9.

Heller, M. A. and R. S. Eisenberg (1998), 'Can Patents Deter Innovation? The Anticommons in Biomedical Research,' *Science*, 280, 698–701.

Henderson, R., L. Orsengio and G. P. Pisano (1998), 'The Pharmaceutical Industry and the Revolution in Molecular Biology: Exploring the Interactions between Scientific, Institutional and Organisational Change,' unpublished manuscript, draft 5.0.

Hobday, M. (1998), 'Product Complexity, Innovation and Industrial Organisation,' *Research Policy*, 26, 689–710.

Hopkins, M. (1998), 'An Examination of Technology Strategies for the Integration of Bioinformatics in Pharmaceutical R&D Processes,' unpublished Masters dissertation, SPRU, University of Sussex.

Houston, J. G. and M. Banks (1997), 'The Chemical–Biological Interface: Developments in Automated and Miniaturised Screening Technology,' *Current Opinion in Biotechnology*, 8, 734–740.

Hughes, T. (1983), *Networks of Power: Electrification in Western Society, 1880–1930*. Johns Hopkins University Press: Baltimore, MD.

Jansen, J. (1996), 'High Throughput Screening as a Discovery Tool in the Pharmaceutical Industry,' *Laboratory Robotics and Automation*, 8, 261–265.

Jones, D. (1999), 'Bioinformatics as a Competency: Its Nature, Development and Use,' unpublished Masters dissertation, SPRU, University of Sussex.

King, R. A., J. L. Rotter and A. G. Motulsky (1992), *The Genetic Basis of Common Diseases*. Oxford University Press: Oxford.

Lander, E. S. (1996), 'The New Genomics: Global Views of Biology,' *Science*, 274, 536–539.

Lazonick, W. (1991), *Business Organisation and the Myth of the Market Economy*. Cambridge University Press: Cambridge.

Leonard-Barton, D. (1995), *Wellsprings of Knowledge: Building and Sustaining the Sources of Innovation*. Harvard Business School Press: Cambridge, MA.

Levin, R. C., A. K. Klevorick, R. R. Nelson and S. G. Winter (1987), 'Appropriating the Returns from Industrial Research and Development,' *Brookings Papers on Economic Activity*, 3, 783–820.

Martin, P. (1998), 'From Eugenics to Therapeutics: Science and the Social Shaping of Gene Therapy,' unpublished DPhil thesis, SPRU, University of Sussex.

McKelvey, M. (1994), 'Evolutionary Innovation Early Industrial Use of Genetic Engineering,' Linkoping Studies in Arts and Sciences no. 107, Linkoping.

—————————————— *Economies of Scale in Experimentation* ——————————————

Mirowski, P. (1989), *More Heat than Light: Economics as Social Physics, Physics as Nature's Economics*. Cambridge University Press: Cambridge.

Mowery, D. and N. Rosenberg (1979), 'The Influence of Market Demand upon Innovation: A Critical Revue of some Recent Empirical Studies,' *Research Policy*, 8, 103–153.

Mowery, D. and N. Rosenberg (1989/1994), *Technology and the Pursuit of Economic Growth*. Cambridge University Press: Cambridge.

Murray Rust, P. (1994), 'Bioinformatics and Drug Discovery,' *Current Opinion in Biotechnology*, 5, 648–653.

Nelson, R. R. (1991), 'Why Do Firms Differ, and How Does it Matter?,' *Strategic Management Journal*, 12, 61–74.

Nelson, R. R. and S. G. Winter (1982), *An Evolutionary Theory of Economic Change*. The Belknap Press at Harvard University Press: Cambridge, MA.

Nightingale, P. (1998), 'A Cognitive Model of Innovation,' *Research Policy*, 27, 689–709.

Nightingale, P. and R. Poll (2000), 'Innovation in Investment Banking: the Dynamics of Control Systems in the Chandlerian Firm,' *Industrial and Corporate Change*, 9, 113–141.

Oppenheimer, J. R. (1956), 'Analogy in Science,' *American Psychologist*, 11, 127–135.

OTA (1993), *Pharmaceutical R&D Costs, Risks and Rewards*. Office of Technology Assessment: Washington, DC.

Parket, S. (1998), 'Biotech Stocks and European Capital Markets: An End of Year Review,' *European Pharmaceutical Review*.

Pavitt, K. (1984), 'Sectoral Patterns of Technological Change: Towards a Taxonomy and a Theory' *Research Policy*, 13, 343–374.

Pavitt, K. (1990), 'What Do We Know about the Strategic Management of Technology?' *Californian Management Review*, 32, 17–26.

Pavitt, K. (1987), 'The Objectives of Technology Policy,' *Science and Public Policy*, 14, 182–188.

Pavitt, K. (1991), 'What Makes Basic Research Economically Useful?' *Research Policy*, 20, 109–119.

Pavitt, K. (1996) 'National Policies for Technical Change: Where Are the Increasing Returns to Economic Research?' *Proceedings of the National Academy of Sciences, USA*, 93, 12693–12700.

Pavitt, K. (1997), 'Technologies, Products and Organisation in the Innovating Firm: What Adam Smith Tells Us and Joseph Schumpeter Doesn't,' *Industrial and Corporate Change*, 7, 433–452.

Petroski, H. (1986), *To Engineer is Human: The Role of Failure in Successful Design*. Macmillan: London.

Pinto, J. K. and O. P. Kharbanda (1995), 'Variations in Critical Success Factors over the Stages of the Product Life Cycle,' *Journal of Management*, 14, 5–18.

Pisano, G. (1998), *The Development Factory*. Harvard Business School Press: Cambridge, MA.

Rich, D., J. Green, M. V. Toth, G. R. Marshall and S. B. Kent (1990), 'Hydroxyethylamine Analogs of the p17/p24 Substrate Cleavage Site are Tight Binding Inhibitors of HIV Protease,' *Journal of Medicinal Chemistry*, 33, 1285–1288.

Rosenberg, N. (1974), 'Science Innovation and Economic Growth,' *Economic Journal*, 84, 333.

Rosenberg, N. (1990), 'Why Do Firms Do Basic Research with Their Own Money?' *Research Policy*, 19, 165–174.

Rosenberg, N. (1992), 'Scientific Instrumentation and University Research,' *Research Policy*, 21, 381–390.

Rosenberg, N. (1994), *Exploring the 'Black Box': Technology, Economics and History*. Cambridge University Press: Cambridge.

Rothwell, R. (1992), 'Successful Industrial Innovation: Critical Factors for the 1990s,' *R&D Management*, 22, 221–239.

——————————— *Economies of Scale in Experimentation* ———————————

Salter, A. J. and B. Martin (2000), 'The Relationship between Basic Research and Economic Performance: A Critical Review,' *Research Policy* (forthcoming).

Scherer, F. M. (1993), 'Pricing, Profits and Technological Progress in the Pharmaceutical Industry,' *Journal of Economic Perspectives*, 17, 97–115.

Searle, J. R. (1995), *The Construction of Social Reality*. Allen Lane: London.

Spellmeyer, D. (1995), 'Measuring Chemical Diversity: Experimental Design of Combinatorial Libraries,' paper presented at the Conference of the Molecular Graphics and Modelling Society: 'Computational Aspects of Chemical Libraries', April 10–12, 1995, University of Leeds.

Stankiewicz, R. (1993), 'The Development of Beta-blockers at Hassle-Astra,' unpublished manuscript.

Stirling, A. (1998), 'The Economics and Analysis of Technological Diversity,' SPRU Electronic Working Paper Series, Paper 28, University of Sussex.

Sykes, R. (1997), 'The Pharmaceutical Industry in the New Millennium: Capturing the Scientific Promise,' Centre for Medicines Research (CMR) Annual Lecture, CMR, London.

Teece, D. J. (1996), 'Firm Organisation, Industrial Structure, and Technological Innovation,' *Journal of Economic Behaviour and Organisation*, 31, 193–224.

Thomas, L. G. (1994), 'Implicit Industrial Policy: the Triumph of Britain and the Failure of France in Global Pharmaceuticals,' *Industrial and Corporate Change*, 1, 451–489.

Turro, N. J. (1986), 'Geometric and Topological Thinking in Organic Chemistry,' *Angewandte Chemie International Edition (English)*, 25, 882–901.

Ugi, I., D. Marquarding, H. Klusacek, G. Gokel and P. Gillespie (1970), 'Chemistry and Logical Structure,' *Agnew. Chemistry International Edition (English)*, 9, 703–730.

Vincenti, W. G. (1990), *What Engineers Know and How They Know It*. John Hopkins University Press: Baltimore, MD.

Weinstein, J. N., T. G. Myers, P. M. O'Conner, S. H. Friend, A. J. Forance, K. W. Kohn, T. Fojo, S. E. Bates, L. V. Rubenstein and N. L. Anderson (1997), 'An Information Intensive Approach to the Molecular Pharmacology of Cancer,' *Science*, 275, 343–349.

Whittle, P. J. and Blundell, T. L. (1994), 'Protein Structure Based Drug Design,' *Annual Review of Biophysics and Biomolecular Structure*, 232, 349–375.

Woodward, J. (1958), *Management and Technology*. Her Majesty's Stationary Office: London.

Zucker, L. G. and M. R. Darby (1997), 'Present at the Biotechnological Revolution: Transformation of Technological Identity for a Large Pharmaceutical Firm,' *Research Policy*, 26, 429–446.

[8]

Does Good Science Lead to Valuable Knowledge? Biotechnology Firms and the Evolutionary Logic of Citation Patterns

Michelle Gittelman • Bruce Kogut

Department of Management and Organizational Behavior, New York University, Stern School of Business,
44 West 4th Street, Tisch 7-11, New York, New York 10012
The Wharton School, University of Pennsylvania, Philadelphia, Pennsylvania 19104, and
Centre de Recherche en Gestion, Ecole Polytechnique, 1, rue Descartes, Paris 75005, France
mgittelm@stern.nyu.edu • kogut@poly.polytechnique.fr

This study looks at the United States biotechnology industry as a community of practice caught between two evolutionary logics by which valuable scientific knowledge and valuable innovations are selected. We analyze the publications and patents of 116 biotechnology firms during the period 1988–1995. In models that link scientific capabilities to patent citations, we show that scientific ideas are not simply inputs into inventions; important scientific ideas and influential patents follow different and conflicting selection logics. Publication, collaboration, and science intensity are associated with patented innovations; however, important scientific papers are negatively associated with high-impact innovations. These results point to conflicting logics between science and innovation, and scientists must contribute to both while inhabiting a single epistemic community. We identify individuals listed on patents and scientific papers and find they effectively integrate science with innovation, leading to more successful innovations. Our findings suggest that the role of the small, research-intensive firm is to create a repository of knowledge; to act as an organizational mechanism to combine the capabilities of versatile scientists within and outside the boundaries of the firm; and to manage the selection of scientific ideas to produce valuable technical innovations.

(*Science; Citations; Patents; Scientists; Epistemic Community; Biotechnology*)

Introduction

One of the important lessons of the sociology of science is that the creation of scientific knowledge is an activity that is institutionally constructed and organized. Until the sixteenth century, scientific endeavors were cloaked in secrecy to withhold knowledge and the powers it conferred from the "vulgar multitude" (David 1998). The institutionalization of science encouraged the validation and diffusion of ideas as open to public scrutiny (Merton 1973). To support these institutions, norms that standardized the language and presentation of results developed under the auspices of academic journals. The careers of scientists were tied to their success in publishing these results in prestigious journals and withstanding subsequent public criticism. As science evolved, it also fragmented into distinct communities, with separate identities, journals, and models of experimentation and validation.

The sociology of science, though rich and variegated, broadly agrees with the view of science as embedded in distinctive communities. As Merton (1973) and economists such as Dasgupta and David (1994) and Stephan (1996) have noted, these norms

Management Science © 2003 INFORMS
Vol. 49, No. 4, April 2003, pp. 366–382

0025-1909/03/4904/0366$05.00
1526-5501 electronic ISSN

create incentives that are efficient insofar as professional ranking is related to effort, excess duplication of effort is contained, and scientists desire to broadly disseminate results to earn respected reputations. A priority-based publication system is an important principle that promotes rapid dissemination of knowledge within scientific communities and organizes individual contributions in the form of a series of races. De Solla Price (1970, p. 6) observes, "scholarship is a conspiracy to pool the capabilities of many men, and science is an even more radical conspiracy that structures this pooling so that the totality of this sort of knowledge can grow more rapidly than any individual can move by himself." At the bench level, science is "manufactured" in laboratories in which scientists seek power and alliances to persuade each other that they occupy important positions (Knorr Cetina 1999, Latour and Woolgar 1979). In this context, a published paper is a legitimate tool of persuasion and a symbol of achievement.

Even though science is "manufactured" in the context of academic communities, firms can usefully apply scientific knowledge to develop new technologies; indeed, the argument that science could drive commercial innovation was a major justification for public support of a nation's scientific infrastructure. The dilemma for firms seeking to profit from scientific knowledge, however, is that science is not available as ready-made inputs, but is produced by scientists situated in these scientific communities. The useful equation of science as an input to technology is problematic when scientific inputs are also seen as *not only* producing technology but also *manufacturing* scientific outputs valued by other scientists.

The importance of a community operating within a well-organized social structure, sharing a strong epistemic culture, has not been addressed in studies on the economics of science and technology, which have been concerned with showing the functional relation between scientific inputs and technological outputs. The focus on the "production function" is seen in recent studies that have taken a more fine-grained approach to understanding this relationship; these have found that better science leads to more technology (Cockburn and Henderson 1998, Henderson and Cockburn 1994). What happens when the inputs,

called professional scientists, care about their perception of what they do, why they do it, when, and for what kinds of rewards?

We propose that the logic of scientific discovery does not adhere to the same logic that governs the development of new technologies, and that these conflicting logics pose potential problems for science-based innovation. The communities in which scientific ideas circulate and the logics by which they are selected mean that value calculations in science and industry are different. Innovation builds on knowledge made in science, but science that is "good" for innovation is propelled by a logic that is different than that employed by a scientific community to determine "valuable" or "important" science. While industry may need scientific insights to resolve technological problems or find new projects, firms do not directly benefit from contributing to important or controversial scientific questions. We use the difference in the processes by which highly cited scientific discoveries are acknowledged by other scientists and by which valuable innovations are selected by market forces to show how different evolutionary logics weaken the science-technology linkage. Consequently, we generate evidence regarding how firms differentially manage the disconnect between the two evolutionary logics by which science and innovations are rewarded. In all, we find evidence for a single epistemic community, for different and conflicting evolutionary logics between science and technology, and for the important role of bridging scientists in the production of valuable innovations.

Science, Epistemic Communities, and Citation Analysis

The community perspective implies a stickiness in the flow of scientific knowledge to firms. The literature on the motives for firms to publish their research in scientific journals has implicitly or explicitly acknowledged the central importance of forming ties to this community, via boundary-spanning "gatekeepers," to access socially embedded knowledge (Allen 1977, Hicks 1995, Lieberman 1978, Tushman 1977). Research in pharmaceutical and biotechnology companies reveals, in particular, the importance of personal links between commercial firms and universities. The fortunes of biotechnology companies are linked to their engagement of "star scientists" and their location in research regions (Zucker et al. 1998). Indeed, because star scientists often maintain university employment, colocation with research centers is a natural and necessary requirement of successful biotechnology firms. In numerous studies in pharmaceuticals and biotechnology, collaboration with researchers external to the firm and an internal science orientation leads to higher research productivity (Cockburn and Henderson 1998, Cockburn et al. 2000, Gambardella 1995, Powell et al. 1996, Zucker et al. 2002). These studies suggest the operation of absorptive capacity to include direct participation in external communities as a means of acquiring knowledge outside the firm's boundaries (Cohen and Levinthal 1990, Rosenberg 1990). In other words, *how* firms access and practice science—in particular, whether they establish credible linkages with the scientific community—matters to the production of valuable innovations.

Of course, investing in science represents a cost to the firm. Stern (1999) argues that research-oriented scientists have an inherent "taste" for research and publication. The skills of these scientists are needed for translating research into product development. The firm allows them to engage in research and publish as a form of payment in exchange for these other activities; these scientists receive lower wages than scientists who are not allowed to publish. Stern's thesis raises the intriguing implication that the relationship of commercial innovations and scientific knowledge is problematic for the scientist whose activities are aimed at commercial outcomes, but whose identity remains embedded in the values and reward systems of a scientific community.

In effect, those scientists working on research for firms engage in commercial innovative endeavors, while operating within a single "epistemic culture," to appropriate Knorr Cetina's (1999) term to differentiate scientific communities. These cultures code for rules by which scientists define their careers, identities, methods of empiricism, and collaboration with others. Experimental physics utilizes large-scale equipment and engages the efforts of hundreds of scientists who publish jointly. Research in the life

sciences relies on smaller teams and is less capital intensive; collaborative publishing, though common, is characterized by fewer authors. Clearly, rules regarding empirical validation and publication are powerful cultural expressions of the collective organization of individual contributions in the development of scientific knowledge.

Citation traces are the bibliometric fossils by which to measure the replication success of an idea. These fossil records permit an investigation of the relative success of an idea to influence subsequent work. Citations to papers are an important way that scientists evaluate their relative standing, by which they exchange gifts, acknowledge prestige, and seek to prevail in their arguments (Crane 1972, Latour and Woolgar 1979). Forward patent citations, that is, citations made by later patents to a patent previously issued, are similarly indicative traces of the importance of commercial innovations, although the process by which they are generated is less deferential to status and reputation effects (Hall et al. 2000). Campbell (1974) contended that academic fields advance by an "evolutionary epistemology" in which favored ideas are promulgated and disfavored ones are lost. The sociology of science shows that articles that are not cited within five years are unlikely to be remembered (Crane 1972). Citation patterns to patents show a similar time frame (Jaffe et al. 1993).

Our analysis relies principally upon citations. We analyze citations in scientific papers to other scientific papers, and also look at citations in patent documents to other patents and to scientific papers. We propose that the difference in evolutionary logics that generate paper and patent citations reflects the difficulty faced by private firms to translate knowledge produced in a scientific setting into valuable technologies. Where the scientific and commercial endeavors diverge is seen in the different citation traces generated by the distinctive rules that govern the logic by which a good paper or a valuable patent is selected and replicated.

How do scientists jointly operate in the distinct communities of science and technology, without damaging their credibility in the former, or their efficiency in producing the latter? To demonstrate first the importance of this question, we turn to an analysis of citations to scientific papers included in other scientific papers and in patents. We put forth the hypothesis that both of these artifacts, papers and patents, will share common antecedents, i.e., they will show similar citation patterns to scientific papers. In other words, there is no ascriptive distinction in the acknowledgments paid by scientists when they publish a paper or when a patent is filed, naming them as the inventor (even if the property right is held by the firm). The citation patterns should reflect membership in the relevant scientific community.

If scientists in universities and those in firms inhabit a single epistemic community, should we not then expect that influential patents should be the product of influential research? At a first pass, it would seem that if citations to scientific research in patents follow the same norms as citations in papers, then influential patents should also cite influential papers. However, the evolutionary logic that selects out "better" patents is different than that which selects the more influential papers. The evolutionary dynamics for patents reflect the joint factors of market demand, technological opportunity, and legal claims on property rights. Market demand increases the efforts to commercialize, and hence patent, more in particular sectors, thus favoring only those related and relevant patents (Mowery and Rosenberg 1982). Second, technologies differ in their opportunities, with some offering a richer set of opportunities than "dead-end technologies" (Kim and Kogut 1996, Stuart and Podolny 1996). Because of the difference in these selection dynamics, we do not expect influential papers to lead to influential patents. Scientists do not reward papers for their market and technological promise; they reward them for reasons proper to their own epistemic community.

Collaborative activity, we would also expect by this argument, reveals the long hand of the sociology of scientific communities. Certainly, firms desire their scientists to engage in external collaboration to improve their productivity and to acquire property rights to research generated in universities and public laboratories. Scientists collaborate on publications, but they do so in accordance within the norms of doing science. Thus, we expect collaboration to help research, and to result in more or better papers, but

we propose that this collaboration will be embedded in the social relations and rituals of the scientific community.

We expect to find firm-level heterogeneity in the relationship between scientific inputs and innovation outputs. We model this heterogeneity as the degree to which firms succeed in integrating the two worlds of science and invention. Because we believe that scientific knowledge is embedded in a community inhabited by scientists, we identify these individual scientists and look at whether they publish or patent. We focus on the intersection of these groups as the set of individuals who inhabit both the world of open science and the world of technology creation, and measure their impact on patenting. We pose the question: Are individual scientists who perform both more productive in producing important patents? In this sense, these scientists are technological gatekeepers, as studied by Allen (1977) and Tushman (1977), but more specifically, as suggested by Lieberman (1978), they help firms patent by bridging the worlds of discovery and innovation.

To summarize, we make the following propositions:

PROPOSITION 1. *Papers and patents exhibit similar citation patterns to scientific papers.*

PROPOSITION 2. *The evolutionary logics that select valuable scientific papers and valuable patents are different, and because of this, influential papers are no more likely to lead to influential patents than other papers.*

PROPOSITION 3. *Collaboration choices are influenced by status in a scientific community.*

PROPOSITION 4. *Scientists who bridge discovery and innovation are able to reconcile these two conflicting logics more effectively than those specializing in either science or technology.*

Science, Publications, and the United States Biotechnology Industry

We address these issues in the context of the United States (U.S.) biotechnology industry, which is characterized by rapid knowledge diffusion and intense technological competition. Biotechnology firms are actively engaged in keeping at the forefront of publishing in the scientific literature (Powell et al. 1996). Because this industry so heavily relies on commercializing on the basis of scientific discovery, the relationship between scientific knowledge and innovation outputs are especially strong.

Biotechnology firms act as organizational vehicles for the private appropriation of knowledge produced in university laboratories and moving it to a commercial marketplace. Indeed, the professional and cognitive divide between individuals engaged in scientific research and those engaged in commercialization is much less sharp in biotechnology than in other technology-intensive industries, e.g., microelectronics. Zucker et al. (1998) show that specialized biotechnology firms are formed with the intent of capturing knowledge held by academic scientists close to the "frontier" of knowledge discovery. A startup company enables them to extract economic value from their valuable knowledge (Audretsch and Stephan 1996).

Biotechnology firms are defined in our study as new firms specializing in the use of molecular technologies to develop new drugs, diagnostic tools, or other novel products. However, their collaboration networks extend beyond their own field. If we were to describe, to use Callon's (1986) terminology, the action networks of a biotechnology firm, they would consist of partnerships with other pharmaceutical firms, research institutes, universities, and, to a lesser extent, other startup companies. We know that these networks have two important properties. First, Powell et al. (1996) showed that firm performance increases with the intensity of interfirm collaboration. Second, Shan et al. (1996) found that networks have a self-replicating property; cooperation reenforces the network by building upon the previous pattern of relationships. Thus, collaboration appears to result in the creation of useful knowledge in a durable web of relationships for the production of patentable knowledge.

In summary, biotechnology startups are characterized by a heavy reliance on scientific knowledge sourced from university and academic laboratories. Their own scientists, while competing to produce valuable technologies, also actively engage them in

the production of that knowledge. Thus, the biotechnology industry is a rich field for an analysis of the relationship between scientific capabilities of firms and valuable innovations.

Modeling Science as an Input to Discovery

Sample of Firms and Data Collection

The first step in the sampling procedure was to create a representative sample of U.S. biotechnology firms. To accomplish this, we made use of an existing database established by one of the authors (Gittelman 2000). This database includes some 14,000 biotechnology patent records, each corresponding to a single invention filed by U.S. organizations during the period 1982–1997. The source of the data is *Derwent Biotechnology Abstracts*, a comprehensive database of biotechnology patents. Patents are restricted to those as classified by Derwent as relating to genetic engineering and/or biopharmaceuticals. Patents relating to plant and agricultural uses, and other industrial applications outside of human health care, are excluded. From the patent data, a number of sources were used to identify which of the assignees were U.S. biotechnology firms. All identifiable biotechnology firms were included, subject to the criterion that the company was based in the U.S., and was granted at least one U.S. patent during 1988–1995. Biotechnology firms that are subsidiaries of other firms, but have maintained an independent identity are included, e.g., Genentech, once partially owned by Roche. Biotechnology divisions of pharmaceutical firms are not included. The primary sources used to identify biotechnology firms include BioScan, a proprietary directory of the biotechnology industry, Ernst & Young annual biotechnology reports, member directories of the Biotechnology Industry Organization, and company sources.

These criteria yielded a sample of 116 U.S. biotechnology firms. We collected four kinds of data for the firms in our sample: (1) data on publications in the scientific literature during the period 1988–1994 (these data also reveal research collaborations with external institutions), (2) patents issued to the firms in the U.S. during 1992–1995, (3) individual scientist data, and (4) data on firm-level characteristics.

370

Publications and Collaborative Research Among Sample Firms

The sample firms produce many more publications than they do patents. Our sample of firms published nearly 7,000 articles in the scientific literature (about 1,000 per year) during the sample period (1988–1994), with 30% of articles published in just 10 journals (including the prestigious publications *Nature, Science,* and *Cell*). The total number of articles published by the firms has been rising at about 10% per year, from 711 articles in 1988 to 1,258 articles in 1994.[1]

The firms were granted some 1,200 U.S. patents, a rate of about 300 per year, during 1992–1995. This is a rough indication that a significant portion of publications did not lead to a patent. The average firm in our sample published 60 articles, was granted 10 patents, was founded in 1984, and employed 262 persons. There is a great deal of heterogeneity among these firms. One firm (Genentech) accounted for 1,400 publications, while some firms had no publications. Size differences are also great, from 3,000 employees (Amgen) to 5 employees (Symbollon and Immunologic Pharmaceuticals).

Collaboration with external organizations is a defining feature of the research activities of the sample firms. We measure research collaboration as represented by articles in which both the firm and an outside organization are listed as institutional affiliations of one or more of the authors. The share of collaborative publications is about 70% of total articles published, and this has remained steady during the sample period. This high proportion is not skewed by firm size.

The great majority of shared research is between a biotechnology firm on the one hand and a university on the other; firm-to-firm collaborations are a small portion of the total. In all, some 1,800 organizations are listed as collaborating institutions with the sample firms. An analysis of the top 200 of these research partners shows that only 15 were other firms; the rest were universities, research institutes, and government labs, U.S. and foreign. Extrapolating from this, it is estimated that 90% of the research partners were universities or other research institutions (government labs, hospitals, or research institutes). The data indicate that copublications allow the firms in our sample to tap into high-quality networks of academic scientists, with prestigious universities and research institutes in the life sciences dominating the population of collaborators.[2] Given the sensitivity to prestige and grounding in scientific practice, it is not surprising that past studies found that many collaborations are not formalized in legal contracts (Liebeskind et al. 1996).

Dependent Variable: Forward Patent Citations

We are interested in exploring whether scientific research impacts the value of a firm's innovations, as captured by its patents. Our dependent variable is the cumulative forward citation frequencies to an individual patent.[3] Forward citations count the number of times a patent (the "cited patent") is included in the prior art of subsequent patents. The evidence strongly supports the conclusion that patent citations contain information about a patent's technological importance, and that they can also be used as a proxy for economic value to the innovator (see Hall et al. 2000 for a review). In biotechnology, where patents are a key means of appropriating returns to innovation, citation rates are more likely than in other fields to contain information about the technological and economic value of a given invention. We count all forward citations received by each patent at of the end of 1999. We call this measure CITES TO PATENT.

---

[1] We use the *ISI Science Citation Index* to collect information on all publications in which the firm is listed as an institutional author during the period 1988–1994.

[2] The top institutional collaborators are (with percent of all collaborative articles): Harvard University (6%), University of California–San Francisco (5%), University of Washington (5%), National Cancer Institute (5%), Stanford University (4%), University of Texas (4%), University of California–Los Angeles (3%), Scripps Clinic and Research Institute (2%), Johns Hopkins University (2%), and University of California–San Diego (1%). In total, the University of California (UC) system accounts for 13% of the collaborations, reflecting the pronounced role of the UC system in fostering a California biotechnology industry and the linkages between UC researchers and scientists at those firms.

[3] Our data give information on the full patent family, comprising the full portfolio of patents issued around the world on a given invention. We utilize the first U.S. patent in the family issued during 1992–1995, inclusive.

Citations may accrue to a patent for reasons that do not reflect its importance but rather its vintage: older patents are likely to be cited more than, but are not necessarily more important than, younger patents. Technological field effects may also influence citation rates: patents in crowded fields may be cited more than patents in sparse fields because there are more citing patents. On the other hand, patents in sparse fields may have higher odds of being cited by subsequent patents because there are fewer cited patents. In these cases, the reasons for citations are likely unrelated to the importance of the patent that we are aiming to measure. We are, therefore, careful to include variables in our regressors that control for patent age and technological field, as well as other characteristics of the firm and cited patent that could affect the frequency with which it is cited.

Our models seek to capture knowledge capabilities at the level of the firm as a whole; to do this, we aggregate various data from all of the firm's patents and publications. For firms with only one patent, the data likely give a poor measure of firmwide capabilities. We, therefore, leave out firms with only one patent in our sample when we estimate our models. This eliminates 15 patents (corresponding to 15 firms). As we are not modeling the *level* of innovation effort, but rather citations to patents, our models should be interpreted as estimating the relative success of the innovative effort, conditional on the firm having a capability to innovate. Additionally, to help ensure that the research effort covered by our patents does not precede the research effort represented by publications, we eliminate 96 patents that were filed prior to 1987, corresponding to an expected 1988 journal publication date, the first year of our publication data.

Independent Variables: The Bibliometrics of Publishing, Copublishing, and Patenting

First, we consider the effect of investing in science on the firm's patents. To measure this effect, we develop several variables from the bibliometric and patent data.

Firm Publish Dummy. This takes a value of 1 if the firm published at least one article up to the year in which the observed patent was filed.

Publication Volume. The total number of firm publications, cumulated up to the year in which the observed patent was filed. This gives an indication of the volume of publishing. The variable is specified in log form to take account of the highly skewed distribution of publications.

Percent of Copublications. Percentage of all publications by the firm that were collaborative publications with an external organization, cumulated up to the year the observed patent was filed.

Science Intensity of Firm's Patents. We measure the closeness of the firm's technologies to knowledge produced in open science. A patent is required to list the prior art that it builds upon: this includes both other patented inventions and publications in the scientific literature that are not patented. Firms that seek to integrate scientific findings into their inventions are more likely to cite scientific findings in their patents, hence have more nonpatent references in their prior art. Science intensity is a "backwards" citation count, measured as the number of times a patent references nonpatented literature in its prior art. Deng et al. (1999) find that science intensity is positively associated with subsequent financial performance for a group of technology- and science-based companies. There is great variation in the degree to which patents in our sample build upon science: the mean is around 40 citations to published works, with up to as many as 1,675 such citations. For the firm, we calculate the mean number of citations to nonpatented literature across all of its patents.

Firm Average Cites to Publications. This variable captures citations to a firm's publications and is our primary measure of firm-level scientific research capabilities. Raw citation counts to each article are normalized by the mean and standard deviation of citations received by all sampled articles in its publication year. Normalizing the raw citations by year allows citations to be summed across years for each firm; the aggregate citation counts are then divided by the number of the firm's publications, to yield an average citation measure for the firm as a whole. Averaging the citations this way removes bias toward large-volume publishers; we have separately estimated the effect of publication volume and here we want to isolate the effect of publication quality. In our models, normalized citations are aggregated up to the year the observed patent was filed. The measure, therefore, represents the relative quality of the firm's stock of scientific knowledge, cumulated from the start of the publication period up to the time of the observed innovation.

Control Variables

We also include a number of control variables, to account for heterogeneity among the firms, as well as to control for age and field effects.

Patent-Level Controls.

Age of Patent. Years elapsed since the patent was filed. This control is particularly important, because we expect citations to patents and to papers to increase with age.

Patent Family Size. This variable is an indicator of the value of the invention to the firm, as evidenced by the number of the patents the firm issued or renewed in different countries (Cockburn and Henderson 1998). We count the total number of patents in the patent family, including patents granted overseas, whose forward citations are captured in our dependent variable. It is costly to maintain multiple patents; this variable, therefore, acts as a fixed-effect control for each invention, allowing for random luck in the innovation process and firm efforts to promote their innovations in multiple markets.

Patent Number of Inventors. We hypothesize that the research effort is associated with the number of people assigned to that effort, and that this is reflected in the number of people listed on the patent. We therefore include this measure as a proxy for the resources invested in the research project that resulted in the observed patent.

Technology Class of Patent (Patent Drug, Patent Test). We wish to control for technology segments that may be inherently more cited than others. We expect that patents that are in drug-related categories may be less cited than technique-based patents, as the former may represent a stopping point in further innovation once a patent has been issued, whereas techniques may spawn a host of incremental innovations. As these patterns would not necessarily reflect the underlying importance of the innovations (and, indeed, may

mask importance in the case of a drug patent), we need to create controls for them, as the classification system does not automatically distinguish patents in this way. Using international patent classification codes, we create two main categories of technologies: Patent drug indicates whether the patent is classified in A61K, medicines and pharmaceuticals; patent test indicates whether the classification is C12Q or G01N, which cover measuring, testing, and immunoassays using genetic materials. These two categories account for 345 patents.[4]

Firm-Level Controls.

Firm Age. Number of years since the firm was founded. Older firms have had more time to accumulate a knowledge base that can be applied across a range of innovations, however, they may represent knowledge of an older vintage than younger firms.

Firm Pharmaceutical Strategy. This is a broad measure of the technological orientation of the firm, to identify those firms that are seeking to develop biopharmaceuticals against firms that are primarily specialized in research tools, tests, instruments, and information-based products and services. From the BioScan (1994) data, we coded for a dummy variable that takes a value of 1 if the firm is involved in research oriented toward discovering new human biotherapeutics. Sixty-seven firms are coded as belonging to this category. We expect that on average investments in science will have a greater payoff for firms engaged in drug discovery.

Table 1 gives summary data for the dependent and independent variables, and Table 2 gives the bivariate correlations. None are high enough to suspect multicollinearity, further confirmed by regression results and additional tests (discussed further below).

Model Specification

Because the data are counts of citation frequencies, we employ a count model that makes use of the information contained in the numerous observations that

---

[4] In models not shown here, we control for all technological subfields by adding dummies for the main (first) patent class listed on each patent, but these controls were not significant and did not add to the power of the model.

Table 1   Summary Statistics

	Mean	Median	Maximum	Minimum	Std. dev.
Cites to patent	12.3	7	463	0	21.5
Patent age	8.88	9	13	5	2
Patent family size	6.14	6	39	1	5.32
Patent number inventors	3.05	3	20	1	2.12
Firm age	16.9	17	25	7	3.98
Firm pharma strategy	0.69	1	1	0	0.46
Patent drug	0.28	0	1	0	0.45
Patent test	0.08	0	1	0	0.28
Firm publication dummy	0.93	1	1	0	0.26
Firm publication volume	139	31	1,395	0	260
Firm % copublication	0.66	0.7	1	0	0.24
Firm % joint patent-publishers	0.6	0.63	1	0	0.21
Firm average cites to publications	−0.29	−0.19	5.63	−3.36	1.12
Firm science intensity	39.9	28.8	373	0	32.7
Patent science intensity	37.9	17	1,675	0	82.8
Patent % joint patent-publishers	0.69	0.86	1	0	0.37

are never cited. Count data are frequently estimated with one parameter Poisson models. Poisson models are nested within the negative binomial model, a two-parameter model that estimates an overdispersion parameter and produces correct standard errors for count data that is overdispersed (Cameron and Trivedi 1998). Because patent citations exhibit a great deal of overdispersion, we estimate negative binomial models. We test for robustness by estimating cluster regression models.

Models of Publishing Effects on Patent Citations
Models 1–4 (see Table 3) include the control variables and add different science investment variables. Model 1 includes only the control variables. As expected, older patents and patents from large families receive more citations. The proxy for cost of the project, number of inventors on the patent, is also positive ($p < 0.01$). There is a small negative effect of firm age but it is not significant. Firms coded as working on drug development do receive higher cita-

Table 2   Bivariate Correlations

Variable		1	2	3	4	5	6	7	8	9	10	11	12	13	14	15
1	Cites to patent	1														
2	Patent age	0.36	1													
3	Patent family size	0.33	0.47	1												
4	Patent number inventors	0.24	0.12	0.26	1											
5	Firm age	0.08	0.17	0.05	0.05	1										
6	Firm pharma strategy	0.12	0.04	0.11	0.11	0.24	1									
7	Patent drug	0.06	0.12	0.22	0.08	0.11	0.25	1								
8	Patent test	0.00	−0.05	−0.07	−0.03	−0.11	−0.22	−0.20	1							
9	Firm publication volume	−0.04	−0.22	−0.06	0.01	0.55	0.28	0.08	−0.04	1						
10	Firm % copublications	0.09	0.05	0.06	0.09	0.15	0.24	0.06	0.09	0.07	1					
11	Firm % joint patent-publishers	0.08	0.05	−0.01	−0.05	0.22	0.26	0.09	−0.16	0.32	−0.09	1				
12	Firm average cites to publications	0.07	0.20	0.11	0.05	0.29	0.33	0.09	−0.07	0.33	0.25	0.14	1			
13	Firm science intensity	0.09	−0.01	0.09	0.10	0.40	0.32	0.12	−0.08	0.47	0.03	0.24	0.19	1		
14	Patent science intensity	0.07	−0.08	0.16	0.14	0.14	0.11	0.11	−0.04	0.18	0.02	0.06	0.07	0.27	1	
15	Patent % joint patent-publishers	0.09	0.07	−0.03	−0.07	0.11	0.06	0.05	−0.09	0.14	−0.10	0.47	0.03	0.07	0.09	1

GITTELMAN AND KOGUT
Does Good Science Lead to Valuable Knowledge?

Table 3   Negative Binomial Models of the Effect of the Quality of a Firm's Science Capability on the Production of Highly Cited Patents

Model	1. Control variables only		2. Firm publication dummy		3. Collaboration and science measures for firms that publish		4. Citation to publications without other firm-level science indicators	
Sample	All firms with > 1 patent		All firms with > 1 patent		All firms with > 1 patent and > 1 publication		All firms with > 1 patent and > 1 publication	
Variable	β	S.E.	β	S.E.	β	S.E.	β	S.E.
Constant	−2.64***	0.34	−2.80***	0.37	−3.67***	0.42	−3.35***	0.40
Log (patent age)	2.02***	0.16	2.05***	0.16	2.38***	0.19	2.19***	0.19
Patent family size	0.05***	0.01	0.05***	0.01	0.03***	0.01	0.04***	0.01
Patent number of inventors	0.06***	0.01	0.06***	0.01	0.08***	0.02	0.08***	0.02
Firm age	−0.00	0.01	−0.00	0.01	−0.02*	0.01	0.00	0.01
Firm pharma strategy	0.21***	0.07	0.21***	0.07	0.23***	0.09	0.33***	0.08
Patent drug	−0.05	0.07	−0.05	0.07	−0.06	0.07	−0.06	0.07
Patent test	0.23**	0.11	0.23**	0.11	0.19	0.12	0.24**	0.12
Firm publication dummy			0.13	0.12				
Log (publication volume)[a, b]					0.05**	0.03		
Firm % copublications[b]					0.23	0.15		
Firm science intensity					0.00***	0.00		
Firm average cites to publications[b]					−0.14***	0.04	−0.11***	0.03
Overdispersion parameter	−0.14***	0.05	−0.14***	0.05	−0.15***	0.05	−0.13***	0.05
N	1,120		1,120		934		934	
Log likelihood	−3,748		−3,748		−3,044		−3,052	

*$p < 0.10$, **$p < 0.05$, ***$p < 0.01$.
[a] A value of 1 is added to each observation.
[b] Cumulated to the year of patent filing.

tions to their patents. On the other hand, drug patents do not receive significantly more citations, although patents in test-related categories patents do. These are not necessarily contradictory results: firms with a pharmaceutical strategy also patent in test-related categories.

Model 2 adds a dummy variable to show whether the firm had any publications up to the file date of the observed patent. It is positive but not significant. This surprising result may indicate that science has no significant impact on innovation outcomes; or it may indicate that there are groups of firms within our sample, and that for some of the firms, science does not impact innovation. Finally, poor specification of the variable is a plausible explanation. Because most of the firms in our sample had at least one publication (only 7% of patents were issued by firms with no publications), this rather crude measure of science invest-

ments may not pick up real differences in research levels and capabilities.

The models that follow use measures taken from each firm's publications; in all subsequent models, we, therefore, only include patents of firms that have published at least one article up to the filing date of the cited patent. Model 3 includes publication volume, percent of collaborative publications, science intensity of the firm's patents, and the average citations to the firm's publications. We consider each in turn.

Publication volume raises the patent citation rate and is significant ($p < 0.05$). This result is counter to the findings of Cockburn and Henderson (1998) and Gambardella (1995) among pharmaceutical firms, who find (respectively) that the volume of publication does not appear important in predicting patent performance, and that only recent publications appear to matter in estimates of patent outputs. However, if we remove Genentech from the analysis shown in Model 3, the variable is no longer significant ($p = 0.14$), while the coefficients and standard errors of all the other variables remain unchanged. Because Genentech is such a large part of our sample—accounting for 10% of patents and 20% of publications—and places such strong emphasis on science-driven discovery, we ran all our models both with and without it. We do not report those separately; in all cases, coefficients and standard errors of the reported variables only slightly change significance levels are unchanged.

The degree to which the firm collaborates is positive but not significant in Model 3 (in subsequent models, it is just significant at the 10% level). Empirical evidence has shown the positive role of collaboration in innovation (Cockburn and Henderson 1998, Powell et al. 1996); the measure here may not be sufficiently fine grained to pick up these effects. Science intensity is positive and highly significant ($p < 0.01$). Patents that build on science are more likely to be cited, hence, more likely to generate further innovations and value. This result supports the notion that integrating scientific research with innovation has a positive impact on a firm's innovation performance, and adds to earlier findings that science intensity is associated with firm value (Deng et al. 1999).

The variable of central importance to our study is firm average cites to publications. It has a negative effect on the citation rate ($p < 0.01$). Highly cited papers are associated with less cited patents. The negative sign implies that the production of high-quality publications actually detracts from the innovation effort. This is a strong result, for it indicates that successful patents and successful papers follow different selection logics, and that these logics are opposing. To confirm this, we construct the scientific quality variable in a variety of different ways. We measure research quality as the firm's percentile ranking among all firms in the sample, ranked according to citations to their publications; by including the citations to the firms' single most highly cited publication across all years in the sample period; and by estimating separate models for citations to firm-only versus collaborative publications. The results are robust to all

specifications of the variable: negative and significant effect ($p < 0.01$) on patent citations.

It is important to note that the bivariate correlation between highly cited patents and highly cited papers is positive at 0.36, and a regression of important papers on important patents shows a significant and positive relationship. However, once age of the patent is included, the relationship turns negative and significant; a likelihood test shows the added variable significantly improves the fit. Theoretically, this is to be expected: older patents cite older papers and both have more forward citations. Thus, the bivariate correlation is spurious, reflecting age, and, hence, disappears once a control is added. As a further check, in Model 4, we regress average citations to publications without any of the other firm science measures. The coefficient is still negative and significant and the standard error is stable, indicating that the relationship is not due to collinearity with the other firm science measures.

Robustness: An Independent Test to Investigate Measurement Error

To test the validity of this important finding, we collect additional data and perform additional tests. We want to see whether patents that build upon highly cited scientific articles are more influential than patents that build on undistinguished scientific articles. If influential patents are associated with highly cited scientific articles, that would contradict our model result of publication citations having a negative impact on patent citations.

We construct two separate samples, ranked sample and firm sample, to measure important science on an absolute scale and on a firm-specific scale, respectively. Each sample includes two groups: highly cited articles and a control group. We first consider the ranked sample. The highly cited group in the ranked sample includes articles falling into the top 0.05 percentile of all articles published by the firms in 1990 (49 articles) or 1991 (57 articles), for a total of 106 highly cited articles. The average publication in this group was cited 470 times; the group includes major scientific findings published in prestigious journals such as Nature, Science, and Cell. It is an indicator of the excellent science carried out by the firms publishing

376

these articles that one-quarter of the articles (26 out of 106) were firm-only publications; the remainder were coauthored with scientists at universities or research institutes. To create the control group, each of these top-ranked articles is matched to 106 randomly sampled articles from our database published in 1990 (49 articles) or 1991 (57 articles) that are not in the highly cited group. The average publication in the control group was cited 42 times, with one-quarter (27/106) as firm-only publications.

The firm sample is constructed to measure science quality on a scale relative to a firm's total scientific outputs. We select the most highly cited article published by the firm in any year to create the highly cited group (total = 93 articles, 1 per firm).[5] The average publication in this group received 321 citations, and 24 were firm-only articles. For the control group, we match the firm's most highly cited article to the median article published by the firm in the same year (total = 93 articles; average citations = 30, 26 firm-only articles). We run separate tests for ranked sample and firm sample.

Table 4 shows that for both the ranked sample and the firm sample, highly cited articles generate many more patents than the control group. This is true for total patents as well as patents by the firm that authored the article (self-patents). Important science attracts innovation efforts, both by firms that generated the findings and by other organizations.

These data establish that firms cluster their innovate efforts around important scientific findings. Interestingly, this is true whether science importance is measured in an absolute sense (ranked sample) or relative to a firm's own scientific research outputs (firm sample).

We next consider whether these patent clusters around important science are more valuable as innovations. We test whether patents that build on highly cited scientific articles receive more citations, through the year 2000, than patents that build on the control groups. We regress the citations received by each patent against a dummy variable, indicating whether

the patent references a highly cited article in its prior art (variable takes a value of 1) or a control group article (variable takes a value of 0). In the ranked sample, only 3 patents cite both a highly cited article and a control group article, indicating that these are different clusters of innovations. In the firm sample, 12 patents cite both a highly cited and a control group article. We remove these patents from our estimations. A positive and significant coefficient for the dummy variable would indicate that patents that cite important articles are more highly cited themselves: valuable science leads to valuable innovations. This finding would contradict our earlier model results.

Table 5 reports the results of these regressions. Both are negative binomial estimations. We include controls for the age of the cited patent, the total number of articles cited in the patent's prior art, and whether or not the patent was a self-patent (assigned to the same firm as authored the article). The results provide strong corroboration for our earlier result. In the ranked sample, highly cited articles are negatively associated with patent citations ($p < 0.01$). In the firm sample, the sign is positive but not significant ($p < 0.68$). The findings are strongly suggestive that highly cited patents do not build upon valuable

Table 4 Number of Patents Citing Articles: Highly Cited Articles Versus Control Groups (Self-Patents Shown in Parentheses)

Group	Ranked sample	Firm sample
Highly cited articles	232 (46)	194 (29)
Control group articles	44 (6)	58 (13)

Table 5 Negative Binomial Models of Citations to Patents Citing Sampled Articles

Variable	Ranked sample Coeff.	S.E.	Firm sample Coeff.	S.E.
Intercept	−2.88***	0.49	−3.17***	0.51
Log (patent age)	2.29***	0.23	1.98***	0.19
Science intensity of patent	0.30***	0.09	0.30***	0.09
Self-patent	0.27	0.27	0.49*	0.28
Cites a highly cited article = 1	−0.81***	0.33	0.11	0.26
Overdispersion parameter	0.67***	0.13	0.38***	0.14
Log likelihood	−452		−444	
N	229		212	

*$p < 0.10$, **$p < 0.05$, ***$p < 0.01$.

[5] This sample is smaller than our full sample of firms, because it is limited to firms that published at least two articles during a given year of our sample period.

science, and, indeed, important science leads to innovations that poorly perform.

The tests of patenting and patent citations indicate that firms cluster their innovation activities around important scientific discoveries, both their own and those made by other firms. However, innovation clusters around highly cited scientific discoveries do not result in highly cited patents, indeed, patent clusters around important science produce less cited patents. This provides support for our earlier finding of conflicting logics between the selection of valuable scientific ideas and valuable technologies.

Bridging Scientists: Joint Patent-Publishers

Why, then, do firms invest in carrying out excellent science, and why do they associate with prestigious scientists? One possibility is that some firms are "captured" by star scientists whose reputations bring prestige and tangible resources to firms. The most skilled practitioners of their art are those who are most wedded to its selection logic yet our results suggest that the process of selecting among ideas comes into conflict with the selection logic of patents. The firm's ability to integrate and mediate these conflicting logics becomes important: as shown by Cockburn et al. (1999), the balancing of incentives based on science with rewards that are market oriented becomes a fulcrum for explaining firm heterogeneity in innovation performance.

We model firm heterogeneity in innovation as the degree to which firms succeed in integrating the two worlds of science and invention at the level of the individual scientist. We identify two overlapping sets of scientists. The first group, called publishers, includes those scientists who are listed on at least one publication in our sample. In total, we identified 19,638 different names of publishers; while our data do not allow us to identify where these individuals actually work, we suspect that a significant proportion are employed by outside institutions, mainly universities. The second group, called inventors, includes scientists involved in developing new technologies, as revealed by their being listed on a patent. On the patents, we identified 2,035 names. This yields a

ratio between publishers to inventors of about 10 to 1. Our data indicate that these two groups of scientists are largely distinct, though they overlap. Among inventors, the overlap with the publishers is relatively large: 57% of inventors are also publishers (1,170 out of 2,035 inventors). We identify these individuals as forming a group of joint patent-publishers. However, only 6% of publishers are also inventors (1,170 out of 19,638). This indicates that the firms are intensely leveraging the contributions of scientists working in academic institutions, because few of this large group ever appear on a patent for the firm.

We expect that this measure of overlap between scientists who publish and patent at the firm level is an important indicator of the degree to which a firm is able to successfully translate research into invention. To capture this heterogeneity, we construct a variable called percent of joint patent-publishers. We calculate this variable first at the patent level, as the percentage of all individuals listed on a patent who are also listed on at least one publication. We then calculate this variable for the firm as a whole, by aggregating all scientists listed on the firm's patents during the sample period. These give measures of the degree to which scientists who patent are also active (or have been active) in scientific research, at the level of the individual project (patent) and at the aggregate firm level.

Model 5 (See Table 6) adds the effect of joint patent-publishers at the firm level to the earlier models of patent citations. The variable is positive and significant ($p < 0.05$) and including it does not affect our previous findings about science intensity and science quality. Model 6 replaces the measure at the firm level to show the percent of joint patent-publishers at the level of the cited patent itself. We also include the science intensity of the cited patent rather than for the firm as a whole. These measures can pick up effects at the more detailed project level rather than the aggregated data for the firm. Both variables are positive and significant ($p < 0.01$). Finally, to show robustness, in Model 7, we re-estimate Model 6 by using a clustering regression that controls for random firm effects. A fixed-effect model, with firm dummies included but not shown, is reported in Model 8; several firms are

378

GITTELMAN AND KOGUT
Does Good Science Lead to Valuable Knowledge?

Table 6   Negative Binomial Models of Effects of Bridge Scientists on the Production of Highly Cited Patents

Variable	5. Effect of firm's joint patent-publishers		6. Effect of patent's joint patent-publishers		7. Model (6) with standard errors adjusted for clustering of firm		8. Model (6) with firm fixed effects (firm dummies not shown)		9. Model 7: elasticities	
	All firms with > 1 patent and > 1 publication		Same as (6)		Same as (6)		Same as (6)		Same as (6)	
	β	S.E.	β	S.E.	β	Robust S.E.	β	S.E.	ey/ex	S.E.
Constant	−3.82***	0.43	−4.0***	0.44	−3.98***	0.61				
Log (patent age)	2.33***	0.20	2.38***	0.20	2.34***	0.12	3.11***	0.34	5.05***	0.63
Patent family size	0.03***	0.01	0.03***	0.01	0.03***	0.01	0.04***	0.01	0.18***	0.07
Patent number of inventors	0.08***	0.02	0.08***	0.02	0.08***	0.02	0.10***	0.12	0.23***	0.07
Firm age	−0.02	0.01	−0.01	0.01	−0.01	0.02	0.33	0.44	−0.21	0.27
Firm pharma strategy	0.20**	0.09	0.25***	0.09	0.25	0.14	−0.52	0.85	0.18*	0.10
Patent Drug	−0.08	0.08	−0.08	0.08	−0.07	0.09	−0.09	0.08	−0.02	0.02
Patent Test	0.22*	0.12	0.21*	0.12	0.21	0.16	0.31***	0.12	0.02	0.01
Log (publication volume)^{a,b}	0.04	0.03	0.04	0.03	0.04	0.22	0.18*	0.10	0.15	0.13
Firm % copublications^{b}	0.29*	0.15	0.26*	0.15	0.26	0.22	0.40	0.36	0.17	0.15
Firm science intensity	0.00***	0.00								
Firm average cites to publications^{b}	−0.14***	0.04	−0.13***	0.04	−0.013***	0.05	−0.15*	0.08	−0.05***	0.02
Firm % joint patent-publishers	0.43**	0.21								
Patent science intensity			0.07***	0.03	0.07*	0.04	0.05**	0.03	0.19*	0.12
Patent % joint patent-publishers			0.25***	0.10	0.25**	0.11	0.21**	0.10	0.18**	0.08
Overdispersion parameter	−0.15***	0.05	−0.16***	0.05	−0.16	0.07	−0.46***	0.06		
N	934		934		934		934		934	
Log likelihood	−3,042		−3,040		−3,040		−2,916			

*p < 0.10, **p < 0.05, ***p < 0.01.
^{a}A value of 1 is added to each observation to take the log.
^{b}Cumulated to the year of patent filing.

dropped due to collinearity. Both indicate robustness of the main results to within-firm variance.

In the final column, we report the elasticities of Model 7. The coefficient to the highly cited publication variable implies an elasticity of −0.05, with a 95% confidence interval ranging from 0.01 to −0.35. These effects are not high, until it is recalled that the citations to the more successful papers (see Discussion below) average around 500. The important effect of patent-publishers, shown here at the project level (with an elasticity of 0.18 for Model 7) supports the interpretation that integrating research and innovation at the level of the individual scientist is more important to the innovation effort than firm-level scientific capabilities as measured by the volume and quality of scientific publications. As a note on the side,

by far the most influential effect on patent citations is age, with an elasticity near five.

Discussion

In a knowledge-based industry, it is reasonable to expect that firms with access to superior knowledge resources or skills should outperform those with weaker resources or skills. Our models do not provide strong support for this hypothesis. Highly cited patents are associated with science intensity and firm effects, but not with cutting-edge science; they are associated with scientists who publish, but only weakly associated with publication volume. Indeed, we find that the ability to produce excellent science has a strong *negative* impact on the patent citation rate. Taken together, the models indicate that investing in scientific research produces mixed results, and the relationship between research and innovation is more complex than a simple human capital story would predict.

Instead of a smooth internal transfer between firm scientific capabilities and innovation, the results indicate the very different processes involved in acquiring scientific knowledge and generating high-impact innovations. The negative relationship between scientific capabilities and the innovation effort points to a problematic disconnect between the scientific knowledge of the firm and its ability to generate high-impact innovations. Scientific ideas are not simple inputs into inventions; important scientific ideas and influential patents follow different and apparently conflicting evolutionary logics. This raises the question: Why do firms invest in scientific research when those investments do not seem to pay off in terms of more highly cited patents?

Two factors emerge as important in predicting patent citations. High-impact innovations heavily build upon the scientific literature and are made by people who both invent and do research. These factors are not independent of one another. Joint patent-publishers may perform the important function of identifying and applying the scientific research that the firm would most profit from in its projects. This function includes identifying as well as accessing external researchers in the field who are likely to bring new or complementary knowledge to the firm. Put another way, bridging the disconnect between scientific knowledge and innovation appears to depend on access to individuals who perform both activities, rather than on the ability to generate valuable scientific knowledge alone. Papers and patents do not follow the same selection logics and, yet, scientists produce both. Firms recruit scientists who can successfully bridge these logics and provide incentives that support their dual activities. In this regard, our findings imply the firm-level properties that Cockburn et al. (2000) found important.

Conclusions

Scientific knowledge and patents are related, but good publications and good patents are not. This can be easily explained by recalling that the two artifacts are not chosen by the same evolutionary logic of selection. In other words, patent citations are filtered by the conjoint influence of technical richness and market impact. These are very different evolutionary criteria than those faced in the world of publications. As long as these heavily cited patents defer to the papers that influenced them, the process will generate a very different selection citation pattern for influential patents than for patents overall.

This filtering of the technologically valuable patents by the selection dynamics among patent citations means that there is a technological and market component to patenting. Namely, because certain patents open richer technological veins, the subsequent advances in related technical knowledge encourage more innovative efforts in that area and, hence, more patents. These, in turn, cite the initial patents that opened this avenue of technological innovation. It is this feedback that carves a trace in the patent patterns. Patent citation patterns do not acknowledge what Merton (1973) called Matthew effects in science of prestige, attracting citations and resources; they reflect perceived technical and market opportunities.

This conclusion has a simple implication for understanding what firms do in biotechnology. On the most basic level, a firm that has excellent capabilities to do scientific research may not succeed well in producing marketable innovations, as indicated by Stern's (1999) analysis of scientists' wages. However, having a reputation for performing "good" science may be necessary to attract the kinds of people the firm needs to innovate. Firm heterogeneity in innovation performance centers on the ability to translate knowledge produced within the epistemic community of science into knowledge that a market will value. Scientists who simultaneously publish and invent are instrumental in bridging the disconnect between scientific knowledge and important technologies. Heterogeneity in innovation performance comes from firms' abilities to access and create the capability to do science, while bypassing the evolutionary logic that selects among its outputs. This role points to potential differences in the capabilities of firms to recruit and manage intellectual capital, as found in the studies by

Cockburn and Henderson (1998) and Cockburn et al. (2000).

Our study concerns an industry with particularly strong linkages between technological innovation and scientific knowledge (Cohen et al. 2002). Although the reliance on public science sets this industry apart, in a wider perspective, these findings are not unique to science- or technology-based industries. They point to the broad claim that knowledge of firms is created within and shaped by occupational and epistemic communities. Individuals embody knowledge that is useful when moving within the firm (Argote et al. 1995) or between firms (Almeida and Kogut 1999, Gittelman 2000). They are also anchored in identities and in what van Maanen and Barley (1984) call "occupational communities" that span across firm boundaries. These communities influence as well the organizing principles that guide the internal structure and the coordination among people and divisions inside the firm. While a resource to the firm, occupational communities pose potential conflicts in directing the exploration and efforts of their members. These results point to the important influence of membership in communities broader than a firm's boundaries that both abet and hinder the search for commercially valuable technological innovations.

Acknowledgments

The authors thank Andreas Noetzel and Mervyn Tan for exceptional research assistance. This paper benefited from comments received from participants in New York University's Management Seminar, and the Harvard Strategy Seminar, as well as from Pierre Azoulay, Lee Fleming, Diana Hicks, Maryellen Kelley, Steve MacMillan, Richard Nelson, Brian Silverman, Olav Sorensen, Brian Uzzi, Andrew Wang. The authors also thank the editors of this special issue, the referees, and associate editor. This research was financed by the National Institute of Standards and Technology and the Reginald H. Jones Center, The Wharton School, University of Pennsylvania.

References

Allen, T. 1977. *Managing the Flow of Technology.* MIT Press, Boston, MA.

Almeida, P., B. Kogut. 1999. Localization of knowledge and the mobility of engineers in regional networks. *Management Sci.* **45**(7) 905–918.

Argote, L., S. L. Beckman, D. Epple 1995. The persistence and transfer of learning in industrial settings. *Management Sci.* **41**(11)

Audretsch, D., P. Stephan. 1996. Company-scientist locational links: The case of biotechnology. *Amer. Econom. Rev.* **86**(3) 641–652.

BioScan. 1994. *Knowledge Access International*, Vol. 63. Mountain View, CA.

Callon, M. 1986. The sociology of an actor-network: The case of the electric vehicle. M. Callon, J. Law, A. Rip, eds. *Mapping the Dynamics of Science and Technology.* Macmillan, London, U.K.

Cameron, A. C., P. Trivedi. 1998. *Regression Analysis of Count Data.* Cambridge University Press, Cambridge, U.K.

Campbell, D. 1974. Evolutionary epistemology. P. A. Schilpp, ed. *The Philosophy of Karl Popper*, Vol. 14-I. Open Court, LaSalle, IL.

Cockburn, I., R. Henderson. 1998. Absorptive capacity, coauthoring behavior, and the organization of research in drug discovery. *J. Indust. Econom.* **46** 157–182.

——, ——, S. Stern. 1999. Balancing incentives: The tension between basic and applied research. Working paper 6882, National Bureau of Economic Research, Cambridge, MA.

——, ——, ——. 2000. Untangling the origins of competitive advantage. *Strategic Management J.* **21** 1123–1145.

Cohen, W., D. Levinthal. 1990. Absorptive capacity: A new perspective on learning and innovation. *Admin. Sci. Quart.* **35**(1) 128–152.

——, R. Nelson, J. Walsh. 2002. Links and impacts: The influence of public research on industrial R&D. *Management Sci.* **28**(1) 1–23.

Crane, D. 1972. *Invisible Colleges: Diffusion of Knowledge in Scientific Communities.* University of Chicago Press, Chicago, IL.

Dasgupta, P., P. David. 1994. Toward a new economics of science. *Res. Policy* **23** 487–521.

David, P. 1998. Common agency contracting and the emergence of "open science" institutions. *AEA Papers Proc.* **88**(2) 15–21.

Deng, Z., B. Lev, F. Narin. 1999. Science and technology as predictor of stock performance. *Financial Analysts J.* **53**(3) 20–32.

De Solla Price, D. 1970. Citation measures of hard science, soft science, technology, and nonscience. C. Nelson, D. Pollock, eds. *Communication Among Scientists and Engineers.* Heath Lexington Books, Cambridge, MA.

Gambardella, A. 1995. *Science and Innovation: The US Pharmaceutical Industry in the 1980s.* Cambridge University Press, Cambridge, U.K.

Gittelman, M. 2000. Mapping national knowledge networks: Scientists, firms and institutions in biotechnology in the United States and France. Unpublished Ph.D. thesis, The Wharton School, University of Pennsylvania, Philadelphia, PA.

Hall, B., A. Jaffe, M. Trajtenberg. 2000. Market value and patent citations: A first look. Working paper 7741, National Bureau of Economic Research, Cambridge, MA.

Henderson, R., I. Cockburn. 1994. Measuring competence? Exploring firm effects in pharmaceutical research. *Strategic Management J.* **15**(Winter Special Issue) 63–84.

Hicks, D. 1995. Published papers, tacit competencies and corporate management of the public/private character of knowledge. *Indust. Corporate Change* **4**(2) 401–424.

Jaffe, A., M. Trajtenberg, R. Henderson. 1993. Geographic location of knowledge spillovers as evidenced by patent citations. *Quart. J. Econom.* **108**(3) 577–598.

Kim, D., B. Kogut. 1996. Technological platforms and diversification. *Organ. Sci.* **7** 293–301.

Knorr Cetina, K. 1999. *Epistemic Cultures: How the Sciences Make Knowledge.* Harvard University Press, Cambridge, MA.

Latour, B., S. Woolgar. 1979. *Laboratory Life: The Construction of Scientific Facts.* Princeton University Press, Princeton, NJ.

Lieberman, M. 1978. A Literature citatiuon study of science-technology coupling in electronics. *Proc. Inst. Electr. Electronics Engr.* **66**(01) 5–13.

Liebeskind, J., A. Oliver, L. Zucker, M. Brewer. 1996. Social networks, learning, and flexibility: Sourcing scientific knowledge in new biotechnology firms. *Organ. Sci.* **7**(1) 428–443.

Merton, R. 1973. *The Sociology of Science.* Chicago University Press, Chicago, IL.

Mowery, D. C., N. Rosenberg. 1982. Technical change in the commercial aircraft industry, 1925–1975. N. Rosenberg, ed. *Inside the Black Box: Technology and Economics.* Cambridge University Press, Cambridge, U.K.

Powell, W. K. Koput, L. Smith-Doerr. 1996. Interorganizational collaboration and the locus of innovation: Networks of learning in biotechnology. *Admin. Sci. Quart.* **41** 116–145.

Rosenberg, N. 1990. Why do firms do basic research (with their own money)?" *Res. Policy* **19**(2) 165–174.

Shan, W., G. Walker, B. Kogut. 1994. Interfirm cooperation and startup innovation in the biotechnology industry. Res. notes and comm., *Strategic Management J.* **15** 387–394.

Stephan, P. 1996. The economics of science. *J. Econom. Literature* **34**(3) 1199–1235.

Stern, S. 1999. Do scientists pay to be scientists? Working paper 7410, National Bureau of Economic Research, Cambridge, MA.

Stuart, T. J. Podolny. 1996. Networks, knowledge, and niches: Competition in the worldwide semiconductor industry, 1984–1991. *Amer. J. Soc.* **102**(3) 659–689.

Tushman, M. 1977. Specialty boundary roles in the innovation process. *Admin. Sci. Quart.* **22** 587–605.

Van Maanen, J., S. R. Barley. 1984. Occupational communities: Culture and control in organizations. B. M. Staw, L. L. Cummings, eds. *Research in Organizational Behavior*, Vol. 6. JAI Press, Greenwich, CT, 287–365.

Zucker, L., M. Darby, M. Brewer. 1998. Intellectual human capital and the birth of U.S. biotechnology enterprises. *Amer. Econom. Rev.* **88**(1) 290–306.

——, ——. J. Armstrong. 2002. Commercializing knowledge: University science, knowledge capture, and firm performance in biotechnology. *Management Sci.* **48**(1) 138–153.

Accepted by Linda Argote, William McEvily, and Ray Reagans; received March 1, 2001. This paper was with the authors 8 months for 2 revisions.

[9]

Introduction

Arguments about the importance of 'market pull' versus 'technology push' are in this sense artificial, since each market need entering the innovation cycle leads in time to a new design, and every successful new design, in time, leads to new market conditions.[1]

Overview

This book arises from a conviction that technological change, or innovations, are crucial processes underlying dynamic change in advanced market economies. They lead to changes in productivity and to economic growth for firms and for national economies. For example, the development and diffusion of agricultural machinery has replaced many manual jobs in agriculture but has instead created jobs in the design and production of those machines and in other newly created sectors. Seen in a longer perspective, economic growth and increased social welfare usually involve major upheavals, including changing patterns of productivity and job creation in different industrial and service sectors. Technological innovation processes are important because they introduce dynamics into economic growth and impact the wider society. It is therefore important to study such processes explicitly, as it is done in this book.

Innovations are novelties which add economic value and may be technological, market, and/or organizational in orientation. Innovations may take the form of a new machine, a new way of doing things, or they may involve redesigning an existing product for a niche market or changing how work is carried out in an organization. Technological innovations can thus add economic value as some combination of changes in goods and services, including production processes and organization. By adding novelty to the economy, innovations thereby change the conditions of competition for firms and often open up room for further technical improvements and profits. Firms in sectors with rapidly changing technologies and dynamic competition have to make continuous improvements and learn.

The resulting technical changes may lead to dramatic changes in the

---

[1] Kline and Rosenberg (1986: 289–90).

2   Introduction

economy, and eventually in individuals' daily lives, or to relatively insignificant changes. The car and surrounding road system has radically changed how modern societies are organized and the structure of production (what is produced), but there are also insignificant changes such as yearly improvements to existing car models. In other words, innovation processes can result in changes which are somewhere between radical and incremental. A few technologies like genetic engineering and information technology are considered more radical than others because they can affect many different aspects of the economy and society.

This book specifically analyses technological innovation processes, that is, how technical novelties of economic value are created. It examines early commercial uses of genetic engineering to see how and why individuals and organizations were able to innovate. The story is much more complex than firms simply imitating what scientists have done. Telling it helps us learn about how and why agents can develop technological innovations. How can they change their vision of a radical innovation into a form useful in the economic sphere? How do they deal with the challenges of developing new science and technology which are in a very fluid phase of change?

The focus here is on innovations which combine economic and technical dimensions in novelty. According to an old dispute, the economic dimension can be seen as 'market pull', which is the idea that innovators make technical improvements in response to perceived market demand. They identify an innovation which is likely to sell and then design a technology to fit what the market will demand. 'Technological push' is the idea that innovators think in technical terms and see obvious opportunities for technical improvements.[2] As the chapter epigraph indicates, this dispute about which is more important is *passé* in modern innovation theory and hence in this book. Economic and technical dimensions are intertwined in technological innovation processes. A technical improvement will change the structure and size of market demand, and the development of an idea based on market demand relies on a series of technical improvements. Economic and technical dimensions directly and irreversibly influence each other in innovations.

Thus, the emphasis here is not on innovations as objects but on the processes whereby innovations are conceived, improved, modified, and used in an economy. Generating such novelty fundamentally depends on knowledge-seeking activities by agents. Agents must conceptualize both economic and technical dimensions. They identify the opportunities and challenges necessary for succeeding with the innovation and try to direct their knowledge-seeking activities to meet them. For radical technologi-

---

[2] See Mowery and Rosenberg (1982) for a discussion of this debate of how market pull and technology push are both necessary for innovations.

Introduction   3

cal changes, many individuals and many organizations contribute. They respond to, and act in anticipation of, a changing environment. Their decisions about where to search for novelty are influenced by market forces, technical opportunities and technical dead ends, and social institutions, among others. Technological change thus depends on the actions of individual agents as well as social interactions among different agents in a socio-economic context.

This book describes parallel technological innovation processes enabling firms to use genetic engineering for the production of pharmaceuticals.[3] More specifically, historical case studies focus on knowledge-seeking activities by firms, universities, and others to develop relevant knowledge and techniques. The following chapters show how and why firms interacted with other agents and informal institutions to develop genetic engineering for the pharmaceutical human growth hormone (hGH). The innovation processes so described centre around two firms which moved into genetic engineering very early—namely the Swedish pharmaceutical firm Kabi and the American biotech firm Genentech. The book thus also deals with innovation processes in two different national contexts.

On the one hand, this book presents original historical material about the challenges and opportunities of commercial uses of genetic engineering in the 1970s and 1980s. This history has been sorely missing in scholarly and popular discussions, which have mostly focused on basic scientists or on venture capitalists. The specific challenges of identifying, modifying, and developing relevant knowledge and techniques within firms has not previously been dealt with in detail. These case studies are thus interesting in themselves but are also indicative of broader trends in the modern biotech industry. The challenges and opportunities facing these firms were often very specific, but they also represent more general categories of problems facing all agents trying to use genetic engineering this way.

So, on the other hand, this book draws upon theoretical approaches to analyse technological innovation processes. It has a multi-disciplinary approach, including perspectives focusing on technological change from a variety of scientific traditions. Each has something to offer. This book thereby opens up new paths and questions for later research and should contribute to ongoing theoretical discussions about technological change and innovations. The argument presented here is that this multi-disciplinary approach clearly illustrates how knowledge, institutions, and

---

[3] Although a distinction is made between the scientific techniques of genetic engineering and the biotech industry *per se*, the terms 'genetic engineering' and 'biotech' are here used synonymously, following common praxis in English. 'Genetic engineering' refers to a set of techniques involving controlled changes to DNA, while the term 'biotechnology' refers to a broad category of biological processes, including beer- and bread-making. Despite this broad definition, the terms 'biotech industry' and 'modern biotechnology' refer to biological processes which are based on the scientific techniques of genetic engineering. See further Chapter 4.

4   *Introduction*

agents coevolve during processes of radical technological change. This interaction underlies—or causes—the pattern here identified as evolutionary. It is important to point out that the concept 'evolution' as used here is explicitly defined in relation to theoretical principles from biology as discussed in Chapter 2. This book provides a fresh approach to understanding the dynamics of technological changes.

Understanding the general characteristics of innovation processes as proposed in this book should benefit both scholars interested in economic and technical change and practitioners involved in actually developing technology. Researchers in various academic disciplines should be interested in the generalizable evolutionary patterns and theoretical arguments in relation to ongoing debates about how to understand technological change. Practitioners involved in developing actual technologies should also be interested but for different reasons. Whether managers of firms, scientific or engineering researchers, or government policy-makers, practitioners can learn about the challenges and opportunities with which others have had to deal. Using the evolutionary metaphor, they can think of the world in terms of individuals and environments, in terms of generation of novelty, retention, and selection. This can help structure their understanding of the opportunities and challenges they face. The conceptual relationships between science and technology (in relation to this evolutionary metaphor) should, for example, help analyse why certain activities are more successful in certain environments than in others.

By focusing on the contributions of multiple individuals and organizations which enabled early commercial uses of genetic engineering, this book provides insights about the complex world in which we live as well as about theoretical discussions of technological change in the economy. It provides a new way of conceptualizing innovation, not as a final object or outcome but as a complex process involving many agents responding to, and creating, environmental conditions.

Situating the Research Questions

This book thus specifically analyses technological innovation processes. The particular importance of innovations to modern market economies was recognized in the early part of the twentieth century by the Austrian economist Joseph Schumpeter. Although often ignored within orthodox economics, Schumpeter's writings have subsequently stimulated many researchers to specify the importance and role of innovations to successful competition in firms and economic growth. Related bodies of work claiming some inspiration from Schumpeter can be found in many scientific fields including evolutionary and institutional economics, geo-

Introduction 5

graphy, management, and history of technology, among others. Schumpeter therefore provides an interesting starting point for analysing the importance of technological change for economic change.

According to the Schumpeterian economist Frederic Scherer (1984: vii), Schumpeter's three major propositions about the relationships between innovations and the economy can be summarized as follows:

1. Innovations, in particular technological innovations, give capitalist economies their dynamic character through the process of 'creative destruction'. Due to major technological changes, old industries and old firms are replaced by new ones. The economic structure is not static; it is in flux.

2. Technological progress in business is one of the most important factors explaining gains in real income per capita. Technological change gives opportunities for productivity gains and is thus one motor of economic development. In this process of stimulating technical change, large firms play a particularly important role.

3. The possibility of capturing a temporary monopoly position stimulates agents to develop technological innovations. Firms have the potential of getting short-term but abnormally high profits if their innovation succeeds. Because the innovator is ahead of the others, s/he gains a temporary monopoly position by increasing price or productivity and hence can capture entrepreneurial returns. Others imitate, driving down prices. The process of innovation drives further economic and technical change. This contrasts sharply with the notion of continuous, perfect market competition. Temporary monopolies reached through innovations create and necessitate disequilibrium situations.

These three propositions will not be tested here. They are instead presented as an introduction to the neo-Schumpeterian perspective found in this broader multi-disciplinary approach to the economics of technical change. Even though much of evolutionary economics has been founded in his name, the extent to which Schumpeter's theory is specifically evolutionary is up for interpretation.[4]

[4] For example, two recent books differ greatly on this issue. Andersen (1994: ch. 1) argues that Schumpeter was indeed an evolutionary economist because of his emphasis on dynamics. In a major review and reinterpretation of the history of economic thought, Hodgson (1993: ch. 10) argues that Schumpeter's view of economic evolution differed greatly from the pattern of change identified in biological evolution and that therefore Schumpeter's theory was not evolutionary. Hodgson argues that Veblen has more specifically evolutionary notions to offer modern economists than Schumpeter (Hodgson 1993: ch. 9). Hodgson (1993: 150) argues that 'despite their current diplomatic convenience and positive ambience, the "Schumpeterian" or "neo-Schumpeterian" [Nelson and Winter 1982: 39] labels are . . . inappropriate for theoretical work of this type.' Hodgson argues that whereas modern evolutionary economists include some analogy for natural selection, Schumpeter saw evolution as development along a given path—as did Marx. This is an 'ultimately unsatisfactory way to reconcile general equilibrium theory with notions of variety of change' (Hodgson 1993: 150).

6   *Introduction*

Whether or not his theories fulfil the criteria for being called evolutionary economics, Schumpeter's perspective stresses dynamic, qualitative change in the economy as well as the importance of major
technological innovations. These two propositions are of central importance to the subject addressed in the current book. Genetic engineering
has involved traumatic economic change as well as radical technical and
scientific novelty.

Although the earlier writings of Schumpeter identified the key role of
individual entrepreneurs in introducing major innovations, his later
writings stressed that firms automated the generation of novelty by
bringing research and development (R & D) in-house, into the firm.
R & D departments brought science into firms and helped institution-
alize processes for introducing novelty into the economy. That novelty is
the basis for future competition. This is why large firms are seen to be
particularly important. In more recent interpretations, Christopher Freeman (1982) has emphasized that major upswings in business cycles are
caused by the exploitation of the productive potential of major new
technologies. Radical new technologies offer more opportunities for
improvements, which increase productivity. Many of these major technologies originate from deliberate activities of searching for novelty,
whether they were specifically scientific or not.

Schumpeterian and evolutionary economics contribute to a wider
body of research on innovations and technical change. One of the major
questions which has been addressed during the past thirty years in this
literature is the relationship between science and technology. It is an
important question quite simply because of the possibility of basic
science to indicate the productive potential of major new technologies
and because of the similarities between search activities in universities,
research institutes, and corporate R. & D. Related questions about the
relationships between science and technology involve the respective
roles of markets and government.

There has been renewed academic discussion to develop an understanding of what science and technology are and how practitioners in
each go about creating novelty. What are science and technology? Do
different organizations tend to specialize in one or the other type of
activity? To what extent do practitioners who utilize these two respective bodies of knowledge interact? What are the cognitive relationships
between these different bodies of knowledge? How do the communities
communicate? Some of this discussion will be taken up below and in
Chapter 3 in relation to the respective roles of firms and governments in
supporting the two communities of researchers.

Another reason for interest in this question has to do with changes in
society. In particular, there has been a massive expansion of government
support for basic research and for universities after World War II. Basic

Introduction 7

science performed by universities and research institutes has long been seen as the origin of major technical changes affecting the rest of society. Many scientists have argued, and policy makers assumed, that government support of science, especially basic research, at non-firm organizations will generate benefits for society in the long run. Medicine is an example where governments support research on diseases in return for new knowledge and treatments which are supposed to benefit the population as a whole. If the potential for profit exists, firms are then assumed to translate science into technology and products. All benefit. Government support of science has often been justified by referring to the indirect but important long-term benefits to society.

The economic arguments for a division of labour between government supporting basic research and firms paying for further technical development has been based on the idea of market failure, especially the difficulties of making a market for information work. According to two early and influential articles by the economists Richard Nelson (1959) and Kenneth Arrow (1962), firms underinvest in long-term research because they have difficulties in evaluating new information and technology. Firms and others face true uncertainty about what will be discovered through R & D activities and about the economic value of that information. Uncertainty means that firms have difficulties in evaluating options, and so firms have difficulties in calculating potential returns on investments in search activities. They must make informed guesses. They also face problems with selling information in a market because it can be difficult to get the buyer to pay for information without disclosing the very information the producer wants to sell. Once the buyer has that information, s/he no longer needs to buy it. The market for information thus differs from markets for commodities.

In addition to the market failure argument, governments are often assumed to have special reasons for being interested in basic research. The results of basic research are assumed to be a public good and, in fact, a special kind of public good where the creation and diffusion of information will lead to significant future benefits to society (the economy).[5] Science produces more benefits for society than for any one individual because information can be used by many without diminishing its value. A physical law like gravity will not change because many utilize it in calculations, but on the contrary, many can benefit from utilizing this physical law in order to design functioning technologies. Because social returns are greater than individual returns, firms will therefore underinvest in the creation of long-term knowledge, e.g. that

---

[5] Bernal (1967) and Bush (1945) both emphasize the long-term benefits of science, albeit from the different perspectives of scientists believing, respectively, in communism and in the ideals of independent basic science.

8   *Introduction*

which is not applicable to fairly immediate problems. Governments therefore ought to support basic research. This division of labour between what governments ought to do and what firms ought to do has often been summarized as science versus technology.

However, both in practice and in theory, it has been shown that it is difficult for firms to develop radically new technologies, even after basic scientific results have been obtained. Radical technologies pose new challenges for firms as they try to translate theory or information into economic practice. Therefore, another reason for this increased interest in the relationships between science and technology is the changing and complex challenges facing firms, particularly strategic and R & D managers and researchers. There is the realization that in many industries, novelty is necessary to survive. At the same time, generating and using novelty is difficult, even when it does not directly entail science. Research has shown that firms' technological innovations often do not directly draw upon contemporary basic science.[6] The division of labour can therefore be not so clear-cut, and corporate knowledge-seeking activities can be challenging.

Even when firms do not directly rely on basic science, interactions between university science and the commercial development of technologies are often important. If nothing else, firms develop and draw upon internal and external bodies of knowledge and techniques to carry out search activities in-house, better known as R & D. Basic scientific research has helped develop, systematize, and pass on such knowledge although other knowledge is specific for the sector, even the firm. Firms are particularly interested in the application of general knowledge and techniques to specific questions, problems, and challenges. Developing such applications may require the development of new types of knowledge, where the firms can work on very different types of challenges from those of university scientists.

Whether or not contemporary basic science is used, major innovations have often been based on previous scientific research or aided by development of new generalizable knowledge. To deal with complexity, firms may therefore need to be able to monitor and understand external scientific research in order to identify opportunities and threats. Doing so often requires special corporate competences. In order to have the capacity to absorb scientific or technical information developed elsewhere, Cohen and Levinthal (1989) have argued that firms must do some research themselves. This may be one explanation for why, in the past century, firms have organized systematic engineering and searching activities.[7] R & D enables firms to access new knowledge

---

[6] Faulkner and Senker (1994).
[7] See Schumpeter (1968), Schmookler (1966), and Freeman (1982).

and to develop longer term strategies and competences beyond the day-to-day running of the firm. Science often indicates future trends, or future technical challenges facing firms. This is often valuable information for making strategic management decisions about the future of the firm. Strategic management decisions must see beyond the present, even though choices will often be based on guesses and visions rather than rational calculations. Corporate R & D also modifies external and internal information into a form useful for the specific context of the firm.

In short, managers and researchers involved with R & D continually have to ask how best to deal with existing and new knowledge in a quickly changing, complex environment. This is a difficult question to answer in general terms because the actions and responses will depend upon firms' own capabilities as well as on local environmental conditions. Firms will have quite different strategies, competences, and abilities to adapt to innovative challenges, including both economic and technical aspects. In different types of industrial and service sectors, managers have begun to recognize the need to be able to incorporate strategies to monitor and carry out technological change as part of the overall firm strategy.

In more recent years, governments have also become increasingly interested in actively aiding the competitive position of national firms. It is argued, for example, that the government should help build up the technical competences of firms in, say, information technologies, in order better to compete in the international market. Technological change has been identified as a key asset in this race. In order to fulfil this goal, governments have been demanding more immediate returns on investment in new knowledge creation.[8] They want to see a return in the form of new jobs; knowledge transformed into new products; competitive positions of firms and economic growth; and so forth. So in addition to basic science, governments have been increasingly interested in providing more direct support for technical development, including joint R & D consortia, specified technical procurement, and encouraging firms to learn to learn. Science no longer reigns supreme over technology; both are expected to deliver benefits. This change in emphasis in government science and technology policy has occurred in parallel with more nuanced academic conceptions of what science and technology are and how they interact.

There are thus many reasons for being interested in relationships between science and technology in connection with questions about firms' innovations, about government policy, and about national economic change. In particular, this perspective opens up research questions

---

[8] See Tyson (1992) for a more theoretical argument about why governments ought to invest in supporting R & D-intensive industries.

10   *Introduction*

about the relationships between innovations and economic change, including the respective roles of markets and governments. Based on this perspective, we can analyse the relative roles of university researchers, of firms, of government agencies, and so forth, in contributing to an innovation process.

Early Commercial Uses of Genetic Engineering

Genetic engineering is one of the most modern, controversial, and dynamic of the science based technologies. Genetic engineering is not an object; it is instead a set of techniques, or way of doing things. The development of these techniques from the 1970s and onwards illustrates changing relationships between research oriented towards science and research towards commercial uses, and between universities and firms. Over time, these two types of activities have been relatively closer or further apart. When activities are relatively closer together, the research itself may be quite similar, and it is easier for individuals to communicate and to move between organizations.

As recently as the early 1970s, scientific research fields like molecular biology and biochemistry, from which genetic engineering techniques were developed, were still considered research mainly of interest within the university world. By the late 1970s and early 1980s, these scientific disciplines were shown to generate knowledge and techniques with potential market value. At this point, there was a small distance between basic university research and economically interesting research and development. This distance was small in the sense that very similar activities could be carried out in universities and in firms, although with different goals in mind.

As will be shown in this book, the distance between different types of activities can change over time. Genetic engineering was initially perceived as important for basic science and far from environments involving market forces. Later, the distance was perceived to be quite small for some activities and further away for others. Luigi Orsenigo (1989) has argued, however, that the main core of university science and of industrial applications in biotechnology quickly developed along divergent paths. Even so, we shall see that firms engaged in scientific activities and university scientists engaged in technological activities, even after divergent trajectories had emerged. The distance between different environments is thus more a matter of perception, e.g. perceptions of scientific, technological, and economic opportunities, than a matter of objective differences.

This book uses comparative case study methodology in order to gather and organize historical material about specific processes of

Introduction   11

technological innovations. The focus is on knowledge-seeking activities enabling early commercial uses of genetic engineering, specifically for production of hGH and to some extent insulin. Following closely after insulin, human growth hormone was the second commercial application of genetic engineering as the basis of production. This book is a tale of the pharmaceutical firm Kabi of Sweden and to some extent Eli Lilly of the USA, which acted very early to access this alternative production technology. It was early relative to the nascent state of basic scientific research. Both contacted the American biotech firm Genentech to carry out industrially directed scientific research, although Lilly had other strategies as well. After a time, Genentech also decided to produce hGH made this way, and so this book is also a tale of Genentech's parallel activities and their relations with university science. The technological innovation processes involving Kabi and Genentech in interactions with others thus form the core empirical material.

Because the technology was so novel scientifically and technically in the 1970s, early commercial uses of genetic engineering to make pharmaceuticals like insulin and hGH presented scientific, technical, and economic challenges to the innovating agents. Initially, both university scientists and corporate researchers and management had common goals in modifying bacteria to build the human protein hGH. Researchers at universities soon moved on to other projects, whereas firms concentrated on solving the challenges of designing an economical and technically functioning production system.

The knowledge-seeking activities of these firms are therefore placed in relation to scientific research, starting with a description of events in the early 1970s. The firms' involvement includes in particular the transformation of knowledge and techniques into products, from funding two published scientific papers in 1979 to approval of recombinant DNA hGH with an extra amino acid (met hGH) in 1985 and then approval of met-less hGH after 1987. This is the period in which genetic engineering techniques became commercially viable because firms invested significant resources in developing technically and economically functioning production processes. After this period, the focus has been on sales, with a world-wide market for hGH of approximately $1 billion.

Although these innovation processes were largely organized within firms, the case studies cannot, in fact, be said to concentrate only on the strategies of specific firms. Instead, a contribution of this particular study is that the innovative activities of firms are placed in relation to other agents' knowledge-seeking activities—whether those agents were involved in basic science or in economic activities—and in relation to changes in the public context. Emphasis is thus placed on how and why different agents, sometimes responding to different environmental conditions, contributed to, accessed, and further developed the knowledge

12   *Introduction*

and techniques necessary for this commercial use of what were initially scientific techniques. Such interactions are a key theme of the book, in the theoretical as well as the empirical chapters.

The challenges and opportunities of commercial use were more evident in the early cases described here than in later cases of genetic engineering. Early applications of genetic engineering can thus provide a useful 'window' to view how radical changes in science and technology are developed and absorbed into firms and nations. Initially, in both the pharmaceutical industry and in basic medical research, only a small number of individuals and organizations used genetic engineering techniques. These were the ones who perceived the potential benefits of engaging in these activities and had competences and resources to develop them. As genetic engineering techniques were developed and proved their worth, more and more people used them, to the point that many are today routinized tools.[9] Therefore, as commercial uses of genetic engineering have become more common, the challenges of innovation have become less obvious.

Gathering this historical but relatively contemporary material about technological innovation processes has involved gathering information which is potentially of economic value to the firms. They may want to use research results and technical specifications in the future to gain competitive advantages, thereby giving them incentives to keep such information secret. A number of interviewees have mentioned that very detailed specifications were secret. Nevertheless, the general course of events and level of detail recorded in subsequent chapters are based on interviews and public information. In fact the firms have also published a number of scientific articles in various scientific and engineering fields. They have incentives to publish such material in order to be seen as participating in scientific communities. This appears particularly relevant for communities of researchers working on technologies close to university research and also for new firms dedicated to commercial use of scientific techniques. Secrecy may thus be balanced by incentives for scientific disclosure, although the level of detail differs.

The degree of secrecy associated with corporate material about technological innovations differs by industry and by whether the technology will be used as a final product or as part of a production process. According to the Yale survey, a large American survey of this question, the amount of secrecy involved depends on how well the firm can safeguard its innovation.[10] Safeguarding means protecting one's own potential economic benefits by trying to prevent competitors from imitating without paying. Protection of the innovating firms' interests can take various forms. Some are directly regulated and embodied in

---

[9] See Clarke and Fujimura (eds. 1992).   [10] Nelson (1990: 201).

institutions such as the patent system, whereas other protection involves a firm's actions relative to competitors. Particularly important are the advantages of moving first, which gives firms the possibility of learning quickly to make a technology more efficient (a steep learning curve) and to gain lead time into the market, which can increase market share. The first-moving firms start developing and using a technology before others and so learn how to use it more efficiently and sell it before others.

The Yale survey indicates that R & D-intensive industries, which include pharmaceuticals, tend to rate patents as more effective in preventing duplication of products relative to other industries. Within the R & D-intensive industries, patents are easier to obtain when 'the composition of the product is relatively easy to define and limit' as in pharmaceutical formulas, in contrast to other R & D-intensive but complex system technologies like aeroplanes.[11] Patents give legal ownership but also require that previously secret information become public knowledge. However, for production process innovations, all industries rated secrecy as more important than patents or the first mover advantages. The pharmaceutical industry rates patents as important for both product and process innovations, but clearly rates patents as more effective for products than for processes.[12]

As this book focuses on developing genetic engineering to manufacture hGH and not on the search process leading to hGH as an effective pharmaceutical, the focus is on new process technologies. The Levin *et al.* survey (1987) indicates that firms would want to keep this type of information secret. In addition to activities to integrate genetic engineering into production, the book also addresses scientific activities to develop genetic engineering and medical knowledge about uses and markets.

Moreover, secrecy in production processes is often used to gain immediate advantages over competitors. For example, secrecy can be used when firms try to develop a product quickly and get it to the market and thereby gain early mover advantages. Secrecy did play a role here when, for example, the firms switched to a new bacteria construction (a new set of genetic engineering techniques and biological materials). In these examples, the information was more valuable to the

[11] Levin *et al.* (1987: 800). Nelson draws out the implications of secrecy and relates firms' responses to empirical data, which shows that the bulk of industrial R. & D. is oriented towards product innovations and not process innovations. In many industries, process innovations are done by upstream firms or suppliers. Nelson argues that the finding is consistent with Eric von Hippel's proposition 'that the locus of inventive activity is determined, in part at least, by where the ability to appropriate returns is greatest' (Nelson 1990: 202). Pavitt (1984) more systematically relates the locus of innovative activity in the economy to ability to appropriate returns. This results in his classification of different types of industries by innovative potential.

[12] Levin *et al.* (1987: 797).

14   *Introduction*

firm immediately when developed, but in that this book describes a
historical process, these secrets are less valuable to the firms now.
Secrecy can also be an important way to continue protecting production process methods over time, when improvements to the techniques
have reduced costs and increased quality. The technical details kept
secret in these case studies were deemed to be so specific that they
were not essential to understanding the broader processes of technological innovation.

There has also been a major patent dispute relating to genetic engineering to produce insulin and hGH. This patent dispute included the
main early agents in the United States, i.e. Genentech, the American
pharmaceutical firm Eli Lilly, and the University of California, and
involved 'numerous contracts and patents that relate to human growth
hormone'.[13] This raises an additional methodological problem about the
interview material. The firms and this university—and hence to some
extent the researchers and managers interviewed or described in written
material—have had a current interest in giving a particular version of
the history before settlement. Giving a particular version could help
them win the patent lawsuit. This is not intended to insinuate that the
interviewees construct a new fable, but it does indicate the difficulties of
weighing different accounts of the same course of events. Attempts were
therefore made to determine the interests and motives of individuals
giving information, and hence the reliability of different sources.[14] Multiple sources of information were also used where possible. Different
strategies were thus used to attempt to deal with the general problem
of investigating technological innovation processes of value for firms.

Reflections on Ensuing Chapters

Chapter 2 explores and develops a theoretical perspective about technological innovation from an evolutionary perspective, while Chapter 3
addresses this question by examining the relationships between science
and technology. These two chapters introduce the main outlines of the
theoretical world within which this book contributes. That theoretical
world contains the partially overlapping contributions of heterodox
evolutionary and institutional economics, management, sociology, and
history of science and technology. The particular perspective developed

---

[13] Raines (1992). Genentech and Eli Lilly settled the eight-year conflict out of court in
January 1995. Eli Lilly agreed to pay '$145 million to settle all claims and counter-claims'
(*Financial Times* 1995: 17).
[14] Historical source criticism provides three pertinent questions. These are: whether the
source is accurate, whether it is relevant, and whether it is representative. See Nilsson
(1973: 179).

Introduction 15

here is of coevolutionary innovation for science based technology, and it is a new combination of theoretical contributions.

In the type of detailed case study work used here, introducing the technology is an important task to help the reader through the historical twists and turns. Chapter 4 therefore provides a layperson's explanation of the use of genetic engineering techniques to make human growth hormone, the scientific basics of these techniques, and how this particular use relates to the larger public debate over genetic engineering. This chapter forms a bridge to move from the highly abstract, theoretical arguments about science, technology, and evolutionary innovation to a very specific, and empirically rich discussion of the historical technological innovation processes involving recombinant DNA techniques to produce human growth hormone.

The detailed historical descriptions in Chapters 5, 6, 7, and 8 chronologically recount the history of these commercial uses of science from the early 1970s to the mid-1980s. Particular emphasis is placed upon knowledge-seeking activities in relation to four environments which influence how and why agents pursue knowledge-seeking activities. For, along with an empirical contribution to our understanding of the internal commercial history of these technical developments, these four chapters also contribute to a further development and discussion of the theoretical issues. Chapters 9 and 10 therefore contain conclusions about innovation processes in relation to an evolutionary perspective on economic change and in relation to science and technology.

Conclusions for Science and Technology

Introduction

This chapter draws some conclusions about science and technology based on the theoretical and empirical discussions in previous chapters. Both scientific and technological activities involve searching for new knowledge and techniques, but it has been proposed that scientific activities are generally about understanding the world whereas technological activities are generally about controlling nature for human purposes. Either type of activity can, however, be carried out under either market or government influence. For example, scientific activities can be financed and expected to give economic advantage, however uncertain and distant it may be.

One advantage of this conceptual framework is that it no longer makes assumptions about which type of activity is carried out at which type of organization. Researchers at different types of organizations—universities, firms, and so forth—are instead assumed to choose which type of knowledge-seeking activities to engage in, and they can make different choices at different points. For example, a university researcher can first compete in the basic scientific environment through scientific activities and then later start a biotech firm to sell scientific-economic knowledge which is perceived to have value in the techno-economic environment. This has been a common pattern in the modern biotech business, where basic researchers start biotech firms to sell R & D contracts to established firms. This perspective thus separates agents and knowledge-seeking activities from the socially constructed environments.

This chapter first addresses how and why agents engage in scientific and technological activities in response to the four environments by analysing the historical material about genetic engineering for human growth hormone and insulin. The focus is on agents, environments, and knowledge-seeking activities from the 1970s and 1980s. The next section discusses cross-stimulus of scientific and technological activities, or interactions in the development of the two bodies of knowledge. This contributes to debates about the similarities and differences among science and technology. The chapter then ends with a discussion of the social nature of selection and concluding remarks about this book.

Agents, Environments, and Knowledge-Seeking Activities

This section addresses the question of whether the agents described in this book engaged in knowledge-seeking activities in response to environmental conditions. Four environments were defined in Chapter 3 as giving incentives for knowledge-seeking activities—the basic scientific, the techno-economic, the scientific-economic, and the techno-government. Each environment consists of informal institutions, incentive structures, and other socially constructed selection criteria which influence the direction of agents' search activities. In addition to these four, the public context has been defined as consisting of debate, informal norms, and formal regulation about the technology and/or final product. It is the larger context which sets parameters for what types of knowledge-seeking activities and results are generally considered acceptable.[1]

Instead of analysing the environments one by one, this section chronologically analyses the main events in preceding empirical chapters. It will indicate how and why agents responded to the incentive structures and informal institutions of various environments. Generally, the agents tended to look for novelty which fitted their perceptions of the environment, although the results could be quite different from those expected. Because learning is a cumulative process, new challenges arose which could not be foreseen. These challenges and opportunities only became evident as different types of knowledge and techniques were integrated into the firms' contexts of use.

The starting-point for these early commercial uses of genetic engineering was the development of relatively practical techniques in the early 1970s. Of particular relevance here, Professors Stephen Cohen and Robert Boyer, working at, respectively, Stanford University and UCSF, developed the recombinant DNA technique, although many other basic researchers were working on complementary approaches. Those who developed such knowledge and techniques did so in reference to the internal basic scientific environment in order to publish papers, compete for future research grants, and so forth. Researchers in these scientific fields had traditionally had little or no contacts with industry or other economic influence. They instead competed for positions within the academic world and for government funding which was distributed by other scientists. These researchers were engaged in scientific activ-

---

[1] It is not an environment in the sense defined here, because it does not give agents specific incentives to engage in knowledge-seeking activities.

282   *Conclusions for Science and Technology*

ities mainly at universities, and they oriented their search to fit the incentive structure (rewards and punishments) of that environment.

In competing to achieve results, they developed both scientific knowledge as well as practical techniques for carrying out experiments and verifying knowledge. Knowledge and techniques for manipulating DNA were thus developed hand in hand over a long time period in this community. Recombinant DNA techniques were, however, radical in giving practical control.

With the development of practical and relatively simple techniques, it was apparent to many basic scientists that genetic engineering could be used for various purposes, both destructive and beneficial. They could also be used to create economic value. Due to the radical implications of possible uses of the technology, however, there arose an internal scientific debate about the risks of genetic engineering, leading to an international, temporary moratorium on certain experiments. Scientists then set guidelines for certain experiments, particularly in the Berg letter, the Berg committee, and finally, in the discussions at the 1975 Asilomar meeting and the publishing of guidelines in *Science*. Scientists based the guidelines on research intended to identify risks as well as acceptable categories of safety precautions. The moratorium and guidelines were voluntary but enforced through strong informal norms.

This informal moratorium was lifted in the mid-1970s, and the initial guidelines were used to develop more formal regulation. In the United States, the National Institutes of Health, the major medical research council and research institute, published guidelines in 1976 based on the Asilomar conference. The Swedish research councils similarly initiated a Genetic Manipulation Advisory Committee, although industrial representatives also participated. The scientists' own debate and decisions about guidelines thus strongly influenced regulation of the technology.

Although the commercial potential of genetic engineering to produce pharmaceuticals was apparent to some, this vision was a long way from reality in the mid- to late 1970s. Even the vision seemed like science fiction to many. In addition to practical difficulties, a question of direct interest to pharmaceutical firms had not been answered. Before investing in developing the new technology, they needed to know whether bacteria could express human proteins in general, particularly the longer proteins relevant as pharmaceuticals. Although genes could now be transferred, experiments had not confirmed that these techniques would work in practice to get bacteria faithfully to reproduce human proteins. Doing so would require a combination of scientific and technological activities. A particularly interesting characteristic of biotechnology is that knowledge-seeking activities to answer such questions were considered relevant for both basic scientists and firms.

By the mid- to late 1970s, genetic engineering thus offered agents a

Conclusions for Science and Technology   283

potential way to compete in the techno-economic and scientific-economic environments as well as the basic scientific one. They could reap benefits from the scientific and technological activities of recombinant DNA techniques. Partly in response to these potential benefits (incentives), some scientists who had previously worked within the basic scientific environment reoriented some activities towards economic potential. This reorientation started in the United States but later spread (to a lesser degree) to scientists in other countries.

The most immediate strategy apparent in this book is the creation of the dedicated biotech firm Genentech in 1975–6 as a combination of venture capital (Swanson) and basic science (Boyer). Once biotech firms like Genentech had been started, they also created a demand for other scientists versed in genetic engineering, either directly through employment or more indirectly through consulting.

From the start of this reorientation towards combined basic scientific and economic goals, scientists trained in fields relevant for genetic engineering could choose from a number of different strategies. This book identifies the following strategies to exploit some of the economic potential of genetic engineering:

1. Stay at a university but engage in multiple activities, where some are directed towards the basic scientific environment and others towards one of the economic environments. This was possible when similar and/or complementary activities had both scientific and technological dimensions due to the radical novelty of techniques. Examples include the research on insulin and hGH by Boyer's (Genentech) and Goodman's (UCSF) respective groups. The results led both to scientific papers and to patentable techniques. Similar experiments could therefore be used to develop scientific knowledge as well as to develop techniques to control nature.

2. Start a dedicated biotech firm to exploit the economic potential of the new technology. Examples include Genentech as well as the large number of biotech firms started by senior researchers from UCSF and elsewhere in the early 1980s and onwards. The researchers who started these firms did not necessarily leave their university. Instead, many senior scientists remained at a university and divided their time between university and firm. These biotech firms in turn hired other scientists.

3. Move from an organization responding mainly to one type of environment to an organization responding mainly to the other. Such moves have generally been from universities to specialized biotech firms or to pharmaceutical firms expanding into these fields. Examples include Kleid, Goeddel, Seeburg, and UCSF post-docs starting work at Genentech.[2]

[2] More recently, some corporate researchers have moved back to universities or else have left the first generation biotech firms to start up new biotech firms in newer fields. Goeddel, for example, left Genentech to start a new biotech firm.

284   *Conclusions for Science and Technology*

Individuals could thus either move physically to organizations mostly oriented to this new environment, or divide their knowledge-seeking activities so that some were towards one environment and some towards the other. This duality of activities is visible in this book, in that scientists at universities and at firms sometimes competed and sometimes co-operated.

As noted, one key question for both basic science and commercial uses of genetic engineering in the mid- to late 1970s was actual expression of human proteins in bacteria. A number of competing university groups engaged in research to answer related questions, but it was university research supported by Genentech which first indicated it was possible in practice. This was the somatostatin research results of December 1977. Boyer and others of Genentech/UCSF carried out this research in collaboration with researchers at the national medical centre City of Hope. This Genentech result for the somatostatin experiment is particularly interesting because it was used to influence the basic scientific and scientific-economic environments as well as the American public context.

The public context was in the process of being shaped in 1977, this time by agents other than scientists. The American public debate over genetic engineering had got underway in a hostile tone, and the US Congress were attempting to impose stricter federal regulation. The national public context would take over regulation from scientists themselves and would more strictly define which types of activities were acceptable. Having decided that this public debate was exaggerating the risks of the technology, these scientists changed tactics. They began to argue that scientific results would offer great benefits to the public, particularly in the medical fields. In particular, this project at UCSF, funded by Genentech, had results showing that the human protein somatostatin could be expressed in bacteria. This, scientists argued, indicated the great potential medical benefits of the technology and the necessity of allowing continued scientific research. No new federal guidelines were imposed.

This was also one of the first times when some scientists indicated more publicly that their activities had moved towards the scientific-economic environment. Their research was financed by a biotech firm, with the underlying understanding that although these activities were scientific, they were also technological. Techniques to control genes, including those enabling protein expression, were crucial for future commercial uses. The newspapers and magazine articles creating the public debate were, however, less clear about the multiple purposes. They rarely if ever simultaneously identified the researchers with both universities and biotech firms; depending on the angle, only one type of organization or the other was mentioned.

Conclusions for Science and Technology   285

This new pattern—for scientific researchers in these fields—of individuals and organizations responding to economic influences also led to changes in the informal institutions. There were implications in the basic scientific environment and the creation of norms for the scientific-economic environment. The first contacts between basic scientists and firms in the USA led to some perceived conflicts of interest and attempts were made to implement clearer rules for interactions. Examples discussed in Chapter 6 include UCSF's Senate hearing about Genentech's relationships with the university and the Pajaro Dunes meeting. The latter brought together firms and the most prestigious research universities in order to discuss the general norms for interactions among universities and firms. In both cases, they discussed new patterns of interaction, or informal institutions regulating interaction. Soon thereafter, the informal institutions of the basic scientific environment more specifically allowed, even encouraged, the university scientists to have more extensive contacts with firms and even to respond directly to the techno-economic environment.

Another consequence of the somatostatin experiment was that the nascent firm Genentech gained scientific prestige which concurrently advertised that they were willing to direct scientific activities in commercial directions. Scientific activities were necessary to address these questions because all experiments to get the bacteria to achieve expression were challenging and required new knowledge and techniques. Partly based on the somatostatin results, Kabi contacted Genentech in late 1978 about hGH and signed an R & D contract in 1978. Eli Lilly also funded similar research at UCSF for hGH and insulin and then later signed an R & D contract for insulin with Genentech. The advantage of biotech firms like Genentech was that they could develop practical and reliable techniques as well as complementary knowledge in order to succeed with commercial developments.

In the long run, Genentech's management saw that genetic engineering went hand in hand with activities involving other types of knowledge. Their vision was that the firm's long term survival would depend on moving more directly into the techno-economic environment. This is reflected by the fact that Genentech's strategy from the beginning was to move into pharmaceutical production and not just sell R & D contracts involving genetic engineering. To fulfil these multiple goals, the organization had to engage in various types of activities. Genentech started by developing its basic scientific potential much more than established firms, but soon hired specialists with other types of competences, such as fermentation, purification, and analytical methods. The same organization, whether university or firm, could thus respond to different environments simultaneously with different sets of knowledge-seeking activities.

286   *Conclusions for Science and Technology*

In exceptional cases of radical science-based change, the same know-ledge-seeking activities could generate novelty relevant to more than one environment. These scientific activities could be made relevant for economic goals, whether more immediately (additional R & D contracts for Genentech) or in the longer-term future (products sold). To produce pharmaceuticals, additional techniques and knowledge had to be devel-oped before the vision could be realized in practice. Even Genentech's early scientific activities however, were, oriented towards commercial use, as illustrated by the comparison of Genentech and UCSF regarding hGH in 1979. The Genentech results were more relevant than UCSF's to engaging in additional technological activities to create economic value. Thus, even when carried out parallel in time and in general orientation, similar knowledge-seeking activities could still be adapted towards different environmental conditions.

Those firms wishing to move into commercial uses early on had to find a way to access the necessary knowledge and techniques. Initially, this required funding knowledge-seeking activities or else licensing very new discoveries. The firm could engage researchers inside the firm (requiring the development of new competence areas) or else contract out the activities to external agents, whether firms or universities. The strategies of these early-moving firms to support or gain access to scientific research directed towards commercial uses included:

1. Giving direct financial support to basic scientists working at uni-versities. Examples include Eli Lilly's support of Baxter and Goodman for research involving insulin and hGH, and Genentech's support of Boyer and City of Hope for insulin and hGH. This strategy involves much uncertainty about what will be developed, partly because the research is basic, and hence involves major unknown factors, and partly because the firm has little control over the scientists and the experiments. It may, however, be the only way to access certain information and monitor future developments.

2. Hiring basic scientists to carry out scientific activities within the firm. Genentech hired, for example, the university scientists Kleid, Goeddel, and Yansura, and Lilly set up a new department with relevant knowledge and techniques. Instead of placing these basic scientists in-house, the firm can also start a specialized biotech firm, as when Kabi and the Swedish government started KabiGen. These activities are some-times relevant for more than one environment. The firms should carry out (basic) scientific activities in-house when it is necessary for being able to monitor (and understand) external research. In addition, doing research in-house can be seen as an admission ticket to being considered a member of the larger scientific community. The firm then contributes as well as draws upon the generally available knowledge.

Conclusions for Science and Technology 287

3. Contacting a dedicated biotech firm to carry out research applicable to a specific commercial goal. These firms have special knowledge and techniques which may not be available in-house in a larger firm. Examples include Kabi contracting Genentech to make hGH in bacteria, and Lilly contracting Genentech to make insulin in bacteria. This also results in a sort of dual-use activity, which might also be relevant to the basic scientific environment. It involves less uncertainty than funding university scientists, because the R & D contracts specify that they will develop practical techniques which are relevant for firms' contexts of use such as level of expression.

4. Buying the right to use the results of previous scientific research by, for example, licensing patents. Examples here include Eli Lilly's attempts to buy Gilbert's patents on insulin techniques (which were sold to Biogen), and Lilly's payment for patent rights held by UCSF and by Genentech. The results already exist, so the firm has some idea of what information and techniques it is buying, but it will often have only a vague idea of the economic value.

The common element in the above four firm strategies is an attempt to gain direct access to the results of activities which could be relevant to the techno-economic environment. The examples given all relate to genetic engineering as the basis of a new production method. The uncertainty in the late 1970s was high because it was not clear what results would be forthcoming, or their technical and economic relevance for the firms. Managers thus had to decide between the risks of losing the money invested and the possibility of high but uncertain future returns. As described in Chapters 5 and 6, knowledge-seeking activities relevant to producing human proteins were oriented towards developing knowledge and techniques of genetic engineering.

From there onward in this book, activities only for genetic engineering in the basic scientific environment are no longer discussed. Instead, the focus is on early commercial uses, and there is therefore a shift to how and why the firms engaged in activities to integrate genetic engineering with biological production systems and to create markets.

Having accessed genetic engineering techniques either in-house or externally, the firms concentrated on using genetically engineered bacteria as the basis of production. Integrating the new and existing technologies and bodies of knowledge sparked new challenges for the specific uses, whose solutions required significant investment in focused and specific knowledge-seeking activities. The firms had to spend much money on R & D to make the process work so that the desired protein(s) could be sold.

In this R & D, Kabi and Genentech reacted to their interpretations of economic criteria for pharmaceutical sales. R & D decisions had to take

288   *Conclusions for Science and Technology*

the different aspects of the integrated production system into account, such as identifying challenges requiring solutions. Firms' internalization of the environment thereby affected the type of solutions they generated as well as which selection criteria were internalized through testing. For example, the yield of pure proteins was initially low largely because the initial set of genetic engineering techniques was not very efficient in getting the bacteria to express proteins. If the firms had kept that level of yield, then the costs of production and hence the selling price for hGH would have been too high. Yield was therefore one of the most important challenges to be solved, both specifically for these products and generally for genetically engineered proteins. The firms therefore had to invest in R & D to try to increase the yield of proteins made in bacteria.

Activities to improve the protein yield initially focused on changing the chosen set of genetic engineering techniques. Genentech had the special competence to do so by trying out different alternative bits of the puzzle, mainly developed in universities. Changing the promoter (start signal) turned out to be particularly effective in increasing yield. When testing these different alternatives, Genentech tried to develop model systems which could be applied to their various products. The way of making changes and the changes themselves were intended to be standardized, because this would reduce costs of additional R & D. Models are thus a way of reducing diversity to a subset which is likely to work. The general corporate goals of reasonable yield, standardization of techniques, high purity, and so forth were based on expected economic and regulatory conditions.

Changes in the specific set of genetic engineering techniques and biological material or in one component technology usually required changes elsewhere as well. Most such changes entailed new challenges for R & D, where solutions also had to be adapted to the specific conditions of use. The amino-acid feed in the fermentation phase, for example, had to be calibrated to the specifics of the bacteria strain and also affected protein purification, making it more or less difficult. Research within different specialist bodies of knowledge and techniques could often offer different solutions to the challenges. Improving yield, for example, could not only be achieved by changing the genetic engineering but also by developing more sensitive parameters for fermentation and by improving the protein purification steps. There had to be decisions about how much improvement could be achieved by making various changes. Doing so also required feedback and communication inside the firms to keep the different specialist research groups heading in the same direction.

In fact, these component technologies and underlying bodies of knowledge were so important that Kabi could contribute even initially in the 1970s to the collaboration with Genentech. In other words, Kabi was not just the recipient of the results of the R & D contract. Initially in 1979, Genentech and Kabi researchers worked together in California to ferment (grow) the genetically modified bacteria, but the resulting solution was then sent to Sweden for analysis and purification. Kabi already had competence and experience in these areas due to their existing product, pit hGH. Before long, however, Genentech built up their competences in these areas by hiring experienced individuals, and Kabi also built up competence in areas relating to genetic engineering, albeit in a separate biotech firm. Thus each firm had identified certain bodies of knowledge and techniques which were crucial for their survival. In the short term, they could rely on others (co-operation; external contracts) but in the longer term, they had decided these areas needed to be represented in-house. They stopped collaborating as well.

The division of labour between Kabi and Genentech can also be related to the public contexts. During 1979 and 1980, the two national public contexts influenced the parameters for research somewhat differently for the two firms. In the USA, the debate had become more positive about the scientific and economic potential of the technology, and restrictions relaxed. In Sweden, public debate arrived somewhat later than the USA and arrived about the time Kabi began grappling with the first experiments. Initially the firm had no permit, then no labs to grow the bacteria; they were instead working with Genentech in California. As in the United States, the Swedish debate over genetic engineering was polarized into opponents and proponents. Partly for political reasons, the Swedish government called for a temporary moratorium on industrial activities. Kabi then shifted some experiments to England, not least because no labs meeting safety specifications were available in Sweden. The Swedish government and Parliament subsequently constructed more formal government regulations for the technology (replacing the scientists' committee), although Sweden followed America's lead in incorporating much of the regulation of genetic engineering into existing laws. These changes to regulation in the public context clearly limited the direction of search and innovative activities considered acceptable. At the same time, however, having stable boundaries assured firms that they knew what would be considered acceptable, and could therefore get on with their business.

In these cases of genetically engineered proteins used as pharmaceuticals, public debate and formal regulation set standards not only for genetic engineering but also for the resulting product. Firms in both countries had to take these regulatory conditions into account when directing knowledge-seeking activities. In other words, research managers had to make decisions about the specific characteristics of the product and market when deciding about R & D. For example, a high yield of proteins had to be correlated with purity because very high purity is a condition for government approval of pharmaceuticals. Firms therefore had to invest resources to solve technical challenges such as impurities arising in the production process (contaminants; variants). Solving such challenges was particularly dependent upon devising more sensitive tests and analytical methods. As these tests and methods became more sensitive and could 'see' more impurities, it became obvious that changes also had to be made to protein purification steps or fermentation parameters. Discussions were necessary to identify what the common challenges were.

As the market for rDNA hGH did not yet exist during this period—although the market for pituitary hGH did—the firms had to make assumptions about the desirable trade-offs between quality and quantity. How much would potential patients and health services be willing to pay, at what quantity and quality? Firms had to make assumptions about conditions in the techno-economic environment, and they made such bets directly dependent both on medical research for hGH and on R & D involving the production process. Better analytical methods were also needed to convince both patients/doctors and regulatory agents that the pharmaceutical product could be used safely. Because genetic engineering represented a novel source of supply, the innovating producer firms engaged in extensive discussions with government regulatory agencies and doctors. In these cases, doctors were not only a professional group representing users but also a group developing scientific knowledge about the efficiency and use of the product.

Doctors have the power to change the boundaries of acceptable uses by, for example, indicating the efficacy of a substance in treating a new condition or in expanding the boundaries for treatment of a known indication. By changing the boundaries, medical researchers increased the potential demand for the product. Changing the boundaries of medical knowledge directly affected market demand when it either expanded the potential group for one use (short but not hGH-deficient children) and/or discovered new uses (retaining muscle mass in the elderly; Turner's syndrome). The firms have therefore had incentives to invest in medical scientific research. On the fringes of the accepted market, a black market for illegal uses (athletes) also sprung up when these users perceived a potential, although medically questionable, benefit from the product. Firms, medical researchers, and government regulatory agencies had a common interest in defining legitimate uses and in patrolling uses and abuses.

This book has shown that environmental conditions shaped the directions of agents' search activities. Solving different challenges could involve scientific and/or technological activities, depending on the current state of knowledge and techniques. The two were often dependent upon each other, so that developing new knowledge required more practical techniques or more sensitive analytical methods. Moreover, agents in one environment often monitored and/or financed knowledge-seeking activities in another, to make general knowledge and techniques applicable to their uses. The contexts of use discussed were mostly commercial, entailing both general and very specific challenges.

Cross-Stimulus of Scientific and Technological Activities

Cross-stimulus means that scientific and technological activities mutually influence each other. Cross-stimulus refers to significant interactions and influences between the two types of knowledge-seeking activities, which is evident throughout the historical chapters of this book. Some technological activities are relevant for science—such as analytical methods—whereas other corporate improvements are too incremental to be of interest in basic science, such as which specific promoter is used by a firm.

There are various reasons why cross-stimulus is possible. One is that both types of activities draw upon common bodies of knowledge and techniques, albeit with different degrees of access even for specialists. The more specialized something is, the fewer there are who understand it. Another reason is that both scientific and technological activities can lead to the development of more general or more specific knowledge and techniques. The more general something is, the more there are who are likely to be interested in it for different reasons.

The history told in this book emphasizes interactions among agents engaging in various scientific and technological activities. These early commercial uses of genetic engineering involved both scientific and technological activities, and it is clear that using basic scientific knowledge and techniques for early commercial uses was usually not easy. It often involved significant challenges and necessitated the development of a different kind of knowledge and techniques. Scientific techniques and knowledge could form the starting-point from which new, commercially relevant activities could be developed. Knowledge-seeking activities relevant for these cases have been classified as involving: (1) genetic engineering techniques, (2) integration of genetic engineering and production systems, and (3) medical knowledge about uses of hGH. However, these new activities had to be specifically oriented in directions relevant for the scientific-economic and techno-economic environments.

Knowledge and techniques relevant for these three types of knowledge-seeking activities could be general for both scientific and technological activities. For example, once university scientists had improved SDS gels through silver staining, their results were available to the community of interested practitioners and could be used for various purposes (generalizable). These initial methods however, were, impractical for corporate goals. If corporate researchers wanted to use silver staining to analyse proteins, then they had to make many additional improvements and changes to make them more reliable, easier to use, and so forth. Those changes could in turn be reincorporated by university scientists (some of whom worked on similar improvements). There could thus develop a general body of knowledge and techniques available to trained researchers in different environments.

Knowledge and techniques relevant for both scientific and technological activities involving rDNA hGH can be classified as more or less available. Some have been shown to be very specialized and only available to a few specialists, whereas other parts are more generally available. Note, however, that the extent to which knowledge is general or specialized within the community of practitioners will change over time. Recombinant DNA techniques were highly specialized and only practised by some scientists in the 1970s but have become increasingly available to a larger number of researchers. The extent to which knowledge and techniques are specialized or more generally available within the community can thus change over time. Its diffusion depends to a large extent on what is taught to new practitioners, either in formal training situations or through experience in firms. When this general body of knowledge does not suffice for solving challenges, then firms have to engage in knowledge-seeking activities which are useful in their local contexts of use. These may nevertheless result in more general solutions of interest to others.

Through the transfer of knowledge, techniques, behaviour, and individuals, the knowledge search activities (and their results) in one environment have been shown to change agents' abilities to engage in search activities in another. Firms can be influenced by basic science and vice versa. As an example, basic scientists can be influenced by firms when they take a firm's very specific problems with growing genetically modified cells in fermenting tanks as the basis of developing model systems useful for understanding general principles.

The cross-stimulus may be based on knowledge and techniques which have been known for a while in the other environment, or which are quite recent. Agents in one environment can perceive the economic and technical potential, or relevance, of knowledge and techniques which are just being established in another environment. When they do perceive such potential, then the agents can choose whether to wait and see what happens or to invest resources to develop that idea or technique further. The relevant comparison here is between those established firms which waited to see how genetic engineering would be developed before moving in the 1980s and the firms described here who actively searched for relevant developments by the late 1970s.

Conclusions for Science and Technology 293

Due to the economic incentive of potential profits, some individuals and firms saw the possibility of developing genetic engineering to make human proteins for sale in the techno-economic environment as soon as the techniques were possible on lab scale in the mid- to late 1970s. Their interest in functioning techniques coincided with the interests of some contemporary basic scientists, who were dependent upon these new techniques for controlling genes and cells in order to develop theoretical knowledge. The ability of basic scientists to do things in practice and to use new knowledge to do new things was one feature which made this basic scientific research attractive to firms. The environments converged in giving incentives for very similar activities at a point when there was a high level of uncertainty and risks involved.

The firms then played a particularly important role in translating what they perceived to be economically possible (market signals) and scientifically possible into functioning technology. Once rudimentary genetic engineering techniques were available for the firms' specific purposes in 1979, then these firms shifted their R & D emphasis to integrating genetic engineering techniques and biological material within a controlled biological production system. As shown in this book, the firms translated scientific knowledge and lab techniques into practical, usable production systems that made a marketable product.

Translating knowledge into practice through the intermediary of technological activities was not, however, a simple proposition. Sometimes it was relatively simple, but sometimes it was quite challenging. Accumulated knowledge, techniques, and equipment formed the starting-point for additional R & D, but how the translation would actually be achieved was often not evident. Instead, a variety of approaches and alternatives were tested, and often, several variants functioned. The firms had to choose among them based on both economic and technological criteria. Firms thus initially faced uncertainty about which technical alternative(s) would function more efficiently technically and economically and which would become dominant in the future. (These two conditions of efficiency and dominance are not necessarily synonymous because an inferior technology can become dominant.) Firms had to make decisions based on their expectations of future selection criteria.

For existing firms, these early uses of genetic engineering required mastery over different scientific knowledge and different organizational and distributional channels for obtaining the raw materials, and posed new problems relating to scale-up, purification, and production. The new technology represented a break with former skills and knowledge for those firms like Kabi which had previously extracted hGH from tissue. They faced a technological discontinuity which could destroy firm competences. Biotechnology thus threatened to make certain knowledge obsolete.

Existing pharmaceutical firms reacted, however, and although the technology changed, both existing pharmaceutical and new biotech firms could compete. Researchers and organizations active in biologically-based pharmaceutical production and R & D could acquire new knowledge and skills. They could compete with new, specialized firms which sold specialized research and which sometimes entered the market for pharmaceuticals. An important reason why there has been room for both types of firms is that the pharmaceutical industry has no single core technology which could be directly threatened by this change. Instead, there are multiple technologies whose integration is important. Genetic engineering required major changes and acquisition of new knowledge, but this new technology is put into the context of other bodies of knowledge and techniques.

For managers, an important implication of cross-stimulus of scientific and technological activities is that those firms involved in science-based innovation need to monitor and carry out knowledge-seeking activities within more than one environment. Genentech's activities as documented here include activities which were considered relevant within the basic scientific, the scientific-economic, and the techno-economic environments. This is true to a lesser extent for the established firms, although each type of firm actively developed in-house competence and experience in crucial fields.

A particularly interesting contribution of this book has to do with the analysis of the integration of genetic engineering with biological production. Both the problems of integration and the importance of these bodies of scientific and technological knowledge and techniques have been neglected. Too much of the analysis of biotechnology only emphasizes molecular biology.

Shifting the perspective in this way has allowed us to see the challenges and creativity of the other fields of science and technology which have contributed to commercial uses. Integrating the new and the old has required the development of a whole body of knowledge and techniques which were general for these commercial uses. Nevertheless, many of these challenges arose because the firm suddenly found local and very specific problems having to do with their particular configuration of the system, such as amino-acid feed or where in the cell the proteins were made. Many specific challenges, such as the low yield of hGH or contaminants 'invisible' to contemporary testing methods, were then identified as representing general categories of problems. The knowledge and techniques were different from those developed in basic science because they had to take into account the specific criteria which interested firms. Some of the challenges encountered by both corporate and university researchers—such as the problems of insoluble proteins and inclusion bodies when bacteria overproduce human proteins—in turn stimulated new scientific research on questions which were particularly important to the firms. There has thus been significant cross-stimulus.

Developing commercially relevant genetic engineering techniques and integrating them into production systems for hGH and insulin thus involved both scientific and technological activities. The two types of activities were concurrently important and stimulated each other. Interaction has increased the diversity of knowledge and techniques which can be used to generate additional novelty. We can therefore say that diversity within each specific environment is increased, compared to a situation of isolated activities in each environment. Overlap enlarges the agents' search spaces. It helps provoke additional questions and solutions, which are sometimes relevant in more than one environment. This may help explain the potential of genetic engineering for productivity gains and for creating new sectors.

Selection as a Social Process

A final discussion relevant for coevolution in scientific and technological activities is the extent to which selection processes are social. Chapter 2 argued that the social nature of selection processes was one of the major differences between biological and socio-economic evolution. Socio-economic selection criteria are created through interactions. This is one of the main reasons why it is so important to try to understand how selection criteria are created in different environments and how they influence agents' knowledge-seeking activities. This social basis of selection has been illustrated in different ways in this book.

First, the definition of environments relies on the idea that agents belong to a community of practitioners when engaging in knowledge-seeking activities. These are practitioners of scientific and/or technological activities although they may work at different types of organizations. Their communality lies in their shared, specialized bodies of knowledge and techniques, but they also often share norms. Sometimes these communities can set the informal institutions and incentive structures defining what is relevant and rewarded in an environment; sometimes these criteria are defined in interactions with non-researchers, such as with managers and others in firms. Individual decisions and actions are positioned and evaluated relative to a social context.

Secondly, acting in response to different environments means directing search activities towards different audiences and towards different reward and punishment structures. To draw on an earlier example, a firm can 'engage in scientific activities', and the results can be of relevance within the basic scientific environment. Nevertheless, the firm will go bankrupt if it only does basic scientific activities unless it is funded by government research grants (and evaluated scientifically). If the main force influencing survival is market allocation, then this firm will have trouble in the long run because of the difficulties of using scientific activities to create economic value. Even if they sell the knowledge and techniques to others, they must translate them into a usable form, which may no longer involve basic science. The difficulties of this translation as described in this book differ from the common understanding that biotech is science-based.

The difficulties have to do with the uncertain and long-term character of scientific activities to understand the world. Therefore, when a firm engages in scientific activities, these may not be of immediate economic interest, in the sense that they might not be saleable, but they had better be of some longer-term economic interest. Longer-term interest can include, for example, building up the knowledge foundation for an existing or future product; building up competences within the firm which make later knowledge-seeking activities possible; and/or being seen as a participating member of the research community in order more easily to access others' research of potential interest.

The fact that selection depends on social processes also leads to a number of additional research questions. For example, even though market forces (supply/demand) are important, one implication is that even market selection should be analysed more thoroughly as a social process. In this book, for example, medical knowledge about the product was necessary to get government approval to sell at all. Another implication is that selection criteria cannot be fully determined in advance. For this reason, a hypothesis that agents try to maximize returns according to known selection criteria cannot be supported. The implications of this perspective for understanding technical and economic change should be further explored.

There are also important questions about how and why the generation of novelty is so closely related to selection in human knowledge processes. It has been argued that agents interpret environmental conditions when and while they generate novelty. This implies that during innovation processes, agents must internalize some conditions into the innovation process. For technological activities, this is done particularly through testing and through the development of models or other generalizable knowledge and techniques. Internalizing environmental conditions means that the firm narrows the range of diversity to a subset considered favourable or most valuable. This argument could be further explored through, for example, analysing whether and why some firms can select a more fruitful range of diversity than others.

As to whether economic or technological criteria are most important for innovation, this book illustrates how agents try to interpret them in tandem. Successful innovations are a combination of economic and technical aspects. For example, firms try to estimate what the conditions of sale and demand are likely to be before actually developing and selling a product. In connection with that, management must make decisions about the trade-offs between, for example, performance and price. However, they can also try to change the current trade-offs by increasing quality at the same price. Another alternative is to try to change environmental conditions if they are apparently unfavourable, as when negative public debate threatened to stop research involving genetic engineering.

Agents, however, are working under conditions of uncertainty about future selection criteria and about the results of their knowledge-seeking activities. They will therefore be more or less successful in interpreting those conditions a priori. For example, it is not surprising that Genentech and Kabi were not the only firms in the world which recognized the economic and technical potential of genetic engineering for pharmaceuticals in the 1970s. Nor is it surprising that some pharmaceutical firms decided not to invest in genetic engineering because they felt the risks and uncertainties to be too great. Firms made diverse calculations of risks and possibilities in a situation of uncertainty. Those firms which did develop early commercial uses of genetic engineering still made different decisions about how to access the necessary knowledge and how to proceed with actualizing their visions. Analysis of both firm-level and economic performance must therefore be based on firms' diversity of knowledge, experience, and strategy, and how their choices corresponded to economic and political conditions. The process of evolutionary innovation rests on social interaction as well as on individual decisions.

Concluding Remarks

The theoretical perspective of evolutionary innovation has provoked a number of research questions about economic and technological change in market economies. The analysis of evolutionary innovation unifies market incentives and institutional structures for understanding agents' decisions and behaviour when engaging in knowledge-seeking activities. By so doing, the book will hopefully provoke further reflections and discussions not only about technological innovation processes but also about the implications of an evolutionary and institutional approach for analysing such phenomena.

The contributions of this book lie in uniting empirical and theoretical domains of enquiry in order to address questions about science-based, also known as high-tech, innovation processes. The book presents original empirical material about genetic engineering and biotechnology and an analysis of how technological innovation processes occur. The current approach has implications for our understanding of the world around us and for additional research.

The conclusions to be drawn should therefore interest readers with a more practical orientation, such as R & D managers, researchers, and policy-makers in the fields of science, technology, and industry, as well as readers with a more theoretical interest in technological change. The latter category includes researchers and students from a variety of disciplines, such as economics of technical change, history of science and technology, management of technology, and sociology of technology. If nothing else, the past thirty years of research on science, technology, and innovation has indicated that technological change is a complex social phenomenon. It does not neatly fit into any one discipline or field of activity. Conclusions drawn on the basis of a multi-disciplinary approach can therefore contribute to discussions among a broad range of readers.

The benefits of this multi-disciplinary approach to science-based technology can therefore be as valid for researchers working specifically on technical change as for practitioners directly influencing technical change in firms and government agencies. An example of the latter case can clarify the argument. Let us assume that management in a firm accepts the idea that the development of technological knowledge includes both R & D activity for specific goals and R & D activity to keep up with new knowledge elsewhere but with no apparent return. In this case, the corporate R & D management can see the necessity of having search activities which apparently do not give returns to investment but which still give benefits in the realm of technological knowledge, such as giving absorptive capacity for identifying new options.

It also indicates that high-tech, or science-based, innovation processes in particular involve extended interaction and communication among researchers in different organizations such as universities and firms. This book clearly shows that researchers located in different types of organizations may be doing similar research for similar or different purposes and may be doing related but quite different research. There should be further analysis of how and why researchers at the different types of organizations communicate and how such shared knowledge can be encouraged through policy.

Multi-disciplinarity, the very core of this book, in itself indicates interesting directions for future research. Having chosen a research problem, it is possible truly to integrate different disciplines. Integration means that the specific questions asked, the methodology used, and the theoretical and conceptual discussions referred to come from more than one scientific discipline.

Conclusions for Science and Technology 299

The current book addresses questions and conceptual issues about technical novelty in firms and economies as raised in the economics of technological change.[3] It draws upon historical methodology and upon discussions in history and in sociology of technology about the cognitive dimensions of science and of technology and engineering. Finally, it develops the notion of coevolution of knowledge, institutions, and agents through an analysis integrating economic, sociological, and techno-scientific factors as modes of explanation of change.

One conclusion to be drawn is, therefore, that an approach which truly integrates economic, sociological, and historical approaches is a fruitful way to address research questions about technological innovation processes. The key to doing so is to select limited research questions dealing with complex issues which can be addressed in parallel. Such an approach is not always necessary or beneficial. There are many times when it is enough to compare parallel theoretical and empirical discussions. Nevertheless, multi-disciplinarity at the very core of analysis can open new perspectives and address existing questions in new ways. Like technological innovation processes, this process involves many dead ends and problems, new challenges and opportunities which do not exist but must instead be created and solutions proposed.

[3] It is important to restate that this heterodox tradition within evolutionary and institutional economics and economic history differs quite substantially from neo-classical economics, not least in its similarity to other social sciences in the conceptualization of individual action relative to context.

References

Andersen, Esben Sloth (1994), *Evolutionary Economics: Post-Schumpeterian Contributions* (London: Pinter Publishers).

Arrow, Kenneth (1962), 'Economic Welfare and the Allocation of Resources for Invention', in *The Rate and Direction of Inventive Activity: Economic and Social Factors* (Princeton, N.J.: Princeton University Press), 609–25.

Berg, Paul; Baltimore, David; Boyer, Herbert; Cohen, Stanley; Davis, Ronald; Hogness, David; *et al.* (1974), 'Potential Biohazards of Recombinant DNA Molecules', *Science* 185: 26 July, 303.

—— Brenner, Sydney; Roblin, Richard III; and Singer, Maxine (1975), 'Asilomar Conference on Recombinant DNA Molecules', *Science* 188: 6 June, 991–4.

Bernal, John D. (1967), *The Social Function of Science* (Cambridge, Mass.: MIT Press).

Bush, Vannevar (1945), *Science: The Endless Frontier. A Report to the President on Postwar Scientific Research* (Washington, DC).

Clarke, Adele and Fujimura, Joan (1992) (eds.), *The Right Tools for the Job: At Work in Twentieth-Century Life Sciences* (Princeton, N.J.: Princeton University Press).

Cohen, Stephen (1982), 'The Stanford DNA Cloning Patent', in Whelan, William, and Black, Sandra (eds.), *From Genetic Experimentation to Biotechnology: The Critical Transition* (New York: John Wiley and Sons).

Cohen, Wesley and Levinthal, Daniel (1989), 'Innovation and Learning: The Two Faces of R & D', *The Economic Journal* 99: Sept., 569–96.

Faulkner, Wendy and Senker, Jacqueline (1994), *Knowledge Frontiers: Public Sector Research and Industrial Innovation in Biotechnology, Engineering Ceramics, and Parallel Computing* (Oxford: Clarendon Press).

Financial Times (1995), 'Eli Lilly and Genentech Settle Dispute' (6 Jan.), 17.

Freeman, Christopher (1982), *Economics of Industrial Innovation* (London: Pinter Publishers).

Hippel, Eric von (1988), *The Sources of Innovation* (Oxford: Oxford University Press).

Hodgson, Geoffrey (1993), *Economics and Evolution: Bringing Life Back into Economics* (Cambridge: Polity Press).

Kline, Stephen, and Rosenberg, Nathan (1986), 'An Overview of Innovation', in Landau, Ralph, and Rosenberg, Nathan (eds.), *The Positive Sum Strategy: Harnessing Technology for Economic Growth* (Washington, DC: National Academy Press).

Levin, Richard; Klevorick, Alvin; Nelson, Richard; and Winter, Sidney (1987), 'Appropriating the Returns from Industrial Research & Development', *Brookings Papers on Economic Activity 3* (Washington, DC: The Brookings Institution).

Mowery, David, and Rosenberg, Nathan (1982), 'The Influence of Market Demand upon Innovation: A Critical Review of some Recent Empirical Studies', in Rosenberg, Nathan, *Inside the Black Box: Technology and Economics* (Cambridge: Cambridge University Press).

Nelson, Richard (1959), 'The Simple Economics of Basic Scientific Research', *The Journal of Political Economy 67*, 297–306.

—— (1990), 'Capitalism as an Engine of Progress', *Research Policy* 19, 193–214.

—— and Winter, Sidney (1982), *An Evolutionary Theory of Economic Change* (Cambridge, Mass.: Belknap Press).

Nilsson, Göran B. (1973), 'Om det fortfarande behovet av källkritik: Jämte några reflexioner över midsommaren 1941', *Historisk tidskrift* 1, 173–211.

Orsenigo, Luigi (1989), *The Emergence of Biotechnology: Institutions and Markets in Industrial Innovation* (London: Pinter Publishers).

Pavitt, Keith (1984), 'Sectoral Patterns of Technical Change: Towards a Taxonomy and a Theory', *Research Policy* 13, 343–73.

Scherer, Frederic (1984), *Innovation and Growth: Schumpeterian Perspectives* (Cambridge, Mass.: MIT Press).

Schmookler, Jacob (1966), *Invention and Economic Growth* (Cambridge, Mass.: MIT Press).

Schumpeter, Joseph (1968), *The Theory of Economic Development* (Cambridge, Mass.: Harvard University Press).

Tyson, Laura D'Andrea (1992), *Who's Bashing Whom? Trade Conflict in High-Technology Industries* (Washington, DC: Institute for International Economics).

Interviews and Personal Communication

Raines, Stephen (1992), Legal Departments, Genentech. Letter to author dated 10 Dec., (South San Francisco).

Part III
New Specialised Biotechnology Firms

[10]

Schumpeterian innovation and entrepreneurs in capitalism: A case study of the U.S. biotechnology industry *

Martin KENNEY

Department of Agricultural Economics and Rural Sociology, The Ohio State University, Columbus, OH 43210, USA

This paper uses the theoretical framework developed by Joseph Schumpeter to examine the first ten years of the U.S. biotechnology industry. The role of the entrepreneur, scientist/inventor, manager and capitalist are distinguished There is a discussion of the obstacles the small firms have had to overcome to create a "New Economic Space" in the marketplace. It is argued that the earlier Schumpeter of *The Theory of Economic Development* and *Business Cycles* emphasizing the role of small firms more accurately describes the U.S. biotechnology industry in 1985, than Schumpeter's later work *Capitalism, Socialism and Democracy* which postulated that the large established firms have preempted the role of small firms in innovation. A discussion of the role of large established firms and the small entrepreneurial firms presents the tensions inherent in the "cooperative" arrangements between these two types of business enterprises. The role of small biotechnology firms in reducing these innovations to practice and in their ability to continue to grow demonstrates that the independent entrepreneur recognized by the early Schumpeter has been very active in the biotechnology industry.

1. Introduction

The creation of the biotechnology industry in the midst of the global slowdown of the 1970s is a striking example of the innovatory capabilities of capitalism. It seems remarkable that the theories of Joseph Schumpeter written in the 1930s would once again be so prescient in predicting the response of capitalist economies to a prolonged cri-

sis. This paper is not a contribution to the debates regarding whether time-series data can establish the existence of long waves [1]. Rather we take up Schumpeter's microanalysis of the role of the entrepreneur in detonating a new upswing through his entrepreneurial activity. We will argue that a case study of the biotechnology industry yields a nearly perfect fit with the observations and hypotheses Joseph Schumpeter [2,3] formulated nearly 50 years earlier regarding the development of new capitalist enterprises. On the other hand, the hypothesis Schumpeter advanced, that large corporations have developed such a strong position that they will be able to stifle small entrepreneurial companies, is questioned in light of the development of the biotechnology industry [4].

The "new" biotechnology industry is a particularly appropriate case study for the following reasons [5]. The size of the market on which biotechnology will impact is enormous – potentially any industries processing or using organic matter of any sort. The important industries affected include agriculture, pharmaceuticals, organic chemicals, hydrocarbon energy, waste processing, and even mining. Estimates of the worldwide market for genetically engineered manufactured products in agriculture alone are as high as $50–100 billion per year by the year 2000 [6]. It seems likely that in 50 years many of the products currently synthesized through the use of organic chemistry will be produced by biosynthesis.

The second reason for examining the biotechnology industry is that the business environment in which the technology will be deployed includes large chemical/pharmaceutical companies aware of its importance. This sets the stage then for a comparison between new entrepreneurial companies and the large established corporations. A fur-

---

* Many of the arguments in this paper are adapted from my forthcoming book *Biotechnology: The Birth of an Industry* (Yale). I would also like to thank Richard Nelson and three anonymous reviewers for their perceptive, thorough, and useful comments. Obviously, any errors remaining are my own.

Research Policy 15 (1986) 21–31
North-Holland

ther fortuitous aspect of the industry is that the process of moving from invention to innovation is very clear. The inventions were made in the university and these were very visibly moved to industry. The Japanese, recognizing biotechnology's importance, have said that, "biotechnology will be the last major technology developed in the twentieth century" [7].

The author explicitly recognizes that the U.S. biotechnology innovation process is to some degree unique because of the role that the small biotechnology firms have played. In fact, a recent U.S. Office of Technology Assessment has specifically pointed out the importance of these companies in creating and maintaining the U.S. léad in biotechnology [8]. However, it is possible that in scale-up and marketing the large established companies may be able to overtake the small companies. The relative advantage in the scale-up and production phase may also shift to the large Japanese chemical, pharmaceutical, and fermentation companies [9]. Very obviously, at this historical juncture it is entirely impossible to predict which companies, countries, or even technologies will be most successful.

The paper first discusses Schumpeter's theory of innovation and the role of entrepreneurs integrating in some of the more contemporary work on innovation. The next section discusses the founding of the industry and the partnership that are formed between an entrepreneur and a university scientist. This is followed by a discussion of the obstacle to innovation that the small companies had to overcome in their effort to create the new industry. The role of the large established companies is also discussed and their relative tardiness in entering biotechnology [10]. The conclusion recapitulates the usefulness of Schumpeter's theories of entrepreneurial activity as a primal force in bringing radical innovations to the market.

2. Schumpeter, innovations, and entrepreneurs

Schumpeter's most extensive discussions of the role of innovation in the growth of capitalist economies are found in *Business Cycles* and the *Theory of Economic Development*. It is important to note that Schumpeter was very careful to separate the concept of inventions from that of innovation, "the making of the invention and the carrying out

of the corresponding innovation are two entirely different things" [3, p. 10]. The social processes involved with producing inventions and innovations belong to different spheres with complex interrelationships and "do not stand in any invariant relationship to each other" [3, p. 11]. Important inventions or scientific breakthroughs can occur without being incorporated into innovations affecting industry [1]. The innovation is the outcome of a process of combining production factors in novel ways to produce old products more efficiently or to create entirely new products. For example, tissue culture has been available for over 20 years as a scientific tool, but only became an innovation as industrial applications have been developed in the last 4 years [12]. Tissue culture was transformed from being an invention to being an innovation by business's application of the technique to production.

The process of innovation is not unproblematic, rather the innovator whom Schumpeter terms the "entrepreneur" or "New Man" is the central actor in transforming inventions to innovations [3, p. 77]. The entrepreneur's motivations fall into three categories: The first is to be successful financially and secure the possibly very large capital gains or entrepreneurial profits. The other important and, perhaps, vital motivation for the entrepreneur is an overwhelming desire to succeed and "conquer." The joy of creating a company is vital to the entrepreneur [2, p. 93]. Schumpeter repeatedly emphasized that his entrepreneur need not necessarily be the inventor nor need he be an investor in the new company.

Schumpeter's entrepreneur is a visionary who has had to struggle heroically against obstacles including: a lack of interest and skepticism among potential capitalists, prohibitions upon the use of the new machinery, a dearth of customers, and inadequately trained labor [3, p. 7]. The vast majority of the entrepreneurs are not capitalists (i.e. holders of large amounts of investable funds) and, therefore, must convince capitalists to provide sufficient capital to purchase the inputs necessary to establish the company. The critical input of the entrepreneurs is not managerial expertise, but it is rather his vision and energy that is so necessary to a fledgling company. In the Schumpeterian scheme entrepreneurial activities are the key to economic development and progress, because they open the economic space into which capitalism will expand.

In his earlier works Schumpeter thought that the "new firms" will grow and eventually displace firms whose growth had been based on earlier innovations. In fact, Schumpeter argues [3, p. 69] that:

> Most new firms are founded with an idea and for a definite purpose. The life goes out of them when that idea or purpose has been fulfilled or has become obsolete, it has ceased to be new. That is the fundamental reason why firms do not exist forever. Many of them are, of course, failures from the start. Like human beings, firms are constantly being born that cannot live.

This model has been termed "Model 1" by Freeman [13,14].

But Schumpeter [3, p. 118] was very careful not to argue that all "old" firms would expire:

> For some of the 'old' firms new opportunities for expansion open up: the new methods or commoditives create New Economic Space. But for others the emergence of new methods means economic death; for still others, contraction and drifting into the background.

Schumpeter's last work, *Capitalism, Socialism and Democracy*, puts forward Model 2 which essentially argues that large capitalist organizations now control innovation. The strength of these companies led Schumpeter to despair of the potential for new entrepreneurs to succeed in establishing new firms in the face of the growing power and research capabilities of the trusts [4, p. 133]. Recent research has indicated that 'Model 1' is not necessarily completely outmoded especially in the electronics and medical instruments industries [14, p. 138]. And, in fact, this paper will argue that Schumpeter's 'Model 1' is very applicable to the biotechnology industry.

The creation of the new firm by the entrepreneur is an individual act, but Schumpeter argued that these innovations tended to cluster in certain historical periods. This is borne out in the recent work by Mensch who postulated that basic innovations clustered in periods of depression [15]. The chronological clustering of innovations occurs in sectors of the economy which were usually related to the originally innovating sector. So, for example, biotechnology was first envisioned as a pharmaceutical technology, but by 1981 other entrepreneurs perceived the market opportunities in industries such as agriculture and waste processing

and founded companies to exploit these markets.

The market potential created by new technologies and possible new products encourages a rush of entrepreneurs into what Schumpeter termed a "New Economic Space." The rush into this new space is not limited to new firms. Older firms that have more farsighted management and have provided internal entrepreneurial space will also expand into this growing new area. These larger established firms have advantages over the new firms in that they have superior facilities and ready access to cash. This reality was the basis for Schumpeter's conclusion that smaller companies had little opportunity to be successful.

3. The birth of the biotechnology industry

The scientific basis for the new biotechnology industry was built in the molecular and cellular biology, virology, and immunology laboratories of the world's research universities. The single most important invention was made by Stanley Cohen of Stanford University, Herbert Boyer of the University of California, San Francisco, and their research teams when they succeeded in reprogramming a bacterium's DNA to express a foreign DNA sequence [16]. This invention would spark a multibillion dollar industrial investment in biology [17].

The idea of commercially exploiting genetic engineering was conceived by Robert Swanson, MIT chemistry undergraduate and Harvard MBA [18]. In 1975 Swanson had been employed by Kleiner, Perkins, Caufield, and Byers, a venture capital fund, managing their investment in Cetus, a company involved in cell selection technology. While performing this function, Swanson became aware of the potentials of genetic engineering and left Kleiner, Perkins to found his own company to exploit genetic engineering [19]. At this early stage, the esoteric skills which are required to use the techniques of genetic engineering were only available in the laboratories of a select few professors in the life science departments of the world's best research universities.

Swanson, after contacting a number of professors, discovered Herbert Boyer, a professor at the University of California, San Francisco, and principal in the Cohen–Boyer gene splicing patent. The two men formed a partnership and launched

Table 1
Startup dates of biotechnology firms: Important companies and total per year [a]

Year	Number	Important Companies
1971	2	Cetus
1972	1	Bioresponse
1973	3	
1974	1	
1975	1	
1976 [b]	2	Genentech
1977	3	Genex
1978	6	Biogen, Hybritech, Collaborative Research
1979	9	Molecular Genetics, Monoclonal Antibodies
1980	18	Calgene
1981	33	Genetic Systems, Integrated Genetics
1982	11	
1983	4	

Source: [42]

[a] The data in this table is based on information that the author believes is accurate. The author did not list every firm, but rather selected the most important companies directly involved in biotechnology.

[b] Genentech is the first company to be entirely devoted to genetic engineering.

Genentech with a $100,000 investment from Kleiner, Perkins [20]. With this and further investments Herbert Boyer's team working in his university laboratory succeeded in 1978 in inducing a bacterium to produce a human hormone, somatostatin. As can be seen from table 1 listing founding dates of small companies, Genentech's success convinced other entrepreneurs to enter the industry. It is interesting to note that until 1980 the large established chemical/pharmaceutical companies made very few investments in biotechnology [8, pp. 100–101].

The four social roles: the entrepreneur, the inventor/technician, the manager, and the capitalist that Schumpeter saw as critical for the creation of new firms have been present in all of the new biotechnology companies, though in certain companies the roles have been combined in one individual (see table 2). In all cases a scientist must be involved in forming the company, but the scientist may also be the entrepreneur with the vision and dedication to go out and seek funding. In these cases, the professor usually approaches a venture capital fund which evaluates the proposed company, its employees and product targets, and decides upon its commercial potential. The capitalist must make his funding decision on the basis of a business plan and the evaluation of the personnel involved, because the proposed company has no assets – a decision fraught with risk [21].

Swanson has called Genentech "a kind of mar-

Table 2
New biotechnology firm personnel classified according to their role and previous positions occupied [a] [d]

Company	Entrepreneur [b]	Scientist [b]	Manager [b]	Capitalist (initial) [b]
Genentech	R. Swanson, ex-venture capital employee	H. Boyer, professor UCSF	G. Kirk Raab, ex-pres., Abbott Labs.	T. Perkin, venture capitalist
Genex	R. Johnston, venture capitalist	D. Jackson, ex-prof. Univ. of Michigan	L. Glick, president Assoc. Biomedic. Sys.	R. Johnston, venture capitalist
Advanced Genetic Sciences	D. Adams, ex-venture capital employee	L. Bogorad, ex-prof. Harvard	T. Dyott, manager, Rohm and Haas	N.A. [c]
Genetic Systems	D. Blech, ex-stock broker	R. Nowinski, ex-prof. Univ. of Washington	J. Glavin, vice pres. Oximetric Inc.	J. Connors, venture capitalist
Hybritech	B. Byers, venture capital principal	I. Royston, professor, UCSF	H. Greene, asst. to president, Baxter Travenol	Kleiner, Perkins [c]
Amgen	Rothschild, Inc. [c]	M. Caruthers, professor, Univ. of Colorado	G. Rathmann, vice-president, Abbott Laboratories	Rothschild, Inc. [c]
Integrated Genetics	D. Anderson, venture capitalist	D. Housman, professor, MIT	R. Carpenter, pres. of Fenwal Division, Baxter Travenol	D. Anderson, venture capitalist

[a] Cetus has been omitted because it was founded before genetic engineering developed.
[b] Many of these roles may include more than one individual, but this table includes only the most important single individual.
[c] Further information unavailable.
[d] The information contained in this table is complete and accurate to the best of the author's knowledge.

riage between science and business" [20, p. 66]. In this partnership, Swanson developed criteria for commercial feasibility and Boyer decided what was scientifically possible. The product goals were the result of an iterative cooperative process between the two men. In nearly all of the companies there is somewhat of a blurring of the entrepreneurial role between the senior researcher(s) who has the scientific vision and an entrepreneur who provides the economic vision and goals that motivate the new firm. The extreme complexity of the technology makes it unlikely that a scientist could also be the entrepreneur. For example, Walter Gilbert formerly a professor of biology at Harvard and then president of Biogen recently resigned as Biogen's chairman possibly because of the enormous losses Biogen was sustaining [22]. At the managerial level, in many cases, a professional manager is hired to undertake corporate day-to-day management, as in the case of Genentech where Robert Swanson turned over the presidency and chief operating office roles to G. Kirk Raab, formerly an executive at Abbott Laboratories in 1985 after Swanson had run the company for eight years [23].

4. The obstacles to innovation

The launching of a company based on a new technology is not a simple task and, in many ways, is more difficult in biotechnology than it was in many other industries. This section briefly discusses the obstacles to innovation that were experienced by the biotechnology industry and which, in large measure, have now been overcome. There can be no doubt that the growth of the biotechnology industry in the U.S. was very much assisted by the precedent of the microelectronics industry.

In few cases do the founding entrepreneurs risk large amounts of personal capital. For example, Swanson and Boyer contributed only $500 each to form Genentech in 1976 [20, p. 66]. Estimates of start-up costs in 1980 for a small genetic engineering research facility were in the range of $6–7 million for the first three years. A larger more econimically viable company would require $10–12 million financing for the first three years [24]. A small hybridoma business venture would be slightly less expensive; approximately $3.5–4 million over three years, the bulk of these expenses would come

in the form of salaries [25]. Clearly, capital investments of this magnitude are usually too large for most individuals and an outside source of capital is mandatory.

In the very early stages of biotechnology (1976–79), neither large corporation nor banks exhibited great interest in investing in such an untested technology. This forced the fledgling companies to seek financing from venture capitalists, that is, financiers who specialize in provding the initial capital to prospective companies in return for significant blocks of equity and seats on the board of directors. Thus, obstacles regarding startup funds were not as serious in biotechnology as in earlier industries.

More important obstacles to the use of recombinant DNA than lack of capital was the public outcry regarding biohazards [26] and, until the 1980 Chakrabarty Supreme Court decision, uncertainty regarding the patentability of living organisms. In this case, the new machinery of production, microorganisms, became the subject of controversy that at various points could have led to stringent regulation or a moratorium on scientific and commercial development. The debate regarding recombinant DNA in the period from 1974 to 1978 discouraged some companies, especially larger established companies, from attempting to enter the genetic engineering field. As Schumpeter presciently observed, these prohibitions, though initially important obstacles, were soon overcome and the fledgling industry went forward.

In the pharmaceutical industry it seems likely that genetically engineered products will be readily accepted as new products are commonplace and adoption is rapid. This, in conjunction with the fact that genetically engineered pharmaceuticals are professionally dispensed, ensures that consumer acceptance is prompt. Indeed, rather than problems of acceptance of genetic engineering, the converse has been true. Acceptance has been prompt, but few products have been produced. Further, even with only a few commercially viable products, the industry has found that capital is easy to acquire. This can be explained by the belief among investors that the products will be easily marketable, thus the lack of customers will not be a significant obstacle. Nevertheless, there still has been some apprehension within the genetic engineering industry regarding public acceptance [27].

The lack of sufficient qualified labor was initially a problem [28]. The primary source of experienced labor was the university, but to lure professors wages have been bid-up to high levels. In the frenzy among both small and large companies to secure scientists in the period from 1979 to 1982 salaries were bid up to "more than $50,000 per year along with fringe benefits and stock options" for "promising relatively inexperienced scientists" [29]. In the current phase of the industry's growth (1985) labor shortages in biotechnology have eased. To overcome the earlier labor shortages companies and public universities developed numerous strategems to enlarge the labor pool. For example, the University of Houston has developed a program to familiarize undergraduates with simple recombinant DNA procedures [30]. The University of Maryland, Baltimore County has initiated courses in applied molecular biology [31]. Industrial response has been to develop gene splicing machines that can displace skilled workers. Finally, with the maturing of many corporate research efforts the types of skills desired have undergone a change. Whereas earlier, immunologists, microbiologists and molecular biologists were in greatest demand, the need has now shifted to fermentation, separation, and biochemical engineers. This reflects the maturation of the technology to the point of scaling-up production to commercial level.

The problems analyzed in this section corresponded to those indicated by Schumpeter. Contrary to his emphasis on the scarcity of capital as the most important limiting factor this was a bottleneck only in the very earliest days of the industry. However, there have been a few bankruptcies and it is clear that in 1985 capital is no longer automatically forthcoming and is almost entirely unavailable for new biotechnology companies. Increasingly, financiers are requiring that the biotechnology entrepreneur have a concrete business plan. Finally, the societal institutions are addressing the scarcity of qualified workers by developing training programs. The surmounting of these obstacles has brought ever greater numbers of entrepreneurs into genetic engineering and is creating a clustering of innovation chronologically and sectorally as the companies search for innovations and less crowded niches within which to survive.

5. The entrepreneur and the venture capitalist

The venture financier, a capitalist whose sole purpose is to invest in new firms is a unique role not envisioned by Schumpeter. These venture capital investors actually contribute capital to the company in hopes of being able to realize a capital gain upon resale of their holdings to a large company or the public through an equity offering. As mentioned earlier venture capitalists hope that their individual investments will bring a 10-fold return or, even better, the jackpot returns that can soar another order or magnitude higher. Thomas Sager of Rothschild, Inc. is quoted as saying, "We're going in for five to seven years." [32]. A long-term investment by most U.S. corporate standards, but in deploying a new technology this is not long-term. Venture capitalists provide the capital that allows the entrepreneur to form a company in the hopes that it will become self-sustaining. For the venture capitalist the new company and its expertise become the commodity to be sold.

The venture capitalist even more than a banker is apt to become involved if the company has difficulties. In some privately held biotechnology companies such as Biologicals or the International Plant Research Institute (IPRI) the corporate founder and president (the entrepreneur) were replaced by better managers due to the founder's inability to properly manage the firm [33]. Another less drastic example is the case of Bethesda Research Laboratory (BRL), a privately held general biotechnology company which was formed by Stephen Turner, an entrepreneur. In late 1981, the company experienced cashflow problems and was forced to seek further financing from venture capitalists. An important aspect of the refinancing was the "transfer of power, no matter how limited, from the entrepreneur to outsiders" [34].

An even more striking case of the potential struggle between entrepreneur and capitalist is that of the new defunct Armos. Frederick Adler, the president of Adler and Co., a venture capital fund, was quoted as saying [34]:

"The people (at Armos) turned down capital. It was the biggest mistake they made, but you can't blame them. Their price was too high due to a wrong perception of the market for investment money." For them to raise money, moreover, would have meant giving up control of the company, something Sheehan and Carlock (the founders) apparently were unwilling to do. "If somebody refuses money they can retain control but lose the company."

The relationship between the entrepreneur and the financier can be strained and at times the financier will bring in new management to replace the entrepreneur.

On the other hand, the relationship between venture capitalists and the entrepreneur is not always one-sided. Genentech, the most successful of the new companies, has used stock offerings, contract research with large companies, and the first royalties on the insulin which Eli Lilly Pharmaceuticals is producing with Genentech-engineered bacteria to move to the next stage in its plan to become a full-fledged company. It has also used offers of research and development partnerships (RDLPs) to raise capital through private placements [35]. RDLPs provide investors a non-equity return, similar to royalties, on any products developed. The RDLP is an important step forward because it reduces the dependence of a fledgling company on an outside company for the development and testing of its products with the result that a greater proportion of the production process is brought in-house. The RDLP is a vehicle that permits the small company to move from a research mode to a production and distribution mode. Success permits the company to become a viable growing corporation that is financially independent.

6. The large companies [36]

There can be no doubt that the genetic engineering companies carved out a new economic space. Initially the strategy of the large companies was to refrain from investing in genetic engineering. However, their response changed as the success of the new companies became apparent and the large companies began to make a number of biotechnology investments. The implication that many observers draw (in opposition to Schumpeter's theories as presented in his first two books, but agreeing with his later ideas) is that the large firms with their massive hoards of capital will soon overtake the small firms. Dow Chemical's director of biotechnology, Dr. Donalds, now president of Collaborative Research Inc., a small biotechnology startup [37], agrees:

The small companies of every new industry start out running circles around the old companies, but they still need size to do certain things. Don't think [recombinant DNA] will revolutionize the way business is done. The technology is revolutionary, but it will get sorted out in the end.

At Dupont similar echoes are heard. For example, Dr. Howard Simmons [37, p. 9] Dupont's director of central research, is quoted as saying:

Recombinant DNA is just a way to synthesize things. As soon as you've inserted the gene, it's identical to what you would do in the chemical industry anyway. [The small firms will] play an important role in the early years, but less so down the line. In the early years, people will buy discoveries, but why buy from somebody and share the profits when you can do it all yourself?

Both of these companies used relatively passive strategies concentrating on developing in-house programs to spearhead their efforts in the biotechnology. Clearly, as Donalds indicates, things will "get sorted out in the end" but which corporations get "sorted out" is not a foregone conclusion. It is interesting to note that Donalds is now the president and chief executive officer of Collaborative Research, Inc., a small biotechnology firm [38] and Dupont lost its director of biological research, Ralph Hardy, to Biotechnica International, Inc., a small Massachusetts company.

There has been a significant amount of investment by large companies in the small biotechnology companies (between $500 million and $1 billion) in the last three years and not every company has a blasé attitude to biotechnology. In contrast to Dupont and Dow, companies such as Monsanto and Schering-Plough have invested large sums in all aspects of biotechnology research and development (see table 3 for major investments by Monsanto and table 4 for Schering-Plough). These companies apparently feel that biotechnology could be very important for their futures.

The relationship between the large established companies and small biotechnology firms is contradictory. The small company has technology the large firm wishs to secure. Conversely, the small company's strategy is thus to attempt to transfer as little technology as possible and to build its in-house skills while working for the large company. Thus, the small companies try to sell marketing rights or even to deliver a recombined organism, but not technology. As the small company gets stronger it may also secure the right to provide a percentage of the large company's product requirements, thus learning to scale-up production. Similarly, the small company may sell overseas production and marketing rights while keeping its domestic ones. Further, the small company is not merely the pawn of one large company, but, in fact, nearly all have contracts with a number of different companies thereby ensuring more flexibility and independence. The motivation of the small company is to maximize its equity value and to do this it must become more than a contract research operation [39].

The large company is motivated by a different set of criteria. First, most large companies did (do) not have access to the best university scientists and contracting with the small companies provided(s) this access. The large companies need access to new products to keep their marketing pipeline full and if these products must be purchased so be it. However, the large companies have also developed in-house research staffs and obviously would like

Table 3
Monsanto's investments in biotechnology [a]

In-house investment
  $185 million invested in biological sciences research center
Biotechnology companies (equity investments and important contracts)
  Collagen – artificial bone materials
  Biogen – tissue plasminogen activator
  Genentech – bovine growth hormone
  Genex – venture capital investment
  Biotechnica International – *B. subtilis* protein expression
University contracts
  Harvard University – biomedical research ($23 million)
  Washington University – biomedical research ($23 million)
  Rockefeller University – photosynthesis research ($4 million)
  Oxford University – sugar chains ($1.5 million)
Drug companies
  G.D. Searle – purchased for $2.7 billion, manufacturer of aspartame and has major biotechnology research facilities in U.S. and Great Britain.
Seed company subsidiaries

Jacob Hartz Seeds	Monsanto Seeds
Hybritech Seed Co.	Farmers Hybrid Co.

Source: adapted from [46].
[a] Information complete to the best of author's knowledge. However, given the secrecy that surrounds the industry all information is subject to error.

Table 4
Schering-Plough biotechnology investments [a]

In-house investment
  Not available
Biotechnology companies (equity investments and important contracts)
  DNAX Research Institute – wholly-owned subsidiary
  Biogen – 12 percent ownership and research contracts
recombinant DNA production facilities
  Interferon plant in County Cork, Ireland – $106 million
Joint ventures
  Takeda Pharmaceutical for interferon production, may also be broadened to other areas.

Source: Compiled by author.
[a] Information complete to the best of the author's knowledge.

to bring the research in-house. There are cases in which large companies have abruptly cancelled contracts when they secured the knowledge they required and were able to do the remainder of the research in-house. The arrangements between the small and large companies are marriages of convenience with both partners needing each other for the movement.

Leslie Misrock, a lawyer specializing in recombinant DNA litigation, has described the environment in the new biotechnology firms thus "The real innovation comes out of the hothouse atmosphere at universities and small companies, but is stifled at large companies" [32, p. 10]. Robert Luciano, a Schering-Plough executive vice-president, said about Biogen, "You just couldn't hire people (Biogen's scientists) like that to work in an industrial setting" [40]. The small biotechnology companies with university-like atmospheres are much more conductive to creative scientific research. For some reason entrepreneurs have been better able to create an internal corporate environment that makes mental laborers more innovative. Further, most large corporatons have been unsuccessful in their attempts to develop small " venture capital" divisions [41].

Schumpeter correctly described the exploitation of a new economic space by the earliest entrepreneurial companies. Large companies and many smaller companies have since moved into this space. Schumpeter's observations that entrepreneurs appear in swarms is certainly borne out; in 1976 there was only one true genetic engineering firm, but by 1982 estimates were that up to 250 firms in the U.S. alone were using the new genetic engineering techniques [42]. Clearly this swarm-like behavior was facilitated by the pioneering successes of the first companies which then prompted other entrepreneurs to launch their own companies.

Regardless of future failures, some small companies will probably survive and prosper. There should also be no doubt that some large companies will also fail. Similarly, the bulk of the smaller firms will not survive if they cannot develop a true competitive advantage. These will either be merged with larger firms or go bankrupt and disappear – an entirely natural process in capitalism [43].

7. Conclusion

This paper has not attempted to answer larger questions concerning the reasons that biotechnological inventions are becoming such an important area of investment. Rather I have tried to describe certain salient surface features of this growing and, in the U.S., now established industry. Because this paper has concentrated on microlevel data, that is, concrete individuals and their activities, it has an obvious affinity with the work of Schumpeter and Mensch. Thus, macrolevel theories such as those relating to capital accumulation or focusing on the demand for labor are not contradicted by my approach [44]. Rather, my focus has been to contribute to a rehabilitation of the sociological aspects of Joseph Schumpeter's work.

I have demonstrated that the entrepreneurial role in genetic engineering has usually been a partnership between an entrepreneur and a scientist. This is due to the very complex nature of genetic engineering. This division of labor reflects the new realities of both science and finance – only a specialist can handle either of these areas. Currently, for all these new companies research is critical. As Genentech's Swanson puts it, "if the research goes well we can handle the rest of the problems of the world [20].

A most intriguing new feature in today's innovation is the role of the venture capital financier, a role that in Schumpeter's times was not yet so clearly defined and demarcated. These capitalists search out innovations with the aim of realizing enormous capital gains when the new company and its innovation achieve sufficient maturity to be either sold to another company or to have their stock sold to the public. The venture capitalists are not merely passive investors, but rather they demand and receive seats on the corporate board of directors. If the fledgling corporation is not meeting benchmarks the capitalist's financial power is such that even the entrepreneur can be replaced.

Venture capital appears to be a systemic response to the behavior of the large U.S. companies, many of which find it difficult to provide an internal environment conducive to radical innovation. These large firms, content to buy small profitable firms, may find the most successful new genetic engineering companies growing so fast that they will become too large and profitable for purchase. Biotechnology companies, thus far, have largely surmounted the problems of shortages of financing, customers, labor and the public's fear of the unknown. A race is underway for all companies, large and small, to ensure survival in this increasingly competitive industry [45].

Schumpeter's analysis of capitalism remains as fresh and timely today as when it was written nearly 50 years ago and provides a powerful tool for analyzing the innovation process in genetic engineering. It is interesting to speculate as to whether the tremendous impact of genetic engineering and, more generally, biotechnology will provide a sufficiently high number of products and cheapen existing production processes enough to become a keystone technology in undergirding another kondratieff upswing. Once again in American capitalism, the role of the unattached entrepreneur has been crucial to commercializing a major new technology. If, as Schumpeter and Mensch theorized, entrepreneurial activity is an indication of innovation and the imminence of an economic upswing, then certainly the current activity in the U.S. biotechnology industry, in particular, and the economy, in general, is a promising sign for the U.S. economy.

Schumpeter's work remains the fundamental model for analyzing the mechanisms through which innovation is introduced in capitalist societies. Though many aspects of the innovation process have changed, the broad outlines of the process of innovation remain almost identical to his observations made 50 years earlier. The concept of the creation of a new economic space is richly suggestive not only of research directions for social scientists interested in innovation, but also public policy initiatives. The lack of historical perspective in much discussion regarding the formation of new companies to exploit genetic engineering or any "high tech" could be remedied by consulting Schumpeter's works. Schumpeter's analysis of the role of the entrepreneur in the rejuvenation of capitalist economies remains as fresh today as when he wrote in the 1930s. The vitality of the Schumpeterian model in providing insights into the process of innovation in capitalist economies has been clearly demonstrated in this paper. In fact, reflection upon Schumpeter's theories could assist U.S. policymakers in designing public policy to favor increased entrepreneurship.

Notes and References

[1] See, for example, C. Freeman, J. Clark and L. Soete, *Unemployment and Technological Innovation: A Study of Long Waves and Economic Development* (Francis Pinter, London, 1982). J. Forrester, Growth Cycles, *De Economist* 125 (1977) 525–543. J.J. van Duijn, *The Long Wave in Economic Life* (Goerge Allen and Unwin, London, 1983).

[2] J. Schumpeter, *The Theory of Economic Development*, (Harvard University Press, Cambridge, MA, 1934).

[3] J. Schumpeter as abridged by Reindig Fels, *Business Cycles* (McGraw-Hill, New York, 1964).

[4] J. Schumpter, *Capitalism, Socialism, and Democracy*, (Harper, New York, 1942).

[5] The term "new biotechnology industry" refers to the new biological production technologies that have been developed in the last 10 years. These techniques include: genetic engineering, tissue culture, hybridomas, DNA probes, etc. It is fully recognized that these new techniques are dependent for their successful deployment on previous "biotechnologies" such as plant breeding, fermentation, and bioprocess engineering. For further discussions of what the term "biotechnology" might mean there are numerous references. For the sake of simplicity this paper confines itself to the recently developed biotechnologies, while clearly recognizing the historical antecedents of biotechnology. For further discussions of the technology see: U.S. Congress Office of Technology Assessment, *Impacts of Applied Genetics* (U.S. Government Printing Office, Washington, DC, 1981). A. Bull, G. Holt and M. Lilly, *Biotechnology: International Trends and Perspectives*, (OEDC, Paris, 1982). P. Dunnill and M. Rudd, *Biotechnology and British Industry* (Biotechnology Directorate of the SERC, London, 1984).

[6] J. Murray and L. Teichner, *An Assessment of the Global Potential of Genetic Engineering in the Agribusiness Sector*, (Policy Research Corporation and Chicago Group Inc., Chicago, 1981).

[7] H. Samejima (Managing Director, Kyowa Hakko, Kogyo Co.), Interview by author (June 1984).

[8] United States Congress Office of Technology Assessment, *Commercial Biotechnology: An International Assessment* (U.S. Government Printing Office, Washington, DC, 1984).

[9] M. Kenney, Biotechnology and the Information Age: Some Observations on the Commercial Competition between the U.S. and Japan, unpublished manuscript. P. Katzenstein and S. Tanaka, Biotechnology: Japan's Industry in Competitive Perspective, paper presented at Policy for High Technology in Japan: An Example for the United States Conference, (New York, March 17–19, 1984).

[10] For a much longer and more in-depth discussion of the points, see M. Kenney, *Biotechnology: The Birth of an Industry* (Yale University Press, New Haven, CT, 1986).

[11] C. Freeman, J. Clark and L. Soete, *Unemployment and Technological Innovation: A Study of Long Waves and Economic Development*, (Francis Pinter, London, 1982), pp. 87–88. J. Schmookler, *Invention and Economic Growth*, (Harvard University Press, Cambridge, MA, 1966). R. Nelson and S. Winter, *An Evolutionary Theory of Economic Change*, (Harvard University Press, Cambridge, MA, 1982).

[12] For discussion of tissue culture as an industrial process, see M. Kenney, F. Buttel and J. Kloppenburg, Jr., Understanding the Socioeconomic Impacts of Plant Tissue Culture Technology on Third World Countries, *ATAS Bulletin* 1 (1985).

[13] C. Freeman, et al., pp. 38–42, see [1].

[14] C. Freeman, *The Economics of Industrial Innovation*, 2nd edn. (Francis Pinter, London, 1982), pp. 211–214.

[15] G. Mensch, *Stalemate in Technology*, (Ballinger Press, Cambridge, MA, 1979).

[16] S. Cohen, *et al.*, Construction of Biologically Functional Bacterial Plasmids *In Vitro*, *Proceedings of the National Academy of Sciences* 70 (1973) 3240–3244.

[17] There are very few calculations of the total capital invested in biotechnology, but Genentech's assets as of June 30, 1984 were $127 million. As table 3 indicates Monsanto has invested in excess of $200 million in biotechnology-related research. It is safe to conclude that total global biotechnology investments are far in excess of $3 billion.

[18] Genetic engineering is a subset of the numerous commercially valuable tools that come under the heading "biotechnology."

[19] E. Sylvester and L. Klotz, *The Gene Age* (Scribner's and Sons Press, New York, 1983).

[20] S. Benner, Genentech: Life Under a Microscope, *Inc.*, May (1981) 62–68.

[21] The venture capitalists are aiming for returns ideally at least ten times the value of the investment, but returns can soar as high as hundreds of times their investment. On the other hand, bad investments can result in loss of the entire investment. However, in aggregate, venture capital funds have outperformed the stock market and inflation.

[22] Dr. Gilbert left Harvard to become president of Biogen full-time, but in December 1984 resigned from Biogen to be replaced by a professional manager. For further discussion, see W. Bulkeley, Biogen's Chief, Walter Gilbert, Quits Top Posts, *Wall Street Journal* (18 December) 22. J. Sterling, Walter Gilbert Resigns as Biogen Chairman, *Genetic Engineering News* 5 (1985) 1, 48.

[23] *Business Week*, Genentech Tries for Growth, *Business Week* (March 4, 1983) 40.

[24] N. Schneider, Prepared statement of Nelson Schneider at a hearing before the Subcommittee on Science, Technology,

and Space of the Committee on Commerce, Science, and Transportation, U.S. Senate, *Industrial Applications of Recombinant DNA Techniques* (May 20, 1980) 45.

[25] M. Treble, Scale-up of Hybridoma Business Ventures: Investment Requirements and Perspectives, *Genetic Engineering News* 2 (July/August 1982) 5.

[26] J. Lear, Recombinant DNA: The Untold Story, (Crown Press, New York, 1978). S. Krimsky, *Genetic Alchemy: The Social History of the Recombinant DNA Controversy*, (MIT Press, Cambridge, MA, 1982).

[27] T. Powledge, Public Education Urged to Counter Biotech Critics, *Bio/Technology* 2 (1984) 8–10.

[28] *Chemical Week*, Wanted: More Genetic Engineers – and Soon, *Chemical Week* (March 29, 1981) 29.

[29] J. Fox, Can Academia Adapt to Biotechnology's Lure?, *Chemical and Engineering News* (March 29, 1981) 40.

[30] R. Reiss, Houston U. Trains Gene Workers, *Genetic Engineering News* (May/June 1981) 17.

[31] A. Adelman, Biotechnology Growth Spawns Generation of New University Technical Training Programs, *Genetic Engineering News* (November/December, 1982) 1, 27.

[32] J. Fox, Biotechnology: A High-Stakes Industry in Flux, *Chemical and Engineering News* (March 29, 1982) 10–15.

[33] Biotechnology Firm IPRI in Financial Trouble, *Chemical and Engineering News* (November 1, 1982) 6. IPRI Founder to Form New Biotech-Computer Venture, *Genetic Engineering News* (July/August 1982) 3.

[34] J. Fox, Armos: Profile of Biotechnology Firm's Failure, *Chemical and Engineering News* (September 13, 1982) 12.

[35] The most recent Genentech RDLP was completed in 1984.

[36] The large companies involved in the genetic engineering industry include: Abbott Laboratories, Allied Corporation, Baxter Travenol, Damon, Dupont, Exxon, General Foods, W.R. Grace, Johnson and Johnson, Monsanto, Merck, Schering-Plough, Stauffer, and many others.

[37] A. Brown, Can the Gene Splicers Survive Commercial Success? *Chemical Business* (July 26, 1982) 16.

[38] Collaborative Research, Inc., *Annual Report*, (Collaborative Research, Inc., Waltham, MA, 1984).

[39] M. Kenney, *Biotechnology: The Birth of an Industry* (Yale University Press, New Haven, CT, 1986 in press).

[40] G. Bylinsky, DNA Can Build Companies, Too, *Fortune* (June 16, 1980) 144–153.

[41] N. Fast, Pitfalls of Corporate Venturing, *Research Management* 24 (1981) 21–24.

[42] *Genetic Engineering News*, Third Annual *GEN* Guide to Biotechnology Companies, *Genetic Engineering News* (November/December 1984) 4–23.

[43] Repeated predictions of an impending shakeout in the industry have proven premature and companies that have not produced a commercial product have successfully secured repeated capital infusions. Only three companies of sufficient significance to warrant any publicity have gone bankrupt (Armos, Southern Biotech, and LEE Biomolecular). These bankruptcies were in every case due to bad management. In the Schumpeterian scheme these were the "failures from the start" and should be considered the normal casualties of the innovation process. It is, however, increasingly clear that there will be more casualties as companies are increasingly judged by their ability to develop products that will generate a cash flow—they can no longer merely consist of a research program. As Brook Byers, the chairman of Hybritech Inc. and a partner in the venture capital firm of Kleiner, Perkins, Caufield and Byers, recently said: "Next year (1983) is the year all good companies must prove themselves. People won't invest in professors anymore" (*Business Week*, Biotechnology is Now Survival of the Fittest, (April 12, 1982) 36–37). This was again premature, but a day of reckoning must come. The innovators must prove that their innovations are economically viable.

[44] Examples of macrolevel capital accumulation theories are: E. Mandel, *Late Capitalism*, (Verso Books, London, 1978). J. Forrester, Growth Cycles, see [1]. Those focusing on labor are best represented by C. Freeman, et al., *Unemployment and Technical Innovation: A Study of Long Waves and Economic Development*, see [1].

[45] For an excellent discussion of a similar process in the semiconductor industry, see G. Dosi, *Technical Change and Industrial Transformation* (Macmillan Press, London, 1984).

[46] F. Buttel, M. Kenney and J. Kloppenburg, From Green Revolution to Biorevolution, *Economic Development and Cultural Change*, forthcoming.

[11]

The Role of Small Firms in
U.S. Biotechnology Clusters

ABSTRACT. The biotechnology industry poses a particular challenge to analysis because its origins were poorly understood or even noticed at the time. The purpose of this paper is to examine and make sense of recent developments in the U.S. biotechnology industry. The pioneers and subsequent followers in biotechnology came from other fields. They altered their career trajectories in sciences like genetics and medical research to enter an unknown and undefined field, which only subsequently became labeled as biotechnology. Those regions exhibiting the greatest success in developing biotechnology clusters also possessed the greatest ability to unleash the potential commercialization latent in those scientists. To generate a successful regional cluster, the existence of world class scientific talent is a necessary condition. However, it is not a sufficient condition. The ancillary or complementary factors must also be available to translate this knowledge into a commercialized product. The complementary factors include the presence of venture capital and other forms of finance, the existence of an entrepreneurial culture, and transparent and minimal regulations fostering the start-up and growth processes.

1. Introduction

The purpose of this paper is to examine and make sense of recent developments in the U.S. biotechnology industry. This paper tries to understand how the American biotechnology industry has evolved, with a particular emphasis on the role of public policy. The biotechnology industry poses a particular challenge to analysis because its origins were poorly understood or even noticed at the time. The pioneers and subsequent followers in biotechnology came from other fields. They altered their career trajectories in sciences like genetics and medical research to enter an unknown

Final version accepted on December 15, 2000

Institute for Development Strategies
Indiana University
SPEA Building
Suite 201, Bloomington
IN 47405, U.S.A.

and undefined field, which only subsequently became labeled as biotechnology. Those regions exhibiting the greatest success in developing biotechnology clusters also possessed the greatest ability to unleash the potential commercialization latent in those scientists.

2. Definitions

The terms *biotechnology* and *biotech industry* are typically used in a casual and imprecise manner. To provide more precision and a common understanding it is important that the following concepts have a more exact definition:[1]

- *Biotechnology*
  A group of techniques and technologies that apply the principles of genetics, immunology and molecular, cellular and structural biology to the discovery and development of novel products.
- *Biotechnology Industry*
  The industry is composed of around 5,000 private companies that apply various biotechnologies to develop commercially viable products. The biotech industry is typically an input in the health care, food and agriculture, industrial processes, and environmental cleanup industries.
- *Pharmaceutical Industry*
  This industry is composed of around one hundred private companies whose products involve the design, discovery, development and marketing of new agents for the prevention, treatment and cure of disease. Pharmaceutical firms have a heavy reliance on scientific research. The industry may be evolving towards a new name – the biopharmaceutical industry – which reflects the heavy reliance of pharmaceutical companies on biotechnology.

Small Business Economics **17**: 3–15, 2001.
© 2001 *Kluwer Academic Publishers. Printed in the Netherlands.*

4 *David B. Audretsch*

- *Medical Research*
  This refers to science-based inquiry, both basic and applied, where the goal is the improvement in health and the eradication or mitigation of disease and disability.

3. Formation and development of regionally based biotech industry

The United States biotechnology industry has grown considerably over the past fifteen years. In 1992 there were 1,358 patent applications by U.S. companies in the United States. By 1997, patent applications were up to 3,014. Between 1990 and 1999, private investment in the biopharmaceutical industry has grown from $10 billion to $24 billion. While the biotechnology industry in the United States has exploded over the past fifteen years, this growth has not been neutral with respect to geographic location. The U.S. biotechnology industry has been regionally concentrated in just a handful of geographic locations. The strongest presence of biotechnology companies has been in California, which accounts for around one-quarter of the biotechnology companies. Massachusetts accounts for the second highest share of biotechnology companies. Nearly one-tenth of American biotech firms are located in Massachusetts. A third tier of states accounts for about five percent of the biotechnology firms, including New Jersey, Maryland, North Carolina and Texas, each with a share of around five percent of U.S. biotechnology companies. In fact, even within these states, the location of the biotechnology industry is geographically concentrated within a very small region. In California, one-fifth of the biotechnology firms is located in the East Bay area, 15% in San Francisco, 18 percent in Santa Clara, 22 percent in Los Angeles, 18 percent in San Diego, and 4 percent in Sacramento (Blakely et al., 1993).

There are strong reasons both why the biotechnology industry remains geographically concentrated in just a few regions, and why these specific regions have emerged as the main *Standort* for the U.S. biotechnology industry. The most important input in the biotechnology industry is specialized knowledge. In most cases, only a few scientists have the potential to acquire this knowledge. In addition, the scientists possessing that knowledge must also have the information about a potential commercial market for viable products resulting from that knowledge. They also must be willing to act on that knowledge by commercializing that knowledge through a biotechnology enterprise. To commercialize requires start-up capital as well as managerial capabilities. For all of these factors to come together has been impossible with the exception of just a few select regions.

Biotechnology is a new industry that is knowledge-based and is predominantly produced by new start-ups and small firms. The relative small scale of most biotechnology firms may be attributable to the diseconomies of scale inherent in the "bureaucratic process which inhibits both innovative activity and the speed with which new inventions move through the corporate system towards the market" (Link and Rees, 1990, p. 25). Zucker et al. (1994, p. 1) provide considerable evidence suggesting that the timing and location of new biotechnology firms is "primarily explained by the presence at a particular time and place of scientists who are actively contributing to the basic science." More specifically, they find that firms are likely to locate in geographic areas where scientists who have published articles on gene sequencing are located.

The early evolution of the biotechnology industry took place in California, and in particular in the San Francisco bay area. This is because the key ingredients identified above were the most prevalent in this region. As Prevezer (1997) explains, there was a strong presence of venture capital already in Silicon Valley, as a result of the boom in high-technology industries such as semiconductors and computers. This also meant the existence of a very entrepreneurial culture. There was expertise in starting and growing high-technology companies, high job mobility, and strong networks. This was also the location of a number of the country's most important universities and government research laboratories.

The two most significant discoveries fuelling the biotechnology industry occurred in the early 1970s, when Cohen and Boyer at Stanford University developed the recombinant DNA technique, and Kohler and Milstein's production of monoclonal antibodies in Cambridge, U.K. The pioneering biotechnology companies, such as Cetus, Genetech and Hybritech, were located in

California near the top scientific centers, and Biogen was located in the Boston area to take advantage of the top scientists at Harvard and MIT. According to Prevezer (1997), the San Francisco bay area became the focal point for the incipient biotechnology industry, relying heavily upon scientific knowledge at the University of California at San Francisco (UCSF), Stanford, and CalTech. While the subsequent explosion of the biotechnology industry has remained concentrated in these regions, other regions combining the demanding requirements of scientific knowledge and a strong and vigorous entrepreneurial culture have emerged in Dallas, Texas, Research Triangle (North Carolina), and in Maryland (between Baltimore and Washington, D.C.).

4. Linkages & cooperation

Linkages and cooperation are particularly important in the biotechnology industry. These linkages occur between entrepreneurial firms, between the scientists involved with the firms, between the firms and universities, and between corporations and biotech firms.

Strategic alliances between large corporations and biotechnology companies have been particularly important for biotech companies specializing in therapeutics. This is because the costs of developing a new drug, complying with the various layers of regulation, manufacturing the product, and then marketing the product, have required a level of finance that far exceeds the budgets of most small firms. Cullen and Dibner (1993) estimate that the cost of bringing a therapeutic drug from basic research to the market is around $250 million. At the same time, the average budget for research and development of a biotech firm is $12.5 million. To close this gap, biotech firms have engaged in a broad range of marketing and licensing agreements. Under these agreements, the biotech firm provides access to cutting-edge technology in exchange of an infusion of capital from their corporate partners.

In documenting the evolution of strategic alliances in biotechnology, Cullen and Dibner (1993, p. 18) conclude that, "The primary strategic goal of small and medium-sized biotechnological companies was to develop products to be marketed by their partners and their primary concern was finding and developing alliances." The obvious advantages of such strategic alliances are that they enable a small, new company to concentrate on its core mission – moving from basic research to commercialization through technological innovation. The strategic alliances also enable the biotech company to reduce financial risks as well as operating costs. In addition, the biotech firm is able to better offset the major liabilities associated with biotech start-ups – acquiring manufacturing capabilities, marketing and sales.

5. The role of established firms towards biotech enterprises

Established firms are generally quite positive and supportive towards biotechnology firms. This is because of the strong complementary nature of biotechnology firms and established firms, particularly in the pharmaceutical industry. There are a number of reasons why such a complementary relationship has evolved between established and biotechnology firms. The first is that the former have recognized that it may be a more efficient structure to engage in an arms length market relationship to obtain new biotechnology products than to produce them internally. The market exchange is apparently more efficient than the internal transaction. The reason for this involves agency problems in doing research that is highly uncertain and asymmetric. In addition, the exposure to legal liabilities resulting from biotechnology research is reduced when that research is done by a small firm with limited assets rather than in a large corporation with massive assets.

Sharp (1999) identifies three main phases in the relationship of established firm and biotechnology companies. The first phase involved the formation and incipiency of the biotechnology industry. Sharp (1999, p. 137) reports that "most of the established pharmaceutical companies were uncertain what to make of the new technology and especially of the hype surrounding its development that grew with the small-firm sector in the U.S." This uncertainty combined with a considerable degree of skepticism resulted in most established pharmaceutical companies distancing themselves from the fledgling biotechnology industry in this initial phase. At the same time, Sharp points out that

6   *David B. Audretsch*

most established companies invested in sufficient scientific expertise to enable them to keep abreast of developments in biotechnology and monitor the industry.

The second phase began in the mid-1980s, when the period of watching and waiting ended. The established pharmaceutical firms recognized that, in fact, biotechnology had a valuable market potential. While strategies pursued by the established enterprises varied, most devised and implemented a strategic biotechnology policy. One common strategy that all companies pursued was to invest heavily to develop an in-house competence in biotechnology. How this was done varied considerably from one company to another. In some companies, scientific teams were assembled. Other pharmaceutical companies acquired such competence through the acquisition of biotechnology firms or, in some cases through mergers. Another strategy was to engage in external linkages with biotechnology companies. As Cullen and Dibner (1993) document, strategic alliance between biotechnology firms and established enterprises exploded in the mid-1980s.

The third phase, which started around a decade ago involves the commercialization of biotechnology products. The first successful biotechnology products reached the market in the early 1990s. As Juergen Drews, head of R&D at Hoffman LaRoche observed in 1993, "While there are some redundancies among the 150 or so novel proteins in development, about 100 represent truly novel substances that have no precedent in medical therapy. Not all of these proteins will reach the market, but it is fair to assume that their attrition rate will be lower than that for small chemical entities because they should cause few unmanageable toxicological problems. A conservative estimate would expect 30–40 of the recombinant proteins now under development to become successfully marketed products over the next 5–6 years. This means that an average of 5–8 novel proteins should become available each year. If we assume an average sales volume for the forthcoming recombinant proteins equal to the average revenues generated by today's recombinant drugs, the portfolio of recombinant proteins now in clinical trials should amount to $10–$20 billion".

In this third phase, the large established companies take the new biotechnology products developed by biotechnology companies and convert them into large-scale marketed products. For example, Intron A was developed by Biogen but marketed by Schering-Plough, resulting in $572 million of sales in 1993. Humulin was developed by Genetech but marketed by Eli Lilly, for $560 million of sales in 1993. Engerix-B was developed by Genetech but marketed by SmithKline Beecham for $480 million. RecombiNAK HB was developed by Chiron but marketed by Merck for $245 million.

In addition, this third phase has experienced a shift by the established companies from the broad learning strategies of phase two and increasingly towards a more focused approach, targeting specific technologies. For example, Ciba Geigy reduced its portfolio of interests in biopharmaceuticals in 1989 in order to focus on the development of just several targeted products. Ciba Geigy subsequently increased its investment in those targeted areas and engaged in a number of research and licensing agreements with biotechnology companies. Similarly, Bayer reduced its biotechnology research in agro-chemicals while concentrating its focus on pharmaceuticals. Hoffman LaRoche similarly pulled out of agro-biotechnology to concentrate its focus on pharmaceuticals.

6. Investor attitudes

Investors come from established companies, venture capital and initial public offerings. A unique feature of biotechnology is that biotechnology firms have no commercial product on which to draw revenues for a substantial time period. This makes evaluating their success or whether to invest in them or not especially complicated. In addition, the science and research upon which they are based is typically so complicated and advanced that only a limited number of experts can hope to evaluate them. The high degree of uncertainty combined with asymmetric knowledge that is prohibitively costly to transact means that investing in biotechnology is different from investing in most other industries.

Since the biotechnology companies typically have no commercial products or production upon which they can be evaluated, potential investors focus instead on the major input of the biotechnology process, which involves the scientific team.

In terms of *bait* to the investment community, biotechnology firms, in the early stages of development, miss no opportunity to signal the abilities of their scientists as well as the science they are doing. It is not uncommon for prospectuses to read like proposals to the National Institutes of Health (NIH). Stephan and Everhart (1998) have shown the importance of scientific reputation at the time a biotechnology company goes public. Based on a sample of 45 firms, they find that the more substantial the reputation of the university-based scientific affiliates, the greater are the proceeds raised from the initial public offering (IPO).

7. Positioning profile of bioregion in global economy

The focus of a region's biotechnology industry generally reflects the scientific strengths at the knowledge source as well as the focus of complementary industries. For example, California's biotechnology industry is largely focused on human-diagnostic and therapeutic products (Blakely et al., 1993). By contrast, considerably less activity in biotechnology is devoted towards agriculture, or plant and animal, and biotechnology suppliers. This reflects that the emphasis on research as product of biotechnology is more pronounced in California than elsewhere inthe United States. It also reflects the relatively weak influence of pharmaceutical companies in California. Most of the pharmaceutical companies are located in New Jersey and New York. The pharmaceutical companies tend to rely on biotechnology companies also located in those regions for new products and innovations. In addition, the presence of the United States Food and Drug Administration (USFDA) on the East Coast has resulted in the location of testing facilities in New York and New Jersey.

The Biotech Industry Research Group reports that, as the result of a massive survey of biotechnology firms located in California, the most important factor determining location is the availability of qualified workers in the geographic area. The relative importance of the presence of this factor varies somewhat across fields. It is relatively more important in therapeutics and agritech,

and relatively less important in diagnostics and for suppliers. The second most important factor in shaping location is the proximity of research universities and other research organizations. The third most important factor is the cost of industrial space. The fourth most important fact is county and city regulations. The fifth most important factor is taxes. The sixth most important fact is wage rates, followed by proximity of major suppliers, proximity to venture capital and financial institutions.

8. Institutional barriers impeding regional development of biotech clusters

A number of regions have tried to develop a biotechnology industry. Many of these attempts have not proven successful; only a handful of regions in the United States have successfully developed a biotechnology industry. For example, the region around Hanover, New Hampshire tried to develop a biotechnology cluster and failed. Atlanta, Georgia has been unable to develop a biotechnology industry. Only recently (1999) have there been signs of an emerging biotechnology industry.

One of the greatest impediments to developing biotechnology is the existence of institutional barriers. Perhaps the most imposing constraint has been an absence of scientific talent in a region. If no cluster of scientific talent is present, either at universities, government research laboratories, or private firms, then it is not likely that a biotechnology industry can be developed. The major reason why the biotechnology industry is so geographically concentrated is that the best scientific talent is also geographically clustered in just a few regions (Audretsch and Stephan, 1996).

However, the presence of a critical mass of excellence in science is a necessary but not sufficient condition for the development of biotechnology. There are numerous examples of clusters of scientific talent in the relevant fields where there is still an absence of biotechnology activity. This is attributable to the presence of other formidable institutional constraints.

For example, Dartmouth College, including the medical center, is the home of a large number of scientists involved in biomedical research. A project to develop a biotechnology cluster failed.

One of the reasons given why this attempt failed is that while the scientific talent was present, there is no base of large companies in the relevant areas, and no existence of an entrepreneurial culture.

In contrast, Atlanta not only has a very strong presence of biomedical researchers in Atlanta, which is the home of Emory University, with one of the top medical schools in the country, and the Center for Disease Control (CDC), but also is rich in private industry. The inability of Atlanta to translate the scientific knowledge into a biotechnology industry may reflect the absence of venture capital and perhaps the fact that the industry has had a tradition of large corporations such as Coca Cola but a relatively absence of high-tech entrepreneurship. This would suggest two other important institutional barriers to developing a biotechnology industry – the availability of finance and venture capital, which tend to be regional and not national, and the presence of an entrepreneurial culture, which is also a regional phenomenon. Venture capital and other forms of informal finance are local because the venture capitalist needs to monitor and interact with the company. The success of the company is highly dependent upon this interaction. Thus, geographic proximity between the venture capitalist and the biotechnology firm is essential for commercial success.

The presence of an entrepreneurial culture is important because this creates the opportunities for a scientist to change his career trajectory and shift it away from research to commercialization. In studying Silicon Valley, Saxenian (1990, pp. 96–97) emphasizes that it is the existence of such an entrepreneurial culture that provides individuals the information and support needed to become entrepreneurs themselves: "It is not simply the concentration of skilled labor, suppliers and information that distinguish the region. A variety of regional institutions – including Stanford University, several trade associations and local business organizations, and a myriad of specialized consulting, market research, public relations and venture capital firms – provide technical, financial, and networking services which the region's enterprises often cannot afford individually. These networks defy sectoral barriers. Individuals move from established firms to startups. And they continue to meet at trade shows, industry conferences, and the scores of seminars, talks and social activities organized by local business organizations and trade associations. In these forums, relationships are easily formed and maintained, technical and market information is exchanged, business contacts are established, and new enterprises are conceived. . . . This decentralized and fluid environment also promotes the diffusion of intangible technological capabilities and understandings."

A fourth type of institutional constraint to the development of a regional biotechnology cluster is the high cost of regulation. Blakely et al. (1993, p. 23) document the high concern that biotechnology firms have for regulation, "Predictable application of regulations concerning environmental and public health and the use of land are crucial to the location of biotechnology firms. Companies must be assured that their development activities will not be hindered by inconsistent applications of regulations."

9. Regional & global networks

Audretsch and Stephan (1996, 1999) document the strong linkages that exist among biotechnology scientists and between universities and biotech firms. These linkages are crucial because biotechnology companies are strongly defined by their scientists. Many of these scientists, particularly senior scientists with strong reputations, do not work for the biotechnology company full time, but instead are members of university faculties. For example, Audretsch and Stephan (1999) show that, of 101 founders of new biotechnology firms in the early 1990s, nearly half (50) are from universities. Of these fifty, 35 remain associated with their universities on a part-time basis, while the remaining 15 founders left the university to work full-time for their biotech firm.

These university-based scientists fulfill a variety of roles within biotechnology companies. Some are founders, others serve as members of scientific advisory boards (SABs), while still others serve as directors. The degree of knowledge provided by university-based scientists varies according to the role played by the scientist. Scientific founders seek out venture capitalists in order to transform technical knowledge into economic knowledge. Scientific advisors provide links between scientific founders and other researchers doing work in the area. They, along with founders, also provide the possibility of out-sourcing research into university laboratories staffed by graduate students and post-docs. The concept of scientific advisory boards also provides the firm the option of having, at minimal cost, a full roster of the key players doing research in the firm's area of expertise.

In addition to providing knowledge to newly formed biotechnology companies, university-based scientists also provide a signal of firm quality to the scientific and financial communities. An effective way to recruit young scientists is to have a scientific advisory board composed of the leading scientists in the field. George B. Rathman, president and Chief Executive Officer of Amgen, attributes much of the company's success to an SAB of "great credibility" whose "members were willing to share the task of interviewing the candidates for scientific positions." Rathman goes on to say that the young scientists that Amgen recruited would not have come "without the knowledge that an outstanding scientific advisory board took Amgen seriously" (Burrill, 1987, p. 77).

Certain roles, such as being a founder of a biotechnology firm, are more likely to dictate geographic proximity between the firm and the scientist than are other roles that scientists play. This is because the transmission of the knowledge specific to the scientist and firm dictates geographic proximity. Presumably scientists start new biotechnology companies because their knowledge is not transferable to other firms for the expected economic value of that knowledge. If this were not the case there would be no incentive to start a new and independent company. Because the firm is knowledge-based, the cost of transferring that knowledge will tend to be the lowest when the firm is located close to the university where the new knowledge is being produced. In addition, the cost of monitoring the firm will tend to be minimized if the new biotechnology startup is located close to the founder.

By contrast, the role of scientific advisor to a biotechnology company does not require constant monitoring or even necessarily specialized knowledge. Thus, the inputs of scientific advisors are less likely to be geographically constrained.

Furthermore, geographic proximity of all major researchers in a particular scientific field is unlikely given the opportunity cost that universities face in buying into a single research agenda. Thus, if firms are to have access to the technical knowledge embodied in the top scientists in a field, they will be forced to establish links with researchers outside of their geographic area. Scientists whose primary function is to signal quality are also less likely to bet local than are scientists who provide essential knowledge to the firm. Their quality signal is produced by lending prestige to a venture they have presumably reviewed – a task that can be accomplished with credibility from a distance.

To identify the links between knowledge sources, the incentives confronting individual scientists, and where the knowledge is commercialized, Audretsch and Stephan (1996) rely upon a data base collected from the prospectuses of biotechnology companies that prepared an initial public offering (IPO) in the United States between March 1990 and November 1992. This includes a total of 54 firms affiliated with 445 university-based scientists were identified during this time period. By carefully reading the prospectuses, it was possible to identify the names of university-based scientists affiliated with each firm, the role that each scientist plays in the firm, and the name and location of their home institutions. Universities and firms were then grouped into regions, which are generally larger than a single city but considerably smaller than a state. Certain areas, for example, metropolitan New York, cross several state lines.

Only 138 of the 445 links observed between scientists and biotechnology companies are local in that the scientist and firm are located in the same region. This suggests that geographic proximity does not play an important role for links between biotechnology companies and scientists in general. However, the geographic link between the scientist and the founder is influenced by the particular role played by the scientist in working with the firm. Most strikingly, 57.8 percent of the scientist-firm links were local when the scientist was a founder of the firm; 42.1 percent were non-local. By contrast, when the scientist served as a member on the SAB, only 31.8 percent of the links were local, while 68.2 percent were non-local.

This disparity suggests that the nature of the knowledge transmitted between the university and the biotechnology firm may be different between scientists serving as founders and those serving on a SAB. Presumably it is the difference in the nature and quality of the knowledge being transferred from the university to the company that dictates a higher propensity for local proximity in the case of the founders, but not for SAB members.

10. Innovation strategies of biotech firms

Biotechnology firms have pursued a broad spectrum of innovation strategies. The diversity of innovation strategies reflects heterogeneity of fields, backgrounds and goals. Still, several key innovation strategies have emerged in the biotechnology industry. One successful strategy has been to develop and exploit close links with university researchers. Such close links provide access to state-of-the art research and knowledge, as well as a source of potential employees. Another important strategy involves marketing. A successful marketing agreement with a large pharmaceutical company can result in the ability for a small biotechnology company to extend its product to a mass-market. Biotechnology companies are generally successful when they devote their resources and competence to their core product – research – and rely on other third-party firms for clinical testing and marketing.

11. Characteristics of newly founded regional biotech enterprises

The biotechnology industry is composed of nearly 1,300 companies that use various biotechnologies to develop products for use in health care, food and agriculture, industrial processes, and environmental cleanup. Most of these companies are very small. Two-thirds of them have fewer than 50 employees. All but twenty of these companies are "unencumbered by revenues" (Rosenberg, 1999). As Rosenberg (1999, p. 69) observes, "The biotechnology industry is a very entrepreneurial industry, but one with relatively few commercial successes."

In order to understand biotechnology enterprises, it is essential to understand the scientists that start these enterprises. This is because the biotechnology firms are an extension of the scientific knowledge and competence embodied in those firms.

Audretsch and Stephan (1999) use a data base drawn from the prospectuses of 60 firms that made an initial public offering (IPO) in biotechnology during the period March 1990 to November 1992 to examine the sources and incentives for commercializing new knowledge. Prospectuses for the offerings were carefully read in order to identify the scientific founders of the new firms. In cases where it proved difficult to identify founders from the prospectuses, telephone calls were made to the firm. In addition, firm histories were checked and confirmed in *BioScan*. Founders having a Ph.D. or a M.D. were coded as scientific founders for the purposes of this research. In addition, several individuals who did not have a doctorate but were engaged in research were included as scientific founders. All told, we were able to identify 101 scientific founders for 52 firms making an initial public offering during this period.

Biographical information was also collected from the prospectuses and was supplemented by entries from standard reference works such as *American Men and Women of Science*. Four types of job experience were identified – academic experience (which includes positions at hospitals, research foundations and the government); experience with pharmaceutical companies; training experiences (as a student, post-doc, or resident), and "other" experience. This information was used to distinguish among five distinct career trajectories followed prior to the founding of the company:

1. The *academic trajectory* describes scientists who had spent all of their time since completing their training employed in the academic research sector;
2. The *pharmaceutical trajectory* describes those scientists whose careers subsequent to receiving training had been entirely spent working in the drug industry;
3. The *mixed trajectory* describes scientists who had worked in both the pharmaceutical industry and the academic research sector;
4. The *student trajectory* describes individuals

who went directly from a training position to founding a biotechnology firm; and

5. The *other trajectory*, which includes scientists who have been employed by non-pharmaceutical firms.

Additional biographical information coded was ascertained concerning date of birth and educational background. Citation counts to first-authored published scientific articles were measured using the 1991 *Science Citation Index* produced by ISI and are used here as an indicator of scientific reputation.

Summary data, presented in Table I, show that fifty percent of the scientific founders' careers followed an academic trajectory; slightly more than 25 percent a pharmaceutical trajectory. Half of this latter group had established their careers exclusively with large pharmaceutical companies such as SmithKline and Beckman; half had come from smaller pharmaceutical firms, some of which, like Amgen, were a first generation biotech firm. Table I also indicates that approximately an eighth of the founders had a mixed career in the sense that prior to founding the firm they had held positions in both a pharmaceutical company as well as a university or non-profit research organization. A handful of founders moved directly from a training position such as a residency or post-doctorate appointment to the startup firm, thereby short-circuiting the traditional trajectories from pharmaceutical firms and/or academe. The

career trajectory of the remaining scientists was either indeterminate or followed another type of path.

The employment status of the founders with the biotechnology company was also determined. Note that 59 of the 101 scientific founders were working full time with the new firm at the time of the public offering; 41 were working part time, and almost all (35) of these had followed an academic trajectory. This means that 70 percent of the academic founders maintain full-time employment with their academic institutions, serving as consultants or members of the Scientific Advisory Boards to the startup firms. Only 15 of the academic founders had moved to full-time employment with the firm by the time the IPO was made. By contrast, all 28 scientists whose careers had been exclusively in the pharmaceutical sector held full-time positions with the firm at the time of the IPO; 9 of the 13 whose careers followed a mixed trajectory were full time.

The evidence from Table I supports the hypothesis that the incentive structure varies considerably between the pharmaceutical founders and the academic founders. Those founders coming from universities and non-profit research organizations have the option of eating their cake and having it too, by maintaining formal contacts with their previous employer, often in a full-time position. Even those from the academic sector who are full time with the new firm are often able to maintain some connection with the non-profit sector as

TABLE I
The age and citation record of biotechnology founders

	N	Birth date			Citations		
		M	SD	N_{known}	M	SD	N_{known}
All scientific founders	101	1943.18	10.20	96	92.13	171.05	99
All academic founders	50	1940.55	10.06	49	149.32	226.51	49
Part time	35	1938.79	10.29	34	172.71	259.03	35
Full time	15	1945.06	8.54	15	72.21	78.70	15
All drug founders	28	1945.61	9.20	28	29.71	46.28	28
Small	14	1945.93	9.84	14	30.30	57.40	14
Big	12	1947.00	7.67	12	34.00	34.41	14
Mixed career	13	1943.80	8.76	13	62.69	57.56	13
Student career	6	1957.00	3.54	5	58.17	83.72	6
All full time	57	1945.64	9.61	57	46.59	60.69	57
All part time	40	1939.42	10.03	37	159.30	245.52	37

Source: Audretsch and Stephan (1999).

adjunct or clinical faculty. By contrast, those scientists who have a career path in pharmaceuticals take full-time positions with the company, at least by the time the company goes public.

There are other differences between those scientists coming from an academic trajectory and those scientists coming from a pharmaceutical trajectory. The most notable is the difference in age at the time the public offering was made. On average, those coming from universities were born approximately five years earlier than those coming from the pharmaceutical sector, a difference which is statistically significant at the 95 percent level of confidence. As would be expected, we also find that those following the academic trajectory have significantly more citations than those coming from a pharmaceutical trajectory.

Of perhaps even greater interest are the differences between the part-time academics and the full-time academics. Academic founders who remain full-time with their institution, working but only part-time for the new firm, were, for example, born more than six years earlier than academic founders who leave their institution to go full time with the firm were. The part-timers are not only older; they are also more eminent, having significantly more citations than academics that go full time with the firm. This suggests that eminence gives these scientists the luxury of hedging their bets; both the firm and their research institution welcome a chance to claim them as affiliates. And, although we have not yet measured the incidence, such individuals often serve as directors and members of Scientific Advisory Boards of additional start-up firms. The full-timers, by contrast, have developed sufficient human capital to be recognized as experts but lack the luster to hold "dual" citizenship. In terms of both citation counts and date of birth they are remarkably similar to their fellow founders who followed a pharmaceutical trajectory.

These preliminary observations suggest that the incentive structure depends upon the career trajectory that the scientist has followed as well as upon whether the scientist has established sufficient eminence to be able to sustain multiple roles. Scientists working in incumbent pharmaceutical firms face the well-known problem of deciding whether to remain with the incumbent firm or start a new firm. Furthermore, the goal of an incumbent

firm to capture their economic knowledge seldom permits a scientist to establish a reputation based solely on publication. Instead, their scientific reputations are typically established in terms of the products they helped to develop and are known primarily to "insiders" in the industry. Scientists in academe, however, face a different incentive structure. They live in a world where publications are essential for the establishment of reputation. Early in their careers they invest heavily in human capital in order to build a reputation. In the later stages of their career, scientists may trade or cash in on this reputation for economic returns. A variety of avenues are available to do this, including the establishment of a new firm.

The data suggest that this *cashing out* pattern is determined in part by eminence. As noted, a number of academic founders have established sufficiently strong reputations as to be able to eat their cake and have it too. They maintain their full-time jobs in academe, while seeking part-time opportunities to gain economically from their knowledge and scientific reputation. The economic returns are tied to the shares they own in the startup companies. A subset of academic scientists, however, go full time with the firm. They, too, hold stock in the firm. But, their rewards are more immediate in terms of the salaries paid to executives in the companies. And, while they have established solid reputations, they are considerably less cited than those academic founders who maintain full-time positions in academe. Although this may be a result of age (they are, after all, about five years younger), it is more likely a characteristic that age cannot alter. Science, as numerous researchers have established, is noteworthy for persistent inequality which age merely amplifies.

12. National biotech promotion programs

There is, in fact, no official program for promoting biotechnology at the federal, or national, level. This, however, does not imply an absence of programs that *de facto* promote biotechnology. Because scientific research and knowledge plays such an important role in biotechnology, policies promoting the underlying scientific research also promote biotechnology. Much of this scientific research has been funded by a national agency, the

National Institutes of Health (NIH). As Penhoet (1999, p. 41) founder and former CEO of Chron observes, "The history of biotechnology to date has involved the commercialization of technologies that were funded almost entirely by the National Institutes of Health (NIH). Partnership, therefore, has been a major theme of the biotechnology story – partnership between government, industry, and universities. Much of the technology commercialized by industry was developed in the university setting using NIH grants."

Rosenberg (1999) concludes that the role of advocacy of research organizations has fueled the rapid growth of the NIH: "Effective lobbying from the Association of American Medical Colleges, Research America, Funded-First, disease groups, and others, have been responsible for the growth in support for federally-funded medical R&D over the past 30 years, and the past 10 years especially." In 1990, the NIH budget was $8 billion; in 1999 it is $14.6 billion, with projected strong growth. This is essentially a doubling of NIH funding within a decade, and there are indications that the Congress will double the budget of NIH again over the next five years.

The NIH has steadily expanded its support of biotechnology startups. For example, in 1998, NIH and private companies entered into 166 cooperative research and development agreements (Cards), which allow for the sharing of compounds and other research materials and results, as well as for the exchange of funds. This is the largest number ever of Cards between NIH and the private sector.

The Small Business Innovation Research (SBIR) program is another federal policy that provides NIH funding of biotechnology firm (Audretsch et al., 1999). The United States Congress enacted the SBIR program in the early 1980s as a response to the loss of American competitiveness in global markets. Congress mandated each federal agency with allocating around four percent of its annual budget to funding innovative small firms as a mechanism for restoring American international competitiveness. A Phase I award provides an opportunity for a small business to establish the feasibility and technical merit of a proposed innovation. The duration of the award is s six months and can not exceed $70,000. A Phase II Award is granted to only the

most promising of the Phase I projects based on scientific/technical merit, the expected value to the funding agency, company capability and commercial potential. The duration of the award is a maximum of 24 months and generally does not exceed $600,000. Approximately 40 percent of the Phase I Awards continue on to Phase II. A Phase III Award is for the infusion and use of a product into the commercial market. Private sector investment, in various forms, is typically present in Phase III.

Through the Small Business Innovation Research (SBIR) program, NIH awarded $266 million in grants to small firms for medical and biopharmaceutical research. It is expected that the SBIR program at NIH will exceed $300 million in 1999.

In addition to the NIH, the United States Department of Defense also uses the SBIR program to fund biotechnology firms. Between 1983 and 1997 there was more than $240 million in SBIR awards for biotechnology companies from the Department of Defense. Phase I accounted for $47 million and Phase II accounted for $194 million.

In addition, the Advanced Technology Program (ATP) funded by the U.S. Commerce Department awarded $29 million to small biotechnology companies in 1999. The ATP, like the SBIR, is expected to grow in the future.

There is compelling evidence that the SBIR program has had a positive impact on developing the U.S. biotechnology industry. The benefits have been documented (Audretsch et al., 2000) as:

• The survival and growth rates of SBIR recipients have exceeded those of firms not receiving SBIR funding.
• The SBIR induces scientists involved in biomedical research to change their career path. By applying the scientific knowledge to commercialization, these scientists shift their career trajectories away from basic research towards entrepreneurship.
• The SBIR awards provide a source of funding for scientists to launch start-up firms that otherwise would not have had access to alternative sources of funding.
• SBIR awards have a powerful demonstration effect. Scientists commercializing research

14   *David B. Audretsch*

results by starting companies induce colleagues to consider applications and the commercial potential of their own research.

13. Regional biotech support programs

While the SBIR is a federally funded program, it also has a regional element. Many or most states have formed programs to assist firms in applying for SBIR grants. States sponsor seminars and hire agents who contact potential recipients of SBIR grants.

In addition, a number of states and cities have developed programs to try to develop a biotechnology cluster. Most of these efforts revolve around transferring the knowledge out of universities via new-firm start-ups. They typically revolve around technology transfer programs and technology parks. As Blakely et al. (1993, p. 12) point out, "State and regional policy makers are taking advantage of the lessons learned from the micro-electronics experience and applying them in recent attempts to revitalize the declining industrial areas with new technologies. Unlike traditional approaches to economic development that are aimed at luring high-technology firms from their established bases, states are becoming increasingly sophisticated in their approach to economic development through biotechnology." In particular, the traditional incentives and instruments used to promote economic development, such as low taxes, cheap land, low-cost labor, and subsidies, are not effective in developing biotechnology. In an assessment of the state's approach to economic development through biotechnology, the California State Senate Office of Research concluded, "The strategies that states are employing in the pursuit of biotechnology are a good deal more sophisticated and better funded than the past economic development strategies, which have been, more often than not, marketing efforts designed to encourage firms to locate new production facilities in their states" (Blakely et al., 1993, p. 12).

For example, the University-Industry Relations (UIR) program at the University of Wisconsin has resulted in the formation of a dynamic cluster of biotechnology companies in Madison, Wisconsin. The size of the grants is generally modest. In 1995, they ranged from $2,500 to $33,000. To date, the program has funded 39 biotechnology start-ups that have received subsequent private investment of $10,723,200.

Similarly, South Carolina recently established the South Carolina Research Grant Program, with one goal being to develop a biotechnology industry. Faculty at the universities was encouraged to submit proposals for funding. Not only were the proposals meant to advance science, but also contribute to the economic development of South Carolina by inducing scientists to commercialize their knowledge via a biotechnology start-up.

14. Conclusions

Because the biotechnology industry in the United States is rapidly evolving it is difficult to make inferences about appropriate policy in the new century. What may have been effective when the industry was incipient may be less effective or even counterproductive today. Still, several major policy lessons emerge from the experience in the United States. First, the biotechnology industry is and remains a local phenomenon. The biotechnology industry not only clusters regionally, but there are only a handful of regions that have successfully generated a viable biotechnology industry. To generate a successful regional cluster, the existence of world class scientific talent is a necessary condition. However, it is not a sufficient condition. The ancillary or complementary factors must also be available to translate this knowledge into a commercialized product. The complementary factors include the presence of venture capital and other forms of finance, the existence of an entrepreneurial culture, and transparent and minimal regulations hindering the start-up and growth processes.

Acknowledgements

I am grateful to the conference organizers and two anonymous referees for their comments and suggestions.

Note

[1] These definitions follow those provided for by Rosenberg (1999) for the United States National Academy of Sciences.

References

Audretsch, D. B., 1995, *Innovation and Industry Evolution*, Cambridge: MIT Press.

Audretsch, D. B., 1996, 'R&D Spillovers and the Geography of Innovation and Production', *American Economic Review* **86**(3), 630–640.

Audretsch, D. B. and P. Stephan, 1996, 'Company-Scientist Locational Links: The Case of Biotechnology', *American Economic Review* **86**(3), 641–652.

Audretsch, D. B. and P. Stephan, 1999, 'How and Why Does Knowledge Spill Over in Biotechnology?', in D. B. Audretsch and R. Thurik (eds.), *Innovation, Industry Evolution, and Employment*, Cambridge: Cambridge University Press, pp. 216–229.

Audretsch, D. B., C. Weigand and J. Weigand, 2000, 'Does the Small Business Innovation Research (SBIR) Program Foster Entrepreneurial Behavior? Evidence from Indiana', in C. Wessner (ed.), *Evaluating the Impact of the U.S. Small Business Innovation Research (SBIR) Program*, Washington, D.C.: National Academy of Sciences.

Blakely, E. J., N. Nishikawa and K. W. Willoughby, 1993, 'The Economic Development Potential of California's Biotechnology Industry', *Biotechnology Review* **1**, 11–27.

Board on Science, Technology and Economic Policy, National Research Council of the National Academy of Sciences, 1999, *Government-Industry Partnerships in Biotechnology and Computing*, unpublished manuscript.

Burrill, G. S. and K. B. Lee Jr., 1992, *Biotech 93: Accelerating Commercialization*, San Francisco: Ernst & Young.

Burrill, G. S., 1987, *Biotech 88: Into the Marketplace*, San Francisco: Ernst & Young.

Cullen, W. C. and M. D. Dibner, 1993, 'Strategic Alliances in Biotechnology: Imperatives for the 1990s', *Biotechnology Review* **1**, 110–119.

Link, A. N. and J. Rees, 1990, 'Firm Size, University Based Research, and the Returns to R&D', *Small Business Economics* **3**(1), 1–38.

Penhoet, E., 1999, 'Biotechnology: Needs and Opportunities', in *Board on Science, Technology and Economic Policy, National Research Council of the National Academy of Sciences*, unpublished manuscript.

Prevezer, M., 1997, 'The Dynamics of Industrial Clustering in Biotechnology', *Small Business Economics* **9**(3), 255–271.

Rosenberg, L., 1999, 'Partnerships in the Biotechnology Industry', in *Board on Science, Technology and Economic Policy, National Research Council of the National Academy of Sciences*, unpublished manuscript.

Sharp, M., 1999, 'The Science of Nations: European Multinationals and American Biotechnology', *Biotechnology* **1**(1), 132–162.

Saxenian, Anna Lee, 1990, 'Regional Nertworks and the Resurgence of Silicon Valley', *California Management Review* **33**(1), 89–112.

Stephan P. and S. S. Everhart, 1998, 'The Changing Rewards to Science: The Case of Biotechnology', *Small Business Economics* **10**(2), 141–151.

Zucker, L. G., M. R. Darby and M. B. Brewer, 1994, 'Intellectual Capital and the Birth of U.S. Biotechnology Enterprises', National Bureau of Economic Research (Cambridge, MA) Working Paper No. 4653.

[12]

Intellectual Human Capital and the Birth of U.S. Biotechnology Enterprises

By LYNNE G. ZUCKER, MICHAEL R. DARBY, AND MARILYNN B. BREWER *

The number of American firms actively using biotechnology grew rapidly from nonexistent to over 700 in less than two decades, transforming the nature of the pharmaceutical industry and significantly impacting food processing, brewing, and agriculture, as well as other industries. Here we demonstrate empirically that the commercialization of this technology is essentially intertwined with the development of the underlying science in a way which illustrates the significance in practice of the localized spillovers concept in the agglomeration literature and of the tacit knowledge concept in the information literature. Indeed we present here strong evidence that the timing and location of initial usage by both new dedicated biotechnology firms (*"entrants"*) and new biotech subunits of existing firms (*"incumbents"*) are primarily explained by the presence at a particular time and place of scientists who are actively contributing to the basic science as represented by publications reporting genetic-sequence discoveries in academic journals.

By quantifying separable effects of individual scientists, major universities, and federal research support we provide specific structure to the role of universities and their faculties in encouraging local economic development through what are conventionally described in the literature as geographically localized knowledge spillovers.[1] Such localized knowledge spillovers may play fundamental roles in both economic agglomeration and endogenous growth (Paul M. Romer, 1986, 1990; Gene M. Grossman and Elhanan Helpman, 1991). However, our evidence, like the other literature cited here, specifically indicates localized effects without demonstrating that they can be characterized as spillovers (or externalities).

Section I lays out our basic hypothesis. The data are described in Section II. Empirical results are reported and discussed in Section III. A summary and conclusions section (Section IV) and Data Appendix complete the article.

I. The Hypothesis

Innovations are generally treated in the growth literature as a nonrivalrous good—freely useable by an unlimited number of potential

* Zucker: Department of Sociology and Institute for Social Science Research, University of California, Box 951484, Los Angeles, CA 90095, and National Bureau of Economic Research; Darby: John E. Anderson Graduate School of Management, University of California, Box 951481, Los Angeles, CA 90095, UCLA Department of Economics, and National Bureau of Economic Research; Brewer: Department of Psychology, Ohio State University, 1885 Neil Avenue, Columbus, OH 43210. This research has been supported by grants from the National Science Foundation (SES 9012925), the University of California Systemwide Biotechnology Research and Education Program, the University of California's Pacific Rim Research Program, the UCLA Center for American Politics and Public Policy, and the UCLA Institute of Industrial Relations. We acknowledge very useful comments on earlier drafts from two anonymous referees, and from David Butz, Harold Demsetz, Robert Drazin, Martin Feldstein, Zvi Griliches, Keith Head, Adam Jaffe, Benjamin Klein, Josh Lerner, Gary Pisano, Jeff Rosensweig, L. G. Thomas, Ivo Welch, and others. We are indebted to a remarkably talented team of postdoctoral fellows Zhong Deng, Julia Liebeskind, and Yusheng Peng, and research assistants Paul J. Alapat, Jeff Armstrong, Lynda J. Kim, Amalya Oliver, Alan Paul, and Maximo Torero. Armstrong was principally responsible for conducting the analysis and cleaning the firm data set and Torero cleaned the scientist data set; comments from both substantially improved the paper. This paper is a part of the NBER's research program in Productivity. Any opinions expressed are those of the authors and not those of their employers or funders.

[1] Zvi Griliches (1992) has surveyed the importance of R&D spillovers as a major source of endogenous growth in recent "new growth theory" models and the difficult empirical search for their existence. Despite these difficulties, there have been a number of articles reporting evidence of geographic localization of knowledge spillovers, including Adam B. Jaffe (1989), Jaffe et al. (1993), and Edwin Mansfield (1995).

users at a zero marginal cost (Richard R. Nelson and Romer, 1996). A complementary literature recognizes that some information requires an investment of considerable time and effort to master. The human capital developed by this investment is seen as earning a normal return on the cost of the investment, both direct costs and foregone earnings. We believe that some innovations, particularly a breakthrough "invention of a method of inventing" (Griliches, 1957), may be better characterized as creating (rivalrous) human capital—intellectual human capital—characterized by natural excludability as opposed to a set of instructions for combining inputs and outputs which can be protected only by intellectual property rights. This natural excludability arises from the complexity or tacitness of the information required to practice the innovation (see Nelson [1959], Kenneth J. Arrow [1962], Nelson and Sidney G. Winter [1982], and Nathan Rosenberg [1982]).

Based on both extensive interviews and empirical work summarized in Zucker and Darby (1996), we believe that, at least for the first 10 or 15 years, the innovations which underlie biotechnology are properly analyzed in terms of naturally excludable knowledge held by a small initial group of discoverers, their coworkers, and others who learned the knowledge from working at the bench-science level with those possessing the requisite know-how. Ultimately the knowledge spread sufficiently widely to become part of routine science which could be learned at any major research university. After the initial 1973 discovery by Stanley Cohen and Herbert Boyer of the basic technique for recombinant DNA—the foundation of commercial biotechnology as well as of a burst of scientific innovation—the financial returns available to talented recombinant-DNA scientists first rose dramatically as the commercial implications became widely appreciated and then more gradually declined as more and more scientists learned the techniques, until knowledge of the new techniques per se earned only the normal return for the time required for a graduate student to master them. Further, mere knowledge of the techniques of recombinant DNA was not enough to earn these extraordinary returns; the knowledge was far more productive when embodied in a scientist with the genius and vision to con-

tinuously innovate and define the research frontier and apply the new research techniques in the most promising areas.

We hypothesize that entry of firms into biotechnology in a given year thus will be determined by the geographic distribution of stars and perhaps others then actively practicing the new science as well as by the geographic distribution of economic activity. Stars are properly viewed as locationally (semi-)fixed since few star scientists who knew how to do recombinant DNA were willing to abandon their university appointments and laboratory teams to pursue commercial applications of biotechnology. The primary pattern in the development of the industry involved one or more scientist-entrepreneurs who remained on the faculty while establishing a business on the side—businesses which, where successful, resulted in millions or even billions of dollars for the professors who acquired early ownership stakes. Thus, we see the university as bringing about local industrial benefits by permitting its professors to pursue private commercial interests while their faculty appointments tie them to the area. In preliminary work not reported here, we tried to develop measures of local economic activity for industries, like pharmaceuticals, specifically impacted by the new technology, but these attempts never added significantly to the measures of general activity used in the empirical work below. The *local* availability of venture capital is widely believed to play a significant role in the birth of new biotech entrants (Martin Kenney, 1986; Joshua Lerner, 1994, 1995); so we also include that variable in our regressions.

II. The Data

Data has been collected in panel form for 14 years (1976–1989) and 183 regions (functional economic areas as defined by the U.S. Department of Commerce, Bureau of Economic Analysis [BEA], 1992b). Frequently, the data are aggregates of data at the zip code or county level.[2] Lagged variables

[2] The BEA's functional economic areas divide all the counties in the United States into regions including one or more cities, their suburbs, and the rural counties most closely tied to the central city.

include data for 1975 in the unlagged form. See the Data Appendix for more details.

A. *Firms*

Our data set on firms was derived from a base purchased from the North Carolina Biotechnology Center (NCBC) (1992) which was cleaned and supplemented with information in *Bioscan* (1989–1993) and its precursor (Cetus Corp., 1988). We identified 751 distinct U.S. firms for which we could determine a zip code and a date of initial use of biotechnology. Of these 751 firms, 511 were entrants, 150 incumbents, and 90 (including 18 joint ventures) could not be definitively classified. By 1990, 52 of the 751 firms had died or merged into other firms.

We then calculated the number of births in each region by year of initial use of biotechnology for all 751 firms as well as for their identified subcomponents of entrants and incumbents. We also have the stocks of surviving firms, entrants, and incumbents by region and year.

B. *Scientists*

Early in our ongoing project studying the scientific development and diffusion of biotechnology, we identified a set of 327 star scientists based on their outstanding productivity through early 1990. The primary criterion for selections was the discovery of more than 40 genetic sequences as reported in *GenBank* (1990) through April 1990.[3] However, 22 scientists were included based on writing 20 or more articles, each reporting one or more genetic-sequence discoveries.[4] In the 1990's,

sequence discovery has become routinized and is no longer such a useful measure of research success. These 327 stars were only three-quarters of one percent of the authors in *GenBank* (1990) but accounted for 17.3 percent of the published articles, almost 22 times as many articles as the average scientist.

We collected by hand the 4,061 articles authored by stars and listed in *GenBank* and recorded the institutional affiliation of the stars and their coauthors on each of these articles. These coauthors are called *"collaborators"* if they are not themselves a star. Some data on the stars and collaborators who ever published in the United States is given on the left side of Table 1, where the scientists are identified by the organization(s) with which they were affiliated on their first-such publication. The higher citation rate for firm-affiliated scientists is explored at length in Zucker and Darby (1996).

Figure 1 illustrates the time pattern of growth in the numbers of stars and collaborators who have ever published and the total number of firms using biotechnology in the United States. There was a handful of stars who published articles reporting genetic-sequence discoveries before the 1973 breakthrough, but even after 1973 their number increased gradually until taking off in 1980. The numbers of collaborators and firms lagged behind the growth in stars by some years.

To identify those scientists clearly working at the edge of the science in a given year, we term a star or collaborator as *"active"* if he or she has published three or more sequence-discovery articles in the three-year moving window ending with that year. As seen in the right side of Table 1, this stringent second screen provides an even more elite definition of star scientists as well as identifying some very significant collaborators. We count for each year the number of active stars and active collaborators who are affiliated with an organization in each region.

The locations of active stars and firms are both concentrated and highly correlated geographically, particularly early in the period. Figure 2 illustrates this pattern for the whole period by accumulating the number of stars who have ever been active in each region up to 1990 and plotting them together with the

---

[3] See Zucker et al. (1993). As will be obvious, much of the time between 1990 and the initial submission of this paper was spent in developing reasonable measures of intellectual human capital and in collecting and coding data necessary to locate the authors of the discoveries reported in the articles in question and to trace the diffusion process.

[4] Scientists advised that some sequence discoveries are more difficult than others and thus merit an article reporting only one sequence. Therefore we included scientists with 20 or more discovery articles to avoid excluding scientists who specialized in more difficult problems.

TABLE 1—DISTRIBUTION OF STAR SCIENTISTS AND COLLABORATORS
WHO HAVE EVER PUBLISHED IN THE UNITED STATES

Organization type[a]	Full data set		Ever active in U.S.[b]	
	Number of scientists	Citations[c]/ scientist/years	Number of scientists	Citations[c]/ scientist/years
Stars:				
University	158	85.5	108	110.8
Institute	44	63.0	26	98.7
Firm	5	143.7	1	694.3
Dual	0	n/a	0	n/a
Total	207		135	
Collaborators:				
University	2901	10.4	369	30.6
Institute	776	13.7	88	35.8
Firm	324	29.2	43	99.1
Dual	3	7.2	0	n/a
Total	4004		500	

[a] The organization type refers to the affiliation listed on their first publication with a U.S. affiliation.

[b] Ever active in the U.S. means that in at least one three-year period beginning 1974 or later and ending 1989 or earlier, the scientist was listed on at least three articles appearing in our data set of 4,061 articles which reported genetic-sequence discoveries and were published in major journals and that the affiliation listed in the last of the three articles was located in the United States.

[c] Citation counts are for 1982, 1987, and 1992 for all articles in our data set (whenever published) for which the individual was listed as an author.

location of biotech-using firms as of early 1990.

C. *Other Measures of Intellectual Human Capital*

Active stars and collaborators may be incomplete measures of location of the scientific base because there are techniques other than recombinant DNA which have played an important role in commercial biotechnology. Some skeptical readers might also think that some simpler measures of regions' relevant academic resources would contain all the information which we have laboriously collected. We found two measures of regional scientific base which entered separately in regressions reported below, but none which were capable of eliminating the effects of the star scientists.

One measure is a count of the number of *"top-quality universities"* in a region where top quality is defined by having one or more *"biotech-relevant"* (biochemistry, cellular/molecular biology, and microbiology) departments with scholarly quality reputational ratings of 4.0 or higher in the 1982 National Research Council survey (Lyle V. Jones et al., 1982). There are 20 such universities in the United States.[5] Our second measure, *"federal*

[5] The 20 universities were: Brandeis University, California Institute of Technology, Columbia University, Cornell University, Duke University, Harvard University, Johns Hopkins University, Massachusetts Institute of Technology, Rockefeller University, Stanford University, University of California-Berkeley, University of California-Los Angeles, University of California-San Diego, University of California-San Francisco, University of

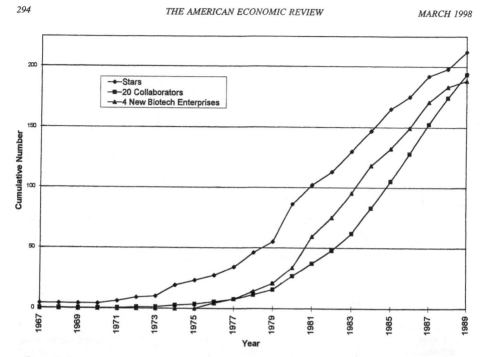

FIGURE 1. CUMULATIVE NUMBER OF U.S. STARS, COLLABORATORS, AND NEW BIOTECH ENTERPRISES, 1967–1989

support,'' is the total number (in hundreds) of faculty supported by 1979–1980 federal grants to all universities in each region for biotech-relevant research.[6] These variables take on the same value for a given region in each year.

D. *Other Variables*

Using listings in Stanley E. Pratt (1982), we measure *"venture capital firms"* as the number of such firms in a region legally eligible to finance start-ups in each year up to 1981. For later years, the number of firms is fixed at the

number in 1981 to avoid possible simultaneity problems once the major wave of biotech founding began.[7] (While great bookstores spring up around great universities, the former should not be counted as causing the latter.)

Since entry of biotech firms would be expected to occur where there is other economic activity, particularly involving a highly skilled labor force, we also include *total employment* in all industries (in millions of persons) and *average wages* (measured by deflated average earnings per job in thousands of 1987 dollars) for each region and year.

Finally, an increase in the (all-equity) cost of capital, as measured by the *earnings-price ratio* on the Standard & Poor's 500 Index would reduce the net present value of entry

---

Chicago, University of Colorado at Denver, University of Pennsylvania, University of Washington (Seattle), University of Wisconsin-Madison, and Yale University.

[6] We also tried a measure of biotech-relevant research expenditures as reported by the universities, but this variable was too collinear with the federal support variable to enter separately and appeared to be less consistently measured across universities.

[7] Instrumental variables would provide a more elegant approach to this problem if suitable instruments had been found.

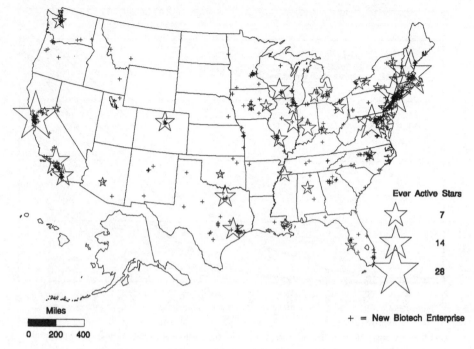

FIGURE 2. EVER ACTIVE STARS AND NEW BIOTECHNOLOGY ENTERPRISES AS OF 1990

and so should have a negative impact on birth of new firms, entrants or incumbents.

III. Empirical Results

We test our hypothesis using both the full panel data and by regressing the geographical distribution of the data in 1990 on values of the independent variables circa 1980. The former more fully exploit the available information while the latter avoid problems of possible simultaneity which might arise after 1980 when commercial biotechnology became a significant economic factor in some regions. All the regressions reported here, as well as an extensive sensitivity analysis noted below, were estimated in the poisson form appropriate for count variables with numerous zeroes using LIMDEP (William H. Greene, 1992 pp. 539–49), with the Wooldridge regression-based correction for the variance-

covariance matrix estimates.[8] The poisson regressions estimate the logarithm of the expected number of firm births; so the signs and significance of coefficients have the usual interpretation. Although OLS regressions are inappropriate for our count dependent variables with most observations at zero and the rest tailing off through small positive integers, we

---

[8] As discussed in Jerry Hausman et al. (1984), the poisson process is the most appropriate statistical model for count data such as ours. In practice, overdispersion (possibly due to unobserved heterogeneity) frequently occurs. Given the problems with resort to the negative binomial (A. Colin Cameron and Pravin K. Trivedi, 1990), Jeffrey M. Wooldridge (1991) developed a flexible and consistent method for correcting the poisson variance-covariance matrix estimates regardless of the underlying relationship between the mean and variance. We are indebted to Wooldridge and Greene for advice in implementing the procedure in LIMDEP.

reported broadly consistent results using that technique in an earlier version of the paper (Zucker et al., 1994).

In our sensitivity analysis, we ran the same poisson regressions for entrants and incumbents defined both exclusive and inclusive of the arguable case of joint ventures. The results were generally very similar to the subcomponent regressions in Table 4. In other unreported poisson regressions, we found that eliminating those regions with no firms and no stars from the sample did not result in qualitatively different results.

A. *The Long-Run Model*

Because of concerns about possible simultaneity biases once the industry became a significant economic force, we begin our empirical discussion with models which relate the number of firms in each region at the beginning of 1990 to the distribution of intellectual human capital and other variables as of about 1980. These results provide something of an acid test of our approach.

In Table 2, we present cross-section poisson regressions across the 183 regions explaining the number of firms in each at the beginning of 1990 when our data set ends.[9] Column (a) restrains the analysis to only the numbers of stars and collaborators ever active in each region at any time up through 1980, while columns (b) and (c) add first their squares and then our other intellectual human capital variables. Column (d) considers alternatively other economic variables which might explain entry, and column (e) combines the variables in (c) and (d). Column (f) adds to this model the number of biotech firms existing in 1980.

Column (a) in Table 2 indicates that the number of stars and collaborators active through 1980 is a powerful predictor of the

geographic distribution of biotech enterprises in 1990, since the log-likelihood increases to -871.9 compared to -1401.7 for a constant alone. It is the star scientists that contribute positively, with collaborators having a much smaller negative coefficient in this regression and most of the other long-run models discussed below. We had expected that the coefficient on collaborators would be much smaller than that on stars, but positive. We do obtain a positive coefficient on active collaborators when the squared terms are added in column (b), but that turns negative again upon addition of other variables in the remaining columns of Table 2.[10] (In the annual regressions discussed below, we generally estimate positive effects of active collaborators, but they are often statistically insignificant.)

We can offer two explanations for the generally negative sign on the number of active collaborators in the long-run regressions: (i) This coefficient reflects two partially offsetting influences; collaborators have a positive direct effect on the entry of firms but reduce the effect of stars who are devoting more of their time to training students and relatively less to starting their own firms. Training collaborators is surely a useful and rewarded activity— particularly for the academic stars—but it may take more of the stars' energy than it is worth if firm birth were the only criteria.[11] (ii) The sign and magnitude of the coefficient on collaborators may simply reflect significant multicollinearity among the intellectual human capital variables in the very early years. This is especially likely since when we examine the

---

[9] In an earlier version of this paper we included an alternative form of Table 2 (available from the authors upon request) in order to forestall interpretations that the results in Table 2 may reflect reverse causality. This alternative table reported regressions which explain the number of firms alive at the beginning of 1990 that were born after 1980. Nearly identical results were obtained, reflecting the fact that bulk of new biotechnology enterprises were founded after 1980.

[10] In column (b) of Table 2 (and Table 3), the negative coefficient on the squared term indicates that as the number of stars or collaborators increases, their marginal contribution diminishes eventually passing through zero. For collaborators, in columns (c)–(f) of these tables the sign pattern reverses so that the partial derivative of the log probability of birth with respect to collaborators starts out negative, and increases as their number increases, eventually becoming positive.

[11] In support of this explanation, we note that in our sensitivity analysis we tried regressions which substituted interaction terms multiplying the numbers of active stars and collaborators for the squared terms. In those regressions, we obtain significant positive coefficients on the numbers of stars and collaborators and a significant negative coefficient on their interaction.

TABLE 2—POISSON REGRESSIONS ON THE STOCK OF BIOTECH-USING FIRMS AT THE BEGINNING OF 1990 BY REGION

	(a)	(b)	(c)	(d)	(e)	(f)
Constant	0.911***	0.644***	0.468***	−2.595***	−2.718***	−2.607***
	(0.014)	(0.015)	(0.033)	(0.086)	(0.256)	(0.345)
Number stars active at any time during 1976–80	0.567*** (0.029)	0.587*** (0.072)	0.466*** (0.090)	— —	0.877*** (0.076)	0.649*** (0.084)
Number collaborators active at any time during 1976–80	−0.076*** (0.012)	0.175*** (0.033)	−0.183** (0.068)	— —	−0.333*** (0.045)	−0.261*** (0.037)
(Number stars active at any time during 1976–80)^2	— —	−0.028*** (0.007)	−0.019 (0.014)	— —	−0.049*** (0.012)	−0.024* (0.012)
(Number collaborators active at any time during 1976–80)^2	— —	−0.005*** (0.001)	0.002 (0.003)	— —	0.007** (0.003)	0.001 (0.002)
Number top-quality universities in the region	— —	— —	1.388*** (0.150)	— —	1.594*** (0.107)	0.442* (0.195)
Number faculty with federal support in the region	— —	— —	0.263 (0.143)	— —	0.752*** (0.088)	0.711*** (0.051)
Number venture capital firms in the region in 1980	— —	— —	— —	0.017*** (0.002)	−0.045*** (0.003)	−0.013** (0.004)
Total employment (all industries) in the region in 1980	— —	— —	— —	0.222*** (0.019)	−0.009 (0.043)	−0.213*** (0.049)
Average wages per job in the region in 1980	— —	— —	— —	0.166*** (0.004)	0.143*** (0.014)	0.139*** (0.019)
Cumulative births of biotech firms during 1976–80	— —	— —	— —	— —	— —	0.300*** (0.025)
Log-likelihood	−871.9	−707.3	−543.2	−753.9	−416.0	−350.7
Restricted log-likelihood	−1401.7	−1401.7	−1401.7	−1401.7	−1401.7	−1401.7

Notes: $N = 183$. Standard errors (adjusted by Wooldridge, 1991 Procedure 2.1) are in parentheses below coefficients.
    * Significantly different from 0 at the 5-percent level.
   ** Significantly different from 0 at the 1-percent level.
  *** Significantly different from 0 at the 0.1-percent level.

full cross-section/time-series results just below we obtain (we think more reliable) zero or positive coefficients on collaborators, so the puzzle largely disappears.

The full "fundamentals" model (excepting the decade-lagged dependent variable) is presented in column (e) of Table 2, where all the coefficients are significant except that for total employment. Leaving aside the question of the negative collaborator coefficient, we note the strong, positive, separate effects of stars, top-quality universities, and federal research

grants at universities on birth of firms in a given geographic region. The intellectual human capital variables alone increase the log-likelihood ratio from −1401.7 to −543.2 [see column (c)], with the final three variables bringing this quantity up to −416.0. As to the last three variables, the quality of the labor force, measured by average wages per job, seems much more relevant than its size. Surprisingly, to some observers, the number of venture capital firms in a region enters, but with a significantly negative sign. We interpret the negative sign as evidence that venture capitalists did play an active role in the formation of entrant firms, but they apparently resulted in fewer, larger firms being born in the areas in which they were more active.[12]

This sign of the coefficient of the number of venture capital firms in a region is robust in sensitivity experiments with other forms (not reported here) except for regressions which exclude the intellectual human capital variables such as in column (d). That regression looks good in terms of significance and expected sign pattern although it has a much lower explanatory power than the intellectual human capital variables alone [column (c)]. Just below, we report very similar results in a cross-section/time-series context. Thus, it is certainly easy to see why the evidence for an important positive impact of venture capital firms on the birth of biotech firms may have appeared stronger in previous work than seems warranted based on fuller models: Since venture capital firms have developed around a number of great universities, their presence proxies for intellectual human capital in the absence of more direct measures; if they are the only variable indicating presence of great universities and their faculties, they enter positively even if their packaging activities result in a negative direct effect on births.

The decade-lagged dependent variable is added to the full fundamentals model in column (f) of Table 2. Doing so primarily has the effect of weakening the significance of the top-quality universities variable (but, see the annual model below) due to significant multicollinearity between the variables.[13] One interpretation of this positive coefficient on the lagged dependent variable is that agglomeration effects strengthen the impact of fundamentals on regional development. However, the statistical properties of poisson regressions with lagged dependent variables are somewhat problematic so such regressions and their estimated standard errors should be viewed cautiously.

In conclusion, the intellectual human capital variables play a strong role in determining where the U.S. biotech industry developed during the 1980's. We have been able to identify particular star scientists who appear to play a crucial role in the process of spillover and geographic agglomeration over and above that which would be predicted based on university reputation and scientists supported by federal grants alone. The strong positive role of venture capital variable reported previously is not supported for firm births. Indeed, the data tell us that there were fewer firms founded, other things equal, where there more venture capital firms. It is left to future research to explore whether firms which are associated with particular star scientists or were midwifed by venture capitalists are more successful than other firms.[14]

B. The Annual Model

We next report analogous poisson regressions exploiting the panel nature of our data set with observations for the 183 regions for each of the years 1976 through 1989. Tables 3 and 4 report poisson regressions for this entire panel.

Column (a) of Table 3 reports the results using only the counts of stars and their collaborators active each year in each region. As with the long-run models in Table 2, examination of the data suggested that these effects— particularly for stars—were nonlinear so we add squared values in column (b). Again, as the number of stars increases, their marginal contribution diminishes eventually passing through zero.

These nonlinearities might reflect the declining value over time of the intellectual human capital as we have measured it. Basically,

---

[12] This hypothesis was derived from anecdotal evidence, but note that the top nine of Ernst & Young's list of top-ten companies by 1993 market valuation (G. Steven Burrill and Kenneth B. Lee, Jr., 1994 p. 54) were located and founded in regions richly endowed with venture capital firms: Boston (3), San Francisco (3), Los Angeles (1), San Diego (1), and Seattle (1).

[13] In the alternative version of Table 2 (see footnote 9 above), the coefficient on the lagged dependent variable was nearly as large as in Table 2, so the significant positive coefficient does not arise from firms born 1976–1980 appearing in both the current and lagged dependent variables.

[14] See Zucker et al. (1994) for our first effort to assess the determinants of success of firms after birth.

TABLE 3—ANNUAL POISSON REGRESSIONS ON THE BIRTH OF BIOTECH-USING FIRMS BY REGION AND YEAR, 1976–1989

	(a)	(b)	(c)	(d)	(e)	(f)
Constant	−1.591*** (0.032)	−1.918*** (0.041)	−2.148*** (0.057)	−4.447*** (0.226)	−4.491*** (0.349)	−4.687*** (0.565)
Number stars active in the region and year	0.157*** (0.020)	0.529*** (0.051)	0.270** (0.088)	—	0.361*** (0.080)	0.282** (0.103)
Number collaborators active in the region and year	0.043*** (0.013)	0.083* (0.035)	0.047 (0.049)	—	0.013 (0.047)	0.032 (0.052)
(Number stars active in the region and year)^2	—	−0.022*** (0.002)	−0.014* (0.006)	—	−0.015** (0.005)	−0.014 (0.008)
(Number collaborators active in the region and year)^2	—	−0.001 (0.001)	0.000 (0.001)	—	0.000 (0.001)	0.001 (0.002)
Number stars active in the region and year × DUMMY1986–89	—	—	−0.219 (0.113)	—	−0.298** (0.102)	−0.245 (0.128)
Number collaborators active in the region and year × DUMMY1986–89	—	—	0.117 (0.067)	—	0.115 (0.064)	0.027 (0.081)
(Number stars active in the region and year × DUMMY1986–89)^2	—	—	0.006 (0.007)	—	0.009 (0.006)	0.007 (0.008)
(Number collaborators active in the region and year × DUMMY1986–89)^2	—	—	−0.001 (0.002)	—	−0.001 (0.002)	0.001 (0.002)
Number top-quality universities in the region in 1981	—	—	0.444*** (0.125)	—	0.472*** (0.095)	0.462*** (0.109)
Number faculty with federal support in the region in 1979–80	—	—	0.625*** (0.093)	—	0.982*** (0.094)	0.930*** (0.093)
Number venture capital firms in the region and year[a]	—	—	—	0.019** (0.007)	−0.028*** (0.006)	−0.024** (0.008)
Total employment (all industries) in the region and year	—	—	—	0.173*** (0.051)	−0.081 (0.048)	−0.117* (0.055)
Average wages per job in the region and year	—	—	—	0.153*** (0.010)	0.125*** (0.016)	0.132*** (0.017)
Earnings-price ratio (Standard & Poors 500) for year	—	—	—	−0.024 (0.016)	−0.026 (0.026)	−0.017 (0.039)
Number firms active in the region at end of previous year	—	—	—	—	—	0.020 (0.013)
Number firms active in all U.S. at end of previous year	—	—	—	—	—	−0.000 (0.000)
Births of biotech firms in the region for previous year	—	—	—	—	—	0.054 (0.034)
Log-likelihood	−1677.0	−1429.1	−1274.3	−1669.5	−1202.3	−1184.6
Restricted log-likelihood	−2238.5	−2238.5	−2238.5	−2238.5	−2238.5	−2238.5

Notes: $N = 2562$. Standard errors (adjusted by Wooldridge, 1991 Procedure 2.1) are in parentheses below coefficients.
   [a] For years after 1981, the number of venture capital firms in a region is held constant at the 1981 level to avert simultaneity problems.
   * Significantly different from 0 at the 5-percent level.
   ** Significantly different from 0 at the 1-percent level.
   *** Significantly different from 0 at the 0.1-percent level.

as the knowledge diffuses we expect that more and more stars will result in less and less payoff to any one of them if he or she were to start a firm, and indeed stars are less likely to result in birth of firms after 1985 than before. This is illustrated in column (c) of Table 3 where we add four interaction terms in which these counts and their squares have been multiplied by a dummy DUMMY1986–89 equal to 1 during 1986–1989 and 0 otherwise, as well as the other intellectual human capital terms. During 1986–1989 the positive effect of stars is sharply reduced while that of collaborators more than triples.[15] Nonetheless, we should view this inference cautiously since the significance values of the interaction terms for stars and collaborators with DUMMY1986–89 fall between 0.10 and 0.05, except for stars in the full fundamentals model in column (e) where the stars interaction term is significant at the 0.01 level.

Thus, we see that (at least during the first decade of this industry) localities with outstanding scientists having the tacit knowledge to practice recombinant DNA were much more likely to see new firms founded and preexisting firms begin to apply biotechnology. There is some evidence that as knowledge about gene splicing diffused and the tacit knowledge lost its scarcity and extraordinary value, the training function of universities became more important relative to the attraction of great scientists to an area. It is interesting that the quadratic term for stars is negative, suggesting diminishing returns (or possibly just proportionately fewer, larger firms) rather than the increasing returns suggested by standard views of knowledge spillovers which posit uninternalized, positive external effects from university scientists.[16] In the same regression in column (c), we see that, beyond the identified stars and collaborators, university quality and federal support are also significant measures of intellectual human capital relevant to firm founding.

Column (d) of Table 3 leads to the same conclusions with panel data as found for the same column in Table 2: The economic variables enter significantly with the expected sign if the intellectual human capital variables are omitted from the regression. However, unlike the previous long-run case, we can now enter the earnings-price ratio.[17] Here this variable enters with the correct sign, but does not even reach the 0.10 level of significance.

Column (e) of Table 3 presents the annual full fundamentals model incorporating the intellectual human capital and other variables. The results for the intellectual human capital measures are robust while the sign of the venture capital variable turns significantly negative as in the long-run model and the employment variable becomes insignificant (and negative).

Column (f) of Table 3, analogously to Table 2, adds a lagged dependent variable to the full fundamentals model. We also included the one-year lagged regional and national counts of firms using biotechnology as dynamic influences reflecting local agglomeration effects and market competition effects, respectively. None of the three dynamic variables enter significantly although their signs are consistent with some geographic agglomeration.

Thus, taken as a whole the results summarized in Table 3 support the strong role of intellectual human capital variables in determining the development of the American biotech industry.

The role of the economic variables, particularly the number of venture capital firms in the region, is explored further in Table 4. This table presents representative results for births in the entrant and incumbent subcomponents of firm entry into biotechnology. We see in columns (a) and (b) that if only the economic variables are introduced we get all the expected signs at appropriate significance [except for employment in (a) and the earnings-price ratio in both], including a result consistent with conventional wisdom that the number of venture capital firms has a signifi-

---

[15] To compute the effects of stars in the 1986–1989 period, we need to add the coefficients of the number of active stars and the coefficient of the same variable interacted with DUMMY1986–89 and then do the same for the two terms involving the squared values of these variables. An analogous approach yields the effect of collaborators during 1986–1989. We examined also interactions with dummy variables for 1976–1980 and with a time trend. Since the coefficients were very small and statistically insignificant for interaction terms involving 1976–1980 dummies, we believe the reported form more accurately reflects the time or diffusion dependence than a negative trend throughout the period.

[16] We are indebted to Jeff Armstrong for this point.

[17] The earnings-price ratio had to be dropped from these analyses because it is available only nationally over time.

TABLE 4—ANNUAL POISSON REGRESSIONS ON THE BIRTH OF BIOTECH-USING ENTRANTS
AND INCUMBENTS BY REGION AND YEAR, 1976–1989

	(a) Entrants	(b) Incumbents	(c) Entrants	(d) Incumbents	(e) Entrants	(f) Incumbents
Constant	−4.726*** (0.284)	−5.798*** (0.563)	−4.843*** (0.409)	−5.673*** (0.902)	−4.928*** (0.669)	−5.228*** (1.285)
Number stars active in the region and year	—	—	0.414*** (0.095)	0.323 (0.165)	0.351** (0.124)	0.242 (0.169)
Number collaborators active in the region and year	—	—	−0.006 (0.053)	0.000 (0.105)	0.012 (0.059)	0.019 (0.101)
(Number stars active in the region and year)²	—	—	−0.016** (0.006)	−0.016* (0.008)	−0.017 (0.009)	−0.015 (0.011)
(Number collaborators active in the region and year)²	—	—	0.001 (0.002)	0.002 (0.003)	0.000 (0.002)	0.001 (0.003)
Number stars active in the region and year × DUMMY1986–89	—	—	−0.227* (0.113)	−0.519* (0.237)	−0.196 (0.147)	−0.456 (0.251)
Number collaborators active in the region and year × DUMMY1986–89	—	—	0.096 (0.071)	0.233 (0.141)	0.011 (0.090)	0.144 (0.153)
(Number stars active in the region and year × DUMMY1986–89)²	—	—	0.007 (0.007)	0.018 (0.010)	0.006 (0.010)	0.015 (0.013)
(Number collaborators active in the region and year × DUMMY1986–89)²	—	—	−0.001 (0.002)	−0.004 (0.003)	0.001 (0.003)	−0.002 (0.004)
Number top-quality universities in the region in 1981	—	—	0.440*** (0.110)	0.479* (0.205)	0.410** (0.126)	0.447 (0.238)
Number faculty with federal support in the region in 1979–80	—	—	0.973*** (0.112)	1.114*** (0.296)	0.932*** (0.107)	1.041*** (0.295)
Number venture capital firms in the region and year[a]	0.023** (0.009)	0.006 (0.013)	−0.029*** (0.007)	−0.027* (0.012)	−0.024** (0.009)	−0.024 (0.013)
Total employment (all industries) in the region and year	0.128 (0.067)	0.296** (0.098)	−0.110 (0.058)	−0.052 (0.098)	−0.149* (0.067)	−0.078 (0.103)
Average wages per job in the region and year	0.156*** (0.012)	0.139*** (0.024)	0.123*** (0.018)	0.113** (0.039)	0.127*** (0.020)	0.114** (0.040)
Earnings-price ratio (Standard & Poors 500) for year	−0.036 (0.021)	−0.033 (0.043)	−0.022 (0.031)	−0.056 (0.070)	−0.016 (0.046)	−0.082 (0.092)
Number firms active in the region at end of previous year	—	—	—	—	0.023 (0.015)	0.024 (0.025)
Number firms active in all U.S. at end of previous year	—	—	—	—	−0.000 (0.000)	−0.001 (0.001)
Births of biotech firms in the region for previous year	—	—	—	—	0.037 (0.041)	0.055 (0.061)
Log-likelihood	−1265.1	−486.3	−945.9	−386.8	−935.8	−382.9
Restricted log-likelihood	−1628.7	−607.9	−1628.7	−607.9	−1628.7	−607.9

Notes: N = 2562. Standard errors (adjusted by Wooldridge, 1991 Procedure 2.1) are in parentheses below coefficients.
  [a] For years after 1981, the number of venture capital firms in a region is held constant at the 1981 level to avert simultaneity problems.
   * Significantly different from 0 at the 5-percent level.
  ** Significantly different from 0 at the 1-percent level.
 *** Significantly different from 0 at the 0.1-percent level.

cantly positive effect on the birth of new firms but an insignificant effect on the birth of sub-units of existing firms which would not normally be financed by venture capital firms. The full fundamentals model is reported in columns (c) and (d) for births of entrants and incumbents, respectively, which is to be compared to column (e) for all firm births in Table 3. Again, in the presence of intellectual human capital the simple economic story does not hold up: the coefficients of venture capital firms and total employment turn negative, significantly so in the former case. Similar results are obtained in the dynamic versions of the full model reported in columns (e) and (f) of Table 4. The robustness of the negative venture capital coefficient remains a puzzle for future work, particularly in view of Yolanda K. Henderson's (1989) evidence that, despite some significant localization, most investments by venture capitalists cross regional boundaries.

IV. Summary and Conclusions

The American biotechnology industry which was essentially nonexistent in 1975 grew to 700 active firms over the next 15 years. In this paper, we show the tight connection between the intellectual human capital created by frontier research and the founding of firms in the industry. At least for this high-tech industry, the growth and location of intellectual human capital was the principal determinant of the growth and location of the industry itself. This industry is a testament to the value of basic scientific research. The number of local venture capital firms, which appears to be a positive determinant when intellectual human capital variables are excluded from the regressions, is found to depress the rate of firm birth in an area, perhaps due to the role of these venture capital firms in packaging a number of scientists into one larger firm which is likely to go public sooner.

We conclude that the growth and diffusion of intellectual human capital was the main determinant of where and when the American biotechnology industry developed. Intellectual human capital tended to flourish around great universities, but the existence of outstanding scientists measured in terms of research pro-

ductivity played a key role over, above, and separate from the presence of those universities and government research funding to them. We believe that our results provide new insight into the role of research universities and their top scientists as central to the formation of new high-tech industries spawned by scientific breakthroughs. By being able to quantitatively identify individuals with the ability both to invent and to commercialize these breakthroughs, we have developed new specificity for the idea of spillovers and in particular raised the issue of whether spillovers are best viewed as resulting from the nonappropriability of scientific knowledge or from the maximizing behavior of scientists who have the ability to appropriate the commercial fruits of their academic discoveries.

DATA APPENDIX

The data used here are generally in panel form for 14 years (1976–1989) and 183 regions (functional economic areas as defined by the BEA). Frequently, the data are aggregates of data at the zip code or county level. Lagged variables include data for 1975 in the unlagged form. These data sets, part of our ongoing project on "Intellectual Capital, Technology Transfer, and the Organization of Leading-Edge Industries: Biotechnology," will be archived upon completion of the project in the Data Archives at the UCLA Institute for Social Science Research. A full description of the data is available from the authors upon request.

Biotechnology Firms

The starting point for our firm data set covered the industry as of April 1990 and was purchased from NCBC (1991), a private firm which tracks the industry. This data set identified 1075 firms, some of which were duplicates or foreign and others of which had died or merged. Further, there were a significant number of firms missing which had exited prior to April 1990. For these reasons, an intensive effort was made to supplement the NCBC data with information from *Bioscan* (1989–1993) and an industry data set provided by a firm in the industry which was also the ancestor of the *Bioscan* data set (Cetus Corp., 1988).

We generally counted entry of firms by adding up for each year and region the number of entrants founded and incumbents first using biotechnology. A few special cases should be noted: Where a firm enters the data set due to the merger of a entrant and another firm, we count it for the purposes of this paper as a continuation of the original entrant and not a new birth (the older entrant if two are involved). If firms already in the data set merge and one continues with the other(s) absorbed, the enterprise is counted as the continuing enterprise and not a new birth.

Scientists

Star scientists and their collaborators were identified as described in the text. Individual scientists are linked to locations through the institutional affiliations reported in their publications in the article data set. The discovery of genetic sequences is recognized by *GenBank*'s assignment to an article of a "primary accession number" to identify each. The 22 additional stars added to the 315 with more than 40 primary accession numbers thus had 20 or more articles with at least one primary accession number and 20–40 primary accession numbers total.

Articles

Our article data set consists of all 4,061 articles in major journals listed in *GenBank* as reporting genetic-sequence discoveries for which one or more of our 327 stars were listed as authors. (A small number of unpublished papers and articles appearing in proceedings volumes and obscure journals were excluded to permit the hand coding detailed below.) All of these articles were assigned unique article ID numbers and collected by hand. For each article, scientist ID numbers are used to identify the order of authorship and the institutional affiliation and location for each author on each article. This hand coding was necessary because, under the authorship traditions for these fields, the head of the laboratory who is often the most prestigious author frequently appears last. Our stars, for example, were first

authors on 18.3 percent of the articles and last authors on 69.1 percent of the 4,031 articles remaining after excluding the 30 sole-authored articles.[18] Unfortunately, only first- and/or corresponding-author affiliations are available in machine-readable sources.[19]

The resulting authorship data file contains 19,346 observations, approximately 4.8 authors for each of the 4,061 published articles. Each authorship observation gives the article ID number, the order of authorship, the scientist ID number of one of our stars and collaborators, and an institutional ID number for the author's affiliation which links him or her to a particular institution with a known zip code as of the publication date of the article.

Citations

We have collected data for 1982, 1987, and 1992 on the total number of citations to each of our 4,061 published articles listed in the Institute for Scientific Information's *Science Citation Index* (1982, 1987, 1992). These citation counts are linked to the article and authorship data set by the article ID number. The citations were collected for articles if and only if they appeared in the article data set; so scientists are credited with citations only insofar as they are to the 4,061 articles reporting genetic-sequence discoveries and published in major journals.

Universities

Our university data set consists of all U.S. institutions listed as granting the Ph.D. degree in any field in the Higher Education General Information Survey (HEGIS), Institutional Characteristics, 1983–1984 (U.S. Department of Education, National Center for Education

---

[18] This positional tradition holds across national boundaries: As a percentage of articles coauthored by their fellow nationals, American stars are 16.4 percent of first authors and 71.2 percent of last authors, compared to 21.2 percent and 63.1 percent, respectively, for Japanese, and 19.7 percent and 69.2 percent for other nationalities.

[19] The *Science Citation Index* lists up to six of the affiliations listed on the paper but only links the corresponding author to a particular affiliation.

Statistics, 1985). Each university is assigned an institutional ID number, a university flag, and located by zip code based on the HEGIS address file. Additional information described in the text was collected from Jones et al. (1982) for those universities granting the Ph.D. degree in biochemistry, cellular/molecular biology, and/or microbiology which we define as *"biotech-relevant"* fields.

Research Institutes and Hospitals

For those U.S. research institutions and hospitals listed as affiliations in the article data set, we assigned an institutional ID number and an institute/hospital flag, and obtained an address including a zip code as required for geocoding. No additional information has been collected on these institutions.

Venture Capital Firms

We created a venture capital firm data set by extracting from the Pratt (1982) directory the name, type, location, year of founding, and interest in funding biotech firms. This information was extracted for all venture capital which were legally permitted to finance start-ups. This latter requirement eliminated a number of firms which are chartered under government programs targeted at small and minority businesses. This approach accounts includes founding date of firms appearing in the 1982 Pratt directory, excluding those firms that may have either entered thereafter or existed in earlier years but exited before the directory was compiled.

Other Economic Variables

Total employment and average earnings per job by region and year are as reported by the Bureau of Economic Analysis based on county level data in U.S. Department of Commerce (1992b): Total employment is from Table K, line 010 (in millions of persons). Average earnings is from Table V, line 290 (wage & salary disbursements, other labor income, and proprietors income per job in thousands of current dollars), deflated by the implicit price deflator for personal consumption expenditures. The annual

data for the implicit price deflator for personal consumption expenditures were taken from U.S. Department of Commerce (1992a p. 247, line 16) as updated in the July 1992 *Survey of Current Business,* (p. 92, line 16). The S&P 500 earnings-price ratio was taken from *CITIBASE* (1993), series FSEXP.

REFERENCES

Arrow, Kenneth J. "Economic Welfare and the Allocation of Resources for Invention," in Richard R. Nelson, ed., *The rate and direction of inventive activity: Economic and social factors*. National Bureau of Economic Research Special Conference Series, Vol. 13. Princeton, NJ: Princeton University Press, 1962, pp. 609–25.

Bioscan, Vols. 3–7. Phoenix, AZ: Oryx Press, 1989–1993.

Burrill, G. Steven and Lee, Kenneth B., Jr. *Biotech 94: Long-term value, short-term hurdles*. Ernst & Young's Eighth Annual Report on the Biotechnology Industry. San Francisco: Ernst & Young, 1994.

Cameron, A. Colin and Trivedi, Pravin K. "Regression-Based Tests for Overdispersion in the Poisson Model." *Journal of Econometrics*, December 1990, *46*(3), pp. 347–64.

Cetus Corp. "Biotechnology Company Data Base," predecessor source for *Bioscan*. Computer printout, Cetus Corp., 1988.

CITIBASE: Citibank Economic Database. Machine-readable database, 1946–June 1993. New York: Citibank, 1993.

GenBank, Release 65.0. Machine-readable database. Palo Alto, CA: IntelliGentics, Inc., 1990.

Greene, William H. *LIMDEP: User's manual and reference guide, version 6.0*. Bellport, NY: Econometric Software, 1992.

Griliches, Zvi. "Hybrid Corn: An Exploration in the Economics of Technological Change." *Econometrica*, October 1957, *25*(4), pp. 501–22.

_____. "The Search for R&D Spillovers." *Scandinavian Journal of Economics*, 1992 Supplement, *94*, pp. 29–47.

Grossman, Gene M. and Helpman, Elhanan. *Innovation and growth in the global economy*. Cambridge, MA: MIT Press, 1991.

Hausman, Jerry; Hall, Bronwyn H. and Griliches, Zvi. "Econometric Models for Count Data with an Application to the Patents-R&D Relationship." *Econometrica*, July 1984, *52*(4), pp. 909–38.

Henderson, Yolanda K. "Venture Capital and Economic Development." Paper presented to the New England Advisory Council, Federal Reserve Bank of Boston, Boston, MA, July 11, 1989.

Institute for Scientific Information. *Science Citation Index*, ISI compact disc editions, machine-readable database. Philadelphia: Institute for Scientific Information, 1982, 1987, 1992.

Jaffe, Adam B. "Real Effects of Academic Research." *American Economic Review*, December 1989, *79*(5), pp. 957–70.

Jaffe, Adam B.; Trajtenberg, Manuel and Henderson, Rebecca. "Geographic Localization of Knowledge Spillovers as Evidenced by Patent Citations." *Quarterly Journal of Economics*, August 1993, *63*(3), pp. 577–98.

Jones, Lyle V.; Lindzey, Gardner and Coggeshall, Porter E., eds. *An assessment of research-doctorate programs in the United States: Biological sciences*. Washington, DC: National Academy Press, 1982.

Kenney, Martin. *Biotechnology: The university-industrial complex*. New Haven, CT: Yale University Press, 1986.

Lerner, Joshua. "Venture Capitalists and the Decision to Go Public." *Journal of Financial Economics*, June 1994, *35*(3), pp. 293–316.

———. "Venture Capitalists and the Oversight of Private Firms." *Journal of Finance*, March 1995, *50*(1), pp. 301–18.

Mansfield, Edwin. "Academic Research Underlying Industrial Innovations: Sources, Characteristics, and Financing." *Review of Economics and Statistics*, February 1995, *77*(1), pp. 55–65.

Nelson, Richard R. "The Simple Economics of Basic Scientific Research." *Journal of Political Economy*, June 1959, *67*(3), pp. 297–306.

Nelson, Richard R. and Romer, Paul M. "Science, Economic Growth, and Public Policy," in Bruce L. R. Smith and Claude E. Barfield, eds., *Technology, R&D, and the economy*. Washington, DC: Brookings Institution and American Enterprise Institute, 1996, pp. 49–74.

Nelson, Richard R. and Winter, Sidney G. *An evolutionary theory of economic change*. Cambridge, MA: Harvard University Press, 1982.

North Carolina Biotechnology Center. *North Carolina Biotechnology Center U.S. companies database*, machine-readable database. Research Triangle Park, NC: North Carolina Biotechnology Center, April 16, 1992.

Pratt, Stanley E. *Guide to venture capital sources*, 6th Ed. Englewood Cliffs, NJ: Prentice-Hall, 1982.

Romer, Paul M. "Increasing Returns and Long-Run Growth." *Journal of Political Economy*, October 1986, *94*(5), pp. 1002–37.

———. "Endogenous Technological Change." *Journal of Political Economy*, October 1990, Part 2, *98*(5), pp. S71–S102.

Rosenberg, Nathan. *Inside the black box: Technology and economics*. Cambridge: Cambridge University Press, 1982.

U.S. Department of Commerce, Bureau of Economic Analysis, Economics and Statistics Administration. *National income and product accounts of the United States, volume 2, 1959–88*. Washington, DC: U.S. Government Printing Office, 1992a.

———. *Regional economic information system, version 1.3*, CD-ROM, machine-readable database. Washington, DC: Bureau of Economic Analysis, May 5, 1992b.

U.S. Department of Education, National Center for Education Statistics. *Higher education general information survey (HEGIS), institutional characteristics, 1983–84*, machine-readable database, ICPSR 8291. Ann Arbor, MI: Inter-University Consortium for Political and Social Research, circa 1985.

Wooldridge, Jeffrey M. "On the Application of Robust, Regression-Based Diagnostics to Models of Conditional Means and Conditional Variances." *Journal of Econometrics*, January 1991, *47*(1), pp. 5–46.

Zucker, Lynne G.; Brewer, Marilynn B.; Oliver, Amalya and Liebeskind, Julia. "Basic Science as Intellectual Capital in Firms: Information Dilemmas in rDNA Biotechnology

Research." Working paper, UCLA Institute for Social Science Research, 1993.

Zucker, Lynne G. and Darby, Michael R. "Star Scientists and Institutional Transformation: Patterns of Invention and Innovation in the Formation of the Biotechnology Industry." *Proceedings of the National Academy of Sciences*, November 12, 1996, *93*(23), pp. 12709–12716.

Zucker, Lynne G.; Darby, Michael R. and Armstrong, Jeff. "Intellectual Capital and the Firm: The Technology of Geographically Localized Knowledge Spillovers." National Bureau of Economic Research (Cambridge, MA) Working Paper No. 4946, December 1994.

Zucker, Lynne G.; Darby, Michael R. and Brewer, Marilynn B. "Intellectual Capital and the Birth of U.S. Biotechnology Enterprises." National Bureau of Economic Research (Cambridge, MA) Working Paper No. 4653, February 1994.

[13]

ELSEVIER

Research Policy 32 (2003) 621–638

research
policy

www.elsevier.com/locate/econbase

Development of SMEs and heterogeneity of trajectories: the case of biotechnology in France

Vincent Mangematin*, Stéphane Lemarié, Jean-Pierre Boissin, David Catherine, Frédéric Corolleur, Roger Coronini, Michel Trommetter

INRA/SERD, Université Pierre Mendès France, BP 47, 38040 Grenoble Cedex 9, France

Received 14 November 2001; received in revised form 19 March 2002; accepted 2 April 2002

Abstract

Biotechnology is an emergent sector based on the creation of research-intensive Small and Medium Enterprises (SMEs). While some SMEs are growing, most of them remain small, even those set-up several years ago. What is the pattern of development of the biotech sector? What are the patterns of development of firms? Studies on the development of high-tech SMEs have focused on a business model, in which entrepreneurs rely on growth forecasts to persuade capital investors (business angels and venture capitalists) to invest in a radical innovation project. Firms aim for a world market to industrialise their innovation, and initial public offering (IPO) enables initial investors to make profits that offset risky initial investment. While this model is appealing, it is simply one of the possible models of biotechnology development. Some firms are not designed to experience exponential growth, and choose to target local markets. Moreover, not all firms have the ambition of being listed on the stock exchange. Based on an in-depth analysis of the business and development of 60 French biotech SMEs, this article identifies two business models. By defining the development trajectories of each of these models, it highlights the temporary nature of the emergent model.
© 2002 Elsevier Science B.V. All rights reserved.

JEL classification: M130; Q130; L210; O30

Keywords: Biotechnology; SMEs; Resources; Innovation; Research; Corporate strategy; Business model; Trajectories of development

1. Introduction

Biotechnology is one of the emergent sectors whose development is largely based on the creation of research-intensive SMEs. In France, the number of new biotech SMEs is skyrocketing, from fewer than 10 new firms per year a decade ago, to more than 30 in 1998. However, while some SMEs are growing, most of them remain small, even those set-up several

* Corresponding author. Tel.: +33-4-76-82-5686;
fax: +33-4-76-82-5455.
E-mail address: vincent@grenoble.inra.fr (V. Mangematin).

years ago. What is the pattern of development of the biotech sector? What are the patterns of development of firms?

Studies on biotechnology development have focused on a business model that emerged along with the development of high-tech SMEs. Based on a radical innovation project, a group of entrepreneurs creates a firm designed to grow fast. These entrepreneurs rely on optimistic forecasts of a promising scientific breakthrough to convince capital investors (business angels and venture capitalists) to fund their technological developments. The firm targets an international market in which it can industrialise its innovations and

622 V. Mangematin et al. / Research Policy 32 (2003) 621–638

generate a comfortable income. Initial public offering (IPO) enables initial investors to make profits that offset risks taken at the outset. While this model is appealing, both in its ability to link up scientific discovery and economic valorisation of science, it is only one of the possible biotechnology development models. Some firms are not designed for exponential growth and target only a local market. Moreover, not all firms have the ambition of being listed on the stock market. Even if going public depends partly on age and on timing (there are periods when the window is open, and periods during which going public is very difficult), fewer than 30% of biotechnology firms are listed in the US and fewer than 7% in Europe. Biotechnology firms, therefore, develop along the lines of several models that have to be described and understood if the evolution of the sector, relations between the actors involved and the effects of public policies are to be understood.

The main features of the biotech sector (science-based, importance of start-ups, and heterogeneity of actors involved) are helpful in understanding its dynamics. Even if they are based on the growth of firms, trajectories and development logics can differ from one company to another. A survey of 60 firms was conducted in the year 2000 to understand the development of SMEs in France. In-depth face-to-face interviews were held during the same year within each of the 60 firms. The results shed light on the diversity of SMEs, which can be described in terms of business models. A business model describes a category of firm in relation to the market it targets, its expected growth, its modes of governance, and the organisation of its activity. The diversity of business models of biotechnology SMEs is a point that is rarely considered in studies on factors promoting the development of these firms (see Section 2). This approach requires not only the differentiation of firms' activities, but also an explanation of their resulting position compared to other actors in the industry and, more generally, the institutional framework around the firm. While the factors facilitating the creation of start-ups are now known, few studies have been made of those facilitating their sometimes-fragile survival and development. The present article highlights the logic of the development of firms and shows that modes of development differ for each business model.

Section 2 discusses the role of SMEs in industrial dynamics and analyses how the diversity of SMEs

has been studied in different contributions. Data and methods are presented in Section 3. The size of the innovation project appears to be a key variable to split the sample of firms into two homogeneous clusters. Networks of the firms (founders, alliances, etc.) are then discussed in Section 4 to describe the Sectoral Innovation System (SIS) in biotechnology.

2. Heterogeneity of SMEs

2.1. Sectoral characteristics of biotech firms

Innovation in biotech firms has common characteristics. Three main features have attracted a great deal of attention in recent years. (1) Development of the biotech sector is based on the entry of a large number of SMEs; the biotech sector is often described as a large and highly turbulent population of innovators; SMEs are a leading force in a science push context, while the role of large firms is mainly to integrate new discoveries into their products after they have been developed by SMEs. (2) Biotech is a science-intensive sector. This characteristic has two consequences (Zucker et al., 1994; Feldman, 1999). First, SMEs are located close to the source of knowledge, i.e. close to the main universities, even if they are not spin-off of universities. Second, most of the founders have a scientific background and a Ph.D. and are members of scientific networks. (3) Strategic alliances progressively appear as central features of the biotech sector. The status of alliances is changing from strategic alliances, as a means to acquire and co-ordinate resources for technological and scientific development, to a new organisational form. Powell et al. (1996) explain the development of SMEs by the inter-organisational collaboration between SMEs and large firms. For established firms in traditional life sciences sectors, such as pharmaceuticals, chemicals, or seeds, the introduction of biotechnology has been a competence-destroying innovation. Because of the novelty of this scientific field and the risks attached to biotechnology, established firms channel their investments in biotech research to SMEs through long-term contracts or by forming joint ventures (Arora and Gambardella, 1990). SMEs, in turn, enter into long-term contracts to obtain complementary assets, such as product testing, commercialisation, or

V. Mangematin et al. / Research Policy 32 (2003) 621–638 623

distribution capabilities (Teece, 1986; Barley et al., 1992). Consequently, the biotech industry is characterised by a network structure of inter-organisational alliances that govern the exchange of complementary assets among the different actors involved: academic teams, large firms, SMEs. The SMEs are seen as a nexus of these networks because of their role between scientists who make the scientific discoveries, and the large established firms that have the capabilities to market the products.

The characteristics of biotech firms transcend the national context; they are similar in all developed countries. The concept of a SIS, defined by Breschi and Malerba (1997), clearly accounts for the dynamics of the biotech sector. A SIS is defined as a "system (group) of firms active in developing and making a sector's products, and in generating and utilising a sector's technologies; such a system of firms is related in two different ways: through processes of interaction and co-operation in artefact-technology development, and through processes of competition and selection in innovative and market activities" (p. 131). The key actors of a SIS are the private firms. Breschi and Malerba (1997) emphasise the fact that competition and selection processes involve firms with different capabilities and capacities to mobilise other actors—large firms, universities, research institutes—and innovative performances.

The approach in terms of SISs highlights the key role of firms in the dynamics of innovation. As shown in our previous work (Mangematin, 2000a), most researchers (Stephan and Everahrt, 1995; Audretsch and Stephan, 1996; Klavans and Deeds, 1997) focus on fast growing SMEs as a key leading mechanism. For this kind of firm, critical resources are access to scientific competencies and techniques developed by academic research, and to capital markets. Their development relies on signals of these scientific competencies, i.e. patent applications, new products and researchers involved in the firm or in its scientific advisory board. Patents, collaborations with academic world, and partnerships with pharmaceutical firms appear to be positive signals of competencies for potential investors when the firm enters into the stock market. Researchers with a high level of scientific visibility often create these firms. The connexion with the scientific network is a condition for growth. It is not sufficient, however, for firms must not only

develop high-tech research, but also transfer and commercialise their results. This often involves research or development contracts with a large company, in which the SME undertakes to provide its partner with specific materials, technologies, know-how, or expertise. Relations are formed on the basis of a specific competency recognised by the large firm. The SME's technological lead depends on the quality of its research, and the launching of the activity relies on capital input for the development of the product or process. We argue, in this paper, that the development of the biotech sector is based on the coexistence of two types of firms, fast growing ones, which will form the elite of European industrial biotech leaders on which the sector could grow and compete with that of the US; and a large number of small firms mainly involved in services to biotech, and which are not expected to become worldwide leaders.

2.2. Heterogeneity among firms within biotech sector and business models

Most studies on the key factors promoting development of biotechnology SMEs focus on this emerging business model. Yet, empirical evidence shows that the strategies of biotechnology SMEs can vary widely, and it seems that all external factors do not influence the different modes of development in the same way. In other words, by treating biotech SMEs as a uniform population (from the point of view of strategies and modes of development), analyses tend to aggregate effects that may be very different from case to case. A first series of studies differentiates between biotech SMEs by using external indicators that could easily be collected by questionnaire or from published data. Based on the technologies used and the targeted domains of application, these studies try to group together firms with similar profiles in this respect (Arundel and Rose, 1998; Saviotti et al., 1998; Lemarie and Mangematin, 2000; Lemarie et al., 2001). From the statistic point of view, their results explain only a small part of the variance. In other words, it is difficult to partition all techniques and areas of application since many combinations of techniques and markets are tested by firms. The analysis also shows that relatively old SMEs have been able to incorporate very recent technologies, thus indicating that combinations of techniques used by each firm can evolve

624   V. Mangematin et al. / Research Policy 32 (2003) 621–638

significantly. In a recent study, Mangematin (2000b) focuses on the composition of shareholding. The very nature of research-intensive firms generates substantial movements in their capital. Three main types can be distinguished: (1) firms owned by individual shareholders; (2) firms owned by institutional investors, i.e. venture capital or listed on the stock exchange; and (3) firms owned by major industrial groups. The analysis shows that firms in the latter two categories are bigger and develop faster, irrespective of the technologies used or the application domains targeted.

A second series groups together studies that analyse firms' development strategies on the basis of monographs. The main monographs published include: Genentech presented by McKelvey (1996), Celltech presented by Dodgson (1991), and Cetus presented by Rabinow (1996) through the history of PCR. Finally, Nilsson (2001) recently analysed biotech firms in Sweden, on the basis of seven monographs. Estades and Ramani (1998) also conducted a comparative study of 20 biotech SMEs based in France and the UK. Simplified versions of monographs are also published in the form of corporate profiles in specialised journals (e.g. Nature Biotech, Biofutur) or the reports of consultancy firms such as Ernst & Young. The analyses in these studies are often very interesting because they point out the different logics of development of the firms under review. However, it is difficult to generalise, given the limited number of firms studied and the heterogeneity of frames of analysis used in the different contributions. It is also difficult to generalise the result of the case studies because of the embeddedness of those firms in their environment. Lastly, the environment also differs from one case study to another, depending on the continent and the year in which the firm is studied.

These studies of biotech firms' profiles constitute a break in research focused on competition (Porter, 1980). Analysis in terms of resources and competencies has been developing since the mid-1980s. This change reflects the shift of interest from external towards internal analysis: organisations are studied from within rather than in relation to their environment. The increasing openness of organisations has gone hand-in-hand with a relative disappearance of boundaries. Thus, the theory of resources, for which the definition of boundaries is less fundamental than it is for approaches focused on competition, seems

particularly rich. It proposes an analysis of the organisation and of its competitive advantages in terms of tangible and intangible resources and competencies (Barney, 1991; Grant, 1991; Hamel, 1991; Barney et al., 1994). Competitive advantage is based on logic of comparative advantages derived from resources and competencies. It is by knowing and controlling them that strategic options can be defined and a competitive advantage created.

Resource-based theory (Penrose, 1959; Peteraf, 1993; Mahoney and Rajendran Pandrian, 1992; Shrivastava et al., 1994; Russo and Fouts, 1997), like the theory of dependent resources (Pfeffer and Salancik, 1978), distinguishes between resources which are inputs into the production process and can be of various kinds (capital, human resources, equipment, co-operative or commercialisation networks, reputation or scientific visibility), on the one hand, and competencies which are related to the use and implementation of those resources, on the other. The sustainable nature of competitive advantage depends on the difficulty another organisation would have imitating the source of the reference organisation's success.

This approach is fruitful for analysing an industry that relies primarily on a combination of resources because few of its products or services have as yet been marketed. The identification of critical resources for each type of organisation helps to understand development logics. Each business model has its own development logic which is coherent with the needed resources—customer and supplier relations, a set of competencies within the firm, a mode of financing its business, and a certain structure of shareholding (Teece et al., 1993). Two main business models have been identified.

1. SMEs that run small projects and target market niches (type A), i.e. small and segmented market in a small geographic area. Although innovation is crucial for these firms, the need to maintain profitability forces them to limit investments in research. In other words, they realise incremental innovations whose value can often be explained by the entrepreneur's early intuition and launching of a research programme to transform that intuition into an innovation. Firms in the A group sell both products and services. For those which sell products, agreements may be made with outside partners who

V. Mangematin et al./Research Policy 32 (2003) 621–638   625

will then perform part of the business (e.g. sub-contracting all or part of the production; granting a license for distribution of the product in a foreign country). Mechanisms for appropriating profits from innovation differ, depending on tacitness and codification of knowledge. Typically, a firm that sells a product protects itself against risks of imitation. If certain activities are outsourced, it takes steps to remain in a position of leadership as regards the products concerned. Appropriation may be achieved by legal means, such as patents, by strategies of the first mover advantage type or, more generally, by the ownership of specific assets. LCA is one of the examples of such firm. It commercialises diagnostic kits to test the soil quality. When a firm sells services, by contrast, secrecy is most often enough to guarantee the fruits of innovation, for the service offered generally provides too little information on adequate techniques. Aquanal can be an example of this kind of firms. It develops methods to detect GMOs in food and more generally to analyse food compositions (mainly for certification).

It is rarely in the interests of SMEs in the A group and their customers to reserve all exclusive rights to the products or services provided. An SME (i.e. the supplier) that reserves exclusive rights will end up relying on certain customers and will be forced to increase its prices to cover fixed costs. If the customer is a firm, the question of exclusivity may arise. We note, however, that in that case the products or services bought generally correspond to factors that are not critical to maintenance of the customer firm's competitive advantage. It is preferable to outsource this activity to ensure better flexibility and possibly put several suppliers in competition.

2. Research-intensive SMEs that target broader markets (type B), i.e. niche markets which cover a large geographic area or large national or international markets. Famous SMEs that have made their mark in the history of biotechnology until now (e.g. Genentech, Chiron, Millenium) are firms with very big research programmes. The profitability of such programmes is credible only if the markets targeted downstream are very broad. This positioning is, therefore, radically different from that of type A firms, which target niches so as not to enter into direct competition with big companies.

Firms in group B develop one of the two strategies: either they are involved in contract research with large firms, or they carry out large research programmes themselves and valorise the product or service in the market. When they undertake large research contracts for big industrial groups, the "client" will have exclusive rights to the results. In this type of contract, the SME is remunerated in instalments and in the form of royalties on innovations resulting from the programme. In other words, these firms do not develop the final product. This type of configuration leads SMEs to adopt a complementary position in relation to major industrial groups. They explore new technological domains that big groups prefer to outsource to remain more flexible. Research is based on the establishment of technological platforms (generally comprising large facilities, software and human capital) that can be amortised on several major research operations.

Depending on the stage of the project, the firm's strategy to acquire and co-ordinate resources for development is not the same: during the years following start-up, the SME has no sales. As it approaches the marketing stage, the SME generates sales licence revenues for use of its product. In the case where heavy investments are needed to move into the industrial phase, the firm will tend to develop the product with outside partners. This is typically the case in the pharmaceutical industry, where the cost of developing a drug is very high, especially in the clinical testing phases prior to marketing. Most biotechnology-based drugs are developed by biotech SMEs and are commercialised by pharmaceutical groups.

As far as large companies are concerned, type B firms have a credible project because they are able to set-up extensive research programmes. Since these research programmes are radically new, the incumbent leaders' experience is not crucial and SMEs are not at a disadvantage. In certain cases, innovation requires additional assets. It is generally in such cases that type B SMEs enter into collaboration with industrial partners.

The French firms, Genome Express and Genset, during the late 1990's, illustrated types A and B. Genome Express has been created in 1994 by Y. Laurent and J.P. Mouret, who worked previously at Perkin-Elmer, a firm specialised in instrumentation for life sciences industries. Genome Express is basically a firm dedicated to high speed and high quality gene sequencing. In 2000, the firm sells mainly services,

626 *V. Mangematin et al./Research Policy 32 (2003) 621–638*

i.e. gene sequences, to a large number of clients, from academic labs to pharmaceutical industry. During this period, Genome Express is clearly a type A firm, which generates turn over from its activity, and which runs small research programs to perform incremental innovations. At the opposite, Genset has been founded in 1989 by two well-known researchers in biotechnology, P. Brandys and M. Vasseur. From the very beginning, this firm develops large research programmes to identify gene involved in important pathologies like obesity or cancer. It also runs a small sequencing activity linked to its research activity. Genset is financed through venture capital and stock market. It belongs to type B firms. Thus, these two firms belong to different business models even if they both generate turn over through the sequencing activity. For Genome Express, the sequencing activity covers the running costs of the firm. For Genset, it covers less than 10% of the running costs which are mainly research expenditures.

To conclude, it seems that a SIS exists in biotech. The central role played by SMEs is underlined by most of the actors. However, it seems that all SMEs do not play the same part. The different case studies show that the activity of the firm has an important impact on the type of firm and on its development logic. The consequences of this heterogeneity is discussed around two hypotheses.

H1. The size of the innovation project determines the two business models with internal coherence for each business model.

The first hypothesis discusses the existence of different business models and introduces a criterion of relative size of innovative project to structure the business models. What percentage of resources is devoted to the innovative project? Most of the resources? A small part of the resources of the firm? H1 divides firms in two categories: firms in which the innovative project represents a small part of the activity of the firm, and those built around the innovative project.

H2. Each firm in a business model requires different resources to run the operational activity. The firm has to convince different kinds of partners to ensure their development.

Hypothesis H2 underlines the fact that firms in each business model target different resources to ensure

their development. To secure these resources, different kinds of alliances are concluded.

3. Survey and method

For analysing the factors stimulating firms' growth and determining business models based on their activities, a sample of 60 firms was selected amongst the 200 biotech firms in France (Lemarie and Mangematin, 2000). Data on each of these firms were drawn from face-to-face interviews,[1] generally lasting 2 h. The interviewee was either the managing director, research director, or financial director. The entire survey was conducted between January and May 2000.

Surveyed firms were selected from a database of 200 French firms, constituted from the results of a short questionnaire survey[2] by the research ministry in 1999. Selection was made with a view to obtain a sample that was as representative as possible of the firms' business. Each of the six French firms listed on the New Market, EASDAQ, or NASDAQ was included in the sample. Finally, to ensure a sufficiently objective view of the firms' development, those created before 1995 were given preference in the survey (only nine firms out of 60 were created in 1996 or later).

On average, the 60 SMEs had 45 employees each and an average turnover of 4.1 M€. They were founded between 1977 and 1998: a quarter before 1989 and the other three quarters since then. These characteristics are similar to the average values observed on the broader sample of 200 French biotech firms.[3] Table 1 presents the general characteristics of the sample.

The main characteristics of the French biotech SMEs, shown in Table 1, are given as follows.

- The size of the firms, measured in terms of turnover or number of employees, was not correlated with the date of creation. Turnover per employee is highly

---

[1] Time constraints forced us to conduct telephone interviews with 10 of the firms.

[2] See Lemarie and Mangematin (2000) for a presentation of the results of this study. Data on the 200 firms covered their legal status, date of creation, main financial indicators, technologies used, and markets targeted.

[3] In this larger sample, 70% of the firms were created after 1990. The average turnover was 32 million francs and the average number of employees was 36 (Lemarie and Mangematin, 2000).

V. Mangematin et al. / Research Policy 32 (2003) 621–638   627

Table 1
General characteristics of the sample

	Average	Minimum	Maximum	Medium	First quartile	Third quartile
Date of creation	1991	1977	1999	1992	1989	1995
Turnover (M €)	4.1	0.0	30.5	1.0	0.3	5.5
Number of employees	45	0	525	16	7	40
Net income (M €)	−0.9	−14.0	5.0	0.0	−0.7	0.1
R&D expenditures (M €)	1.7	0.0	33.1	0.2	0.1	1.3
Share capital (M €)	2.2	0.0	25.2	0.3	0.1	1.7
Annual growth rate of employee numbers (%) 1997–1998				13	0	50
Total number of alliances	155					
Total number of founders	113					
Total number of shareholders	257					

(1) All the financial data concern the years 1999 (when available) or 1998. The increase in capital is calculated on the firm's entire life cycle. Growth of number of employees is calculated between 1997 and 1999. (2) Alliances: firms have been asked to record the number of alliances and the type of partner for the alliance. Alliances with academic teams and alliances with other firms have been spilt. The localisation of the partner is also important. Is it a local, national, or international partner? When several partners are involved in the same alliance, they are taken into account in each category.

variable (between 0 and 0.5 M € per person) and is not necessarily a good indicator of the firm's health.

- Biotech SMEs are recent. Firms are less than 10 years old on average. Biotechnology is a new sector and firms were set-up recently.
- Biotech SMEs are small in terms of employees and turn over. The biotech sector is composed of small firms with 45 employees on average, and generate 4.1 M € of turnover.
- Biotech SMEs are not profitable in the period studied. They are at the beginning of their life cycle and need large investments before generating enough turnovers to cover running costs.
- Biotech SMEs are science-based. On average, R&D expenditures account for over 40% of the turnover. These SMEs obviously belong to a high-tech sector, where 76% of the founders have a scientific background and 14% are well-known scientists.
- The 46 firms, which answer to the question, are involved in 155 alliances. Thus, partnerships are one of the main characteristics of this sector. They are also means for small firms to acquire resources. Biotech SMEs are involved in partnerships for gaining access to scientific knowledge and scientific platforms. Most partnerships with a scientific team remain local and are mainly based on existing relationships. Biotech SMEs contract mainly with large and multinational firms. See Table 6 for further details.

- Investments in capital are one of the modes of funding the firm's research activity. The level of capital is high and the presence of venture capitalists in the shareholding reveals the importance of capital for biotech SMEs. Compared to other SMEs in the traditional sector, the number of SMEs listed on the stock exchange is high.
- The most interesting characteristic revealed by Table 1 is heterogeneity. The survey focuses on one type of firm (SMEs) in one given sector (biotechnology). The SMEs present a high degree of diversity, in their financial results (turnover, net income—some of them have negative results), in their research activity (from 0 to 33.1 M € of R&D expenditures)[4] and in their size (from one to 525 employees). This diversity suggests that the activities of firms differ and that the logics of development may also be different. Because of heterogeneity, all data are analysed through their medium and quartile rather than mean and standard deviation.

The heterogeneity of resources (human, scientific, and financial) mobilised by firms to guarantee their survival and development relates to various strategies. During interviews with managing directors, the

---

[4] Even if it is rare for science-based firms, some firms have no R&D expenditures, i.e. no formal R&D structure. They are mainly service oriented.

628 *V. Mangematin et al. / Research Policy 32 (2003) 621–638*

relationship between the firm's activity and the ambition of its innovation project appeared clearly. The distinction between firms in groups A and B is based on three variables taken from the interviews: (1) the firm's business strategy and the positioning prompted by its leaders. Will the firm develop a product or technology that caters for a local or an international market? Are patents national, European, or international? (2) Does the firm invest essentially in R&D, or does it primarily develop an offering of products and services that generates turnover? (3) Is the firm's main activity research, or does it also develop sales and production? A qualitative analysis of the interviews enabled us to the divide firms into these groups, according to their business strategy and activity. The following section analyses the coherence of firms in each group and the development logics in which they are grounded.

4. Results

4.1. Firms that mobilise different resources

Basically, firms have been clustered in two groups: group A, consisting of those which have a small innovation project and target market niches; and group B, consisting of those which have a radical innovation project. In our sample, 37 firms are in type A and 23 firms in type B. Table 2 summarises the main characteristics of each type.

The analysis of firms' income and expenses clearly indicates the difference between traditional firms that target market niches (type A) and research-intensive firms (type B). The distribution of firms differs in respect of all characteristics except age and turnover. It, thus, appears that the creation of each type of firm is relatively homogeneous in time and that the appearance of a particular type of firm is not related to a particular period. In general, type A firms are smaller (average number of employees is 10, compared to 30 for type B), generate a turnover that covers running costs (including R&D spending), and grow more slowly than firms in group B. One of the reasons for which turnovers are not significantly different in the two groups is that our sample groups together firms at different stages of their life cycle, in both groups.

Each group has particular characteristics. For type A SMEs, the turnover is always clearly positive and

correlated with the number of employees. Average turnover per employee is equal to 110 K € per person, the average current value in many sectors. When this indicator is studied for each firm, we note a clear link between this type of indicator and the age of the firm: the six SMEs with the highest average turnover per employee were all created before 1990, whereas the six SMEs with the lowest average turnover per employee were all created after 1994. This result is consistent with growth rates that are as high as 30% (on turnover) and 13% (on staff).

These firms, all invest in R&D, but their projects are of a reasonable size compared to the size of the firm. Note, however, that the levels of intensity of R&D of type A firms are high (10% on average) compared to other sectors. To sum up, type A SMEs have similar financial results. These firms have a net growth of their turnover and have to maintain a balanced accounting system. Maintaining this balance is especially difficult during the early years, before the firm has developed its first products. At that stage it is rarely possible to maintain a research team and most of the research is conducted by one or two people.

Type B firms have very different characteristics from those in the preceding category. Some of the firms are at the beginning of their life cycle and therefore focus on research without selling. That explains the high standard deviation on the turnover. For those which sell research through contract research, the average turnover is much higher. On the other hand, when considered in proportion to the number of employees, the turnover is lower than that of type A firms (90K € as opposed to 110K €).

The importance of research programmes is clearly apparent in R&D investments. Some of the B type firms have R&D investments that are far higher than their turnover. In general, the average intensity of R&D generally exceeds 50%. Type B firms fund their activity through capital.

Finally, the analysis of income and expenses enables us to draw a coherent picture of the development of the two types of firm. The development of type A firms is based on their activity. They sell products or services to clients. Their development depends on the size of the market; its expansion leads to a rapid development of the firms. By contrast, type B firms are based on their research and innovation projects. To be

V. Mangematin et al. / Research Policy 32 (2003) 621–638

Table 2
Basic characteristics of the two business models

	Mann–Whitney	A			B			Full sample		
		Medium	First quartile	Third quartile	Medium	First quartile	Third quartile	Medium	First quartile	Third quartile
Number of firms	–	37			23			60		
Date of creation (year)	NS ($P = 0.2703$)	1992	1987	1994	1992	1989	1996	1992	1989	1995
Number of employees	*** ($P = 0.0035$)	10	5	22	30	13	108	16	7	40
Turnover (K€)	NS ($P = 0.8912$)	1007	396	2679	1264	162	7287	1041	336	5482
Net income (K€)	*** ($P = 0.000$)	16	−6	91	−1098	−4369	−141	−5	−714	55
R&D spending (K€)	*** ($P = 0.0001$)	82	37	305	1313	770	2672	229	65	1313
Share capital (K€)	*** ($P = 0.0021$)	175	61	694	1650	257	3011	297	88	1730
Annual growth rate of employee numbers (%) 1997–1998	** ($P = 0.0446$)	11	0	35	37	2	60	13	0	50

If $P < 1\%$ then ***; if $1\% < P < 5\%$ then **; if $5\% < P < 10\%$ then *; if $10\% < P$ then NS (non-significant); The Mann–Whitney test has been used to calculate these data. The Mann–Whitney test is a non-parametrical test which can be performed on data to compare two independent samples. Gaussian distribution is not required to perform this test. All the financial data concern the years 1999 (when available) or 1998. The increase in capital is calculated on the firm's entire life cycle. Growth of number of employees is calculated between 1998 and 1999.

Table 3
Capital contributions

Type of business	Mann–Whitney	A			B			Full sample		
		Medium	First quartile	Third quartile	Medium	First quartile	Third quartile	Medium	First quartile	Third quartile
Cumulated capital contributions on the firm's entire life cycle (M €)	*** ($P = 0.000$)	0.1	0.0	1.4	10.1	2.2	26.0	1.1	0.1	8.9
Cumulated contribution by venture capital on the firm's entire life cycle (M €)	*** ($P = 0.009$)	4.2	1.0	9.7	54.9	13.0	152.8	13.8	3.5	84.5

The contribution of venture capital is calculated as an average for firms who use it. Only those cases in which venture capitalists are the only parties involved at the round table are taken into account.

V. Mangematin et al./Research Policy 32 (2003) 621–638   631

able to achieve their goals, the CEOs[5] of these firms
have to convince investors to invest in their capital,
since capital investors fund their operating activity.

To sum up, the size of the innovative project splits
the sample into two business models. Thus, hypothesis H1 can be accepted. All the firms in the survey
belong to the same sector. They have common characteristics: acquiring and developing scientific capabilities through internal research and alliances with both
firms and academic teams. They also require funds to
develop their activity. However, the degree to which
these resources are needed and the nature of the partnership to acquire these resources are different in types
A and B firms.

4.2. Musicians playing the same score in different orchestras

4.2.1. Types of business, structure of shareholding and movement of capital

Table 1 suggests that start-ups accompany their development by increasing capital. Table 3 presents the
sum of capital invested in firms during the different
round tables, by business model, while Table 4 identifies the different categories of shareholders. It ex-
hibits the heterogeneity of firms in the capital needed
and collected. In type A, small inputs suffice, and can
be made by the founder of the firm alone or by family members or partners, or can be drawn from profits. Capital contributions (Table 2) are a key factor in
the financing of type B firms, even if some of them
have a high turnover. Since their creation, these firms
have managed to raise an average of between 15 and
30 M € from shareholders (see following section for
an analysis of shareholder structure). These capital
contributions explain the high values of share capital, even if only part of the contribution is channelled
into share capital. In other words, the older the firm,
when the capital contribution is made, the higher the
price will be for a given proportion of capital. These
capital contributions are transformed into liquid assets that the firm progressively uses up. Ownership
of these liquid assets results in interest received that
partially compensates losses linked to the firm's business. This is probably the main reason why these firms
have a higher net income than their operating income

---

[5] CEO: chief executive officer.

Table 4
Current structure of shareholders

Number of shareholders	264	A	B	Chi-square test
Identity of shareholding				
Founders	42	30	12	** (P = 0.018)
Other SME	17	9	8	NS (P = 0.382)
Large firms	18	7	11	** (P = 0.018)
Venture capital	34	15	19	*** (P = 0.001)
Public	6	0	6	
Others	8	1	7	
Firms surveyed (60)		37	23	

If $P < 1\%$ then ***; if $1\% < P < 5\%$ then **; if $5\% < P < 10\%$
then *; if $10\% < P$ then NS (non-significant). Shareholders in the
year 2000 are described in six categories. Some shareholders are
present in several firms, making the number of shareholders higher
than the sum of the shareholders by identity. Only "active" shareholders have been mentioned by the CEO, so that other shareholders, who represent a small minority, may have been omitted. The
different categories are mutually exclusive: founders who created
the firm and are still shareholders; individuals, family, friends or
business angels; other SMEs, firms which have participation in
the capital; large firms or multinationals which have participation
in the capital; venture capitalists who have invested in SMEs; the
public, when firms are quoted on the stock exchange. Table 4 is
expressed in terms of presence or absence of each type of shareholder for a given firm. A shareholder that finances two firms is
counted twice. On the other hand, for a firm with several shareholders of the same kind, e.g. several venture capitalists, the type
of shareholder, and not the number of shareholders will be counted.

(Table 2). The opposite result is observed for type B,
probably because of financial expenses.

When firms are listed in the stock exchange, the expected profit is high. As pointed out in Le Monde,[6]
the price earning ratio (PER) of the firms listed in the
NASDAQ is about 144, compared to 20 for the blue
chips which compose the Dow Jones. Type B firms are
those in which the expectation of growth and of future profits is the highest. Thus, institutional investors
(mostly business angels and venture capitalists) fund
the starting steps of these firms. If the firm is successful, business angels and venture capitalists can stay in
the shareholding of the firm to increase the value of
their shares[7] or they are able to sell out their shares
when the firm makes an IPO.

---

[6] De Tricornot, A., L'insaisissable valeur du Nasdaq et des
entreprises de la nouvelle économie, Le Monde, 27 February, p. 20.
[7] The business angels and venture capitalists have lots of stocks,
and it takes time for the market to generate the required liquidity
to buy out all the stocks (at a good price for the shareholders).

The current structure of capital clearly reflects the history of these firms (Table 4). Here, we simply identify the different types of actor present in the capital. Since setting up a business initially involves a few people, natural persons are often shareholders. In the case of type A firms, the founder almost always owns a substantial proportion of the shares. Genome Express, during the late 1990s, is a good example of this situation. By contrast, this is the case in only half of type B firms. When B type firms have been set-up only recently, their founders are more likely to be large shareholders, but their share of capital diminishes when the need for funds increases. The firm still has to convince other actors to invest in it.

The presence of major industrial groups is consistent with the given analysis of relations between SMEs and big companies. Type A firms focus on market niches and are therefore of little interest to large groups. By contrast, type B firms are large groups' potential partners or rivals in their core business. These groups, therefore, frequently acquire shares in firms in these categories when they reach a certain level of maturity. Since at the beginning of their life cycle type B firms' research results are still too hypothetical to interest industrialists, these investors prefer to wait and pay more to gain access to less risky projects.

Venture capital firms are virtually indispensable partners for type B firms, since over 19/23 use them. This proportion is over twice as high as that of type A firms. Moreover, the analysis of the average contribution of venture capital firms per biotech SME also shows substantial differences between types A and B. Very high contributions in the case of new type B firms reflect the significant development of this type of institution in France. More detailed analyses on the origin of venture capital firms reveal substantial differences between different types of SME: most capital contributions for type A firms were from regional venture capitalists, while foreign venture capitalists intervene primarily in type B firms.

Venture capital firms are not intended to remain shareholders in the long run. Their involvement in a high-risk project is based on expectations of large financial returns. Type A firms pose several problems from this point of view: (1) they generally run less risky, and also less profitable projects; (2) in many cases, the founder controls the firm and would like to maintain that control, thus leaving less latitude to any

new shareholders to influence management; (3) finally, these firms rarely reach the stock market which is the most favourable capital outlet for venture capitalists. This type of stumbling block is generally not found in type B firms, and shares can be resold either to big industrial groups or to the public at the time of flotation.[8]

The last issue, concerning shareholding, concerns the evaluation of the firm and the signal that investors look at. Type A indicators are traditional: turnover and profit margins are the main indicators. These firms have no specific characteristics in addition to others in the sector (high level of R&D expenditures, partnerships and capital investments). By contrast, investments in type B firms are mainly driven by expectations of profit rather than existing or even planned profits. Thus, investors have to believe unusual and non-financial signals. As profit expectations are based on science, signals are scientific: curriculum vitae of founders, scientific committees, and tangible commitments like patents or publications. Investors have to follow and believe such signals and indicators to invest large amounts of money in promising, but highly fragile, firms. The concentration of the shareholding is the main consequence of such a situation. It implies as follows.

- The number of actors is reduced. Even when they are public, the shareholding of these firms is concentrated amongst a limited number of shareholders, mainly institutional investors. Up-to-date scientific knowledge is necessary to be able to interpret signals like publications or scientific trends (Boissin and Trommetter, 2001). As a result, the shareholdings of these firms are mainly composed of persons involved in biotechnology. The liquidity of such firms, when they are listed, may be reduced compared to other sectors that are easier to understand for a large public of investors.
- Actors are specialised. Compared to other sectors in which traditional financial indicators can be interpreted with general knowledge of the sector, biotech firms require specific investments to understand the logic of the sector. It is necessary to be involved in both the scientific community and the financial

---

[8] In most cases, flotation calls for additional capital by institutional investors (banks and insurance companies) during the pre-float stage. These investors, largely absent from the databases studied here, have been put into the 'public' category in Table 4.

V. Mangematin et al. / Research Policy 32 (2003) 621–638

community. Even institutional investors need to be specialised and only few actors are involved in biotech firms which require large investments. That is one of the reason why venture capitalists and institutional investors like pension funds are specialised.

- The notion of "good results", which is quite common in traditional sectors (high profit rate, high profit prospects, low level of debt, etc.), has to be redefined and adapted. Thus, both shareholders and CEOs need to agree about what "good results" are in a general framework of financial losses. Patents, contract research, or contract licences with large pharmaceutical companies are usually considered as favourable indicators, as are phase III products.

- Science and shareholding are international. To be able to generate high returns on investments, type B firms have to focus on international markets. Thus, the advance of science is no longer national, but international. This characteristic has two consequences. (1) The decision-making process in type B firms depends upon international criteria and decisions. Before IPO, B firms have international venture capitalists who invest in firms in Europe and the USA. European firms compete with American ones to acquire resources. After IPO, they compete on the same stock exchange (NASDAQ, Euronext, etc.) to attract shareholders. (2) Shareholders have to be more and more professional in their investment because of the internationalisation of the competition and the production of science.

All in all, when indicators are transient, and when the amount of money needed for development is high, investors become less numerous, more professional, and more international. Indicators and their significance require agreement in the community of investors. They are difficult for outsiders to understand and risks of manipulation of prices are high. Since this is a new and transitory situation (because everyone expects to have traditional firms which generate turnover and profit as soon as possible), the community of investors remains small.

4.3. Business models, founder's characteristics and acquisition of resources

As shown in Table 5, most of the founders have a scientific background. But the significance of the

Table 5
Characteristics of the founders

Number of founders	113	A	B	Chi-sqaure
Founders' experience				
Scientific experience				
Scientists	28	23	5	*** ($P = 0.000$)
Scientific researchers	30	13	17	** ($P = 0.014$)
Famous scientists	11	1	10	*** ($P = 0.000$)
Managerial experience				
Junior manager	19	16	3	*** ($P = 0.004$)
Senior manager	20	4	16	*** ($P = 0.000$)
Firms surveyed (55)		32	23	

The number of founders describes the number of persons involved in the creation of the firm. When available, their initial training and professional experience have been recorded. Two professional experiences are recorded: scientific experience and managerial experience. The scientific experience is ranking in three levels: (1) junior scientists are persons who have a scientific background, engineers, or Ph.D.s; (2) senior scientists are at least post doc, and have research experience after the Ph.D. (some of them had a position in an academic team before creating the firm); and (3) famous scientists are those who have already published a lot and have a high level of scientific visibility. In our sample, they are those who have published more than 15 papers recorded in the Science Citation Index. Founders are described as having a managerial experience if they have either a managerial training or a scientific background and a managerial experience in a large firm or in the creation of a biotech SME. Those who have little managerial experience are considered junior. Those who have already been in positions of leadership in a major industrial group or banking institution are senior. Table 5 is expressed in terms of presence or absence of each type of founder for a given firm. If a founder has two competencies—scientific and managerial—she/he will be counted twice. By contrast, if the firm has several scientists which similar competencies, they will be counted only once. The founder's experience is determined on the basis of the last job held. The differences in the experiences of founders are significant between types A and B. For each line, a Chi-square test has been performed with the default entries are junior scientist or junior manager, depending on whether the person is from an academic or industrial environment.

presence of a scientist as a founder is not the same when that person is a junior scientist who is able to operate one technique, and when she/he is a famous scientist who brings with her/him a large network of collaboration and relationships. Thus, scientists play different roles, depending on their experience. Junior scientists can run experiments and do the job themselves. Based on a Poperian analysis of scientific production, Shinn (1988) shows that a division of work exists in laboratories. Senior scientists are able to co-ordinate more complex research and divided work and tasks. They are also able to generate new hypotheses and models. Junior scientists have fewer capacities to co-ordinate complex projects and are oriented more towards the implementation of specialised tools.

Moreover, two main types of founder have been distinguished: academics and managers (Catherine and Corolleur, 2000). In each of these categories a distinction is made on the basis of the person's experience and responsibilities. The analysis reveals a very clear connection between the founder's characteristics and the type of activity of the firm. Founders with extensive experience are almost systematically involved in the creation of type B firms. By contrast, founders with little experience more often appear in type A firms. This correspondence probably stems from constraints relating to the firm's mode of capitalisation. To be able to raise large amounts of capital the founder has to be able to provide guarantees as to the credibility of the proposed project.

As shown previously, the visibility of the scientist in the academic community plays a key part as regards venture capital. It gives credibility to the scientific project of the firm. A similar conclusion can be drawn for alliances. All biotech SMEs enter into partnerships to gain access to critical resources. But the nature of the critical resources differs in types A and B firms. Type A firms mainly need access to technological platforms and to scientific teams with general knowledge. Thus, scientists in the firm, who are generally the founders, usually maintain contacts with their previous team (from their Ph.D. or post doc) to have

access to certain technological facilities or to scientists in their speciality. They also manage contracts with local or national firms to secure access to markets. By contrast, type B firms are more globalise, even when they are small. They conclude international partnerships with academic teams specialised in a particular field and with large firms which valorise their specific competencies (Table 6).

In general, biotech SMEs have more than two partnerships in average. Partners with whom these alliances are concluded have few distinctive characteristics, except for co-operative agreements with local academic teams for type A firms and with international groups for type B firms. These characteristics confirm the relations between founders and the business model. Junior scientists set-up their firm near the laboratory in which they did their Ph.D. or spent the first years of their scientific career, and maintain close ties with the team in that lab. By contrast, type B firms are founded by senior researchers who often have good international visibility that enables them to convince international groups of the quality of research conducted in their firm. The analysis of the founders of a firm, of the partnership and of the way in which the firm finances its activity reveals that types A and B firms require different resources to run operational activities. Depending on the business model, different indicators are analysed by partners and potential investors. Thus, empirical data supports hypothesis H2. The development of types A and B firms is not based on the same resources. Type A firms have to convince local or national businesses

Table 6
General characteristics of the sample

Number of alliances	155	A	B	Chi-square test
Type of alliances				
With a local academic team	27	19	8	** (P = 0.055)
With a national academic team	15	7	8	NS (P = 0.249)
With a international academic team	6	2	4	NS (P = 0.176)
With a local firm	19	9	10	NS (P = 0.191)
With a national firm	21	12	9	NS (P = 0.845)
With a international firm	24	9	15	*** (P = 0.002)
Other	5	4	1	–
Firms surveyed (46)		27	19	

Table 6 is expressed in terms of the presence or absence of each type of partnership in a given firm. A firm that has two partnerships with an international academic team will be counted only once. By contrast, a partnership in which several types of partner are involved will be counted as many times as there are partners.

V. Mangematin et al./Research Policy 32 (2003) 621–638   635

to buy their product or service if they are to generate a turnover enabling them to gradually expand. Their contacts with local and national teams afford them access to scientific breakthroughs and sometimes also to technological platforms. Capital needs are small, and are often met by local venture capitalists linked to the public authorities. In type A firms, partnerships with firms are alliances for production or commercialisation. The resources that the firm exchanges with other partners are mainly products and services as opposed to turnover, and profits as opposed to investments. These results are consistent with Mansfield's ones (Mansfield, 1995), when he finds that firms draw on local universities for applied research, but for basic research they are connected to the most appropriate labs wherever it is located. Type B firms exchange credibility, signals of excellence, and expectations of potential profit as opposed to capital investments to finance research. Thus, firms are not really competing to convince the same partners, clients or network. The driving force of development is therefore also different. Although, types A and B firms belong to the same sector, the two types of firm do not really compete for resources. Each type is a specific case, with type A firms competing for markets and clients and type B for investors. They are playing the same score, but in different orchestra, and in different theatres. To what extent is a firm locked in a specific business model?

4.4. Dynamics of the transition from one type of business to another

Characterisation of types of business has made it possible to explain the financial constraints weighing on a firm's development and, in the case of type B firms, the constraints weighing on the founder's profile due to the need to raise large amounts of capital. Business models enable us to understand the internal coherence of firms and to characterise the resources they mobilise. This representation also helps to explain certain important mechanisms governing each type of activity. The emergent business model characterising type B firms can only be temporary. The longevity of this type of firm involves a transition towards another model. Four types of evolution can be identified.

- *Disappearance of the firm*: Several reasons of a structural and conjunctural nature can explain this

type of evolution. Biotech firms develop an innovative scientific project and explore hypotheses. Some of these hypotheses may prove to be scientific dead-ends, which nullifies investments made and leads to bankruptcy. Other, more conjunctural causes, can also be the cause of bankruptcy, since these firms are highly dependent on venture capital investments. A lack of liquidity of stock markets related to an unfavourable context, as in the first half of 2001, for example, can lead to financial delays or deficiencies that are fatal for firms. Likewise, discussions on intellectual property rights on living organisms, which strongly influence patenting strategies, can lead certain firms to ruin if they have not anticipated new laws. The fall of biotech values following the Blair/Clinton declaration[9] attests to firms' sensitivity to their political and legislative environment. Finally, a discovery made more quickly by another research team in a rival firm can strongly reduce the value of a firm's research. Thus, whether for internal reasons, or reasons related to the environment, certain firms can disappear.

- *Focus on a niche*: The firm's strategy of positioning in international markets may prove fruitless, especially due to the scientific competition to which it is exposed. It may then redeploy its investments, at least partially, in market niches where it valorises specific competencies. This strategy corresponds to a withdrawal to a less ambitious project. Its viability depends on the state of the competition and investments made. Some firms, whose shareholders include venture capital companies that have remained present for many years despite the business not taking off, seem to be in this case. Such firms are not as successful as anticipated, and venture capitalists have difficulty withdrawing their capital since the firms concerned are not quoted on the stock market.

- *Buy-outs*: Some firms are bought out by industrial groups after proving themselves. In this category we find companies such as Appligene (founded in 1985, with an 80% buy-out by the US Oncor in 1995 and then by the Canadian Quantum). Systemix (founded

---

[9] On 16 March 2000, Bill Clinton and Tony Blair advocated free access to the human genome, asking for the results of sequencing of the human genome to be put into the public domain and for patents to be limited to their industrial and medical exploitation. Le Monde, 15 March 2000.

in 1988 and bought out by Sandoz in 1992), and Agrogene (founded in 1989, gradually bought out by Limagrain and then taken over by Perkin-Elmer). These firms differ from subsidiaries created ex nihilo, in so far as they started off independently. This evolution is not specific to type B firms. The event must be considered as a stage in the development of the SME and not as a sign of failure. The buy-out may be explained by problems experienced by the SME (problems securing access to markets, incomplete technology base) but also by the group's wish to acquire it. The buyer's aim is either to complete its technology base with the patent portfolio or technical competencies of the SME, or to use the SME to gain a foothold in the country. In the latter case, the acquirer is often a foreign group. The firm then acts as a bridgehead in France to transfer technologies or products developed by the parent company. The buy-out is followed by changes in the SME. Its research may be concentrated with that of the group to avoid duplication.

- *Coming out on top*: For some firms the gamble pays off. They market radical innovations that generate huge cash flows and make them unavoidable actors. These firms, most of which are in the human health sector, are the ones that, either alone or in partnership with a pharmaceutical firm, produce active principles or new drugs that capture a large slice of the market (what in the pharmaceutical industry are known as blockbusters). If the firm chooses to commercialise its product itself, it will have to acquire the competencies of pharmaceutical firms and act like a firm in that sector. If it sells its innovation to a pharmaceutical firm, it will remain in the biotechnology sector to valorise its research in other directions.

The emergent business model is temporary and the evolution of firms corresponding to this development model vary widely. Possible trajectories of type A firms are more classic. Three possibilities can be identified.

- *Development of an ambitious project based on the firm's competencies*: With a scientific and technological base valorised in existing products and services, some firms can run ambitious projects that take them closer to the type B model. Transition from types A to B is probably rare. We have

seen that it is not easy for founders with limited experience to directly set-up a type B firm. Some entrepreneurs interviewed in the survey, nevertheless, clearly planned to progressively move towards a type B business. In other words, going through type A firms is a way of building up additional experience and credibility in the biotech business. This experience will not, however, be equivalent to that of the founders of type B firms because it will have been acquired in very different organisations. Moreover, the founder will have to agree to concessions regarding control of the company. Thus, the question is to study how type B businesses can be developed when they have a past that differs substantially from that previously observed. The sample studied did not include that type of case except for Kappa biotech which was created by the founders of Biovector-Therapeutics and Genome Express, two persons formerly employed by a large firm which sells biotech equipment. Kappa has chosen to valorise its innovations in cosmetics and is developing research capabilities to move into pharmaceutical markets. It is clearly a type A firm, but the ambition of the founders is to set-up a type B firms as soon as possible. Genome Express began by doing sequencing and functionalisation. It is now setting up large research programmes involving international teams to develop new sequencing and functionalisation techniques.

- *Geographic expansion to conquer new markets*: Founded on the basis of a local or national market, firms with recognised competencies in a market niche can expand by conquering new markets, either in sectors neighbouring those in which their competencies are recognised (application to animals or plants of competencies acquired in human health, as in the case of genomics or bio-computing, for example), or in countries which lack their specific competencies, i.e. countries with a less advanced biotech sector, as in eastern Europe and southern Europe. This strategy often implies external growth operations, such as joint ventures or the buy-out of local partners. It can also take the form of direct investments abroad in the form of creation of subsidiaries.

- *Maintaining the "craft" character of the business*: Some firms have no growth strategy and generate a business enabling them to live.

V. Mangematin et al. / Research Policy 32 (2003) 621–638 637

The analysis of business models enables us to highlight the development prospects of biotech firms. It shows the eminently labile character of the emergent business model that represents only a stage in the firm's development. While it is legitimate for investors and actors in the sector to have their eyes fixed on the short-term evolution of firms, it is legitimate for the public authorities to support the middle-term development of these firms.

5. Conclusion

Based on a study of the development of the biotechnology sector and of biotech SMEs in France, this paper underlines the diversity of firms involved in this sector even if the two main business models share common characteristics: science base, leading role of SMEs, and resource acquisitions through alliances. Our analysis is based on the construction of two business models, each corresponding to a viable position for SMEs in the industrial environment. Within a SIS, firms do not form a homogeneous group. Development of the different types of firm is based on different dynamics, and different actors play a role in those dynamics. The main differences appeared between models A and B. When an SME focuses on a market niche and conducts small research programmes, it will experience steady growth if it is able to reach financial equilibrium fairly quickly. By contrast, when SMEs embark on large research programmes in partnership or competition with major companies in the sector, development is possible only with outside capital and the participation of venture capital firms. The founding members' experience is then a key factor if the SME is to enter into certain partnerships. On the basis of these results, we were able to see how certain business models that are less risky, but probably make less use of the founders' knowledge, can appear as gateways in the establishment of a final business model.

The heterogeneity of the firms studied here affords a different view compared to existing studies of biotechnology SMEs. We mentioned, for instance, some results of the study of factors favouring the growth of biotechnology SMEs, but which do not differentiate the characteristics of these firms. For example, Audretsch and Stephan (1996) highlight the importance of scientific networks in the capacity of

firms to raise capital at the IPO stage. This result is totally consistent with results obtained here for model B, but does not apply to model A. This confirms the more general idea that it is necessary to develop appropriate economic indicators when the firms studied have different characteristics.[10]

Finally, this research provides important elements for an analysis of the advantages of public policies focused on the development of biotech SMEs. The differentiation of business models helps to explain the sometimes very variable effects of certain public policies, and can facilitate the definition of appropriate instruments for each type of business. It finally underlines the crucial of public policy in funding, basing science (through subsidies, venture capital, or other ways) developed in type B firms.

Acknowledgements

Results presented in this article are based on a research contract financed by the Commissariat Général du Plan. The data collected were drawn from interviews with company directors, whom we would like to thank. We also wish to thank Myriam Carrere, Mounia Benjedar and Nadine Mandran for their valuable assistance in the compilation and processing of databases. We would like to thank the participants of the EUNIP conference (Tilburg, December 2000), Paula Stephan, and the anonymous referees of the journal 'Research Policy' for their helpful comments on the previous version of this text. Any error in this paper are entirely our responsibility.

References

Arora, A., Gambardella, A., 1990. Complementarities and external linkages: the strategies of the large firms in biotechnology. Journal of Industrial Economics 4, 361–379.

Arundel, A., Rose, A., 1998. Finding the substance behind the smoke: who is using biotechnology? Nature Biotechnology 16 (7), 596–597.

Audretsch, D., Stephan, P., 1996. Company scientist locational links: the case of biotechnology. American Economic Review 86 (3), 641–652.

[10] See Paranque et al. (1997) for a more in-depth study of problems relating to business indicators in varied sectors.

Barley, S.R., Freeman, J., Hybels, R.C., 1992. Strategic alliances in commercial biotechnology. In: Nohria, N., Eccles, R. (Eds.), Networks and Organizations. Harvard University Press, Boston, MA, pp. 311–348.

Barney, J.B., 1991. Firm resources and sustained competitive advantage. Journal of Management 17, 99–120.

Barney, J.B., Spender, J.C., Reve, T., 1994. Does Management Matter? On Competencies and Competitive Advantage. Lund University Press, Lund Bromley, UK.

Boissin, J.P., Trommetter, M., 2001. Contextes Et Pratiques Du Gouvernement Des Entreprises De Biotechnologie. Troisième forum de la moyenne entreprise, Caen, IAE.

Breschi, S., Malerba, F., 1997. Sectoral innovation systems: technological regimes, schumpeterian dynamics, and spatial boundaries. In: Edquist, C. (Ed.), Systems of Innovation. Pinter Press, London.

Catherine, D., Corolleur, F., 2000. Pme Biotech Et Géographie De L'innovation, Des Fondateurs À Leur Modèle D'entreprise. XXXVIe colloque de l'Association de Science Régionale de Langue Française, Trans Montana (CHE). INRA/SERD/IREPD, Grenoble.

Dodgson, M., 1991. The Management of Technological Learning: Lessons from a Biotechnology Company. Walter de Gruyter, Berlin.

Estades, J., Ramani, S., 1998. Technological competence and the influence of networks: a comparative analysis of new biotechnology firms in France and Britain. Technology Analysis and Strategic Management 10 (4), 483–495.

Feldman, M., 1999. The new economics of innovation, spillover and agglomeration: review of empirical studies. Economics of Innovation and New Technology 8 (1), 5–25.

Grant, R., 1991. The resource-based theory of competitive advantage: implications for strategy formulation. California Management Review 33 (3), 114–135.

Hamel, G., 1991. Competition for competence and inter-partner learning within international strategic alliances. Strategic Management Journal 12, 83–103.

Klavans, R., Deeds, D.L., 1997. Competence building in biotechnology start-ups: the role of scientific discovery, technological development and absorptive capacity. In: Sanchez, R., Heene, A. (Eds.), Strategic Learning and Knowledge Management. Wiley, New York.

Lemarie, S., Mangematin, V., 2000. Biotech firms in France. Biofutur 32–42 (special issue).

Lemarie, S., Mangematin, V., Torre, A., 2001. Is the creation and development of biotech SMEs localised? Conclusions drawn from the French case. Small Business Economics 17 (1/2), 61–76.

Mahoney, J., Rajendran Pandrian, J., 1992. The resource-based view within the conversion of strategic management. Strategic Management Journal 13, 363–380.

Mangematin, V., 2000a. Competing business models in the French biotech industry. In: de la Motte, J., Niosi, J. (Eds.), The Economic and Social Dynamics of Biotechnology. Kluwer Academic Publishers, Boston, pp. 181–204.

Mangematin, V., 2000b. What Business Model for Biotech SMEs. Statcan, Ottawa, Canada. INRA/SERD, Grenoble.

Mansfield, E., 1995. Academic research underlying industrial innovations: sources, characteristics and financing. Review of Economics and Statistics 75 (1), 55–65.

McKelvey, M., 1996. Evolutionary Innovations—The Business of Biotechnology. Oxford University Press, Oxford.

Nilsson, A., 2001. Biotechnology firms in Sweden. Small Business Economics 17 (1/2), 93–103.

Penrose, E., 1959. The Theory of the Growth of the Firm. Basil Blackwell, Oxford.

Peteraf, M., 1993. The cornerstones of the competitive advantage: a resource-based view. Strategic Management Journal 14, 179–191.

Pfeffer, J., Salancik, G.R., 1978. The External Control of Organizations: A Resource Dependence Perspective. Harper & Row, New York.

Porter, M., 1980. Competitive Strategy. Free Press, New York.

Powell, W.W., Koput, K.W., Smith-Doerr, L., 1996. Inter-organisational collaboration and the locus of innovation: networks of learning in biotechnology. Administrative Science Quarterly 41, 116–145.

Rabinow, P., 1996. Making PCR: A Story of Biotechnology. University of Chicago Press, Chicago.

Russo, M.V., Fouts, P.A., 1997. A resource-based perspective on corporate environmental performance and profitability. Academy of Management Journal 40 (3), 534–559.

Saviotti, P.P., Joly, P.B., Estades, J., Ramani, S., De Looze, M.A., 1998. The creation of European dedicated biotechnology firms. In: Senker, J.V.V., Cheltenham, R. (Eds.), Biotechnology and Competitive Advantage: Europe's Firms and the US Challenge. E. Elgar, pp. 68–88.

Shinn, T., 1988. Hiérarchies des chercheurs et formes de recherche. Actes de la Recherche en Sciences Sociales 74, 2–22.

Shrivastava, P., Huff, A.S., Dutton, J.E. (Eds.), 1994. Resource-Based Views of the Firm: Advances in Strategic Management, Vol. 10a. JAI Press, Greenwich, CT.

Stephan, P., Everahrt, S., 1995. The Changing Rewards to Science: The Case of Biotechnology. Department of Economics and Policy Research Center, Atlanta.

Teece, D., 1986. Profiting from technological innovation: implications for integration, collaboration, licensing and public policy. Research Policy 15, 285–305.

Teece, D., Rumelt, R., Dosi, G., Winter, S., 1993. Understanding corporate coherence: theory and evidence. Journal of Economic Behaviour and Organization 22.

Zucker, L., Darby, M.R., Armstrong, J., 1994. Intellectual Capital and the Firm: The Technology of Geographically Localized Knowledge Spillovers. NBER, p. 4946.

Part IV
Reaction and Adaptation of Large Incumbent Companies

[14]

Louis Galambos and Jeffrey L. Sturchio

Pharmaceutical Firms and the Transition to Biotechnology: A Study in Strategic Innovation

During the twentieth century, the pharmaceutical industry experienced a series of dramatic changes as developments in science and technology generated new opportunities for innovation. Each of these transitions forced existing firms to develop new capabilities. The authors examine the most recent such transition, the shift to molecular genetics and recombinant DNA technology (1970 to the present), and explain how and why this transformation differed from the previous ones in pharmaceuticals. Small biotech startups played an important role in this transition, and the large pharmaceutical firms that began to enter the field had to develop new strategies for innovation. Two major strategies were adopted by the early movers, all of which created various kinds of alliances with the small biotech businesses. By the mid-1990s, the leading pharmaceutical manufacturers had established significant capabilities in the new field, but they were continuing to work with specialized biotechs in order to innovate across a broad range of therapeutic categories.

The transition to a new technology, science, market or political regime is difficult for any organization, public or private, to manage. In the twentieth century, the pace of these transformations appears to be accelerating and the resulting pressure to change mounting steadily. Indeed, one of the central themes in the history of modern capitalist enterprises is provided by their efforts to develop appropriate means of understanding and responding to rapidly changing environments. In the American case, robust global competition since

LOUIS GALAMBOS is professor of history and editor of *The Papers of Dwight David Eisenhower* at Johns Hopkins University.
JEFFREY L. STURCHIO is a historian of science and technology, and executive director of public affairs, Europe, Middle East & Africa, at Merck & Co., Inc.

Business History Review 72 (Summer 1998): 250-278. © 1998 by The President and Fellows of Harvard College.

A Study in Strategic Innovation / 251

the late 1960s has had a dramatic impact on these efforts to remain innovative and profitable, efforts that have produced some startling failures as well as impressive narratives of success.

This setting has aroused among scholars in business history, management studies, the history of technology, and economics a new interest in the transitions of large-scale firms attempting to adapt to new competitive, technological, and political conditions. The work in these different disciplines overlaps and is, we believe, coalescing into a coherent body of knowledge about the management of change.[1] That emerging paradigm provides us with our context for analyzing the strategic responses of large pharmaceutical firms to the advent of molecular genetics and recombinant DNA technology.

The pharmaceutical industry provides ample opportunities to study both success and failure in the recent past and, for that matter, through the entire twentieth century. The industry has experienced a series of decisive changes, starting with the development and gradual acceptance of the germ theory of disease at the turn of the century and accelerating during the chemo-therapeutic revolution of the 1930s and 1940s. Synthetic organic chemistry and soil microbiology generated significant opportunities for pharmaceutical innovation, but to take advantage of those situations, firms had to make strategic commit-

---

[1] See Michael J. Piore and Charles F. Sabel, *The Second Industrial Divide: Possibilities for Prosperity* (New York, 1984); Thomas K. McCraw, ed., *America versus Japan* (Boston, 1986); Richard S. Tedlow, *New and Improved: The Story of Mass Marketing in America* (New York, 1990); Joseph Morone, *Winning in High-Tech Markets: The Role of General Management* (Boston, 1993); William Lazonick, *Competitive Advantage on the Shop Floor* (Cambridge, Mass., 1990); Richard R. Nelson, ed., *National Innovation Systems: A Comparative Analysis* (New York, 1993); Amy Glasmeier, "Technological Discontinuities and Flexible Production Networks: The Case of Switzerland and the World Watch Industry," *Research Policy* 20 (Oct. 1991): 469-85; Margaret B. W. Graham, *RCA and the VideoDisc: The Business of Research* (New York, 1986); Richard S. Rosenbloom and Michael A. Cusumano, "Technological Pioneering and Competitive Advantage: The Birth of the VCR Industry," *California Management Review* 29 (Summer 1987): 51-76; Richard S. Rosenbloom, Donald N. Sull, and Richard S. Tedlow, "Managerial Commitments and Technological Change in the US Tire Industry," Harvard Business School Working Paper (1997); Rebecca Henderson and Kim B. Clark, "Architectural Innovation: The Reconfiguration of Existing Product Technologies and the Failure of Established Firms," *Administrative Science Quarterly* 35 (1990): 9-30; Rebecca Henderson, "Underinvestment and Incompetence as Responses to Radical Innovation: Evidence from the Photolithographic Alignment Equipment Industry," *Rand Journal of Economics* 24 (Summer 1993): 248-70; David Teece and Gary Pisano, "The Dynamic Capabilities of Firms: An Introduction," *Industrial and Corporate Change* 3:3 (1994): 537-56, this entire issue of the journal deals with organizational capabilities in different contexts. For guides to the economics literature on transitions, see Lynne G. Zucker, Michael R. Darby, and Marilynn B. Brewer, "Intellectual Human Capital and the Birth of U.S. Biotechnology Enterprises," *American Economic Review* 88:1 (1998): 290-306; and Lynne G. Zucker and Michael R. Darby, "Present at the Biotechnological Revolution: Transformation of Technological Identity for a Large Pharmaceutical Firm," *Research Policy* 26 (1997): 429-46.

Galambos and Sturchio / 252

ments to entirely new capabilities in research and development, manufacturing, and marketing. In the 1940s and 1950s, advances in virology provided another set of new opportunities for entrepreneurship, followed shortly by a new wave of breakthroughs in microbial biochemistry and enzymology, breakthroughs that provided the basis for a new style of targeted pharmaceutical research and development. In each of these transitions, there were losers as well as winners.

The next, partially overlapping wave of innovation—the subject of our paper—was grounded in recombinant DNA technology and molecular genetics, and is generally referred to as the biotechnological or "biotech" revolution.[2] Seen from the perspective of the large pharmaceutical firm, the biotech transition was unusual in several regards. It was, for instance, the first twentieth-century transition in this industry in which the initial stages of applied research and commercial development were centered in small, startup companies rather than the large, well-financed organizations that have for many decades been the primary innovators in pharmaceuticals (see Table 1 for a comparison of the size of the pharmaceutical and biotech firms in 1997). During the 1930s, 1940s, and 1950s, for example, it was large, vertically integrated companies that led the industry into the golden age of medicinal chemistry. The later transition into microbial biochemistry/enzymology followed a similar pattern.[3]

In the 1970s, however, small biotech firms led the way, followed by the large pharmaceutical corporations, which developed alliances of varying sorts with one or more of the biotech companies. The large companies exchanged financial support and established organizational capabilities in clinical research, regulatory affairs, manufacturing, and marketing for the smaller firms' technical expertise and/or patents. Frequently, the biotechs developed "niche technologies"—capabilities using a narrowly defined, often proprietary technological tool. While

---

[2] Sheldon Krimsky, *Biotechnics & Society: The Rise of Industrial Genetics* (New York, 1991), 1-13, provides a useful set of definitions and a brief historical sketch of the major developments in molecular genetics (in the 1950s and 1960s) that led (in the 1970s and 1980s) to a new form of biotechnology—one that manipulated the genes that govern the internal functions of the cell. See also Horace Freeland Judson, *The Eighth Day of Creation: Makers of the Revolution in Biology* (New York, 1979). Robert Teitelman, *Gene Dreams: Wall Street, Academia, and the Rise of Biotechnology* (New York, 1989), esp. 4-10, views the phenomenon from its Wall Street connections. For a longer historical perspective, see Robert Bud, *The Uses of Life: A History of Biotechnology* (Cambridge, U.K., 1993), which deals with the past century.

[3] For an excellent discussion of this latter transition, see Alfonso Gambardella, *Science and Innovation: The US Pharmaceutical Industry during the 1980s* (Cambridge, U.K., 1995), esp. 23-32. See also Rebecca Henderson, "The Evolution of Integrative Capability: Innovation in Cardiovascular Drug Discovery," *Industrial and Corporate Change* 3:3 (1994): 607-30.

A Study in Strategic Innovation / 253

Table 1

1997 Sales in Dollars

Top Pharmaceutical Companies		Top Biotechnology Companies	
Glaxo Wellcome°	11.6 billion	Amgen	2,303 million
Merck°	11.4	Chiron	1,313
Novartis	11.0	Genentech	967
Bristol-Myers Squibb	9.3	Genzyme	536
Johnson & Johnson	8.7	Alza	466

°According to the audited figures in *Scrip World Pharmaceutical News*, Merck ranked first ($11.296 billion) and Glaxo Wellcome second ($10.870). See also *Scrip*, 24 July 1998.

Source: *The Economist*, 21 Feb. 1998, 57. The total industry sales for biotech products were estimated to be $13 billion. *The Economist* was optimistic about the prospects for biotech firms, but *Business Week*, 26 Sept. 1994, 84-92, reported that about half of the companies at that time had only enough money to stay in business two years. This situation was especially threatening since it took an average of seven to ten years and at least $100 to $150 million (according to *Business Week*) to bring one of these new drugs to market.

innovation at large pharmaceutical companies had always been linked to complex networks of public and nonprofit institutions, the relationships most important to the pharmaceutical companies had been with individual scientists, with public institutions (such as the U.S. National Institutes of Health and public health organizations), and nonprofits such as research universities. The growing web of biotech contractual ties added a new element to the networks that have traditionally sustained innovation in this industry.[4]

It is unclear whether or not this change marks a significant, enduring transformation of the process of innovation in pharmaceuticals. The biotech contracts may have been merely a phenomenon of the transition, a fleeting accommodation to particular circumstances in the 1970s and 1980s, owing more to the structure of capital markets than to a fundamental change in the nature and sources of innovation. They may, on the other hand, have represented a decisive turning point

[4] For a discussion of these networks, see Louis Galambos, with Jane Eliot Sewell, *Networks of Innovation: Vaccine Development at Merck, Sharp & Dohme, and Mulford, 1895-1995* (New York, 1995). See also Louis Galambos and Jeffrey L. Sturchio, "The Pharmaceutical Industry in the Twentieth Century: A Reappraisal of the Sources of Innovation," *History and Technology* 13 (1996): 83-100. Volume 20 (Oct. 1991), of *Research Policy* is devoted entirely to "Networks of Innovators," by which the editors and authors mean the type of contractual relationships that accompanied the transition to biotechnology. For an excellent overview, see Chris DeBresson and Fernand Amesse, "Networks of Innovators: A Review and Introduction to the Issue," 363-79. For a general historical evaluation and taxonomy of strategic alliances, see David C. Mowery and David J. Teece, "Strategic Alliances and Industrial Research," in *Engines of Innovation: U.S. Industrial Research at the End of an Era*, eds. Richard S. Rosenbloom and William J. Spencer (Boston, 1996), 111-29.

Galambos and Sturchio / 254

when specialization became more important to the entrepreneurial function than economies of scale and scope or transactions costs.[5] In that case, one would expect the small biotechs to continue to generate most of the initial innovations in the industry, leaving to their large partners the tasks of clinical and regulatory development, process research and engineering, manufacturing, and distribution.[6]

We explore these questions in order to develop a better understanding of the responses of the large pharmaceutical companies to this transition. There is an extensive literature on the small biotech firms. But much less research has been done on the large firms, most of which adopted one of two strategic pathways into biotechnology: the more common strategy was to start by developing highly specific expertise and then attempting to generalize it across a range of different therapeutic categories. One important variation on this strategy was to use the new techniques of recombinant DNA as "enabling technologies" that strengthened existing capabilities in drug discovery and development. The second strategic alternative was to skip the first stage and attempt to acquire and build upon general capabilities very early in the process of establishing licensing, research, and equity relationships with biotech enterprises. Whichever path they selected, by the early 1990s all of the large corporations had found it prudent to extend their traditional networks by establishing new types of contractual ties with biotech firms, a development that invites historical analysis.[7]

[5] Alfred D. Chandler, Jr., *Scale and Scope: The Dynamics of Industrial Capitalism* (Cambridge, Mass., 1990); Oliver E. Williamson, *The Economic Institutions of Capitalism: Firms, Markets, Relational Contracting* (New York, 1985); for an update on transactions cost analysis, see the same author's "Hierarchies, Markets and Power in the Economy: An Economic Perspective," *Industrial and Corporate Change* 4:1 (1995): 21-49.

[6] These issues are discussed, with various conclusions, in the following sources: Alfonso Gambardella, *Science and Innovation*, 61-81; Ashish Arora and Alfonso Gambardella, "The Changing Technology of Technological Change: General and Abstract Knowledge and the Division of Innovative Labour," *Research Policy* 23 (1994): 523-32; Gary P. Pisano, Weijian Shan, and David J. Teece, "Joint Ventures and Collaboration in the Biotechnology Industry," in *International Collaborative Ventures in U.S. Manufacturing*, ed. David C. Mowery (Cambridge, Mass., 1988), 183-222. See also Bengt Holmstrom, "Agency Costs and Innovation," *Journal of Economic Behavior and Organization* 12 (1989): 305-27. For an excellent quantitative map of the trends, see Joseph A. DiMasi, Elaine M. Healy, and Louis Lasagna, "Trends in the Introduction of New Drugs by Pharmaceutical Firms" (mss. courtesy of the authors).

[7] For an important exception to the rule about the analysis of large-firm responses, see Lynne G. Zucker and Michael R. Darby, "Present at the Biotechnological Revolution," *Research Policy* 26 (1997): 429-46. We provide a description and analysis of what Zucker and Darby refer to as the "generalizability of the transformation experience" (440).

A *Study in Strategic Innovation* / 255

The Pharmaceutical Setting for the Biotech Revolution

When recombinant DNA technology first became possible in the 1970s, the pharmaceutical industry was still in the early stages of mastering drug development by design, applying across a broad front the molecular insights provided by microbial biochemistry and enzyme inhibition.[8] The pipelines at several of the leading firms already contained a number of unusually promising drug candidates. At Smith Kline & French, Sir James Black and his colleagues were about to bring to conclusion a decade of research by producing the first H2-antagonist anti-ulcer drug, cimetidine (marketed under the trade name *Tagamet*).[9] Glaxo was following a similar path toward ranitidine (*Zantac*, also an ulcer treatment), and Squibb researchers were developing a compound, captopril, that would inhibit the overproduction of angiotensin II and thus prevent increases in blood pressure. In 1975, Merck was just launching the biochemical research program that would yield two major products, enalapril maleate (an antihypertensive) and lovastatin (a cholesterol-lowering agent).[10]

Many of the largest pharmaceutical companies were preoccupied with these programs—for good reason. Indeed, the immediate economic results appeared to confirm their decisions to focus resources on their existing programs. By the end of the 1980s, most of the world's leading pharmaceutical products (in sales) had emerged from the new style of research in microbial biochemistry and enzymology.[11] Most were treatments for ulcers, hypertension, or arthritis. Several were recording global sales approaching or exceeding a billion dollars a year. Meanwhile, traditional forms of screening and synthetic organic chemistry were also still producing important new therapies.[12]

---

[8] By blocking the normal function of a particular enzyme, the scientists could prevent a specific biochemical sequence from taking place. This sequence might deal with such biological mechanisms as the human body's production of cholesterol or the process by which a virus multiplies.

[9] *Tagamet* blocked the acid-secreting cells that were triggered by histamine (the H2 receptor). On *Tagamet*, see Herdis K. M. Molinder, "The Development of Cimetidine: 1964-1976: A Human Story," *Journal of Clinical Gastroenterology* 19:3 (1994): 248-54. See also the commentary in Arthur M. Louis, "SmithKline Finds Rich Is Better," *Fortune*, 30 June 1980, 63-66. Showing the type of restraint one expects in popular articles, Louis concluded: "With Tagamet, SmithKline has become the most glamorous drug company of them all."

[10] The first Merck product was *Vasotec*, see *Merck World* 7 (Mar. 1986): 4-8; the second, was *Mevacor*, see *Merck World* 8 (Oct. 1987): 1-13. See also Matthew Lynn, *Merck v Glaxo: The Billion Dollar Battle* (London, 1991), 191-94.

[11] IMS International, *The Pharmaceutical Market World Review 1989* (1990), lists the 75 leading products in Table 85. See also IMS AG (U.K. Branch), *The Pharmaceutical Market World Review 1990* (1991), Table 85.

[12] The companies which were continuing to develop new antibiotics were using the same

Galambos and Sturchio / 256

In addition to these important innovations, executives in pharmaceuticals during these years were attempting to position their firms to achieve greater economies of scale and scope.[13] One of the objectives was to conduct research and development across a broader range of therapeutic categories and to take advantage of the many new opportunities available in biomedical science. Even the largest firms were finding it difficult to fund both basic and developmental research across the entire range of these opportunities, and this problem, as well as the necessity to achieve scale economies in manufacturing and distribution, prompted a major wave of mergers, acquisitions, and experiments with strategic alliances.[14] In the United States, the new antitrust policy that emerged in the 1980s made it easier for large companies to form such combinations.[15] This new public policy environment made it easier for large pharmaceutical companies to establish collaborative arrangements with their competitors, as well as with small biotech firms.

The Biotech Startups and the University Setting

Most of the modern science-based industries in the United States, Germany, Britain, and elsewhere had significant ties to universities,

---

style of research and development that had yielded penicillin, streptomycin, and tetracycline in the 1940s and 1950s. The firms screened tens of thousands of samples (of soil, for instance) to discover evidence of anti-microbial activity. Then, they attempted to isolate and analyze the active chemical so that they could use it as the basis for a new therapy; usually, this followed further chemical modification to improve its bioavailability and mitigate its side-effects.

[13] Chandler, *Scale and Scope*, provides the best historical treatment of the role of economies of scale and scope in the industries of the United States, Great Britain, and Germany from the late nineteenth century to the early 1960s. See F. M. Scherer and David Ross, *Industrial Market Structure and Economic Performance* (Boston, 1990), for an analysis of the evidence from the vantage point of industrial organization theory.

[14] *Financial Times*, 23 July 1992, 21-23; 27 Oct. 1992, 20; 23 Mar. 1994, I-VIII; Francis H. Spiegel, Jr., "Strategic Alliances, With Care and Creativity," *Financial Executive* 9 (Mar./Apr. 1993): 28-31; *Wall Street Journal*, 7 Mar. 1996, 1, 12; M. L. Sharp, "Pharmaceuticals and Biotechnology: Perspectives for the European Industry," in *Technology and the Future of Europe: Global Competition and the Environment in the 1990s*, eds. Christopher Freeman, Margaret Sharp, and William Walker (London, 1991); Natasha Alperowicz, "Transforming Hoechst," *Chemical Week*, 22 May 1996, 22-26. Hoechst had acquired Marion Merell Dow. Sharon L. Oswald and William R. Boulton, "Obtaining Industry Control: The Case of the Pharmaceutical Distribution Industry," *California Management Review* 38 (Fall 1995): 140-42. Consolidation was taking place at all levels of the industry. Between 1963 and 1993, the number of wholesalers fell from four thousand to seventy-four. Ibid., 146-47.

[15] The last two structural cases involved IBM and AT&T. The Department of Justice (DOJ) abandoned the IBM case, and in 1982, DOJ settled with AT&T when the company signed the agreement that broke up the Bell System. Between 1982 and 1998, there were no major structural cases; the Clinton Administration seems now to have changed that policy by bringing suit against Microsoft.

A *Study in Strategic Innovation* / 257

but biotech was different in several regards. For one thing, the universities were now directly involved in entrepreneurial activity through contracts, grants, and patents.[16] Moreover, the scientific entrepreneurs who founded many of the early biotech firms frequently continued to hold university positions while they were getting their enterprises underway. Academic scientists had usually played less important roles in the ventures to which they had provided technical expertise.[17] In biotechnology, however, they were often both inventors and entrepreneurs, line and staff, president as well as head of the labs.[18]

Following the discovery of techniques for genetic recombination in 1973, new biotech firms had proliferated in the United States.[19] Recombinant DNA (rDNA) gave molecular biologists the means to produce in bacteria large quantities of particular, desired cellular proteins, an innovation that sparked a growing interest in the commercial potential of the new technology but also widespread fears about the potentially destructive effects of genetic engineering.[20] Uncertainty as to whether these fears would result in stifling government controls and ambiguity about the ability to patent genetic innovations probably persuaded some pharmaceutical executives to adopt a "wait and see" approach in the 1970s. But these imponderables discouraged neither the entrepreneurial scientists in molecular genetics nor their venture capitalist partners.

[16] Luigi Orsenigo, *The Emergence of Biotechnology: Institutions and Markets in Industrial Innovation* (New York, 1989), esp. 72-98. On the unusual case of Genetics Institute and Harvard University, see *Bio/Technology* (Dec. 1983): 84-85; and Kenney, *Biotechnology: The University-Industrial Complex* (New Haven, 1986), 78-83.

[17] See David F. Noble, *Forces of Production: A Social History of Industrial Automation* (New York, 1984), for some interesting exceptions to the norm.

[18] This was particularly true during the early phase of development, from 1971 through 1980. Mark D. Dibner, "Commercial Biotech's Founding Fathers," *Bio/Technology* 5 (June 1987): 571-72; Kenney, *Biotechnology*, 28-72, 90-106.

[19] See Figure 2.1 for the "Aggregate Growth of New Biotechnology Enterprises, 1973-1987," in Krimsky, *Biotechnics & Society*, 31. There were biotech companies—for example, Cetus (1971)—established before scientists developed (1973) the new technique for cutting, splicing, and reproducing DNA by using plasmids.

[20] Stanley N. Cohen, "The Manipulation of Genes," *Scientific American* 233:1 (1975): 24-33. On the mounting public and professional fear, see ibid., 32-33; Bud, *The Uses of Life*, 174-77; Stephen Hall, *Invisible Frontiers: The Race to Synthesize a Human Gene* (Redmond, 1987), 41-68; and *Bio/Technology* (Oct. 1983): 644. The latter publication reported that as late as 1983 "nearly two-thirds of the American people appear to believe society should exercise caution in proceeding with genetic engineering...." Susan Wright, *Molecular Politics: Developing American and British Regulatory Policy for Genetic Engineering, 1972-1982* (Chicago, 1994), provides an excellent comparative history of the rise and fall of concern about and controls on genetic engineering.

The public concern and the policy issues did not go unnoticed in commercial circles, see *Wall Street Journal*, 13 Aug. 1974; 28 Feb. 1975; 21 Mar. 1975; 28 Sept. 1976; 1 Oct. 1976; and 13 July 1984; *Business Week*, 9 Aug. 1976, 66-67; and *The Economist*, 25 Jan. 1975, 25; 8 Nov. 1975, 18; 10 July 1976, 79; and 28 Aug. 1976, 13.

Galambos and Sturchio / 258

In its early stages of development Genentech was typical of the enterprises that sprang from this technical breakthrough. Organized in 1977 by Robert Swanson, a venture capitalist, and scientist Herbert Boyer, co-discoverer of the rDNA technique, the company set out to produce somatostatin, a human brain hormone, in bacteria; later the company successfully used rDNA technology to make human insulin.[21] Genentech was the first of the biotech firms to go public, a 1980 event that was so successful it helped launch a series of initial public offerings (IPOs) and a wave of Wall Street interest in the new industry.[22]

Ready access to capital during the prosperous 1980s combined with an evolving science/technology base to foster rapid expansion of biotechnology along lines reminiscent of Silicon Valley.[23]   Indeed, California was one of the four leading areas of the country for startups, followed by Massachusetts, the New York City area, and Maryland. Access to relevant research universities dominated the location decisions, just as access to capital markets controlled the pace of expansion of the businesses.[24] In addition to rDNA technology, the ability to produce large quantities of monoclonal antibodies prompted entrepreneurial ventures in biotech and enthusiasm in the stock markets.[25]

---

[21] *Business Week*, 12 Dec. 1977, 128, 132.   Maureen D. McKelvey, *Evolutionary Innovations: The Business of Biotechnology* (Oxford, 1996), esp. 91 and following, provides an insightful description and analysis of Genentech and its links with the Swedish firm Kabi and with Eli Lilly.

[22] On Genentech, see Robert Teitelman, *Gene Dreams: Wall Street, Academia, and the Rise of Biotechnology* (New York, 1989). See also Teitelman's *Profits of Science: The American Marriage of Business and Technology* (New York, 1994). As late as 1979, there had still been concern about the nature of the controls the federal government might impose on rDNA research; see *Business Week*, 24 Sept. 1979, 64, 68. The election results in 1980 appear, however, to have erased those fears in the United States.

[23] On Silicon Valley, see AnnaLee Saxenian, *Regional Advantage: Culture and Competition in Silicon Valley and Route 128* (Cambridge, Mass., 1994), 11-27. On biotech scientific developments, see, W. French Anderson and Elaine G. Diacumakos, "Genetic Engineering in Mammalian Cells," *Scientific American* 245 (July 1981): 106-21. On urban financial support, see James R. Murray et al., "The Chicago Model for Developing Biotechnology Enterprise," *Bio/Technology* (July 1983): 407-10. See also the review of state, regional, and national programs in the supplement to *Bio/Technology* (Dec. 1983); and the information on the U.S. Small Business Innovation Research Program in *Bio/Technology* (Jan. 1984): 22. For a later, more reserved evaluation, see Mark Ratner, "Regional Development: The Role and Effectiveness of Governmental Agencies," *Bio/Technology* (July 1989): 671-72, 678-81.

[24] Lynne G. Zucker et al., "Intellectual Human Capital and the Birth of U.S. Biotechnology Enterprises," *American Economic Review* 88:1 (1998): 290-306. R&D limited partnerships—in which investors bought a stake in a specific research project—became popular in the United States in the 1980s, just as biotech venture funds did in Britain. Direct government financing was substantially more important in Europe than in the United States. *Bio/Technology* (Sept. 1983): 546-48, 552, 554-59.

[25] Monoclonal antibodies are homogeneous, biological molecules produced from a single cell line; they recognize and react against a specific chemical structure; their specificity and relative ease of manufacture made them promising candidates as therapeutic agents against

A *Study in Strategic Innovation* / 259

Between 1981 and 1983, the biotechs launched twenty-two public stock offerings. Some of these had disappointing results, but after the stock market boom of 1983 loosened the purse strings of investors, money flowed into such promising enterprises.[26]

The biotechs were dependent upon Wall Street because they were, with few exceptions, unable as yet to produce their own working capital. Often founded on just the promise of a new technology, they were usually trying to balance licensing revenue against the "burn rate" of capital being expended on their scientists and laboratories. Genentech was the exception that proved the rule.[27] Like nuclear energy, space exploration, and other areas of post-World War II science, biotech suffered from the hyperbole that normally accompanies entrepreneurship and government grant-seeking.[28] New breakthrough products were slow to appear, and the great promise of such drugs as interferon—projected initially as an almost miraculous cure for cancer—faded when subjected to the rigorous tests of clinical research.[29]

---

a variety of diseases. Alberto Cambrosio and Peter Keating, *Exquisite Specificity: The Monoclonal Antibody Revolution* (New York, 1995); *Business Week*, 18 May 1981, 147-48, 150, 154, 156. *Bio/Technology* (Apr. 1983): 197-98; (June 1983): 313-14; (Sept. 1983): 534, 536; and (Nov. 1983): 736. See also Roland Carlsson and Cristina Glad, "Monoclonal Antibodies into the 90s: The All-Purpose Tool," *Bio/Technology* (June 1989): 567-68, 570-73.

[26] Kenney, *Biotechnology*, 134, 156.

[27] *Business Week*, 18 Sept. 1978, reported Genentech's success in producing insulin in bacteria by means of rDNA. By 1982, Genentech's R&D expenditures were thus only 111 percent of its operating revenues—a low figure for the biotech companies. U.S. Office of Technology Assessment [herafter OTA], *Commercial Biotechnology* (1984), 271. Arthur Klausner, "And Then There Were Two," *Bio/Technology* (July 1985): 605-12, offers a comparative analysis of Genentech and Cetus. See Barry Werth, *The Billion-Dollar Molecule: One Company's Quest for the Perfect Drug* (New York, 1994), for an exciting blow-by-blow account of Vertex Pharmaceuticals, a biotech closer than Genentech to the norm.

[28] The government as well as the business press added to the hype. See, for instance, *Business Week*, 22 Oct. 1979, 160, 164, 168, 172. A scientist with the OTA suggested modestly that biotech opened up "the possibility of building a sustainable future based on renewable resources." See also *Business Week*, 30 June 1980, 48-9; 28 July 1980, 71.

As the 1980's boom got underway, the percentage of company founders coming from industry, as opposed to academia, steadily increased. Mark D. Dibner, "Commercial Biotech's Founding Fathers," *Bio/Technology* (June 1987): 571-72.

[29] See, for instance, Gene Bylinsky, "DNA Can Build Companies, Too," *Fortune*, 16 June 1980, 144-146, 149, 152-53. This article focused on the startup firm Biogen (organized in 1978) and its production of interferon. "Looking further ahead," the author commented, "the possibilities seem almost boundless." See also Gene Bylinsky, "Future Drugs That Will Be Lifesavers for the Industry Too," *Fortune*, Dec. 1976, 162. On interferon, see Suzie Rubin, "Biotechnology and the Pharmaceutical Industry," *Cancer Investigation* 11:4 (1993): 451-57.

While commercial outcomes were disappointing, the science/technology was progressing rapidly. For advances in sequencing DNA, for example, see Allan M. Maxam and Walter Gilbert, "A New Method of Sequencing DNA," *Proceedings of the National Academy of Science* 74:2 (1977): 560-64; and F. Sanger, S. Nicklen, and A.R. Coulson, "DNA Sequencing with Chain-Terminating Inhibitors," *Proceedings of the National Academy of Science* 74:12 (1977): 5463-67. Also, the polymerase chain reaction (PCR) provided biotech companies

Galambos and Sturchio / 260

After the stock market collapse of 1987, investors became even more cautious than the U.S. regulatory agency, the Food and Drug Administration (FDA), about the potential for new biotech wonder drugs.[30]

The Large Pharmaceutical Companies Move into Biotechnology

More accurate than Wall Street as a prognosticator of biotech's future were the investments that large pharmaceutical firms were making to establish close links with the emerging molecular genetic networks. In the mid-1970s, the large, multinational pharmaceutical companies, most of which were still making heavy investments in microbial biochemistry and enzyme inhibition, had not been ready to pump major resources into molecular genetics and rDNA.[31] The former invest-

---

with the ability to make quickly and efficiently an almost unlimited number of copies of a particular DNA molecule. Dr. Kary B. Mullis discovered this process while working at Cetus Corporation in the mid-1980s. Kary B. Mullis, "The Unusual Origin of the Polymerase Chain Reaction," *Scientific American* (Apr. 1990): 56-65; and Paul Rabinow, *Making PCR: A Story of Biotechnology* (Chicago, 1996).

[30] Compare Arthur Klausner, "Public Markets Favor Biotech Offerings," *Bio/Technology* (June 1987): 548, with Klausner, "Stock Crash Could Squeeze Biotech Firms," *Bio/Technology* (Dec. 1987): 1253-54. See also Lee Rauch, "Surviving the Funding Gap," *Bio/Technology* (May 1989): 435-36, 440.

[31] See Gambardella, *Science and Innovation*, esp. 61-81, 146-58, on microbial biochemistry and enzyme inhibition. We have drawn heavily upon this excellent study, even though our approach is somewhat different than Gambardella's. First, we distinguish between the transition to biochemistry/enzymology and the partially overlapping transition to molecular genetics/rDNA, whereas Gambardella describes these two developments as part of a single transformation in the industry. Second, we distinguish between the strategic alliances and the move into biotech, although we realize that these two phenomena had some similar outcomes. Third, we give less emphasis than Gambardella does to the relationships between firms and universities; these links were common elements in each of the industry's prior transitions, and we are trying to isolate what was particular to the most recent shift into biotech.

See also Pisano et al., "Joint Ventures and Collaboration," 183-222. Despite the title of the book, the authors of this essay deal with collaboration as such, whether international or not. The authors conclude (193) that "The competitive advantage of NBFs [New Biotech Firms] is concentrated in product R&D. For established firms, in general, it lies in downstream activities (manufacturing, marketing) where they have many years of experience." Pisano et al. analyze the collaborations from the perspective of appropriability and transfer costs.

See Kenney, *Biotechnology*, esp. 190-216, for a description that emphasizes the large chemical companies rather than pharmaceutical firms. Kenney, whose book is substantially less analytical than Gambardella's volume, observes (190) that "the large, relatively ponderous corporate bureaucracies (with a few exceptions) have had difficulty responding quickly to biotechnology's potential." Alice M. Sapienza, "Technology Transfer: An Assessment of the Major Institutional Vehicles for Diffusion of U.S. Biotechnology," *Technovation* 9 (1989): 463-78, concludes that: "Pharmaceutical investment in biotechnology was prompted by declining technological performance." The author comments on the "bureaucratic stolidity" of the large companies.

A *Study in Strategic Innovation* / 261

ments remained too promising, the latter investments still too questionable to justify so sudden a change in research strategy.[32] The new investments would be substantial. What the pharmaceutical firms needed were not merely scientists working with nucleic acids, but scientific leaders with diplomatic skills and links to the relevant networks that would enable them to build the teams and productive programs necessary to sustain biotech R&D over the long-term.[33]

In the late 1970s and early 1980s, many large companies began making these investments and building biotech capabilities. Most of the pharmaceutical firms adopted a narrow-front strategy, first building capabilities associated with specific products that they had in market or had targeted for research and development. A few companies bypassed this stage and attempted to acquire general biotech capabilities very quickly, usually through acquisition. Whichever strategy they implemented, the pharmaceutical companies had to manage their way through a transition that was sometimes painful for their existing personnel in R&D and in other parts of the organization. There were "transition costs."[34]

These costs help explain the preference for an incremental transition, as do the relationships between biotech and the pharmaceutical firms' existing product lines and capabilities. For some, biotech was an unusual technological opportunity and for others, a serious threat to their position in the market. Eli Lilly was fast off the mark, for instance, when its dominant position in the U.S. market for insulin appeared to be threatened by the advent of recombinant DNA production. Genentech's success in using rDNA to produce insulin convinced Eli Lilly that it should promptly complete a multi-million dollar, long-term manufacturing agreement with the San Francisco-based biotech company.[35] Lilly used its new form of insulin to protect its U.S.

---

[32] See the sources cited in n. 17 for background on the political and social uncertainties of the mid-1970s. It was unclear at that time whether companies could patent genetic innovations and it seemed possible that the federal government would introduce stringent, perhaps stifling, regulations on industrial uses of rDNA.

[33] On these aspects of innovation, see Galambos, *Networks of Innovation*, esp. 53-77.

[34] The best treatment of the "transition costs" (although they do not use this expression) is in Michael Williams et al., "Biotechnology in the Drug Discovery Process: Strategic and Management Issues," *Medicinal Research Reviews* 13:4 (1993), which discusses Upjohn (426-31) and Berlex Laboratories (the U.S. subsidiary of Schering AG; 431-35). For some of Merck's transition costs see Lynne G. Zucker and Michael R. Darby, "Present at the Biotechnological Revolution," *Research Policy* 26 (1997): 429-46; see also Galambos, *Networks of Innovation*, 197-205; and Orsenigo, *The Emergence of Biotechnology*, esp. 99-165.

[35] The best study is McKelvey, *Evolutionary Innovations*. Also see *Business Week*, 18 Sept. 1978, 31; 29 Oct. 1979, 134-45. In 1980, Lilly announced that it had committed $40 million to build new facilities for the production of human insulin using recombinant technology. Ibid., 4 Aug. 1980, 22. Later, Lilly acquired Hybritech. *Wall Street Journal*, 18 Dec. 1985, 43.

Galambos and Sturchio / 262

market share and to push forward in European markets for the first time.[36]

Merck & Co., Inc., had similarly specific needs when it moved into involvement with biotech in the late 1970s. The company was making its new hepatitis B vaccine (*Heptavax-B*) from plasma, which was difficult to obtain and expensive; it was also fear-provoking once the AIDS pandemic began. Dr. P. Roy Vagelos, President of the Merck Research Laboratories (MRL), guided the company's research team to Dr. William Rutter (University of California, San Francisco) and rDNA technology in an effort to obtain the antigen it needed. While the first efforts failed, Merck was finally able to obtain the right antigen from a combined effort by Rutter's lab, Chiron (the biotech firm of which he was a co-founder),[37] Dr. Benjamin Hall's laboratory at the University of Washington, and Merck's Virus and Cell Biology department. Substantial research and development remained to be done in Virus and Cell Biology before the company had in hand a commercial product, but by 1986, Merck was able to market *Recombivax HB*, the world's first genetically engineered vaccine for humans.[38]

In the course of their work on *Recombivax HB*, the scientists at Merck gradually enhanced the firm's biotech capabilities. The laboratories acquired the services of a specialist in yeast genetics after it became apparent that the desired antigen could not be "expressed" in *E. coli*.[39] Leading these efforts was Dr. Edward Scolnick, a molecular geneticist whom Vagelos had recruited from NIH to be his successor. Scolnick, who had worked in Marshall Nirenberg's laboratory, had in the course of his own research defined the specific genes associated with human malignant disease. He now personally guided *Recombivax HB* to completion, while steadily extending MRL's reach in gene-related research. But initially, the firm's capabilities were—like those of Eli Lilly—specific to the particular task at hand.

[36] *Wall Street Journal*, 19 Dec. 1984, 33. For the impact of the competition in Europe, see *Wall Street Journal*, 8 Mar. 1985, 35.

[37] On Chiron, see Arthur Klausner, "Chiron: Looking Good," *Bio/Technology* (Feb. 1987): 121-22, 124, 126-27.

[38] Galambos, *Networks of Innovation*, 172-73, 181-205. While Chiron held the patent and Merck licensed the vaccine, Merck's Virus and Cell Biology team devoted many months of research to the improvements needed after receiving the vectors from Chiron. Interview, Edward M. Scolnick, 17 July 1995. M. R. Hilleman, R. E. Weibel, and E. M. Scolnick, "Recombinant Yeast Human Hepatitis B Vaccine," *Journal of the Hong Kong Medical Association* 37:2 (1985): 75-85.

[39] The antigen-antibody response is a basic function of the human immune system. Antigens are substances not normally present in the body that stimulate an immune response. In this case, the antigen was being "expressed" in (that is, being produced by) the cellular machinery of the microbe *E. coli*.

A Study in Strategic Innovation / 263

Other leading pharmaceutical firms adopted a similar approach,
edging their way into the new field. SmithKline—a firm some analysts
had considered one of the weaker research organizations in the industry—used part of the profits from its blockbuster ulcer treatment,
Tagamet, to push into new areas of immunology and into the field of
recombinant DNA vaccines.[40] SmithKline was able to bring out a
recombinant hepatitis B vaccine in 1986 and was meanwhile working
with Damon and Amgen on other biotech therapies.[41] Johnson &
Johnson used research contracts (with Immunomedics) and joint projects (with Amgen) as its bridge into genetic research, and by 1988,
Pfizer was collaborating with four different biotech enterprises
through licensing agreements, research contracts, and joint projects.[42]
After consolidating and expanding its in-house programs, the Upjohn
Company also began to develop external links to biotechnology in the
1980s.[43]

Many of the strongest European pharmaceutical firms adopted
this strategy, seeking to develop their own capabilities across a relatively narrow front while working with biotech startups. The small,
university-linked biotechs were at first almost entirely an American
phenomenon, and this gave the large U.S. pharmaceutical companies
an initial advantage, as did the federal government's support for basic

---

[40] *Bio/Technology* (Feb. 1987): 103-04. *Wall Street Journal*, 11 Mar. 1985, 22. Arthur M.
Louis, "SmithKline Finds Rich is Better," *Fortune*, 30 June 1980, 63-66; this article, which
was on *Tagamet*, did not mention biotech. At this time, other pharmaceutical firms were collaborating with biotechs in an effort to develop competing hepatitis B vaccines.

[41] The Belgian government approved Engerix-B, SmithKline's new vaccine, in 1986. By
the time the U.S. FDA approved the vaccine in 1989, SmithKline had merged with a British
pharmaceutical company and become SmithKline Beecham. Under a 1988 licensing agreement, SmithKline paid Biogen royalties from the sales of the new vaccine. See *Nature*, 11
Dec. 1986, 506; *Journal of the American Medical Association* (hereafter *JAMA*) 261:22
(1989): 3278-81; Mark D. Dibner, *Biotechnology Guide U.S.A.: Companies, Data and
Analysis* (New York, 1988), 302. SmithKline continued its work on vaccines, in conjunction
with Cetus. Luigi Orsenigo, *The Emergence of Biotechnology: Institutions and Markets in
Industrial Innovation* (London, 1989), 135.

[42] *Wall Street Journal*, 24 Mar. 1995, B2; Dibner, *Biotechnology Guide*, 301-02.

[43] For an excellent discussion of Upjohn's entrance into biotechnology, see Williams et al.,
"Biotechnology in the Drug Discovery Process," 426-31. See also *Bio/Technology* (Dec.
1983): 837. Dibner, *Biotechnology Guide*, 302. Squibb Corporation bought an equity interest in Cetus Corporation and launched a major joint venture in 1987. *Bio/Technology* (Aug.
1987): 752. For a good review of the collaborations in 1988-89, see Mark Ratner, "Dealing
with Large Companies," *Bio/Technology* (Oct. 1989): 1013-14, 1018-19. For the collaborations as of 1992, see The Boston Consulting Group, "The Contribution of Pharmaceutical
Companies," 111. The financial problems of the biotechs gave the large pharmaceutical
companies an advantage in establishing collaborative and/or equity relationships at this time.
When asked what the key business issues of the next decade would be, 36 percent of the
biotechs identified financial questions. G. Steven Burrill, "The Promise of the Next
Decade," *Bio/Technology* (Oct. 1989): 1023.

Galambos and Sturchio / 264

research in molecular genetics.[44] But soon, European governments were attempting to close the widening biotech gap, hoping that government support would make up for the venture capital that was not available to their potential innovators.[45] One of the interesting international hybrids was Biogen, a startup that was built on American science and that quickly expanded through licensing arrangements with several leading pharmaceutical firms. Biogen soon had operations in Germany, Switzerland, and Belgium. Schering-Plough, an American pharmaceutical company, collaborated with Biogen in the race to capture the anticipated global markets for interferon. To enhance and protect its access to Biogen's products and processes, Schering-Plough bought a substantial equity position in Biogen, a practice that became increasingly popular in pharmaceuticals.[46] In the final heat of the race

---

[44] OTA, *New Developments in Biotechnology: U.S. Investment in Biotechnology* (1988), 3-4. On the resistance in Europe to biotechnology, see Marlis Buchmann, "The Impact of Resistance to Biotechnology in Switzerland: A Sociological View of the Recent Referendum," Robert Bud, "In the Engine of Industry: Regulators of Biotechnology, 1970-86," and Sheila Jasanoff, "Product, Process, or Programme: Three Cultures and the Regulation of Biotechnology," in *Resistance to New Technology: Nuclear Power, Information Technology, and Biotechnology*, ed. Martin Bauer (Cambridge, U.K., 1995), 207-23, 293-309, and 311-31 respectively.

[45] John Gurnsey, "Biotechnology in Europe," *Bio/Technology* (Sept. 1983): 561-64. The lack of venture capital in the early 1980s made the biotech startups dependent upon corporate support, government subsidies, or the U.S. financial markets. See also Susan Wright, "Molecular Politics in a Global Economy," and Herbert Gottweis, "The Political Economy of British Biotechnology," in *Private Science: Biotechnology and the Rise of the Molecular Sciences*, ed. Arnold Thackray (Philadelphia, 1998), 80-104 and 105-30 respectively. Michael Stone, "European Community Approves Second Stage of Biotech Program," *Bio/Technology* (Dec. 1983): 825-26; Stephanie Yanchinsk, "Biotechnology Taking Root in West Germany," in ibid., (Apr. 1984): 291-92; and Roger Longman, "Germany's Biotech Boomlet," *Start-Up* (Windhover Information, Inc., Feb. 1998): 12-25. Longman noted that there were still "cultural and managerial obstacles" to the biotechs. On developments in France, see *Bio/Technology* (May 1983): 291-92, and (Apr. 1984): 145; on Holland, see Douglas McCormick, "Holland is Busy in Biotechnology," ibid., (Sept. 1987): 911-13; on Sweden, see McKelvey, *Evolutionary Innovations*; Alice M. Sapienza, "R&D Collaboration as a Global Competitive Tactic—Biotechnology and the Ethical Pharmaceutical Industry," *R&D Management* 19:4 (1989): 285-95.

It was not easy to close the gap: see Richard L. Hudson, "With a Small Budget, Britain's Celltech Fights to Stay in the Biotechnology Race," *Wall Street Journal*, 19 Aug. 1985, 19. See also Bernard Dixon, "More Collaborations in British Biotech," *Bio/Technology* (Feb. 1987): 112. On European startups, see ibid., (Nov. 1987): 1108. A number of American biotechs moved into this gap, establishing operations in Europe. See Nigel Webb, "Taking American Biotechnology Across the Atlantic," ibid., (Mar. 1987): 222, 224, 226, 228; and Stephen M. Edgington, "Germany: A Dominant Force by the Year 2000?" ibid., (Aug. 1995): 752-54, 756. Other articles in the same issue consider developments in France (757-9), where there were still serious problems; in the Balkans (760-61), where the development was fragile; and in Scandinavia (763-64), where the outlook was promising. For recent developments, see David Firn, "Continental growth pushes European biotech towards the 21st century," *Pharmaceutical Business News*, 11 May 1998, 18-20.

[46] Schering-Plough was also working with Genex on other products. Dibner, *Biotecnology*

A Study in Strategic Innovation / 265

to interferon, Schering-Plough's chief competitor was Hoffmann-La Roche, a Swiss pharmaceutical company that allied itself with Genentech.[47]

After creating a complex network of licensing and research agreements, Hoffmann-La Roche embraced a new strategy that quickly moved the firm more deeply into the biotech field.[48] In effect, Roche decided to transplant a generalized biotech capability through acquisition. It began by buying equity stakes in the biotechs with which it was collaborating, a relatively common element in the large firm/small firm alliances in this industry. But next, it broke the mold by purchasing a controlling (60%) share of the most successful of the biotech startups, Genentech.[49] Meanwhile, the Swiss firm was spending between $130 and $140 million a year on its in-house capabilities in the new field.[50]

Japan's largest pharmaceutical firms adopted a more cautious strategy, one similar to that of most large American pharmaceutical companies. Because there was in Japan no tier of biotech startups and university research in molecular genetics was relatively weak, Japanese firms created a web of contractual relationships with U.S. biotechs, emphasizing specific therapies. For the most part, Japanese pharmaceutical companies did not establish equity positions in the U.S. com-

---

Guide, 302. On Biogen, see Lawrence M. Fisher, "The Rocky Road From Startup to Bigtime Player," *Strategy & Business* (Booz, Allen & Hamilton), Third Quarter, 1997. Gene Bylinsky, "DNA Can Build Companies, Too," *Fortune*, 16 June 1980, 144-46, 149, 152-53; and Tabitha M. Powledge, "Biogen in Transition: From Research Specialist to Manufacturer," *Bio/Technology* (July 1983): 398-405. For a table of equity investments in biotechs, 1976-80, see Wright, *Molecular Politics*, 88-90. See also *Business Week*, 22 Oct. 1979, 160; *Bio/Technology* (Dec. 1983): 840; Dibner, *Biotechnology Guide*, 300-2, lists equity relationships as of 1987. For a thorough review of the continuing importance of these links between biotechs and their pharmaceutical partners, see Roger Longman and Kevin Roche, "Biotech Deals by the Numbers," *In Vivo: The Business and Medicine Report* 15:8 (1997): 16-20.

[47] Arthur Klausner, "And Then There Were Two," *Bio/Technology* (July 1985): 605-12; Gambardella, *Science and Innovation*, Table 3.3. The race ended in a de facto tie. See *Wall Street Journal*, 24 Feb. 1984, 20; 17 May 1985, 6; 3 June 1985, 6; and 24 July 1985, 16.

[48] On the development of the various agreements, see Gambardella, *Science and Innovation*, Table 3.3. *Wall Street Journal*, 18 Jan. 1984, 34; and Gene Bylinsky, "Science Scores a Cancer Breakthrough," *Fortune*, 25 Nov. 1985, 16-21.

[49] Gambardella, *Science and Innovation*, 104. See also *Wall Street Journal*, 5 Mar. 1984, 56, on Hoffmann-La Roche. Boehringer Ingelheim International G.m.b.H. had purchased 4.9 percent of Genentech's shares in 1985 (*Wall Street Journal*, 1 Mar. 1985, 8), and by that time was already marketing Genentech's gamma interferon and its tissue-type plasminogen activator. By 1985, when Genentech had the second largest rDNA manufacturing facility in the world, it was threatening to become the first biotech to succeed in becoming a fully integrated pharmaceutical producer. Eli Lilly's dedicated insulin plant was the world's largest rDNA production facility. *Bio/Technology* (July 1985): 606.

[50] OTA, *Biotechnology in a Global Economy* (1991), 89. In 1987, Biogen sold its entire Geneva, Switzerland, research operation, including the contracts with the scientific staff, to Glaxo. *Bio/Technology* (Sept. 1987): 864. On the European industry's responses to biotech, see Margaret Sharp, "Pharmaceuticals and Biotechnology," in ibid., esp. 222-25.

Galambos and Sturchio / 266

panies.[51] But they used their network of contractual ties as a foundation for building in-house capabilities that had the potential to be generalized across a broad range of therapies.[52]

In the United States, DuPont followed a course closer to the Roche model. The American chemical giant had been exploring pharmaceuticals in the 1960s and 1970s without making a strong commitment to the industry. In the 1980s, however, DuPont made a decisive move into pharmaceutical development and it did so initially without building a foundation of therapy-specific contractual ties to biotech firms. As DuPont's Chairman, Irving S. Shapiro, explained: "We're placing a large bet on life sciences for the future."[53] The first "bet" involved an effort to depend almost entirely on internal resources—the approach DuPont had used in the successful development of nylon. Later, when DuPont went outside the firm to build capabilities, it followed the Roche strategy, acquiring New England Nuclear Corporation, a firm with expertise in monoclonal antibodies.[54]

After a few years of experience, however, DuPont began to shift its strategy again. Like most of the other large firms, it began to develop therapy-specific ties to biotech companies, complementing the sub-

[51] See, for instance, OTA, *New Developments in Biotechnology, 4, U.S. Investment in Biotechnology* (1988), 91-2, Table 5-6. Arthur Klausner, "Today's Trends: Doing Business with Japan," *Bio/Technology* (Oct. 1987): 1019-20, 1024, 1026; in the same issue see Mark D. Dibner, "An Analysis of Partnerships," 1029, 1031-32. One of the exceptions was the joint venture between Amgen and the Kirin Brewery Co. of Japan. *Wall Street Journal*, 14 May 1984, 16.

[52] Mark D. Dibner, "Biotechnology in Pharmaceuticals: The Japanese Challenge," *Science*, 20 Sep. 1985, 1230-35. See also Christopher G. Edwards et al., "Japan Taps Into New Biotech," *Bio/Technology* (Apr. 1984): 307-21. The trade in biotechnology was substantial enough to support brokers who arranged transactions between U.S. universities and Japanese firms. See *Bio/Technology* (Feb. 1984): 115-16. Because of the widespread concern in the 1970s and 1980s about U.S. industrial competitiveness and the rise of Japanese industry and finance, many of the reports and articles were framed in terms of "The Japanese Challenge." See, for instance, *Business Week*, 4 Aug. 1980, 30-31. OTA, *Commercial Biotechnology: An International Analysis* (1984), 7-8: "Japan is likely to be the leading competitor of the United States," the OTA concluded. By 1991, however, the OTA decided: "The efforts of MITI to promote biotechnology as a key technology ... clearly has [sic] been less successful than many anticipated." See OTA, *Biotechnology in a Global Economy, Summary* (1991), 18. William O. Bullock and Mark D. Dibner, "The Changing Dynamics of Strategic Alliances between US Biotechnology Firms and Japanese Corporations and Universities," *Trends in Biotechnology* 12 (1994): 397-400, describes the trend in Japan toward developing in-house capabilities.

[53] *Business Week*, 24 Nov. 1980, 86-9, 92, 96, 98. David A. Hounshell and John Kenly Smith, Jr., *Science and Corporate Strategy: DuPont R&D, 1902-1980* (New York, 1988), 589-90.

[54] For DuPont, this was a return to the strategy the firm had implemented with great success in the 1920s. Fearful that the government would prosecute the firm for violating the antitrust laws, DuPont's leaders had abandoned this strategy in the 1930s. Ibid., 119-221. On the acquisition, see *Business Week*, 18 May 1981, 150.

A Study in Strategic Innovation / 267

A biologist at the Merck Research Laboratories in West Point, Pennsylvania, works on the development of an influenza vaccine based on recombinant DNA technology. DNA that codes for proteins of the influenza virus is used directly to elicit an antiimmune response instead of using whole virus or viral proteins as a vaccine. Courtesy Merck & Co., Inc., Whitehouse Station, N.J., U.S.A.

Galambos and Sturchio / 268

stantial in-house capability the company had already acquired.[55] Later, DuPont moved its entire pharmaceutical operation into a joint venture with Merck to take advantage of Merck's strengths in drug development and marketing in the United States and Europe.[56]

Whether they adopted an incremental approach or the Roche strategy, by the 1990s most of the large pharmaceutical firms had made substantial progress in acquiring new capabilities in biotechnology and molecular genetics.[57] Some of the firms had therapy- or product-specific capabilities. Others had acquired basic research strengths that enabled them to employ their new capabilities across a broad range of scientific and technical areas of drug discovery. At Merck, research management had made a conscious decision to focus its use of biotechnology on supporting its core competencies in developing small organic molecules as drugs.[58] Rather than adopting an acquisition strategy, Merck concentrated on acquiring general capabilities in the new science and technology.

The premium for successfully internalizing those capabilities was high. If the pharmaceutical companies did not have to license innovations, they could eliminate license fees, the one element in their unit costs that did not decline as production got up to scale. Even when licensing was necessary, the pharmaceutical companies had to have in-house the scientific and technical capabilities to transform a licensed compound from the laboratory to the pilot-plant and then to full-scale manufacturing.[59]

[55] Dibner, *Biotechnology Guide*, 300-01.

[56] Merck & Co., Inc., *Annual Reports*, 1989-1992; Wayne Koberstein, "Joseph Mollica," *Pharmaceutical Executive* 11 (May 1991): 24-32. Recently, DuPont purchased Merck's half of the joint venture.

[57] The Swiss firm Ciba-Geigy adopted the Roche strategy in 1994, acquiring Chiron for $2.1 billion. *Wall Street Journal*, 9 Dec. 1994, 2. *Bio/Technology* (Feb. 1996): 135. The Ciba-Geigy-Chiron relationship began with a series of joint ventures, and the Swiss firm also established biotech research programs in Basle and in North Carolina at the Research Triangle Park. *Bio/Technology* (Feb. 1987): 122, 124; (June 1987): 530. The acquisitions continued. American Home Products bought 60 percent of Genetics Institute, Inc. In 1991, Sandoz (Switzerland) bought 60 percent of Systemix, a cell-therapy firm, and four years later, Sandoz acquired Genetic Therapy Inc., for $295 million. See *Business Week* (2 Mar. 1992): 69, 73. *Bio/Technology* (Sept. 1995): 945-47. The financial problems of the biotech firms once again created conditions that favored large pharmaceutical companies seeking alliances of various sorts. See *Business Week*, 26 Apr. 1993, 86; and *Investors Business Daily*, 2 Dec. 1993. On Schering AG's acquisition of Codon and Triton Biosciences through its U.S. subsidiary (Berlex Laboratories), see Williams et al., "Biotechnology in the Drug Discovery Process," *Medicinal Research Reviews* 13:4 (1993): 431-35.

[58] The distinction between proteins and small molecule drugs was important. The former could only be given to patients by injection; the latter could in most cases be taken orally. For patients as well as the pharmaceutical firms, this was a significant difference.

[59] For an exploration of these issues, see Ashish Arora, "Evaluating technological infor-

A Study in Strategic Innovation / 269

A chemist loads reactors with polymers for high-speed synthesis of new chemical compounds in a combinatorial synthesis laboratory. Courtesy Merck & Co., Inc., Whitehouse Station, N.J., U.S.A.

Galambos and Sturchio / 270

By the mid-1990s, it was becoming apparent that various kinds of collaborative arrangements between biotechs and pharmaceutical companies would continue to be an important feature of the current long cycle of innovation in this industry.[60] As Glaxo's director of corporate development explained, "No emerging or established pharmaceutical company is large enough, or smart enough to meet all of its knowledge needs in isolation."[61] The front across which change was taking place in the biomedical sciences was so broad that even the largest pharmaceutical firms could no longer bring in-house all of the research capabilities they needed.[62] Indeed, the "knowledge needs" were so pressing that they had given rise to a new subdiscipline, "bioinformatics," that combined genomic information with computer technology in order to make data more widely available to scientists.[63]

---

mation and utilizing it," *Journal of Economic Behavior and Organization* 24 (1994): 91-114. It was also necessary for pharmaceutical companies to have substantial capabilities in order to sift out the many proposals for collaboration that they received from biotechs. *Chemical Marketing Reporter*, 8 Apr. 1996, 18. By 1996, Merck was receiving about four to five hundred such proposals every year; Eli Lilly received about two thousand a year (of which twenty resulted in collaboration). For a thorough analysis of the impact of these changes on process development in pharmaceuticals, see Gary P. Pisano, *The Development Factory: Unlocking the Potential of Process Innovation* (Boston, 1997).

[60] "More Deals than Ever," *Scrip Magazine* (Jan. 1996): 44-46, lists the mergers, acquisitions, alliances and collaborations for 1995. See also Longman and Roche, "Biotech Deals by the Numbers."

[61] K. H. George, "US Biotechnology United," *Trends in Biotechnology* 6 (1993): 221-22. There had been a distinct upward trend in the number of licensing agreements and acquisitions in the late 1980s and early 1990s. By 1992, there were 215 alliances between biotechs and pharmaceutical companies. Ibid., 220-21. The Boston Consulting Group, Inc., "The Contribution of Pharmaceutical Companies: What's at Stake for America" (Sept. 1993): 110.

[62] This was true despite the increased spending on R&D by pharmaceutical companies. *Hospitals & Health Networks*, 20 Mar. 1996, 12. For a list of "Products in the U.S. Pipeline" in 1991, see *Bio/Technology* (Oct. 1991): 947-49. For a similar list as of the end of 1992, see John P. Santell, "Projecting Future Drug Expenditures—1994," *American Journal of Hospital Pharmacy*, 15 Jan. 1994, 183. For a list of the top ten biotech drugs as of 1995 (in U.S. sales), see *Bio/Technology* (July 1995): 649. See also Pharmaceutical Research and Manufacturers of America, "1996 Survey: 284 Biotechnology Products in Testing," and "Approved Biotechnology Drugs and Vaccines"; J. Drews, "Intent and Coincidence in Pharmaceutical Discovery," *Arzneimittel-forschung/Drug Research* 45:8 (1995): 934-39.

[63] Tefko Saracevic and Martin Kesselman, "Trends in Biotechnology Information and Networks: Implications for Policy," in *Biotechnology: R&D Trends*, ed. George T. Tzotzos (New York, 1993), 135-44. Diane Gershon, "The Boom in Bioinformatics," *Nature*, 18 May 1995, 262. Also, W. Bains, "Bioinformatics in Europe—The Federation Strikes Back," *Trends in Biotechnology* 6 (1993): 217.

By the mid-1990s, the new technologies of combinatorial chemistry and high throughput screening were also changing the information landscape in drug discovery. Combinatorial chemistry enabled researchers "to make a large number of chemical variants all at 1 time, to test them for bioactivity (binding with a specific target or other functional property), and to isolate and identify the most promising compounds for further development." Donald F. Phillips, "Making New Drugs via Combinatorial Chemistry," *JAMA*, 6 June 1996, 1624-26. Also, Robert F. Service, "Combinatorial Chemistry Hits the Drug Market," *Science*, 31 May 1996, 1266-68; Aris Persidis, "Enabling Technologies and the Business of Science,"

A Study in Strategic Innovation / 271

The pharmaceutical companies that were successful in the 1990s had already acknowledged this aspect of the current transition in the industry by seeking mergers, acquisitions, and strategic alliances.[64] Collaborations with small biotech companies were an extension of this basic strategy of innovation, an additional affirmation that their scientific and technological opportunities were expanding more rapidly than they had at any time in the industry's history.

In that sense, the division of labor in innovation that had begun to take shape in the 1970s appeared to be of lasting significance.[65] Some biotech firms were already attempting to go it alone, integrating downstream from research and development into manufacturing and marketing a limited number of biological products. Others were specializing even more in the application of new technology "platforms" in combinatorial chemistry or genomics, marketing their proprietary techniques to a stable of corporate partners interested in adding new approaches to their basic research and development capabilities.[66]

Bio/Technology 13 (Nov. 1995): 1172, 1174-76; Roger Longman, "Combinatorial Chemistry's Challenge," *Start-Up* 2:5 (1997): 28-34; and Kenneth G. Krul, "The Commercialization of High-Throughput Screening Technology," *Spectrum, Pharmaceutical Industry Dynamics,* 20 Mar. 1997, 1-16.

On genomics—which included gene therapy and diagnostics—see, for instance, *Science,* 22 Oct. 1993, 502-04. *Bio/Technology* 13 (June 1995): 534-35. Alan R. Williamson, Keith O. Elliston, and Jeffrey L. Sturchio, "The Merck Gene Index, A Public Resources for Genomics Research," *The Journal of NIH Research* 7 (Aug. 1995): 61-63. Rebecca S. Eisenberg, "Intellectual Property at the Public-Private Divide: The Case of Large-Scale cDNA Sequencing," *The University of Chicago Law School Roundtable* 3:2 (1996): 557-73. On the business developments flowing out of DNA sequencing, see John Carey et al., "The Gene Kings," *Business Week,* 8 May 1995, 72-78. *Wall Street Journal,* 28 Sept. 1994, B1; 10 July 1995, B8. *Merck World* (Feb. 1996).

[64] On the continued merger movement, see Claudio Casadio Tarabusi and Graham Vickery, "Globalization in the Pharmaceutical Industry, Part II," in *International Journal of Health Services* 28:2 (1998): 281-303; Karen J. Partridge and Michael A. M. Partridge, "The Year of the Megadeal: Analysis of 1994 Changes of Ownership and Alliances," *Spectrum* (1995); see also in the same issue Hemant K. Shah, "Horizontal Integration: Three Major Pharmaceutical Companies Pursue Comparable Business Strategies." One of the recent structural changes involved the merger of diversified companies, followed by the spinning off of the pharmaceutical business line. See Hassan Fattah, "A Shrinking Drugs Galaxy," *Chemical Week,* 10 Apr. 1996, 31-32. On Schering-Plough's approach to alliances, see *Financial Times,* 23 Mar. 1994, VI-VII.

[65] Economic conditions in the biotech industry during the 1990s continued to favor the large pharmaceutical firms seeking collaborative research or marketing agreements. Elisabeth Hefti, "Pharmaceutical Investment in Biotechnology: Trends, Issues, and Strategies," *Spectrum,* 30 Mar. 1995, 3-5. *Scrip,* 30 Sept. 1992, had reported on the "growing uncertainty" in biotech financial markets. By that date, total product sales had reached nearly $6 billion, but a survey of 1,231 companies indicated that on balance they had lost $3.4 billion in the previous year. *Chemical & Engineering News,* 5 Oct. 1992, 48.

[66] On combinatorial chemistry and genomics, see note 63 above. On recent developments in genomics involving the effort to sequence the entire human genome, see Eliot Marshall and Elizabeth Pennisi, "Hubris and the Human Genome," *Science,* 15 May 1998, 994-95.

Galambos and Sturchio / 274

Table 2

10 Top-Selling Prescription Drugs, Ranked by Worldwide Sales, 1997

Drug	Therapeutic Use and Firm
Prilosec	Antiulcer, Astra, Astra/Merck
Zantac	Antiulcer, Glaxo Wellcome
Zocor	Cholesterol-lowering, Merck
Renitec (Vasotec in the U.S.)	High blood-pressure lowering, Merck
Prozac	Antidepressant, Eli Lilly
Norvasc	High blood-pressure lowering, Pfizer
Ciproxin	Antibacterial, Bayer
Augmentin	Antibiotic, SmithKline Beecham
Voltaren	Antirheumatic, Novartis
Zovirax	Antiviral, Glaxo Wellcome

Source: *IMS World Review, 1997. Epogen*, a red blood cell booster (Amgen) was the highest ranked biotech product in global sales (13th), followed by Amgen's *Neupogen* (14th), an immunostimulant. Samantha S. Cheng, "Biopharmaceutical Sales and Forecasts to 2002," *Spectrum*, 9 Dec. 1997, 1-2.

By the 1990s, however, it was also evident that the division of labor was less significant than it had appeared to be a decade earlier. The dominant products in world markets were still those developed by the large firms in pharmaceuticals (see Table 2).[67] Moreover, the pharmaceutical organizations had—by acquiring biotech firms, in-house scientific personnel, and new research leaders who were effective both within and outside the organization—internalized significant capabilities in molecular genetics and rDNA.[68] One measure of the progress they had made was the success several leading firms achieved in developing new therapies for HIV/AIDS.[69] The most important of these

[67] See also Vivian Lee, "Lessons from the Latest Biotech Product Failures," *Spectrum Express*, 3 May 1997, 1-14. The author nevertheless estimated that "more than 100 new biopharmaceutical products will reach the market over the next several years."

[68] For different interpretations, see Martin Kenney, "Biotechnology and the Creation of a New Economic Space," in *Private Science*, ed. Thackray, 131-43; and William F. Hamilton, Joaquim Vila, and Mark D. Dibner, "Patterns of Strategic Choice in Emerging Firms: Positioning for Innovation in Biotechnology," *California Management Review* (Spring 1990): 73-86. See Gerard Fairtlough, "A Marriage of Large and Small: R & D for Healthcare Products," *Business Strategy Review* 7:2 (1996): 14-22, for an optimistic projection of small-firm potential.

[69] Another indicator of the change that was taking place was provided by the number of new patents issued in biotechnology. As early as 1986, large pharmaceutical companies were acquiring more such patents than were small biotech businesses. *Bio/Technology* 5 (Mar. 1987): 204. In 1992, the list of firms receiving U.S. patents was headed by Merck, followed by Abbott Laboratories, and Boehringer Mannheim. *Chemical Week* (10 Feb. 1993). While this was happening, new biotech company foundings dropped off and then increased sharply in the 1990s. Mark D. Dibner, "Tracking Trends in U.S. Biotechnology," *Bio/Technology* 9 (Dec. 1991): 1337. Elizabeth Hefti, "Pharmaceutical Investment in Biotechnology: Trends, Issues, and Strategies," *Spectrum*, 30 Mar. 1995, 3-5.

A Study in Strategic Innovation / 275

therapies are the protease inhibitors, three of which were approved by the U.S. FDA in late 1995 and early 1996.

To develop the HIV protease inhibitors, Merck, Hoffmann-La Roche, and Abbott had to have a blend of traditional and new skills in pharmaceutical research. Synthetic organic chemistry was important in developing molecules that would be effective against the virus' ability to replicate and also be orally bioavailable.[70] It was important too in devising production processes that would ensure a sufficient supply, first to conduct clinical trials and later to satisfy the demand for the new therapies. Molecular genetics and enzymology played crucial roles in analyzing the process of viral replication; crystallography was employed in determining how the viral protease enzyme might be inhibited. Recombinant DNA technology was used to clone the protease enzyme.[71]

Typical in several regards of the new pattern of R&D was the experience of the Merck Research Laboratories (MRL). While pursuing in-house research, the firm also worked with two biotech companies on alternative approaches to HIV prevention with a vaccine or treatment. One of the biotechs was Repligen, a Cambridge, Massachusetts, firm (founded in 1981) that specialized in efforts to develop treatments for cancer and inflammation, as well as AIDS. Later, Merck collaborated with MedImmune, Inc., a Maryland biotech, in an attempt to use that firm's monoclonal antibodies as a means of preventing HIV infection. The Merck/Repligen combination at first produced some promising results, but neither the vaccine research nor the explorations of monoclonal antibodies proved fruitful.[72] Meanwhile, MRL's in-house

---

[70] Michael L. West and David P. Fairlie, "Targeting HIV-1 Protease: A Test of Drug-Design Methodologies," *TiPS* 16 (Feb. 1995): 67-74. Agouron, a biotech, has developed a fourth protease inhibitor that the FDA licensed for marketing in the United States in 1997. Agouron's experience typifies the type of cooperative innovation process that has become common in this industry. In bringing Viracept (nelfinavir) to market, Agouron received substantial financial support from Japan Tobacco and regulatory and marketing support from Roche.

[71] Jonathan Greer, John W. Erickson, John J. Baldwin, and Michael D. Varney, "Application of the Three-Dimensional Structures of Protein Target Molecules in Structure-Based Drug Design," *Journal of Medicinal Chemistry*, 15 Apr. 1994, 1035-54. Noel A. Roberts and Sally Redshaw, "Discovery and Development of the HIV Proteinase Inhibitor Ro 31-8959," in *The Search for Antiviral Drugs*, eds. Vincent J. Merluzzi and Julian Adams (Boston, 1993), 129-51. Bennett M. Shapiro, "Biotechnology and Drug Discovery at Merck," address to the World Chemical Congress (San Diego, Calif., Sept. 1992). For a somewhat different interpretation of the relationship between biotechnology and the other biomedical sciences, see Pisano, *The Development Factory*, 69.

[72] *Bio/Technology* 5 (July 1987): 658; and (Nov. 1987): 1118. David T. Karzon, Dani P. Bolognesi, and Wayne C. Koff, "Development of a vaccine for the prevention of AIDS, a critical appraisal," *Vaccine* 10:14 (1992): 1039-52. Emilio A. Emini et al., "Antibody-Mediated

Galambos and Sturchio / 276

research was successful in developing a novel antiretroviral therapy, *Crixivan* (indinavir). This drug and the Roche and Abbott protease inhibitors were indicators of the degree of progress the leading pharmaceutical companies had made by the 1990s in developing capabilities in molecular genetics and rDNA technology.[73]

Reflections

The transition to biotechnology was, then, an historically unique situation—one in which there were two, very formidable, overlapping transitions taking place in the biomedical sciences and technologies. That situation, the early concerns about whether legal restrictions would prevent further research in biotechnology, and the "transition costs," help explain the pace with which large pharmaceutical firms responded to this new set of opportunities. Unquestionably, they were slow. And they left the window of opportunity open for small biotech startups.[74]

In the 1970s and 1980s, they paid a price for leaving the window of opportunity in biotech open as long as they did. A comprehensive cost/benefit, counterfactual analysis is beyond the objectives of this paper. But we can offer a rough estimate of the loss in revenues. In 1994, all kinds of biotherapeutics—including drugs, diagnostics, and delivery mechanisms—logged worldwide sales exceeding $8 billion. About half of these revenues went to large pharmaceutical companies under licensing agreements. If these same large firms had invested in molecular genetics and rDNA technology earlier, they would probably have been supplying virtually all of those markets. By that reckoning,

---

In Vitro Neutralization of Human Immunodeficiency Virus Type 1 Abolishes Infectivity for Chimpanzees," *Journal of Virology* 64 (Aug. 1990): 3674-78; Emilio A. Emini et al., "Prevention of HIV-1 Infection in Chimpanzees by GP120 V3 Domain-Specific Monoclonal Antibody," *Nature*, 20 Feb. 1992, 728-30; Jon Cohen, "Are researchers racing toward success, or crawling?" *Science*, 2 Sept. 1994, 1373.

[73] Jon Cohen, "Protease Inhibitors: A Tale of Two Companies," *Science*, 28 June 1996, 1882-83; Charles C. J. Carpenter et al., "Antiretroviral Therapy for HIV Infection in 1996," *JAMA*, 10 July 1996, 146-54; in the same issue see also Anthony S. Fauci, "AIDS in 1996," 155-56. Part of the information about Merck is drawn from a study in progress (by Louis Galambos and Jane Eliot Sewell) of the research that led to the development of *Crixivan*. By 1996, the world market for HIV therapies had reached $1.5 billion. Michael A. M. Partridge, "Combination Therapy Drives HIV Market Expansion," *Spectrum*, 5 Dec. 1997, 1-16. The U.S. FDA approved Agouron's *Viracept*, a nonpeptide protease inhibitor, in Mar. 1997.

[74] Our conclusions on this point support and elaborate those advanced in Lynne G. Zucker and Michael R. Darby, "Present at the Biotechnological Revolution," *Research Policy* 26 (1997): 441-443.

A Study in Strategic Innovation / 277

they lost about $3 to $4 billion in total revenue as a result of this transition to a new scientific and technological paradigm.[75]

They had not been slow because they were large and bureaucratic. After all, the firms that responded most effectively to the opportunities for innovation provided by microbial biochemistry and enzyme inhibition in the 1970s and 1980s were already very large and bureaucratic. On balance, companies such as Squibb, SmithKline, Glaxo, and Merck had done an excellent job of coping with that earlier transition.[76]

Moreover, once they began to read and respond to the biotech developments, most of the early movers were able to establish appropriate strategies and capabilities in the new science and technology fairly quickly. All utilized cooperative arrangements with biotech companies. Many of the pharmaceutical firms initially stressed a narrow range of therapies and built up specific internal capabilities as they went along—the Lilly and Merck strategies. Others quickly built more elaborate networks of contractual relations and sought through acquisition the generalized biotech capabilities they needed—the Hoffmann-La Roche strategy. Both strategies created significant problems for management and both involved substantial transition costs.

By the early 1990s, however, a number of the early movers had managed their way through the transition and established the general biotech resources they needed.[77] By the time that happened, several of the leading biotechs had collapsed or been acquired by other firms.[78]

---

[75] In a more formal analysis, this figure would be reduced by the revenue lost when large pharmaceutical firms shifted resources from the pre-biotech drugs in their pipelines. The companies would also have incurred more risk, esp. in the mid-1970s. See also Marilyn E. Gosse and Michael Manocchi, "The First Biopharmaceuticals Approved in the United States: 1980-1994," *Drug Information Journal* 30 (1996), 991-1001; and *New Biotechnology Drugs in Development: 1998 Survey* (Washington, D.C., 1998), lists the fifty-four drugs and vaccines cleared for U.S. marketing by the FDA through 1997.

[76] On this point, see Rebecca Henderson and Iain Cockburn, "Scale, Scope, and Spillovers: The Determinants of Research Productivity in Drug Discovery," *Rand Journal of Economics* 27 (Spring 1996): 32-59.

[77] By 1995, *Nature Biotechnology* estimated that pharmaceutical companies were spending about $3.5 billion to acquire biotechs, approximately $1.6 billion for R&D or licensing agreements with biotechs, and from $1.2 to $7.5 billion on in-house biotech R&D. Sylvia Davidson, "Hidden Biotechnology Worth over $7.5 Billion a Year," *Nature Biotechnology* 14 (May 1996): 564.

[78] Elizabeth Hefti, "Pharmaceutical Investment in Biotechnology: Trends, Issues, and Strategies," *Spectrum*, 30 Mar. 1995, 3-5. On the hard times, see: *Scrip*, 28 Sept. 1993, 17; Alice M. Sapienza, "The Dilemma of Biotechnology," *Drug Development Research* 29 (1993): 171-78; and Marion Leopold, "The Commercialization of Biotechnology: The Shifting Frontier," in *Biotechnology: R&D Trends*, ed. Tzotzos, 214-31. The U.S. market for biotech stocks did not begin to recover until 1995-1996. *Fortune*, 1 Apr. 1996, 40. See also Ann M. Thayer, "Market, Investor Attitudes Challenge Developers of Biopharmaceuticals," *Chemical & Engineering News*, 12 Aug. 1996, 13-21.

Galambos and Sturchio / 278

Similarly, some of the largest pharmaceutical firms in the industry during the 1960s and 1970s were no longer around to see the payoff on molecular genetics and biotechnology.[79] But those businesses able to navigate the strategic transitions first to biochemistry/enzymology and then to biotechnology and genomics were positioned to take advantage of what promised to be a very long cycle of innovation. "This is the transistor," one of the industry's leading scientists said, "not the light bulb."

By 1996, a new pattern of innovation had emerged in this industry. Biotech-pharmaceutical collaboration now played a significant role in that process and was likely to remain important for many years to come.[80] Given the improved biotech capabilities of the largest pharmaceutical manufacturers, however, it appears that they will continue to be dominant players in the industry's global markets. Their success with protease inhibitors for HIV/AIDS provides one indicator of how and why that will be the case. Unlike America's tire companies and large steel firms—the corporate dinosaurs of recent decades—the successful pharmaceutical companies have retained the ability to develop and implement new strategies of innovation suited to a rapidly changing scientific and technological environment.

[79] On some of the problems pharmaceutical firms were having, see Thomas M. Burton, "Bad Chemistry, Populism, Profit Slide Forced Bryson Out as CEO of Eli Lilly," *Wall Street Journal*, 28 June 1993. Lilly sold Hybritech at a great loss in 1995. *Wall Street Journal*, 2 Oct. 1995, B4. But soon, Lilly was back in the hunt, spending nearly $45 million under an agreement with Millennium Pharmaceuticals. *Bio/Technology* 13 (Nov. 1995): 1149. Clifford J. Levy, "Drug Maker Reshuffles at the Top," *New York Times*, 22 Sept. 1993, observed (quoting an analyst) in regard to Marion Merrell Dow: "Anything that could go wrong has continued to go wrong for this company." On Warner-Lambert, see *Business Week*, 9 Dec. 1991, 91, 94. On Upjohn, see *Wall Street Journal*, 29 Dec. 1994.

[80] The president of Hoffmann-La Roche International R&D, Jürgen Drews, M.D., recently predicted that the top fifty pharmaceutical companies would generate fewer "interesting compounds" than they needed in order to sustain their growth. The compound gap, Drews says, will have to be filled by the biotech firms. *The Impact of Cost Containment on Pharmaceutical Research and Development* (Surrey, U.K., Centre for Medicines Research, Tenth CMR Annual Lecture, June 1995). See also Jürgen Drews and Stefan Ryser, "Innovation Deficit in the Pharmaceutical Industy," *Drug Information Journal* 30 (1996): 97-108.

[15]

Available online at www.sciencedirect.com

SCIENCE @ DIRECT®

ELSEVIER

Research Policy 33 (2004) 1041–1057

research
policy

www.elsevier.com/locate/econbase

Understanding company R&D strategies in agro-biotechnology: trajectories and blind spots

Joanna Chataway [a,*], Joyce Tait [b], David Wield [c]

[a] *Development Policy and Practice, Technology Faculty, Open University, Walton Hall, Milton Keynes MK7 6AA, UK*
[b] *ESRC Centre for Social and Economic Research on Innovation in Genomics (INNOGEN),*
The University of Edinburgh, High School Yards, Edinburgh EH1 1LZ, UK
[c] *ESRC INNOGEN Centre, Development Policy and Practice, Technology Faculty, Open University, Walton Hall, Milton Keynes MK7 6AA, UK*

Received 30 May 2003; received in revised form 6 April 2004; accepted 21 April 2004

Available online 24 June 2004

Abstract

Companies innovating in agriculture-related biotechnology currently confront a complicated and controversial policy environment. Using analytical frameworks of technological trajectories and paradigms this paper examines R&D strategy in large companies. A large research project found that R&D related decisions taken by managers reflect company distinctiveness and can be characterised as cumulative in important respects. 'Economising' and 'strategising' strategies combine in different ways. But in many instances managers did not sufficiently recognise the importance of complex interactions between public policy and public opinion and failed adequately to incorporate public policy into strategic R&D decision making. This blind spot compounded initial difficulties in bringing products to market and has had significant impact on the rate and direction of innovation in this area, including contributing to the demise of the idea of an integration of agro and health sectors based on life sciences. The paper raises important questions about the relationship between technological paradigms, industry and sector level trajectories and decision making in firms.
© 2004 Elsevier B.V. All rights reserved.

Keywords: Biotechnology; R&D strategy; Public policy; Innovation management; Politics of innovation

1. Introduction

During the 1990s large companies developing new agro-biotechnology products were faced with difficult choices about how to develop new technology over the medium and long term whilst catering to the demands of shareholders in the shorter term. A study of large agro-biotechnology companies, the Policy Influences on Technology for Agriculture (PITA) study,[1] undertook a comprehensive review of how major companies were responding to the challenge of reorienting their R&D strategies to incorporate the new technologies and in particular how policy environments impacted on their strategies. The study aimed to integrate analy-

* Corresponding author. Tel.: +44 1908 652103;
fax: +44 1908 654825.
E-mail address: j.c.chataway@open.ac.uk (J. Chataway).

[1] Policy Influences on Technology for Agriculture was funded under the EU TSER programme (European Commission, Project No. PL97/1280) and involved 8 European partners in five different countries. Reports from the study can be found at http://www.technology.open.ac.uk/cts/projects.htm#biotechnology.

0048-7333/$ – see front matter © 2004 Elsevier B.V. All rights reserved.
doi:10.1016/j.respol.2004.04.004

1042   J. Chataway et al./Research Policy 33 (2004) 1041–1057

sis of policy, governance and regulation and company strategy. Rather than in-depth breakdown of any one of these areas, the value of the work undertaken lies in its investigation of the dynamics between them. This paper follows that approach.

Section 2 of the paper discusses company approaches to biotechnology. Companies positioned themselves differently with regard to the new technology and adopted a range of R&D strategies for changing the technological base of their businesses. These differences relate both to companies' distinctive technological trajectories and organisational features and to the understanding managers had of how best to build on company strengths. Companies combined 'economizing' and 'strategizing' (Teece et al., 1997) approaches but with different emphasis. Yet, despite the differences in individual companies there were some striking similarities. Whilst most initially embraced it, for large chemical companies in the study the dream of a life science that spanned pharmaceutical and agricultural sectors was short lived. The reasons for the collapse of structures that promised to combine sectors are discussed later in this section.

Chemical and biotechnology based agricultural technology is highly regulated and controversial. In Section 9, we contend that the very complicated politics surrounding biotechnology and the highly volatile policy process were initially given relatively little attention by managers. Although companies did have a relatively coherent stance about risk regulation, their position often seemed to ignore the political complexities associated with the discussions of risk and risk regulation. Kaplan et al. (2003) have shown that senior managers' sense-making of strategic options has been important in pharmaceutical firms' response to biotechnology. Our data in turn suggest that managers seemed to have blind spots about politics and policy and the interplay between the two, which are to some extent reflected in conceptualisations of technical change. We argue that public policy and political factors are likely to become more important to innovating firms and that managers will increasingly have to incorporate them in their innovation plans, as will those who research and conceptualise innovation management.

The data and analytical framework presented here raise some important questions about commonly held assumptions amongst innovation theorists. The evi-

dence as presented here suggests that the relationship between technological paradigms on the one hand and firm and sector level technological trajectories on the other is complex and may not be as entirely distinct as a Dosian analysis might suggest (Dosi, 1988). For example, we argue that strategic R&D choices taken by Monsanto have had a profound impact on the rate and direction of broader scientific progress and technological innovation. Thus whilst it seemed clear to many in the late 1980s and early to mid 1990s that a new technological paradigm was being established and firms seemed to be moving into a new life sciences based era, the picture by the late 1990s and early 2000 looked quite different.

This analysis differs from the approach taken in much innovation research and teaching. For example, Kaplan et al. (2003) follow a widely accepted tradition, particularly dominant in the US perhaps, of taking technological trajectories as a given. The key strategic issue is one of agency and how senior managers deal with changing external and 'objective' environments. Our analysis, however, highlights the way in which industry and sector level trajectories are influenced by firm level decision making. A central question then emerges: Given that trajectories are not fixed how should firms move along them?

From strategic, policy and innovation perspectives, a number of issues related to this central question are clearly important. We hope that what follows in this paper shows that assuming that trajectories are fixed (as many managers seemed to do with agri-biotechnology) can lead to blind spots and mistakes. Although psychologists have often made use of the concept of blind spots to our knowledge it has not been used in management or innovation literature.

This paper also highlights the fact that strategic thinking does not amount to one vision per firm. Firms differ in how they deal with external environments and different constituencies within firms will have different views and take different approaches. Strategic thinking often involves the process of promoting one view and suppressing others. This paper highlights the importance of the management processes and structures involved in evaluating different perspectives within firms and aligning them and evaluating them in the light of complex political as well as commercial, economic and technological realities and dynamics.

J. Chataway et al./Research Policy 33 (2004) 1041–1057   1043

2. Company approaches to agro-biotechnology

As the agro-chemical sector became increasingly mature, multinational corporations (MNCs) searched for a new R&D trajectory. As a senior manager in a large agro-chemical company put it in the mid 1980s 'We are overdue for another big splash of revolutionary change.... Our creativity needs new outlets.... Biotechnology will drive us into the next golden era" (Fernandez, 1985). In 1998 and 1999, we interviewed managers[2] in 14 chemical and seeds MNCs as part of the PITA project.[3] For all the major multinational companies interviewed, strategic R&D planning involved the construction of new-agro-biotechnology based trajectories[4] – a combination of chemical and biotechnology developments with varying degrees of synergistic interaction. Companies increased their technological diversity (Granstrand et al., 1997) as a means to introduce improved products and new products. None of the companies envisioned a future which did not include biotechnology. Moreover, in all cases companies defined themselves as moving towards products based on biotechnology, not merely using biotechnology to develop better chemicals. A new technological paradigm was seemingly in the making. We identify three distinct strategies amongst the companies examined:

- Monsanto (from a very early date) and DuPont (later with its take-over of Pioneer) invested major amounts of shareholder funds in acquisitions – a strategy described by other companies as 'buying the channel to market' as well as investing heavily in building up their technology base.

- Other companies (AgrEvo, Zeneca, Novartis, Rhone Poulenc, Dow) tried to capture value in a different way – they invested a great deal of money in the technology and made some acquisitions to give them a reasonable 'route to market' but not as far downstream as Monsanto and DuPont.

- BASF and Bayer were late starters at the time with different strategies for patents, technology and route to market and looked for benefits from their agro-chemical businesses to help them to buy into biotechnology bypassing the earlier innovation phase of other companies.

Many managers thought that of the top ten agro-chemical companies, four to six of them would eventually emerge as leading biotechnology companies in the future.

We look at three companies in some detail in this paper.[5] From the first group of companies, we look at Monsanto. At the time of the research, Monsanto was widely considered to be at the forefront of developments in the new technology. Towards the end of 1999 when the fieldwork was ending, Monsanto was the centre of the storm of negative public opinion about GM crops. The company, whose share price had fallen dramatically, was sold to Pharmacia, a European pharmaceutical company that shortly announced its intention to sell the agro-chemical division but was unable to find a buyer. From the second group we look at Novartis and Zeneca who merged towards the end of the fieldwork period in 2000. Here, we look at the companies separately.

---

[2] Managers interviewed included middle and senior ranking R&D, marketing, divisional and regulatory managers.

[3] Separate studies were made of the following companies: Advanta, AgrEvo, BASF, Bayer, Cebeco, Danisco, KWS, Limagrain, Monsanto, Novartis, Pioneer, Rhone-Poulenc Agro, Seminis, Zeneca. The ten leading chemical MNCs were subsequently reduced to seven including four with a turnover of more than US$3 billion: AgrEvo and Rhone Poulenc merged to form Aventis Crop Science; BASF acquired American Cyanamid; and the agrochemical divisions of Novartis and Zeneca merged to form Syngenta.

[4] The concepts of technological trajectories and paradigms are widely used in innovation theory literature and relate back to conceptual frameworks developed by Dosi and colleagues (Dosi, 1988). A good early account of these concepts in relation to biotechnology is developed by Orsenigo (1989).

[5] We do not look in detail here at companies from the third category (BASF and Bayer). One reason is that access to these companies proved more difficult than with companies in the first two categories. To give a flavour of one company, BASF, it was a latecomer to biotechnology. Until 1998, it continued to invest strongly in agrichemicals, with massive economies of scale and scope through its Verbund system of efficient backwards integration. BASF described itself as a 'fast follower' in 1998 when it moved into plant biotechnology, committing itself to a focus on agbiotechnology. Its biotech R&D has grown rapidly, with commitment to 'leapfrog' into 'second' and 'third' generation products, for example to specialty products such as vitamins, fatty acids and enzymes. Its strategy involves large investments over 10 years from 2000, including significant joint ventures with university departments and public research institutes (Wield, 2001). BASF did not want their strategies followed in too much detail, it seems, since it was not prepared to be interviewed.

Companies were changing R&D strategies in complex market environments and managers were aware of the need to balance the adoption of new technologies and maintaining viable business profiles. As one Novartis document described it:

"The agribusiness industry is in a state of upheaval and rapid change. Low farm commodity prices and depressed farm income have impacted on sales. Margins have eroded, putting pressure on financial results and the distribution channels. Restructuring in the agribusiness industry has created a more aggressive competitive environment. New technologies, including genetically modified crops and precision agriculture, are challenging traditional farming practices. Moreover, farmers and growers are increasingly influenced by other players in the food chain, from food and feed processors and food companies right down to supermarkets and consumers".[6]

External conditions meant that companies had little alternative but to change. What is interesting is that each company had a unique approach relating to its own organisational, strategic and technological distinctiveness.

3. Monsanto

Monsanto was the early leader in developing agricultural applications of biotechnology. The company threw itself wholeheartedly behind the technology making very significant investments in R&D facilities, biotechnology based product development and acquisitions. The scale and nature of Monsanto's investment is detailed in Chataway and Tait (2000) and Lemarie (2001).

It was the unequivocal leader in input trait[7] led crop development such as herbicide resistant maize and soybean and Bt cotton. The company was also focused on second generation 'output' traits. These

could include soybean, rapeseed oils and plant derivatives that are good for heart and blood vessels but these would not be on the market in the short term. During the 1990s the company defined itself as a 'technology provider'. As one manager put it 'We have an almost unshakeable faith in our technology'. Managers interviewed stressed that technological leadership was the cornerstone of strategy.

The company had a marketing strategy that reflected its technological vision, licensing its technology to seed companies and farmers. Designing products to appeal to farmers was identified by managers in interviews as the strategy most likely to yield a better return for shareholders.

Decisions to pursue the new technology were very much related to Monsanto's particular technological and product history. The company had a very narrow technological base with a large proportion of its profit coming from its glyphosate herbicide, Round-Up. The R&D strategy of developing GM crops resistant to glyphosate, and also developing GM induced insect resistance based on Bacillus Thurigeinisis (Bt) genes fitted well with a company that had significant herbicide market share but little presence in insecticides. However, Monsanto's first biotechnology related product was Bovine Somatotrophin (BST) designed to increase milk yields in diary cattle. This technology related to the company's previous work in animal productivity enhancers and somatotrophins so that a great deal of accumulated knowledge laid the foundation for the work (Chataway, 1992). The product was a technological success but Monsanto's efforts to introduce it in Europe coincided with a period of milk surpluses, and the product was widely seen as redundant and unpopular.

Monsanto's vision of being a technological leader was also reflected in its acquisitions strategy. The company spent lavishly on acquisitions, mostly of seeds and biotechnology companies. In 1998 alone, "Monsanto acquired DEKALB Genetics Corp., Cargills international seed business, and Plant Breeding International Cambridge Ltd (PBIC), and entered into an agreement to merge with Delta and Pine Land Co. Monsanto's total investment in 1998 was more than $4 billion" (Monsanto Annual Report, 1998, p. 11).

In total from 1996 to 1999, Monsanto spent over US$ 8 billion in acquisitions (Fortune, 29.03.99). Monsanto had pursued an ambitious acquisitions

---

[6] http://www.seeds.novartis.com.news_article   Backgrounder: New Agribusiness Strategy (Basel, 9/15/99).

[7] Input traits such as herbicide, insect pest and disease resistance aimed increasing production efficiency were based on relatively simple technology which had been the subject of scientific investigation for a considerable time and could be implemented fairly rapidly.

J. Chataway et al. / Research Policy 33 (2004) 1041–1057   1045

strategy, in some cases paying top prices. Monsanto had a distinctive method for judging a company's value to its business. The distinguishing characteristic of this method was that the value of potential acquisitions was measured by the perceived contribution it could make to Monsanto's overall strategy rather than its commercial value. A manager explained it in the following terms:

> "When we valued PBI we actually valued it... from a number of criteria – on its current business, what the current business was worth, over a period of time, using net present value classic methodologies to value businesses – so we decided how much it was worth on that basis, that was current business. We looked at what additional value having access to PBI and all its assets would bring to our high-bred business – so there's value there. We looked at... how its influence on our highbred business would bring value to our biotechnology traits... and then we also looked at PBI in terms of what it gave us, from the point of view of what we call option value – option value is a term or concept which comes mainly from the oil industry... its used where, if you own a particular asset, you then have an option to invest further in that asset to gain extra value. So there was an option value associated with PBI as well... (Monsanto senior manager)".

Some managers within Monsanto strongly disagreed with this way of valuing companies and took the view that Monsanto had acquired too much, too fast on the basis of projected gains from one technology. An important aspect of Monsanto's acquisition strategy, according to one manager, was that it reflected the company's belief in technology. This technology-led ethos ran deep in the company. It was an overall goal of the company to pursue a technology-led strategy. Whilst Monsanto employed a sophisticated decision making tool called Strategic Decision Analysis (SDA), the outcomes of this process were often overridden. Monsanto's aim was to stay a world leader on the technology front not necessarily through making strictly commercially-based decisions. The discrepancy between the analysis-based and vision-based decision approaches was particularly apparent over the issue of acquisitions. In notable cases, SDA would indicate that acquisition might not be the best strategy. Still, Monsanto decided to

go ahead with acquisitions. This reflected the overall vision which drove the company and which was fostered by the CEO. It meant that details were not always fully taken into account and that sometimes "sub-optimal decisions" were made. There were clear tensions between maintaining technology at the centre of strategy and making short-term financial decisions.

In interviews about R&D strategy, Monsanto managers stressed the importance of positioning in terms of technology and markets. Although some were eloquent on the ways in which Monsanto's organisational culture and pathways had led to success they were less inclined to see organisational processes and pathways as key to strategic success. One manager likened the dominant organisational style to a shoal of fish with strong leaders (who could emerge throughout the ranks of the company) providing vision and leadership. In later interviews, however, some saw that the way in which Monsanto had organised itself and its organisational culture had contributed to the company's problems in that checks and balances were lacking.

Monsanto's strategy can thus be understood in terms of previous capabilities and core strengths. Monsanto's radical push in new biotechnology was based on its desire to transcend its weaknesses and narrow technological base but the importance of the vision driving strategy was paramount.

4. Novartis

Another major player in both agro-chemicals and the new biotechnologies, Novartis, had a different technological and industrial trajectory to Monsanto. Its strategy was based on integrating new developments into a broadly based core business, strong both in crop protection and seeds. Novartis was formed in 1996 from a merger of Sandoz and Ciba-Geigy creating what were at the time the world's largest agrochemical company and second largest seed company.[8] The merger was seen as a strategic focus on life sciences.[9] Ciba-Geigy and Sandoz both had their ori-

---

[8] These positions have been altered by subsequent mergers and acquisitions among the multinational companies active in these areas.

[9] The merger involved rationalising and integrating the Group's Global Operations including the disposal of a substantial portion of the former Sandoz US and Canada Agribusiness activities and

1046   *J. Chataway et al. / Research Policy 33 (2004) 1041–1057*

gins as producers of industrial chemicals. Ciba-Geigy and its precursors began pesticide production in 1935 and bought seed companies in the 1970s and 1980s. It established a special biotechnology unit in 1980 and was the first company to market genetically engineered hybrid corn seed in 1995. Sandoz began pesticide production in 1939 and also bought seed companies in the 1970s and 1980s. It began to invest heavily in plant biotechnology in 1990. With the 1996, merger, therefore, Novartis was constituted with a broad commercial and technological reach. In its first year of existence, Novartis as a whole achieved an increase in sales of 19% on a comparable basis. The business portfolio was successfully streamlined and focused on Life Sciences.[10]

In 1997, Novartis completed the US$ 910 million acquisition of the crop protection business of Merck & Co. Inc. In the agribusiness area, this strengthened its profile in the high value acaricide/insecticide and fungicide markets.[11] In the same year, Novartis announced the sale of its world-wide spray-on Bacillus thuringienisis. The spray-based version of Bacillus thuringiensis is one of the few insecticidal products available to organic farmers. Nevertheless, the press release emphasised Novartis' continued commitment to integrated pest management (IPM), based on the optimal combination of cultural, chemical and biological pest control management, and refers to the insecticides abamectin, emamectin and pymetrozine (the first two being part of the Merck acquisition), giving them a unique portfolio well suited to a sustainable agriculture.

1999 was described by Novartis as the year in which it took further steps to focus its business portfolio, 'moving from a Life Sciences company to a pure Healthcare company'.[12] In December, the Boards of Novartis and AstraZeneca agreed to spin off and merge Novartis Agribusiness (crop protection and Seeds Sectors) with Zeneca Agrochemicals to create the world's first dedicated agribusiness company (Syngenta) with

pro-forma combined sales of approximately US$ US 7.9 billion (1998 figures).

To deal with a fast changing technological and economic environment, in June 1999, the Novartis Agribusiness Sector initiated the New Agribusiness Strategy (Project Focus), with the objectives of growth, fitness and sustainable leadership, to provide a framework for the future of the Seeds and Crop Protection sectors. The project was designed to strengthen leadership and cost savings by implementing new product priorities, improving the product mix, optimising asset utilisation, reducing purchasing costs and also by eliminating approximately 1100 jobs world wide. Research projects were prioritised to focus on the most promising products and to reflect the growing importance of new technologies and output traits in the face of new market dynamics characterised by lower agricultural subsidies and commodity prices.[13]

The new course charted by the Agribusiness Strategy focused on:

- the food and feed chain and crop solutions;
- integrating breakthrough technologies;
- exploiting joint synergies between Crop Protection and Seeds.

Global Crop Teams, operating jointly between Crop Protection and Seeds, were formed to define overall strategies for 'pillar crops' (corn, vegetables, cereals and rice) and 'important crops' (oilseeds, sugar beet, cotton, fruits and grapes).

The key components of the agribusiness strategy, including Global Crop Teams, brought together staff from agrochemicals and biotechnology areas with the aim that they would rapidly become indistinguishable in their approach despite their different backgrounds (Tait and Chataway, 2000a). Novartis agribusiness strategy was based on a structuring of the world industry players into two clusters of companies:

- those with both a strong and viable seed base and significant crop protection activities; and
- the 'pure players', mainly in crop protection, along with a small number of players who are still purely seeds based.

Novartis was in the first category and saw itself as being in a very favourable position to gain lever-

---

the demerger of the speciality chemicals division of Ciba and construction chemicals of Sandoz.

[10] Novartis Financial Review 1997 and Novartis Operational Review 1997.

[11] http://www.seeds.novartis.com.news_article   Backgrounder: New Agribusiness Strategy (Basel, 9/15/99).

[12] Novartis Operational Review, 2000.

[13] http://www.cp.novartis.com/e1e14_con.htm.

J. Chataway et al. / Research Policy 33 (2004) 1041–1057   1047

age from chemical/biotechnology synergies, but felt it was not achieving this as effectively as some of its competitors, particularly Monsanto. The 'pillar crop' strategies arose out of this analysis.

The need to maintain a flexible approach to the direction of the agribusiness strategy in the future was partly a result of the uncertain policy and political environment for GM crops in Europe. As a result, the focus on input characteristics would be extended gradually and opportunistically to genomics, the development of marker systems and output traits.

The main emphasis of the agribusiness strategy then was on long-term balance and on keeping the core business running while managing new developments. The overall aim for each crop and region was for a low impact, high output agriculture, considering in a broad way which mix of technologies could achieve this most effectively. Chemical and GM strategies were thus considered in parallel, in the full realisation that some GM strategies would be a threat to chemical-based strategies and others would create opportunities. This aspect of the strategy was still causing difficulties for some staff who saw themselves as 'chemists' or 'biologists' and who were in the process of making a transition to a more holistic perspective.

Novartis managers talked a great deal about organisational processes and pathways in interviews. Novartis' ability to manage knowledge flows and interdisciplinary work was seen as a fundamental strength.

5. Zeneca

Zeneca Plant Sciences was formed in 1993 as part of the de-merger of ICI (a major chemicals conglomerate) and included ICI seeds. At the time of the interviews, Zeneca was the third largest producer of agrochemicals, after Novartis and Monsanto.

Like Novartis, Zeneca saw good opportunities for new chemical based technology in future markets, for example its fungicide product Amistar has a superior environmental and toxicological profile compared to competing products, is applied at lower rates and has less impact on the aqueous environment and non-target species. The company's emphasis was on building a range of technological options around key crops and remaining flexible. As managers noted,

growth in value has to come from innovation and if a company had new and better technology there would be no problem in getting market share. On the other hand, although new products are not necessarily more expensive to begin with, older competing products tend to become cheaper when faced with improved technology and this could cause problems for all pesticide companies in their attempts to market products with a better environmental and toxicological profile. On the whole, where a product had environmental advantages, Zeneca considered that the policy and regulatory environment was more supportive than before in most countries.

The question of how to continue to extract value from off-patent products is a major issue for multinational companies. Zeneca had scrutinised its product portfolio for chemicals with a poor 'return on net assets' and had made major savings by withdrawing altogether from insufficiently profitable markets or by sales of some product rights to other companies. By 2003, 85% of the agrochemicals business was expected to be in approximately 12 products, compared to 50% in 40–42 products in 1994/5. Support was only given to off-patent products if they fitted into the overall portfolio at certain stages of development.

Zeneca thought in terms of waves of product development from biotechnology, 5–10 years from now and 10–15 years from now. A portfolio of products was in the final stages of research and still needed a significant amount of development to be launched beyond 2005. A major planning exercise was under way looking at the value of the biotechnology business over the next 20 years and the necessary time scale for investment, and this was expected to lead to an increased Zeneca commitment to biotechnology.

Zeneca had invested in some seed companies in order to acquire the germplasm base for GM crop development, some gene effects and patent rights, and hence the access to market. However the price of these investments had been relatively lower than investments made by some other companies such as Monsanto because of the timing of the purchases. It was more characteristic of Zeneca's approach to seek collaborations with other companies, as had been done in the development of the GM tomato paste product, protecting the technology base and the channel to market by means of legal contracts with other players.

In developing GM crops, managers were mainly focusing on output rather than input characteristics, e.g. nutritional characteristics of cereal products and incorporation of effects beneficial to health. Input and output-related crop characteristics were seen as different businesses, the former having the grower as the customer while the latter went right along the market chain. However, on crops with output traits, farmers would also want crop protection elements, either genetic or chemical.

Output traits change commodity crops into specialities which can then be segmented in the market place, providing an 'engine of value'. The company would then get its rewards in two ways: for the trait itself and also for protecting a more expensive crop. Farmers were expected to prefer to have this combined in a single package so that output traits will be complemented by both agrochemicals and input traits, hence the combined agro-biotechnology strategies noted below.

Zeneca's delayed ripening GM tomato was a good example of a set of development decisions where the consumer did see a benefit. The product was labelled as GM and people did buy it. In developing this product, Zeneca learned a lot about working with food companies, retailers, consumer and grower groups, and what needed to be done to launch and get acceptance for a GM product. The same degree of close attention was not given by other companies, for example to the launch of soya bean in Europe, and Zeneca's tomato product has suffered as a result (Tait and Chataway, 2000b).

For other GM products, the benefits will be mainly agronomic, for example animal feed tailored to meet the nutritional needs of different species, or 'low phytate' products with the ability to reduce the environmental impact of the feed by reducing the phosphate content of effluent from intensive animal husbandry.

Zeneca considered the development of 'terminator' technology possible but it would be expensive and, given public worries and third world issues, it was not considered a technology worth pursuing.[14] Hybrid

seed would give similar benefits and some crops such as soya and maize are currently grown as hybrids because the yield and quality are better. However the same is not yet true of cereals in Europe.

Zeneca also had projects based on chemical switching to regulate the terminal output of the genes. For example, in anti-sprouting potatoes a 'switch' placed in front of the gene is regulated chemically or by some other means. For ware potatoes, the gene remains switched off; for seed potatoes the sprouting gene can be switched on when needed. This technology seemed to managers to be logical, legitimate and defensible but it also had connotations of attempting to control farmers' use of the crop. As a second example of the use of switching technology, on Bt cotton, rather than creating refugia to prevent the spread of insect resistance to Bt, the gene could be part of a controllable system, with a chemical spray used to activate the gene to give better resistance management.

6. Combined agro-biotechnology strategies

Strategic planning of research and development in all the nine multinational agrochemical companies surveyed combined both chemical and biotechnology developments with varying degrees of synergistic interaction. No company's senior management envisaged a future without biotechnology, including *products* based on biotechnology as well as using biotechnology to develop better chemicals. The predicted overall value of the agrochemical/biotechnology sector was US$ 75 billion by 2020, compared to the current agrochemicals world market of US$ 30 billion. The exact way in which companies went about integrating new and old technologies depended on their distinctive sets of perspective and experience.

For Monsanto, development of products based on biotechnology was most important. Whilst Monsanto continued to innovate in chemicals and improve them with biotechnology techniques, they were primarily oriented towards biotechnology-based products. Zeneca described the strategies of Monsanto and Du Pont as 'buying the channel to market', investing major amounts of shareholder funds in acquisitions as well as investing heavily in building up their technology base. Its own approach on the other hand aimed to capture value by investing in the technology base,

---

[14] This is unlike some other companies. Monsanto for example did pursue this technology in large part because it offered potential IPR solutions and abandoned it later largely because of pressure from NGOs and influential policy groups.

J. Chataway et al./Research Policy 33 (2004) 1041–1057   1049

making more targeted acquisitions to give a reasonable route to market but relying more on partnering than acquisitions for downstream developments of the technology. Novartis also aimed to balance the development of chemical and biotechnology products. However, Novartis' aim was to innovate both in products based on biotechnology and chemicals in a balanced fashion according to commercial and technological criteria.

Zeneca in making decisions about GM crop development selected at two levels of detail. Some of the more interesting and creative thinking was done at the broader level, by global teams, looking at where a crop as a whole is going, what the outputs are used for, how farming is changing, what are the trends in the crop world wide, what competitor companies are developing. A different group then considered more detailed targets – candidate products, benefits, cost effectiveness and the crop/pesticide/GM mix for a range of markets.

The important long term question for managers in this area was, "How will mainstream food be produced?"

- using pesticides, along with non-GM crops (this seems to be the current publicly acceptable approach);
- using pesticides along with GM crops (prognosis unknown); or
- organic crops (without pesticides or GM crops).

Subsidiary questions then were:

- how many of these sectors are there and how big are they;
- what types of GM are acceptable (e.g. extra carotenoids in food);
- does the company have a product, process or capability which provides a recognised benefit.

7. Technological trajectories and understanding the difference between companies

So far, a number of issues are apparent. PITA results strongly support innovation theories that hold that change is a cumulative activity and what companies have done in the past heavily impacts of what they do in the future (Dosi, 1988; Teece et al., 1997).

Even in the context of what Monsanto managers perceived to be radical change, there were strong links to technological and product strategies of the past. In fundamental ways, companies were both limited and influenced in their strategic options by what they had done in the past. Thus, even where companies perceived developments to be fundamental in the technological paradigms in which they worked, change was cumulative and incremental in important respects. For example, Monsanto's radical technology- and vision-led strategy had its roots partly in its previous technology and product base.

The PITA research therefore supports the notion that a firm's resources influence patterns of innovation in that firm. In a 1997 paper, Teece et al. developed a notion of 'dynamic capabilities' as a firm's ability to 'integrate, build and reconfigure' internal and external routines associated with a firm's distinctive activities (Teece et al., 1997). Teece et al. term resource-based and dynamic capabilities based perspectives as economising strategies. Two other schools of thought attenuate competitive forces and strategic conflict, and Teece et al. call these strategizing approaches.

All the companies in this study combine strategizing and economising strategies but with very different emphasis and awareness of the importance of the various elements of strategy. Monsanto's vision and technology led approach whilst related to an assessment of its resources clearly owes a great deal to strategizing approaches[15,16] and relatively less attention paid to its resource and organisational base. For example, Monsanto's determination to rapidly develop new technology based products left it with some key partnerships that became seriously problematic. One of Monsanto's key partners in developing GM crops was Delta and Pine Land. The collaboration set up with Delta and Pine Land ran into problems as Monsanto became increasingly determined to own the company outright. Acrimonious acquisition negotiations left Monsanto unable to acquire the company. Whilst

---

[15] Indeed it is probably the case that strategies based on radical technology based vision may always have more in common with strategizing routes.

[16] John Kay has commented on the threats to its existence faced by a company drive 'not by an assessment of what it is but a vision of what it would like to be' (Kay, 1993).

Monsanto blamed the US Justice Department for the failed deal, many thought the company's burgeoning debt load and increasingly troubled outlook were to blame. In the aftermath of the failed acquisition attempt, Delta and Pine Land sued Monsanto for breach of agreement.

According to some Monsanto managers the partnership between the companies was not well thought through or developed. A senior manager talking about the relationship with Delta and Pine Land noted:

"... an immense amount of the phenomenal relationship- we had with Delta Pine... was down to the personality and integrity of the individuals involved... every time the contract, or whatever it was... was reached for because there was an impasse or a problem, in our working together, this [key] individual would be summonsed to deal with it, and in the end of course, that's unsustainable for a big corporation..."

Another senior manager speaking when it looked as though the deal might be successfully resolved, disagreed with the assessment that Monsanto needed to acquire Delta Pine Land at the time an offer was being made and said:

"I really don't see why we had to buy it... when I think back to 1993 or something we could have paid $30 million for it and when we bought it we paid $1.8 billion or something like that... we ended up... paying back every penny we made".

One factor that PITA researchers observed was the difference in approach taken by US and European firms in relation to more incremental approaches versus large scale rapid change. Teece et al. propose that US companies are often more prone to exaggerate the importance of strategizing rather than economising perspectives in their approaches to R&D strategy.

"... we speculate that the dominance of competitive forces and the strategic conflict approaches in the United States may have something to do with observed differences in strategic approaches adopted by some U.S. and some foreign firms" (Teece et al., 1997, p. 529).

Teece et al. refer to a study carried out by Hayes "which noted that American companies tend to favour 'strategic leaps' while, in contrast, Japanese and German companies tend to favour incremental, but rapid, improvements" (Hayes, 1985, p. 529). It seems likely that cultural patterns might account for some of the differences between firms here[17] and might indicate that it is not simply coincidence that the two firms which have taken the most obvious 'strategic leaps' into biotechnology in this study were US based firms. Additionally, managers in Zeneca and Novartis appeared to us to be more aware of the importance of more immediate organisational opportunities and constraints and less driven by a fixed vision of the future.

The European firms differed in their strategies (and we have already explained some of this difference in terms of trajectories) but it seems clear in the case of the companies examined here that the European firms placed more emphasis on trying to build on accumulated and distinctive organisational difference. Technological trajectories and organisational pathways intrinsically had a major influence over what companies did and awareness of the importance of distinctiveness of technology and organisational patterns made a difference in the way managers approached change.

Despite the differences in R&D strategies, one thing that was common to many of the MNCs looked at in the study was the failure of an initial view of the synergies which drive the organisation of agro-chemical and pharmaceutical companies. This we believe is connected with the cross-company failures to deal with complex policy environments and political realities.

8. The Life Sciences experiment

One key aspect of life sciences innovation during the 1980s and 1990s was the move to integrate pharmaceutical and agro-biotechnology divisions of multinational companies. The hope was that synergies could be built on the basis of the new technology. However, by the late 1990s and early 2000s, most companies had again split their pharmaceutical and agro-chemical divisions. Some writers viewed this experiment about the potential for synergies between agro-biotechnology and pharmaceutical innovations

---

[17] We could also speculate that European firms were more influenced by EU emphasis on the precautionary principle, sustainability and environmental protection.

J. Chataway et al./Research Policy 33 (2004) 1041–1057   1051

with scepticism and noted amongst other things the wide disparities in profit margins (Niiler, 2000). An analysis of longer term shifts by Walsh and Lodorfos (2002) emphasises the importance of technological and economic differences between the two areas. Referring to unpublished work done by Tampubolon they say

"...there is a significant knowledge gap, for example, between what is known about the human genome compared with, say, the rice or wheat genome; the pathways are better understood in drug than agro-food biotechnology; multiple trait genetic modification is still a major problem; and ecological effects and genetic crossing are far from well understood" (Walsh and Lodorfos, 2002, p. 292).

Whilst some of these factors are undoubtedly important, our findings painted a slightly different picture of both the synergies and the reasons for subsequent separations (Tait et al., 2002). Our findings implied a change in, rather than total demise of, the life science concept. Companies were not envisioning a future without biotechnology and, although as we have noted companies had chosen different strategic routes, they expected products to be based on biotechnology and genomics rather than simply using biotech to improve chemical products. Managers, particularly from companies such as Zeneca and Monsanto who were working on combined chemical and biotechnology based product development strategies confirmed the importance of synergies between the two areas.

Early interpretations of the term 'life sciences' assumed that, by using biotechnology to gain a better understanding of the functioning of cells across a wide spectrum of species, there would be useful cross-fertilisation of ideas between the development of new drugs and new crop protection products.

Managers referred to the problem of slow growth in pesticide markets, alongside the need to continue to invest in research to sustain their position in pesticides and also to invest in biotechnology, which would not generate returns for 10 years. Taking a 'life sciences portfolio approach', with its economies of scope as well as scale, and investing R&D money into new technology across the combined pharmaceutical and agrochemical portfolio as part of regeneration of the business, seemed to be one way to resolve this dilemma. This was seen as part of a give-and-take relationship between divisions, with the flow mainly, but not entirely, from pharmaceuticals to plant sciences.

The vision here was of synergy at 'discovery' level, where a better understanding of genomics and cell processes, made possible by fundamental knowledge gained in life sciences, can lead to new products: drugs, pesticides, GM crops and genetic treatments for disease. At this stage, the researchers often do not have any particular application in view and a close association between agrochemicals/seeds and pharmaceutical divisions can allow the company to extract more value from the original research than would otherwise be the case.

However, the original conception of a life science based industry structure is now being reinterpreted. The discovery-level synergy works well where both partners are interested in sources of *chemical* novelty, but not in the *gene* area. Functional genomics, as a platform technology, can help both sides to invent novel and profitable chemicals but the major commercial opportunities in the creation of GM crops have no parallel in pharmaceuticals. And according to managers economic forecasts heavily influenced by the political factors had the major influence. Although experience in the USA and other countries has indicated that GM crop development is potentially very profitable, the negative public reaction and difficult policy environment in Europe has created potential conflicts of interest between agricultural and health-based sectors.

Synergies between pharmaceutical and agricultural areas of biotechnology may emerge again in future, for example, through medicinal benefits from crop plants but by that time the managers we interviewed expected that there would no longer be a link between the agrochemical and pharmaceutical divisions of companies. Managers of the agricultural arms of biotechnology companies saw alignment with other agribusinesses, fighting for agricultural investment, as a more robust strategy than competing for pharmaceutical investment.

The result of this reinterpretation of the relationships between agro-biotechnology and pharmaceutical sectors of companies can be seen in the changed pattern of mergers and de-mergers among life science companies. The splitting off from their pharmaceutical divisions of the agrochemical and seeds divisions of Novartis and Zeneca to form Syngenta was, in one manager's words, 'the first ever pure play in ag-biotech'. In the merger of Zeneca with Astra in 1999, Zeneca had argued for a life sciences model which included agrochemicals but there was no apparent design for the agrochemicals unit in the merged company. Novartis as we have seen split its healthcare and agri divisions in 1999. This pattern has since been repeated in most of the agro-biotechnology MNCs included in our research sample.

Thus, beyond the fundamental discovery stage, there is currently little synergy between the downstream product development processes of the agrochemical and pharmaceutical sectors and when it comes to near-market development, managers interviewed for this project commented that their natural partners were now in the food processing and retailing sectors rather than in pharmaceuticals. Instead of being a broad life sciences vision it is a narrower agricultural one (Tait et al., 2002).

A principal argument of this section is that while a range of factors were important the original life sciences based technological trajectory envisioned by the industry did not come about largely as a result of political forces. It is too soon to tell how policy and political factors will affect further the new life sciences trajectory.

So far we have provided an overview of some of our findings as they relate to R&D trajectories and strategies and their impact on industry structures. Next, we relate this analysis to the political context in which agri-biotechnology emerged. Agri-biotechnology is a contested technology and there is multifaceted disagreement about many aspects of its developments. In the following section, we aim to relate this analysis to the political context in which agri-biotechnology emerged. Agri-biotechnology is a contested technology and there is multifaceted disagreement about many aspects of its development. For the focus of this paper, a number of aspects of the debate are particularly relevant. The first are the ongoing disputes about the ways in which GMOs should be regulated. Whilst it is now broadly accepted, in Europe if not the United States, that GMOs should be subject to additional regulations, based on the novelty of the technological process used to create products, this was not always the case. During the late 1980s and early to mid 1990s, most companies argued vociferously that there should be no additional risk regulation for

products based on biotechnology and that existing regulations designed for agricultural and pharmaceutical products were adequate. Some companies even argued that companies should self-regulate activities and products (Levidow and Carr, 1996).

The 'process versus product based regulation' debate evolved into equally heated disputes over the 'precautionary principle'. The precautionary principle, which holds that risk regulation should be based on broader perception of risk than narrow scientific evidence can address,[18] is currently the focus of disagreement about transboundary regulation of GMOs (Tait and Bruce, 2004). Very broadly, the EU and many developing countries favour the precautionary principle and US negotiators oppose it. For some, the adoption of the precautionary principle underlies the inability of the EU to overcome the moratorium which has unofficially been in place for the past 3 years. Although companies are now much more reticent to air their opinions, they were vocal in their opposition to the precautionary principle during the 1990s.

The 1990s also saw considerable controversy over segregation between GM and non-GM products and over labelling of GM products. Whilst some companies, notably Zeneca, were happy initially to label GM products, Monsanto and other companies bitterly opposed both segregation and labelling.

At the same time, during the late 1980s and 1990s a number of companies undertook high profile public relations campaigns. These started with ICI's 'Feed the World' campaign which promoted biotechnology on the basis of overcoming world hunger. This was widely dismissed as hopelessly technocractic. And ending with Monsanto's advertising and public relations campaign in the late 1990s which ostensibly involved an effort to engage people in debate but was widely thought of as propaganda. These contextual factors are important for understanding the ways in which political disputes have impacted on the evolution of biotechnology, as we see below.

Although controversy about GM and concerns about implementing more precautionary approaches may on the surface appear to be a peculiarly European phenomenon it is important to recognise that

---

[18] The precautionary principle is itself contested, with little agreement about the degree and nature of scientific evidence needed to rule a product an environmental or health risk (Bail et al., 2002).

J. Chataway et al./Research Policy 33 (2004) 1041–1057   1053

broader dynamics are at play. First, as Wiener and Rogers (2002) and others have shown, there is no simple division between EU precautionary and US non-precautionary approaches. Whilst the rhetoric sometimes suggests that might be the case in reality the US has often taken substantially more precautionary approaches than the EU, over the risks of blood contamination with BSE for example. The triggers for concerns may differ[19] but the complex political dynamics involved in innovation are more pervasive.

9. How public policy issues impacted on company decision making and how companies responded to the policy environment

Firms studied had similar understandings of the way in which public policy impacted on innovation. But they had different ways of relating to the policy process. Their R&D strategies whilst not explicitly and directly related to policy will give them different options in terms of responses to future policy environments. Zeneca and Novartis will have more flexibility than Monsanto to respond with product mix in uncertain environments. Yet, given the scale of investment and the strong view that biotechnology would be key to future developments, the way companies viewed and dealt with political factors and policy was perhaps surprising.

In very general terms managers in companies had the following two sets of overriding beliefs about the direction of policy:

- Liberalisation of markets and reduction in CAP would continue and this would be good for future product development.[20]
- Although regulatory environments were uncertain and this posed great difficulties it was broadly felt that tough evidence-based regimes were difficult but could be lived with.

Overall, public policy influences on companies are subtle and in many cases, had a much more indirect,

background influence on companies' product development strategies. At the time of the research, it was not possible to detect direct influences of policies on product development decisions. Of course, directives and regulations were obeyed and lived with, but they were not the drivers used in company 'foresighting' and detailed thinking about company strategy. Some managers recognised the problem, that whilst they had been active in rather opportunistic lobbying this had not amounted to a more strategic approach to integrating policy and political factors into decision making. Senior managers in Novartis noted that if they wanted to create opportunities in Europe they have "...to position themselves in the context of the long term strategies of European society". Managers saw that a "paradigm shift" in companies was necessary because the previous situation where people never had any doubts about the societal value of their products had changed. Their previous 'tactical' engagement with policy and the political process would be insufficient.

Whilst managers had little sense that policy impacted directly on their R&D strategy, they were all concerned about public opinion. In Zeneca, R&D strategies did to some extent reflect this concern. Managers accepted from an early date that consumers, rather than producers, would be key to success in marketing GMO technology. Investment in delayed ripening technology in tomatoes reflected this but this example aside there was little appreciation of the complex interaction between policy and public opinion.

Monsanto was beginning to recognise that the high profile campaign it had launched to try to counter demands for more precautionary types of regulation and to convince the public of the benefits of GM technology had backfired. In the campaign, the company emphasised the continuity between new biotechnologies and traditional plant breeding. Rather than consider precautionary approaches the company pushed for 'appropriate regulation'. Monsanto and other companies argued against segregation of GM and non-GM crops and actively supported WTO attempts to force the EU to take GM crops unlabelled.[21]

Fieldwork interviews took place during a period when GM technology was the focus of considerable

---

[19] In the US concerns over the moral implications of stem cell research has led to a more precautionary approach than in Europe for instance.

[20] Although a minority of managers expressed concern that CAP reforms would make farmers feel less wealthy and use fewer crop protection inputs.

[21] Monsanto managers note that opposition to segregation was partly due to Monsanto's lack of control over what farmers did with bought seed.

media attention and when tabloid headlines regularly referred to 'Frankenstein foods'. Shortly after this fieldwork, shareholder confidence in Monsanto hit rock bottom and the company reached crisis point and was subsequently acquired by Pharmacia.[22] A de-facto moratorium on the commercialisation of GM crops in the EU was instigated, and increasingly belligerent discussions took place between EU and US trade negotiators. Although companies were not linking policy directly to R&D strategy, they were beginning to reflect on the weaknesses of the way they were thinking about policy and public opinion. Thinking here though was in general not well developed.

The conflict over the risks of biotechnology that have already had a role in shaping the industry were related to the relative lack of attention to the complexities of public opinion and policy and to the inept PR efforts of companies and were not inevitable. A range of issues might have been handled differently (Tait and Chataway, 2000c). A few points are particularly relevant to this paper:

1. The agro-biotechnology industry presented a united front in its opposition to the use of the precautionary principle (PP), largely on the grounds that it undermined the scientific basis for risk regulation and opened up the possibility of regulatory decision making based on pressure group preferences rather than sound science. This opposition allowed public interest pressure groups to present industry as pursuing commercial interests while ignoring environmental and health risks for which no precautionary testing had been undertaken (Tait, 2001). However, recent events in Europe where environmental pressure groups seem to have an unlimited list of testing requirements and also to be opposed to such tests taking place confirm industry's fears that the PP would produce lengthy delays which would make companies' investments in the technology non-viable. It is an open question whether continued espousal of a less extreme version of the PP in Europe by industry throughout the 1990s would have avoided the current

more extensive demands now being made in its name.

2. Although several companies were aware of the advantages of developing initial products that provided a clear consumer benefit, most did not state this aim until recently. Zeneca's delayed ripening tomato, one of the first GM crops, sold well but was not sufficient to have more of an impact beyond its niche market. The lack of a political perspective in early R&D decisions has had a profound impact on the way biotechnology has been viewed and subsequent strategic and policy decisions which will shape the technology.

3. Much of European public reaction to GM crops can be traced to concerns about globalisation and vertical integration of food production systems so that they are perceived as being not publicly accountable (Tait, 2001). The role of The World Trade Organisation (WTO) in promoting greater freedom in international trade as part of this globalisation process has been represented by pressure groups as supporting multinational companies in ways which exploit and undermine the needs and desires of ordinary people and damage the environment. This perception has been exacerbated by attempts by US-based companies to market GM seeds and crop products in Europe despite public opposition. US-based companies could perhaps have done something to avoid the worst fall-out from this issue by being more conciliatory to European consumers, particularly on the issue of labelling GM foods.

4. The rapid uptake of GM crops by farmers in North America would have amply justified the technology and marketing strategies of the companies involved, but for the corresponding negative reaction from the European (and perhaps also US) public. Given that two of the major crops involved were maize and soya, GM products became very widely distributed in world food supplies and public interest pressure groups were able to claim that the public had not been consulted and had been given no choice in what was represented as a major change in the food supply. The companies' response on the other hand was that there was no need for consultation or choice as there was no substantive difference between GM crop products and the parent plant variety.

---

[22] Pharmacia announced its intention to sell the ag-chem and biotech side of the company back but has still not succeeded in finding a buyer.

J. Chataway et al./Research Policy 33 (2004) 1041–1057   1055

10. Conclusion

Early choices made by companies about the development of biotechnology based products and the marketing of those products have had weighty consequences for the way in which the technology has subsequently been shaped. Those early choices we have argued were made on the basis of particular company trajectories and strategic orientation. Monsanto with its emphasis on a radical 'strategising' approach carved out a powerful route into a new biotechnology based era but at the same time triggered negative reactions from sections of the public and policy-makers.

Biotechnology is the basis for a very different set of technological options from those previously faced by agri-biotechnology companies, but the foundations for the resulting technological paradigm are weak and the shape of that paradigm is uncertain. Had the importance of public opinion and policy decisions and the complex interaction between the two been judged differently by these companies the move towards agricultural systems based on a combination of chemical and biotechnological inputs may have been more rapid and more pervasive. The initial life science-based vision, uniting pharmaceutical and agro-chemical sectors, may also have remained dominant and given biotechnology a systemic cross-sectoral reach.[23]

One response to the arguments presented here might be that it is easy for us as detached researchers to be wise in retrospect. Our aims here however are to carry forward an analysis initiated in 1993 about the complexities of decision making in biotechnology and the interaction between company strategies and public policy (Chataway and Tait, 1993) and to develop this analysis in ways which can be applied to future trajectories in the bio-pharmaceuticals and nanotechnology sectors.

This paper has analysed the evolution of biotechnology strategies drawing on the concepts of paradigms and trajectories and a number of more theoretical issues could be explored based on the analysis presented in this paper. Alan Russell has critiqued the thinking behind technological paradigms on the grounds that it does not give enough weight to social and political factors and that this limits its appeal to a particular

school of economists.[24] He relates his critique to a problematic of structure-agency which acts as a constraint on the usefulness of the concept. He says

> "A technological paradigm therefore represents a structure, or enabling framework, within which the work of individual researchers, technologists or firms proceeds and the form of the paradigm is influenced by the collective activities and choices of agents over time. In explaining technological progress, however, the point of deploying the concept of paradigm is to suggest that it shapes the work of the agents within it. . . .This dichotomy goes to the heart of the debate over agency-structure determinism. The concept of technological paradigm needs revision along these lines to extend its usefulness beyond the concerns of innovation economists (Russell, 1999, p .242)".

There is certainly a set of fascinating questions about the relationship between strategic decisions taken by individual firms and the emergence and establishment of new technological paradigms. The complexity of that relationship is of course much more obvious in cases where there does seem to be a shift from one paradigm to another. Whether or not there is a fundamental problem with thinking associated with technological paradigms and trajectories, there does seem to be a particular problem of perspective that needs addressing so that politics attracts more attention as a factor influencing the shape and direction of change.[25]

---

[23] We are not making value judgements here, rather observing the causes and possible impacts of R&D strategies.

[24] By noting this critique we are not challenging the usefulness of the concept of paradigms and trajectories to the understanding of ex-post technological change. We are simply reflecting on the potential relevance of this case, i.e. that in the case of agri-biotechnology whilst economic and technological factors seemed to favour the emergence of a new paradigm, a set of political judgements by key actors and political conditions seem to have blocked it, at least temporarily.

[25] A recent paper by Parayil discusses the importance of a political awareness of the importance of debates around patents and IPR will be important in influencing the direction of a biotechnology based technological trajectory. He says 'The technological trajectory will be affected significantly by the success in mediating between the public and private domains by de-centring the primacy of IPR regime in biotechnology research. Ways must be found to adapt the IPR regime such that innovations in agro-biotechnology could be stimulated through new institutional frameworks that encompass broader social goals than patents and private gain (Parayil, 2003).

In another but related context (the importance of constructive technology assessment (CTA) for innovation in firms), Coombs (1995) addresses the deficit between the economic and the social analyses on technology, whilst arguing that it needs closing if firms are to come to grips with increasing public unease over new technologies. He contrasts the more quantitative, technology promotion focus of evolutionary economics, with the more qualitative, 'quality of life', control of technology emphasis of social studies of science analysts. He notes on the one hand the concepts 'technology trajectory', 'technological paradigm' and 'decision routine' vocabulary of those seeking to understand firm strategy and, on the other, the more social constructive, actor-network perspectives, with their recognition of the emergent processual and political character of strategic discussions within organisations. He argues that these two discourses have not been able to engage satisfactorily with each other and 'have even been seen as to some extent pursuing conflicting objectives' (p. 332). Thus, 'the recent intellectual frameworks for understanding technical change have not made a major impact on policy practices' (p. 334).

The deficit or 'blind spot' about the political impact of strategic R&D choices seems to exist as much for those who research and teach innovation management as those who practice it.[26] Even in the literature on strategic management, where the emphasis is usually on 'actors' rather than structures, there is a lack of attention paid to politics, policy and public opinion.[27] These factors often tend to be subsumed under the classification of the 'external environment' and discussion is limited to how firms react to it. For example, in a recent well known textbook on strategic management, the section most relevant to these points is entitled 'The Environment School: Strategy formation as a reactive process' (Mintzberg et al., 1998). But as this paper has shown, the reality is more complex. Firms do act, and react, within a political environment and in the case of biotechnology, a highly politicised environment. And pressures for managers to become strategic in the way they incorporate pol-

icy and politics into R&D and other decisions are likely to intensify.[28] Vernon (1998) relates the need for MNCs to develop new forms of political awareness to tensions between processes of globalisation and the nation state. Firms increasingly need to legitimise themselves in broader societal terms as well as financial terms and risks to reputation are becoming a new area of interest to investors (Brunsson, 1989; Larkin, 2003). Firm level technology strategy in this new environment requires much stronger linkage with social and public networks than is currently the case.

This study confirms that managers did not fully recognise the relationship between R&D decisions and policy environments and often seemed to be acting in ways that had negative impact on the firms themselves and had dramatic unintended consequences on the rate and direction of innovation in the sector.

References

Bail, C., Falkner, R., Marquand, H., 2002. The Cartegna protocol on biosafety. Reconciling Trade in Biotechnology with Environment and Development? Earthscan, London.

Brunsson, N., 1989. The Organisation of Hypocrisy: Talk, Decision and Actions in Organisations. John Wiley and Sons, Chichester.

Chataway, J., Tait, J., 2000. Monsanto Dossier. PITA Project Report. http://www.technology.open.ac.uk/cts/projects.htm# biotechnology.

Chataway, J., 1992. The making of biotechnology: a case study of radical innovation. PhD thesis, Open University.

Chataway, J., Tait, J., 1993. Risk regulation and strategic decision making in biotechnology: the political economy of innovation. Agriculture and Human Values 10 (2), 60–67.

Coombs, R., 1995. Firm strategies and technical choices. In: Rip, A., Misa, T.J., Schot, J. (Eds.), Managing Technology in Society: The Approach of Constructive Technology Assessment, Pinter, London.

Dosi, G., 1988. In: Dosi, G., et al. (Eds.), The Nature of The Innovative Process in Technical Change and Economic Theory. Pinter Publishers, London.

Fernandez, L., 1985. Splashes and ripples: the chemical industry in the new millennium. Chemistry and Industry 23, 787–789.

Granstrand, O., Patel, P., Pavitt, K., 1997. Multi-technology corporations: why they have 'distributed' rather than 'distinctive core' competences. California Management Review 39 (4), 8–25.

Hayes, R., 1985. Strategic planning: forward in reverse. Harvard Business Review 63 (6), 111–119.

---

[26] There are obvious links between the two!

[27] Although of course there are a range of perspectives offered which might promote the incorporation of these concerns. See for example Mahdi (2003).

[28] Meyer-Krahmer and Reger have analysed innovation strategies in the context of complex policy environments from the perspective of lessons for technology policy makers (Meyer-Krahmer and Reger, 1999).

Kaplan, S., Murray, F., Henderson, R., 2003. Discontinuities and senior management: assessing the role of recognition in pharmaceutical firm response to biotechnology. Industrial and Corporate Change 12 (2), 203–233.

Kay, J., 1993. Foundations of Corporate Success. Oxford University Press, Oxford.

Larkin, J., 2003. Strategic Reputation Risk Management. Palgrave Macmillan, Hampshire.

Lemarie, S., 2001 How will US-based Companies Make It In Europe? An Insight from Pioneer and Monsanto. http://www.agbioforum.org/vol14no1/vol14no1ar13lemarie.htm.

Levidow, L., Carr, S., 1996. Biotechnology risk regulation in Europe. Science and Public Policy 23 (3), 133–200.

Mahdi, S., 2003. Search strategy in product innovation process: theory and evidence from the evolution of agrochemical lead discovery process. Industrial and Corporate Change 12 (2), 235–270.

Meyer-Krahmer, F., Reger, G., 1999. New perspectives on the innovation strategies of multinational enterprises: lessons for technology policy in Europe. Research Policy 28, 751–776.

Mintzberg, H., Ahlstrand, B., Lampel J., 1998. Strategy Safari: The Complete Guide through the Wilds of Strategic Management. Financial Times/Prentice Hall, Pearson Education. London.

Monsanto Annual Report 1998. Monsanto Company, St. Louis.

Niiler, E., 2000. Demise of the Life Science Company Begins. Nature Biotechnology 18 (January), 14.

Orsenigo, L., 1989. The Emergence of Biotechnology. Institutions and Markets in Industrial Innovation. Pinter Publishers, London.

Parayil, G., 2003. Mapping technological trajectories of the Green Revolution and the Gene Revolution from modernization to globalization. Research Policy 32 (6), 971–990.

Russell, A., 1999. Biotechnology as a technological paradigm in the global knowledge structure. Technology Analysis and Strategic Management 11 (2), 235–254.

Tait, J., 2001. More faust than frankenstein: the European debate about risk regulation for genetically modified crops. Journal of Risk Research 42 (2), 175–189.

Tait, J., Chataway, J., 2000a. Novartis Dossier. PITA project Report: http://www.technology.open.ac.uk/cts/projects.htm#biotechnology.

Tait, J., Chataway, J., 2000b. Zeneca Dossier, PITA Project Report: http://www.technology.open.ac.uk/cts/projects.htm#biotechnology.

Tait, J., Chataway, J., 2000c. Technological foresight and environmental precaution—genetically modified crops. In: Cottam, Harvey, Pape, Tait (Eds.), Foresight and Precaution, Balkema, Rotterdam.

Tait, J., Bruce, A., 2004. Global change and transboundary risks. In: McDaniels, T., Small, M. (Eds.), Risk Analysis and Society: An Interdisciplinary Characterisation of the Field. Cambridge University Press, pp. 367–419.

Tait, J., Chataway, J., Wield, D., 2002. The life science industry sector: evolution of agro-biotechnology in Europe. Science and Public Policy 29, 253–258.

Teece, D., Pisano, G., Shuen, A., 1997. Dynamic capabilities and strategic management. Strategic Management Journal 18 (7), 509–533.

Vernon, R., 1998. In the Hurricane's Eye: the Troubled Prospects of Multinational Enterprises. Harvard University Press. Cambridge.

Walsh, V., Lodorfos, G., 2002. Technological and organizational innovation in chemicals and related products. Technology Analysis and Strategic Management 14 (3), 273–298.

Wield, D., 2001. BASF: Agbio Fast Follower', Agbioforum 4 (1), 58–62. http://www.agbioforum.org.

Wiener, J.B., Rogers, M.D., 2002. Comparing precaution in the United States and Europe. Journal of Risk Research 5, 317–350.

[16]

THE JOURNAL OF INDUSTRIAL ECONOMICS   0022-1821
Volume XLVI   June 1998   No. 2

ABSORPTIVE CAPACITY, COAUTHORING BEHAVIOR, AND THE ORGANIZATION OF RESEARCH IN DRUG DISCOVERY*

Iain M. Cockburn† and Rebecca M. Henderson‡

We examine the interface between for-profit and publicly funded research in pharmaceuticals. Firms access upstream basic research through investments in absorptive capacity in the form of in-house basic research and 'pro-publication' internal incentives. Some firms also maintain extensive connections to the wider scientific community, which we measure using data on coauthorship of scientific papers between pharmaceutical company scientists and publicly funded researchers. 'Connectedness' is significantly correlated with firms' internal organization, as well as their performance in drug discovery. The estimated impact of 'connectedness' on private research productivity implies a substantial return to public investments in basic research.

I. INTRODUCTION

THE IDEA that the inability of profit-maximizing firms to fully appropriate returns from R&D is likely to lead to under-investment in research relative to the social optimum remains an uncontroversial basis for public support of R&D in general, and basic scientific research in particular, in most developed economies. Yet it is far from clear how much publicly funded research is actually utilized downstream, through which channels, and at what cost. Though efforts to measure the rate of return to public investments in science suggest that it is quite high, these results are dogged by a variety of difficult practical and conceptual problems (Griliches [1979], [1994]; Jones and Williams [1995]; Stephan [1996]).

In this paper we examine the interface between for-profit and publicly

---

* We gratefully acknowledge financial support by the Alfred P. Sloan Foundation through the NBER Project on Industrial Technology and Productivity, as well as by the UBC Entrepreneurship Research Alliance (SSHRC grant # 412–93–0005). We would also like to express our appreciation to those firms and individuals who generously contributed data and time to the study, and to Gary Brackenridge and Nori Nadzri, who provided exceptional research assistance. Lynn Zucker and Michael Darby provided many helpful comments and suggestions. The usual disclaimers apply.

† Authors' Affiliation: Iain M. Cockburn, Faculty of Commerce and Business Administration, University of British Colombia, 2053 Main Mall, Vancouver, BC, Canada V6T IZ2 and NBER, USA.
email: iain.cockburn@commerce.ubc.ca

‡ Rebecca M. Henderson, MIT Sloan School of Management E52–543, MIT Cambridge, MA 02139, and NBER USA.
email: rhenders@mit.edu

158   IAIN M. COCKBURN AND REBECCA M. HENDERSON

funded research in the pharmaceutical industry, one of the most science-intensive sectors of the economy and one where public support for research has been very substantial. We begin with the observation that upstream basic research, though difficult to appropriate, is not a free input to downstream market-oriented R&D. Rapid advances in the life sciences over the past four decades have indeed been accompanied by high rates of product innovation in the pharmaceutical industry, but, arguably, these have been achieved through costly investments in complementary in-house basic research and organizational innovation. As Cohen and Levinthal [1989] and others have noted, firms wishing to take advantage of research conducted outside their organizational boundaries may need to invest in 'absorptive capacity' in the sense of accumulating the knowledge, skills, and organizational routines necessary to identify and utilize externally generated knowledge. These investments may be particularly significant if firms are to take advantage of upstream advances in fundamental science, and there is certainly evidence that this is an important issue in pharmaceuticals (Gambardella [1992], [1995]). However our findings suggest that, at least in this industry, it may be necessary to expand upon this idea. While it is certainly necessary to invest in basic research inside the firm, we believe that it is also important for the firm to be actively connected to the wider scientific community.[1]

The paper begins by drawing on an extensive program of qualitative research to develop the concept of 'connectedness' to the community of open science, to suggest that it is a key factor in driving a firm's ability to recognize and use upstream developments, and to hypothesize that it has a large impact on research productivity by changing the way in which research is managed inside the private firm.

We then move on to a quantitative analysis of pharmaceutical firms' 'connectedness' to the public sector, using data on coauthorship of scientific papers across institutional boundaries. 'Connectedness' in this sense is closely related with a number of other factors that we believe also increase the productivity of privately funded pharmaceutical research, including the number of 'star' scientists employed by the firm and the degree to which the firm uses standing in the public rank hierarchy as a criterion for promotion. Linking these data with measures of research productivity we find that 'connectedness' and research performance are correlated across firms and over time. While any estimate of this type must be treated with great caution, our results suggest that differences in the effectiveness with which a firm is accessing the upstream pool of

---

[1] We thus build on an important stream of work by Zucker, Darby and their collaborators, who have shown that in the case of biotechnology, both rates of firm founding and rates of new product introduction are a function of connections to 'star' university scientists (Zucker, Darby and Brewer [1997]; Zucker, Darby, and Armstrong [1994]; Zucker and Darby [1995]).

knowledge correspond to differences in the research productivity of firms in our sample of as much as 30%.

One interpretation of this result is that it represents a lower bound estimate of the impact of public sector research, since by definition it excludes the impact of any publicly generated knowledge that can be costlessly accessed across the industry. However the fact that 'connectedness' is likely to be correlated with other hard-to-observe organizational practices that improve research productivity, as well as other important sources of unobserved heterogeneity across firms (such as the quality of human capital) makes us hesitate to assign the result too much weight. Rather we draw two more general conclusions. Firstly, they suggest that Cohen and Levinthal's stress on the importance of absorptive capacity is well placed, but that at least in the case of pharmaceuticals the development of absorptive capacity requires rather more than simply conducting a certain amount of fundamental research in house. Our results are consistent with the hypothesis that the ability to take advantage of knowledge generated in the public sector requires investment in a complex set of activities that taken together change the nature of private sector research. In the second place, they raise the possibility that the *ways* in which public research is conducted may be as important as the *level* of public funding. To the extent that efforts to realize a direct return on public investments in research lead to a weakening of the culture and incentives of 'open science,' our results are consistent with the hypothesis that the productivity of the whole system of biomedical research may suffer.

II. QUALITATIVE EVIDENCE: EXPLORING THE INTERACTION BETWEEN THE
PUBLIC AND PRIVATE SECTORS

As a first step in understanding the role that the need to take advantage of publicly generated knowledge plays in the management of privately funded pharmaceutical research, we focus on two types of qualitative data: first, 'case histories' of a number of important drugs, and second, a series of interviews with academic research scientists, senior pharmaceutical company scientists, and research managers at a number of pharmaceutical firms.[2]

Between 1970 and 1995, public funding in the US for health related

---

[2] In the first phase of our research we interviewed over fifty individuals at 10 large pharmaceutical firms. Interviews were semi structured in that each respondent had been provided with a list of key questions before the interview, and each interview lasted from one to three hours. These data were complemented with interviews with a number of industry experts, including senior academics in the field. In the second phase of the research we returned to three of the firms in the original data set, and we also conducted interviews at three large firms that were not represented in the original sample.

research increased nearly 200% in real terms, to $8.8 billion, or 36% of the non-defense Federal research budget, an amount roughly equal to the total research expenditure of all the US pharmaceutical firms. In general the private sector draws from the public throughout the lifecycle of a drug: as a source of fundamental biological and chemical knowledge that can lay the groundwork for the discovery of new drugs, as a source of clinical knowledge for the design of tests for drugs once they have been discovered, and as a source of insight into additional indications for drugs once they have been approved (Comroe and Dripps [1976]; Raiten and Berman [1993]; Maxwell and Eckhardt [1990]; Ward and Dranove [1995]).

Table I illustrates the changing nature of the interaction between public sector and private sector pharmaceutical research. The table summarizes detailed case histories which we constructed of the discovery and development of 21 drugs, identified by two leading industry experts as 'having had the most impact upon therapeutic practice' between 1965 and 1992.[3] While this is a selective and not necessarily representative sample of new drugs introduced in this period, and it is certainly the case that many potentially important drugs arising from more recent discoveries are still in development or have been just introduced, our qualitative work suggests that the general trend identifiable in these data is a plausible summary of recent trends in the relationship between the public and private sectors in the industry.

In the first place, the table confirms the important role that the public sector plays in providing fundamental insights in basic knowledge as a basis for drug discovery.[4] Only 5 of these drugs, or 24%, were developed with essentially no input from the public sector.[5] In the second place, these data are consistent with the hypothesis that public sector research has become more important to the private sector over time. Table I groups the drugs into three classes according to the research strategy by which they were discovered: those discovered by 'random screening', those discovered by 'mechanism-based screening' and those discovered through fundamental scientific advances. Broadly speaking, the degree of reliance on the

---

[3] The two experts were Professor Louis Lasagna (Tufts University) and Professor Richard Wurtman (MIT). The sample includes 5 of the drugs (cyclosporine, nifedipine, captopril, cimetidine and propranolol) that Maxwell and Eckhardt [1990], who used a similar criterion but focused on an earlier time period, included in their sample.

[4] For purposes of general comparison we list a 'date of key enabling discovery' for each drug. The choice of any particular event as the 'key enabling discovery' is bound to be contentious, since in pharmaceuticals, as in many fields, discovery usually rests on a complex chain of interrelated events. In the case of drugs discovered through screening we give the date of first indication of activity in a screen. In the case of 'mechanism' based drugs, we give the date of the first clear description of the mechanism. Dates for the third class are only broadly indicative, and all should be used carefully.

[5] This contrasts with Maxwell and Eckhardt's finding that 38% of their sample of older drugs were developed with no public sector input.

TABLE I

HISTORY OF THE DEVELOPMENT OF THE 21 DRUGS WITH 'HIGHEST THERAPEUTIC IMPACT' INTRODUCED BETWEEN 1965 AND 1992

Generic name	Trade name	Indication	Date of key enabling discovery	Public?	Date of synthesis of compound	Public?	Date of market intro	Lag from enabling discovery to market introduction
'Old Fashioned' or 'random' drug discovery: screening of compounds in whole or partial animal screens								
Cyclosporine	Sandimmune	Immune suppression	NA		1972	N	1983	7
Fluconazole	Diflucan	Anti-fungal	1978	N	1982	N	1985?	67
Foscarnet	Foscavir	CMV Infection	1924	Y	1978	Y	1991	19
Gemfibrozil	Lopid	Hyperlipidemia	1962	N	1968	N	1981	16
Ketoconazole	Nizoral	Anti-fungal	1965	N	1977??	N	1981	12
Nifedipine	Procardia	Hypertension	1969	N	1971	N	1981	21
Tamoxifen	Nolvadex	Ovarian cancer	1971	Y	NA		1992	
'Mechanism driven' research: screening of compounds against a very specific known or suspected mechanism								
AZT	Retrovir	HIV	Contentious	Y	1963	N	1987	16
Captopril	Capoten	Hypertension	1965	Y	1977	N	1981	29
Cimetidine	Tagamet	Peptic Ulcer	1948	Y	1975	N	1977	18
Finasteride	Proscar	BPH	1974	Y	1986	N	1992	30
Fluoxetine	Prozac	Depression	1957	Y	1970	N	1987	28
Lovastatin	Mevacor	Hyperlipedimia	1959	Y	1980	N	1987	11
Omeprazole	Prilosec	Peptic Ulcers	1978	N	1982		1989	34
Ondansetron	Zofran	Nausea	1957	Y	1983	N	1991	19
Propranolol	Inderol	Hypertension	1948	Y	1964	N	1967	35
Sumatriptan	Imitrex	Migraine	1957	Y	1988	N	1992	
Drugs discovered through fundamental science								
Acyclovir	Zovirax	Herpes	?		?	Y	1982	
Cisplatin	Platinol	Cancer	1965	Y	1967	Y	1978	13
Erythropoietin	Epogen	Anemia	1950	Y	1985	N	1989	39
Interferon beta	Betaseron	Cancer, others	1950	Y	Various	N	Various	

Source: authors' compilations.

© Blackwell Publishers Ltd. 1998.

public sector for the initial insight increases across the three groups, and as the industry has moved to a greater reliance on the second and third approaches, so the role of the public sector has increased. The very long lags apparent in the table between fundamental advances in science and their incorporation in marketed products may be shortening as the public and private sectors draw closer together, but it is difficult to draw strong conclusions from this small sample. Thus the data outlined in Table I suggest that public sector research has indeed had an impact on the private sector. However they give little insight into the process by which this occurred. On this front our interview data proved to be much more informative.

The interviews confirmed the hypothesis that public research was important to the industry. All of the researchers with whom we spoke believed that it was critically important to keep abreast of the results of publicly funded research. They also confirmed the importance of the possession of 'absorptive capacity' in the classical sense. All our respondents reported that it was indeed critical to conduct leading edge research inside the firm in order to be able to take advantage of publicly generated research. But our qualitative work also uncovered a number of other aspects of the relationship between the public and private sector that have been less well explored in existing research.

In general our respondents stressed the importance of three additional factors: hiring the best possible people, rewarding them on the basis of their standing in the public rank hierarchy, and encouraging them to be actively engaged with their public sector counterparts, or to be 'tightly connected' to the public sector. Moreover they suggested that this complex of factors have two quite distinct (although complementary) functions. First, they improve access to public sector research and second, they improve the quality of the research conducted within the firm.

The intuition behind this assertion is quite clear in the case of the first two factors and is also consistent with the results of our prior research. Increasing the quality of the human capital in the firm almost certainly improves internal research productivity, although substantive difficulties in measuring the quality of human capital make it difficult to estimate this effect precisely (Henderson and Cockburn [1996]). In previous work we have also shown that firms that are 'pro-publication' in the sense that they promote researchers on the basis of their standing in the scientific community are significantly more productive than their rivals, all other things equal (Henderson and Cockburn [1994]). Using this incentive mechanism appears to improve productivity for several reasons. First, it is efficient, in that it allows the firm to monitor and reward researchers relatively cheaply. Since monitoring researchers in private organizations is notoriously costly, this is no small benefit (Aoki [1990]). Second, it forces researchers to publish and to remain in close touch with the state of knowledge in their field, which is important if the firm is to take prompt advantage of the latest scientific advances. Lastly it acts as a powerful recruiting tool, since the highest quality scientists in a field are often reluctant to work for private firms if they will not be able to publish and thus maintain their personal scientific reputations.

Here we focus our attention on the 'connectedness' factor: the suggestion that in order to access the results of publicly funded research it is not enough simply to hire world class scientists, allow them to do fundamental research and promote them on the basis of their standing in their field: it is also important that these researchers be active collaborators with public sector researchers. Reading the journals, attending conferences, even being an active player on the informal network of information transfer within the industry are insufficient. If they are to be optimally productive, our respondents suggested that a firm's researchers need to be active participants in the construction of publicly available research results, despite the issues of appropriability that such active collaboration raises.

Such collaboration is clearly complementary to some of the other factors that improve research productivity. It probably makes it easier to recruit and retain world class researchers, for example. Our respondents also stressed that active collaboration with public researchers sometimes had the effect of increasing the level and quality of communication within the firm. But nevertheless it is a striking assertion that raises a number of fascinating questions.

In the first place, it highlights the fact that the relationship between public and private sectors is not at all well characterized as one of active donor to passive recipient. In reality, the relationship between the two sectors is very much a bidirectional one, characterized by the rich exchange of information in *both* directions. This observation is well established in the historical literature, where several researchers have observed that knowledge or observations gained by experience and experiment in practice made major contributions to advances in fundamental scientific knowledge in many industries. In the case of steel, for example, major advances in the techniques of steel refining provided the impetus for the first serious studies of metallurgy, and advances in practical knowledge of dyestuffs helped to lay the foundations of modern chemistry (Walsh [1984]).

In the case of pharmaceuticals, on several occasions the development of novel therapies by the private sector or advances made in clinical treatment have *preceded* major advances in fundamental knowledge. Brown and Goldstein's Nobel prize winning work on the structure of the LDL receptor, for example, occurred simultaneously with the discovery of the first effective HMG CoAse reductase inhibitors, and the recognition that stomach ulcers are bacterial in origin flowed from the pioneering work of physicians working in the clinic rather than from basic scientific research. Even apparently 'straightforward' cases such as the discovery of AZT appear, on closer examination, to have a fine grained structure that reflects a bi-directional flow of knowledge rather than the simple transmission of research results or new ideas from the public to private sectors.

In the second place, it highlights the complexity inherent in the concept of 'absorptive capacity.' The work of Zucker, Darby and their collaborators has already focused attention on the ways in which the transfer of knowledge from the public to private sectors in the case of biotechnology is mediated by much more than simple investment in leading edge science. Their results suggest that factors such as geographical proximity and the nature of the contractual relationship between public researchers and private firms also have a very important effect on the efficiency with which information is transferred from public to private sectors. Our respondents seemed less concerned with issues of geographic proximity or of contractual structure.[6] But they did confirm that staying current with leading edge research was a complex process that required more than simple investment in basic science.

Quantifying these factors is far from straightforward, and presents a variety of interesting measurement problems. In prior work we established a statistical relationship between the observed research productivity of some of these firms and variables coded from interview transcripts which attempted to measure important aspects of their internal organization and incentives. Here we supplement these measures with data on the publication activity of scientists employed by pharmaceutical firms. Publication in the open literature generates a paper trail which can be used to quantify some aspects of 'connectedness' in an objective, albeit limited, way. In the remainder of the paper we draw on these data to explore a number of the hypotheses suggested by our qualitative work.

III.   QUANTITATIVE EVIDENCE ON PUBLIC-PRIVATE INTERACTION

The publication patterns of pharmaceutical company scientists present a rich source of data about the nature and extent of interaction between private

---

[6] Biotechnology and conventional small molecule drug discovery are quite different in this respect. Biotechnology, at least in its early days, had an enormous tacit component that may well have required physical proximity for its transferal, and the possibility of high personal gains to public sector researchers probably placed greater stress on the nature of the formal contractual relationships between firms. In small molecule discovery our interviews suggested that physical proximity was less important, and financial relationships between individual scientific investigators and private sector firms were largely confined to cash transfers in the form of research grants, awards, and fee-for-service consulting, rather than the equity holdings, stock options and other high-powered incentives characteristic of the biotech sector, at least in the period covered by our data (1975–1990).

sector researchers and their publicly funded colleagues. We focus here in particular on coauthorship of papers by researchers working in different organizations. As prior research has established, pharmaceutical companies publish heavily, with annual counts of papers comparable to, and sometimes exceeding, the output of similarly sized universities and research institutes (Koenig [1982]; Hicks [1995]). Publication counts are an important indicator of research activity, particularly of the investments in 'doing basic science' which are often interpreted as building absorptive capacity (Gambardella [1992], [1995]). But they are relatively uninformative about *interaction* between researchers at different institutions, in the sense of sustained, direct, and meaningful intellectual collaboration. Since our qualitative work suggests that it is not simply publication that reshapes private sector problem solving but *participation* in the intellectual life of the wider scientific community, a focus on interactions is particularly important.

Following pioneering work by Zucker, Darby and their collaborators we look to collaboration as evidenced by coauthorship of papers to document this type of interaction (See Zucker and Darby [1995]; Zucker, Darby and Brewer [1997]; Zucker, Darby and Armstrong [1994], and Liebeskind, Oliver, Zucker, and Brewer [1995]). While this work focused on biotechnology, and on the technology of gene sequencing in particular, we look at the conventional 'small molecule' sector of the industry that is dominated by the major pharmaceutical firms.

Another approach would be to study cross-citation. It would be interesting, for example, to see whether private sector scientists preferentially cite public sector researchers and vice versa. Citation patterns have been used very fruitfully to trace interaction among researchers and across organizational boundaries, and to identify research communities (See, for example, Penan [1996] and Narin and Olivastro [1992]). However analysis of citations presents a number of difficulties. Citation is often highly ritualized, occurs with variable and often very long lags, and may represent negative as well as positive acknowledgement of previous research. Furthermore, as is often pointed out, citation in the age of the wordprocessor and computerized databases is extremely cheap and easy. By contrast, as many researchers can testify from personal experience, joint authorship is *costly* in terms of effort as well as other resources. In order to be willing to collaborate on writing a paper all of the authors must be willing to incur these costs, which makes an instance of coauthorship a very different datum from a citation.[7]

---

[7] Analysis of citations also presents some very serious practical difficulties and resource requirements for the kind of analysis we perform in this paper: without better access to machine-readable source data, tracking down the institutional affiliations of all of the authors of the 20 or more papers typically referenced by each of the 68 000 papers under consideration is a formidable and very expensive proposition.

166   IAIN M. COCKBURN AND REBECCA M. HENDERSON

We believe that coauthorship also evidences a qualitatively different kind of interaction than does citation. Joint authorship often reflects joint research, which is an important opportunity for the exchange of tacit knowledge. By contrast citation may be seen as an acknowledgement of the exchange of codified knowledge. Citation also refers to old knowledge, whereas coauthorship reflects generation and exchange of new or current knowledge. We note also that citation typically takes place impersonally, and can be done at a distance, whereas (notwithstanding the internet) joint authorship typically entails face-to-face interaction. Most significantly, we see coauthorship as evidence of opportunities for extensive discussion, debate, exchange of ideas, and joint problem solving: something much more than reading the journals, and something that represents a much more significant investment on the part of the firm.

Naturally there are difficulties with this interpretation. Clearly much important interaction of this type takes place at arms length, or otherwise outside the realm of joint authorship of papers: for example researchers circulate working papers, read each others work, correspond informally, listen to conference presentations, serve on professional committees together etc. Coauthorship may also reflect a variety of things other than exchange of information and joint problem solving. It may be offered as a quid pro quo for supplying information or resources, for example money or research materials such as candidate compounds, reagents or cell cultures. Coauthorship can serve as an important means of resolving disputes about priority. It may also be as a way to acknowledge of intellectual debts (for example to a thesis advisor or other mentor). As is well known, it is fairly common in physical and biological sciences to list Laboratory Directors or other senior project leaders as authors on papers which they may have had very little direct involvement in writing. Finally, the presence of some authors on the front page of a paper may simply reflect efforts by the 'real' authors to attain legitimacy, or admission to networks of other researchers.[8]

Notwithstanding these issues, we proceed on the assumption that coauthorships represent evidence of a significant investment on the part of the firm in developing connections to publicly funded research. This interpretation is consistent with our interview data, and moreover coauthorships have the virtue of being a consistently observable quantitative measure.

III(i).   *Data Sources and Methods*

We collected bibliographic information on every paper published in the

---

[8] The interviews that we have conducted with both researchers in both the public and private sector, however, lead us to believe that even these kinds of interaction are characterized by a quite rich exchange of scientific knowledge.

journals indexed in the Institute for Scientific Information's *Science Citation Index* between 1980 and 1994 for which at least one author's address was at one of a sample of twenty major research-oriented pharmaceutical firms. This sample comprises the ten firms which we have studied in prior work, and for which we have very detailed data on their research performance, plus a further ten informally selected to capture the industry's leading R&D performers and to obtain world-wide geographical representation. Obviously, these firms do not constitute a random sample, but we believe them to be reasonably representative of the industry as a whole.[9] For purposes of comparison we also repeated the data collection effort for the National Institutes of Health (NIH).

As in Zucker, Darby, and Brewer [1997] (and related papers) we use authors' mailing addresses to identify the institutions involved in a research collaboration. We measure instances of 'coauthorship' by counting the number of different addresses listed by *Science Citation Index* for each paper. This does not exactly correspond to coauthorship in the usual sense of the word, since the number of addresses on each paper does not necessarily equal the number of authors. For the 20 firms in the sample, there were on average 1.86 addresses per paper and 4.4 authors per paper. This is because *Science Citation Index* only records a maximum of six distinct addresses per paper, while as many as 27 authors may be listed, and identical addresses are not repeated. Though more than 50% of the papers in the sample have only one address listed, and these averages are skewed by the 4.5% of papers which have 10 more authors, this means that these counts of 'coauthorship' should strictly be interpreted as 'instances of institutional contact' rather than coauthorship between individuals in the conventional sense.

To trace interaction across different types of organizations, we then classify each address according to its 'TYPE': SELF, UNIVERSITY, NIH, PUBLIC, PRIVATE, NONPROFIT, HOSPITAL, and a residual category of UNCLASSIFIED. Because this procedure drives much of our analysis, it is worth describing in some detail. We start by searching *Science Citation Index* for all papers listing our 20 firms (and obvious permutations and variations of their names). Papers for which the only addresses listed are for one of these firms (or their divisions or subsidiaries) are classified as SELF. UNIVERSITY addresses are self-explanatory (note though that this classification includes medical schools). NIH is an address at any of the National Institutes of Health. The PUBLIC category comprises government affiliated organizations such as National Laboratories, Departments of Public Health, government departments and ministries, national research organizations etc. HOSPITAL

---

[9] Table III lists the sample of firms.

TABLE II
SAMPLE SOURCE RECORD

Authors:	Felten-DL Felten-SY Steececollier-K Date-I Clemens-JA
Title:	Age-Related Decline in the Dopamine Nigrostriatal System - The Oxidative Hypothesis and Protective Strategies
Source:	ANNALS OF NEUROLOGY, 1992, Vol 32, Iss S, ppS133-S136
Addresses:	UNIV-ROCHESTER, SCH MED, DEPT NEUROBIOL & ANAT, BOX 603, 601 ELMWOOD AVE, ROCHESTER, NY 14642, USA
	OKAYAMA-UNIV SCH MED, DEPT NEUROL SURG, OKAYAMA 700, JAPAN
	ELI-LILLY-&-CO, LILLY RES LAB, INDIANAPOLIS, IN 46285, USA
Cited References:	....
	....

contains hospitals, clinics, community health centers, HMOs, disease treatment centers etc. NONPROFIT is a category made up of research centers, foundations, institutes and other not-for-profit but not government-affiliated organizations. PRIVATE comprises for-profit private sector organizations, principally pharmaceutical and biomedical firms. Organizations which we were unable to classify are given the designation UNCLASSIFIED.

An example may be helpful. Table II reproduces salient information from a sample record from *Science Citation Index*, one of more than 4700 that would be found by searching for papers published between 1980 and 1994 by scientists employed by Eli Lilly. The record for this paper lists five authors, and three addresses, and we interpret it as two 'coauthorships' between Eli Lilly and other organizations: one UNIVERSITY coauthorship with the University of Rochester, and one UNIVERSITY coauthorship with Okayama University. For reasons of economy, we collapse multiple addresses from the same institution, thus a record with four authors and three Eli Lilly addresses (e.g. Lilly Research La Jolla, Eli Lilly GMBH, and Eli Lilly Corporate Center Indianapolis) is treated as *one* instance of coauthorship with SELF. Similarly, a record with two Eli Lilly addresses (Eli Lilly Indianapolis and Lilly Research La Jolla) and two University of British Columbia addresses (UBC Department of Biochemistry and UBC Faculty of Pharmaceutical Sciences) is treated as one instance of coauthorship with UNIVERSITY.

It is important to note that while we made great efforts to count coauthorships and classify organizations consistently, these procedures and definitions are necessarily somewhat arbitrary. One potentially important issue is that it is not always clear to which category some organizations belong. For example it is difficult to draw a clear distinction between teaching hospitals affiliated with medical schools (classified as UNIVERSITY) and non-teaching hospitals (classified as HOSPITAL), or to classify some institutions as either NONPROFIT or PUBLIC. Nonetheless, we were able to classify almost all of the more than 14 000 distinct addresses in the data, with only 1044 falling into the residual category of UNCLASSIFIED. Weighted by their frequency of occurrence in the data, these UNCLASSIFIED organizations constitute less than 1.5% of the observations. We are reasonably confident about the success of the classification procedure. Due to the very large number of observations (and organizations) tinkering with the definition of the categories has little impact on the overall results: the great majority of the observations are attributable to institutions whose TYPE is unambiguous.

We make no attempt here to control for the quality of each publication. Weighting instances of coauthorship by the ranking of the journal, or by the number of citations the paper subsequently receives, may well be worth doing in further work.

For ten of these 20 firms, these data on publications and coauthorship are supplemented by an extensive data set collected for previous studies from confidential internal company records. This data set extends from 1965 to 1990 and includes discovery and development expenditures matched to a variety of measures of output including important patents, INDs, NDAs, sales and market share, as well as variables constructed from interview data which capture the nature of the firms' decision-making procedures and incentive systems. These data are described in more detail in previous work (Cockburn and Henderson [1994]; Henderson and Cockburn [1994], [1996]). Although for reasons of confidentiality we cannot identify these firms, we believe that they are not markedly unrepresentative of the industry in terms of size, or of technical and commercial performance.

III(ii).   *Analysis*

Publication activity of the private sector.   The descriptive statistics for our data on publication and coauthoring activity provide some interesting evidence on the scale and nature of public-private interaction. The first point to be made is that private sector scientists publish extensively. For these twenty firms alone, our working data set contains 68 186 papers published during the 15 year sample period: on average just over 227 papers per firm per year. These statistics conceal considerable variation across firms and over time: publications rates range from a minimum of only 2 per firm per year, to a maximum of 958 per firm per year.

These papers appear in leading journals: among the most frequent

170   IAIN M. COCKBURN AND REBECCA M. HENDERSON

outlets for these papers are *Abstracts of Papers of the American Chemical Society, Federation Proceedings*, the *Journal of Medicinal Chemistry*, the *Journal of Antibiotics*, and the *Journal of Biological Chemistry*. The list of top 50 journal outlets for the twenty firms is similar to that for the 81 574 'public-sector' papers in our NIH sample (obtained by searching for 'NIH' in *Science Citation Index* instead of 'Company X') with some interesting exceptions. There are more pharmacology journals in the top 50 list for the 20 firms, which is perhaps to be expected. But while prestigious journals with a clinical orientation,[10] such as the *Journal of the American Medical Association* and the *New England Journal of Medicine* are among the most frequent outlets for NIH researchers (*JAMA* is ranked 15th and the *New England Journal of Medicine* 12th) they appear very far down the list for the 20 firms (140th and 177th respectively.) This reflects one of the puzzles in these data: pharmaceutical company scientists appear to publish relatively few clinical papers. Searching the Medline database, which covers a somewhat different and smaller journal set but is exhaustively indexed, reveals relatively few papers where keywords for clinical research such as 'HUMAN SUBJECTS' appear in the index fields and a pharmaceutical firm is listed as an affiliation of one of the authors. To the extent that private sector organizations can be expected to publish less the closer the research is to the marketplace, this makes economic sense. But we hesitate to classify clinical research as being necessarily hard-to-appropriate or 'pre-competitive' and we continue to investigate this issue. One factor responsible for this phenomenon may be that editorial policy at some top journals prohibits publication of studies whose authors may get immediate financial rewards from the results.

Turning to the publication behavior of individual researchers, we obtain results similar to many previous bibliometric studies. By using the papers with only one address, we were able to compile a list of researchers affiliated with one of the 20 firms, and then to count the number of times these individuals appeared as an author on one of the papers in the dataset. This procedure excludes individuals who only appear as authors on papers with multiple addresses, but this is a fairly small number for most of the firms. The distribution of annual publication counts for these researchers is highly skewed to the left: of our 48 525 observations on annual publication rates of individuals, 56% are at one paper per year. Just as in public sector research organizations, in some firms an elite of leading researchers publishes very heavily, at a rate as high as 20 papers per year. Though there is substantial variation across firms and over time, it is nonetheless interesting that these private sector institutions exhibit the

---

[10] Great care should be taken in labelling journals as 'basic' or 'applied', or 'clinical' versus 'laboratory', as their content typically spans the whole spectrum of basic and applied research, as well as many different disciplines.

heavily skewed distribution of publications per researcher and disproportionate share of 'star' researchers characteristic of publicly funded research communities.

Patterns of Coauthorship. Our coauthorship measure of interaction with researchers outside the firm reveals extensive collaborative activity. Of the 68 186 papers in the sample, 49.6% involved collaborations across the boundaries of the 20 firms. Moving to the metric of coauthorships rather than publication counts, we observe a total 89 280 instances of co-authorship for our sample of 20 firms, counting papers authored entirely within the firm as one instance of coauthorship with SELF.

Breaking these coauthorships down by the TYPE of coauthoring institution, reveals some interesting information. For the sample as a whole, SELF accounts for 43% of all coauthorships, with UNIVERSITY being by far the largest external category at 34% of the total. HOSPITALs were just under 10% of the total, other PRIVATE institutions 5%, and NONPROFIT and PUBLIC accounted for a further 3% each. Curiously, NIH accounted for only 1% of the total. This appears not to be an artifact of the particular sample of firms: the breakdown of over 112 000 such coauthorships in our NIH sample is not markedly different, with the great majority of coauthorships being with SELF and universities, and less than 4% with private sector institutions. While many university researchers are supported by NIH grants, and thus should perhaps be re-classified as 'NIH', it is still significant that linkages between the private sector and the NIH are via this indirect channel.

Some interesting trends over time are apparent, both in amount of coauthoring activity and in the mix across different types of institutions. The rate of publication has been rising over time, from an average of 150 papers per firm per year in 1980 to more than 350 per firm per year in 1994. An even steeper trend is apparent in the number of coauthorships, which rose from 170 to 510 per firm per year over the same period.[11] The number of coauthorships per paper rose from 1.13 in 1980 to 1.42 in 1994. Over these 15 years the fraction of UNIVERSITY coauthorships rose steadily from 24% of the total to 38%, largely at the expense of SELF. The fraction of coauthorships with other private sector organizations also rose, but no significant trends in the aggregate share of the other types of coauthorships are apparent.

Links to Public Sector Research and Private Sector Research Productivity. These data on 'coauthoring' document significant linkages between private

---

[11] Care must be taken in interpreting these trends, both because we cannot normalize by the number of scientists employed by each firm, and because the journal set covered by *Science Citation Index* changes over time.

172   IAIN M. COCKBURN AND REBECCA M. HENDERSON

sector research and 'upstream' public sector activity. But the import of
such linkages is unclear. Does more participation in the wider scientific
community through publication or coauthoring give a private sector firm a
relative advantage in conducting research? In prior work we found
substantial and sustained variation in the research performance of
pharmaceutical firms, even after controlling at a very detailed level for
economies of scale and scope in the research laboratory, competitive
activity, differences in the composition of the research portfolio across
areas of technological opportunity and 'horizontal' spillovers of knowledge between related research projects within and between firms.[12] As
suggested above, we believe that these performance differentials may be
driven to a significant degree by differences in firms' linkages to the
'upstream' research community.

As Table III indicates, this sample of firms shows marked differences
in publication and coauthoring behavior, and the types of institutions they
collaborate with. The average fraction of coauthorships with SELF ranges
across firms from 31% to 68%, with HOSPITALs from less than 1% to
14%, and with other PRIVATE sector firms from 2% to 9%. There is also
substantial variation across firms in these evolution of these variables over
time. Formal tests strongly reject homogeneity across firms in the distribution of their coauthorships over TYPE, even after controlling for a time
trend. ($\chi^2_{133} = 5141$)

The question then arises as to whether these differences in coauthoring
activity are related to firms' research performance. For a subset of 10
firms, we are able to match the coauthoring data to our data on their
research performance and other measurable characteristics of their
research programs. We then look for a statistical relationship between
these variables and the extent and nature of coauthoring activity in two
ways, firstly by establishing a statistical link between firms' choice of type
of coauthor and the ways in which they organize the research function,
and secondly by testing for an association between these choices (or
other publication-based measures of the degree to which firms employ
the incentive mechanisms of the public sector) upon research
performance.

In prior work we constructed a number of measures of these ten firms'
internal incentive systems and decision-making procedures from structured
interviews. The extent to which each firm was 'pro-publication', in the
sense of using publication in the open literature or an individual's
reputation and standing in the wider scientific community as a basis for

---

[12] We use the term 'horizontal spillovers' to refer to the unpriced impact of research results
from one research project upon the productivity of a related research project at the same stage
of development. By contrast 'vertical spillovers' refers to their impact upon research projects
at different stages of the development process.

TABLE III
PATTERNS OF COAUTHORSHIP BY TYPE OF COAUTHOR AND FIRM

The first column identifies sample firms. The second column gives the total number of papers published by that firm appearing in *Science Citation Index* 1980–1994. The third column gives the total number of coauthorships by each firm. Entries in the rest of the table are the fraction of total coauthorships by each firm with different categories of coauthoring institution.

FIRM	Number of papers	Number of coauthorships	Fraction of coauthorships with:							
			SELF	Hospitals	Public	Private	Non-Profit	Universities	NIH	Unclassified
Abbott	3410	4894	0.40	0.10	0.04	0.04	0.03	0.36	0.02	0.01
Beecham	1139	1286	0.63	0.09	0.02	0.02	0.02	0.21	0.00	0.01
Bristol-Myers	4115	6177	0.37	0.10	0.03	0.06	0.02	0.39	0.02	0.01
Ciba Geigy	7656	10114	0.43	0.08	0.04	0.06	0.04	0.32	0.00	0.02
Fujisawa	1012	1159	0.62	0.04	0.02	0.04	0.00	0.27	0.00	0.01
Glaxo	3944	5517	0.33	0.14	0.02	0.06	0.02	0.41	0.01	0.01
Hoechst	3763	4900	0.41	0.07	0.03	0.10	0.04	0.32	0.00	0.02
Hoffman LR	4137	6061	0.32	0.09	0.04	0.05	0.03	0.41	0.05	0.03
Lilly	4761	6115	0.50	0.08	0.02	0.04	0.01	0.33	0.01	0.01
Merck	9291	11912	0.48	0.08	0.03	0.04	0.02	0.32	0.02	0.01
Nor. Eaton	232	262	0.63	0.05	0.01	0.07	0.00	0.23	0.00	0.01
Pfizer	2963	3927	0.42	0.08	0.02	0.07	0.02	0.36	0.01	0.02
Sankyo	1050	1133	0.68	0.01	0.02	0.03	0.01	0.25	0.00	0.00
Searle	1170	1573	0.38	0.05	0.02	0.05	0.02	0.46	0.01	0.00
Smithkline	3470	4766	0.40	0.08	0.05	0.07	0.03	0.35	0.01	0.01
Squibb	899	1071	0.58	0.11	0.01	0.03	0.00	0.25	0.00	0.01
Takeda	1793	2263	0.52	0.06	0.02	0.02	0.01	0.35	0.00	0.01
Upjohn	4982	6733	0.43	0.06	0.02	0.04	0.02	0.42	0.02	0.01
Wellcome	7840	11780	0.31	0.11	0.04	0.03	0.04	0.45	0.01	0.01
Yamanouchi	559	624	0.57	0.05	0.02	0.04	0.02	0.30	0.00	0.00
NIH	81574	112338	0.39	0.08	0.08	0.04	0.04	0.37	0.00	0.01

Source: *Science Citation Index*, various years, authors' computations.

174   IAIN M. COCKBURN AND REBECCA M. HENDERSON

promotion and compensation was rated on a 5-point Likert scale. A similar variable was constructed to measure the extent to which firms use a 'dictator' to allocate research resources versus a peer-review committee system. These ratings vary over time as well as across firms.[13] Table IV presents results from using these two variables, as well as a time trend and a measure of the scale of the firm's research effort, to model the choice of TYPE of coauthor in a multinomial logit framework. (To avoid computational problems caused by very small numbers of observations in some cells, and to facilitate presentation of the results, NIH is subsumed in PUBLIC, and UNCLASSIFIED is dropped.) The first panel of the table reports the estimated coefficients of the model, while the second panel gives the estimated marginal effects upon the probability of choosing each category.

Encouragingly, we find effects consistent with our hypothesis that variation across firms and over time in coauthoring choices is related systematically to differences in the ways in which they organize their research function. Conditional on publishing, being more 'pro-publication' is negatively associated with the probability of coauthoring only with researchers in the same firm, and positively associated with the probability of coauthoring with external institutions. The largest such association is with the probability of coauthoring with a university. Increasing use of a 'dictator' is associated with an increased likelihood of coauthorship with SELF, and a decreased the likelihood of publishing with external coauthors. Goodness-of-fit measures are hard to interpret in this kind of model, but we note that comparing the estimated model and the restricted model with constants gives a likelihood ratio test statistic of $\chi^2_{20} = 660.7$, strongly rejecting the hypothesis that the estimated coefficients are all zero. Obviously we can infer nothing about causality from these results, but they are at least suggestive of a meaningful link between coauthoring activity and firms' choices about managing research.

We turn therefore to the relationship between research performance and 'connectedness' to the public sector. Table VI presents results from estimating simple research productivity regressions, using explanatory variables derived from the bibliometric data, as well as other controls for factors that our previous work has shown to be related to research productivity such as the scope of the research portfolio. Research output is measured by 'important patents', where 'importance' is defined by the patent being granted in at least two out of three major world markets — Japan, the US and Europe — and research inputs are measured by expenditure on the 'discovery' phase of R&D, excluding

---

[13] See Henderson and Cockburn [1994] for more detail on the interview process and the construction of these variables.

THE ORGANIZATION OF RESEARCH IN DRUG DISCOVERY   175

TABLE IV
MULTINOMIAL LOGIT COEFFICIENTS
DEPENDENT VARIABLE: TYPE OF COAUTHOR INSTITUTION
SUBSAMPLE OF 10 FIRMS, 1980–1988 DATA: 26501 OBSERVATIONS

Category: TYPE of coauthor institution		Explanatory Variables			
	Time trend	Degree to which firm is 'pro-publication'	Degree to which decisions are made by 'dictator'	Size of firm's drug discovery effort in $m	Constant
HOSPITAL	0.029 (0.023)	0.009 (0.044)	0.059 (0.038)	−0.008** (0.002)	−1.220 (1.783)
NON-PROFIT	0.001 (0.027)	0.176** (0.061)	0.075 (0.049)	−0.011** (0.002)	−0.726 (2.275)
PUBLIC incl NIH	−0.066** (0.022)	0.313** (0.053)	−0.103 (0.041)	−0.006** (0.002)	5.155** (1.838)
SELF	−0.062** (0.018)	−0.131** (0.040)	0.163** (0.034)	−0.003* (0.001)	8.089** (1.526)
UNIVERSITY	0.014 (0.019)	0.078* (0.040)	0.085** (0.034)	−0.008** (0.002)	1.018 (1.545)

Notes: Reference category: PRIVATE. BHHH standard errors in parentheses. Log-likelihood function = −24987. Restricted log-likelihood function = −25317.

MARGINAL EFFECTS (IN PERCENT) AT THE SAMPLE MEAN ON PROBABILITY OF
COAUTHORING WITH:

Category: TYPE of coauthor institution		Explanatory Variables		
	Time trend	Degree to which firm is 'pro-publication'	Degree to which decisions are made by 'dictator'	Size of firm's drug discovery effort in $m
PRIVATE	0.078 (0.055)	0.038 (0.119)	−0.035** (0.010)	0.017** (0.004)
HOSPITAL	0.516** (0.094)	0.185 (0.193)	−0.486** (0.169)	−0.026** (0.007)
NON-PROFIT	0.059 (0.049)	0.446** (0.011)	−0.082 (0.090)	−0.015** (0.004)
PUBLIC incl NIH	−0.224** (0.068)	1.739** (0.189)	−0.645** (0.132)	−0.002 (0.004)
SELF	−1.719** (0.151)	−5.443** (0.333)	2.409** (0.289)	0.110** (0.012)
UNIVERSITY	1.292** (0.149)	3.040** (0.326)	−0.850** (0.276)	−0.010** (0.001)
Sample means of covariates	84.99	4.05	3.08	50.57

** indicates significant at 5% level   * indicates significant at 10% level
Source: authors' computations.

© Blackwell Publishers Ltd. 1998.

176   IAIN M. COCKBURN AND REBECCA M. HENDERSON

TABLE V
DESCRIPTIVE STATISTICS FOR REGRESSION SAMPLE

$N = 82$	Mean	Min	Max	Std. Dev.
Important Patents	39.71	2	141	30.58
Research expenditure (constant 1986 $million)	38.59	5.81	>100	26.48
Fraction of coauthorships with universities	0.23	0.00	>0.4	0.07
Papers per research dollar	11.23	0.99	108.84	20.14
Pro-publication	3.26	1.00	5.00	1.39
'R&D Dictator'	2.90	1.00	5.00	1.64
Fraction of publications by top 10% of authors	0.37	0.22	0.64	0.08
SCOPE: Number of substantial research programs	9.58	2.00	19.00	4.44
Stock of past patents	197.76	2.40	585.19	131.78

clinical development.[14] We use a log-log specification, with the relationship between research expenditure and patents mediated by the other explanatory variables. The fraction of coauthorships in each year with universities is a measure of the degree to which the firm is linked to the public sector.[15] Total publications per research dollar per year can be thought of a measure of investment in doing basic science, that is absorptive capacity in the classic sense. A time trend is included in all regressions to control for the strong aggregate trend of declining patents per research dollar in these data. We also calculated the fraction of the firm's publications in each year attributable to the top 10% of its scientists (ranked by number of publications) as a proxy for the presence of a 'star' system within the firm, a key aspect of the incentive system of 'open science' practiced in the public sector. Table V gives descriptive statistics for the sample.[16]

These regressions are presented as a reduced form representation of

[14] See Henderson and Cockburn [1996] for details on the sources and construction of these data. We note that 'important patents' are a very limited measure of research output, and subject to a number of problems of measurement and interpretation. In particular the economic value of patents is highly skewed, and propensity to patent may vary systematically across firms. On the other hand, patent counts are at least consistently measured across firms and over time, and they are closely linked to research activity. Other output measures, such as sales, or regulatory approvals are difficult to attribute to specific research expenditures because of the very long and variable lags in the drug development and approval process.

[15] Collapsing TYPEs by e.g. aggregating UNIVERSITY, NIH, PUBLIC, NON-PROFIT has little effect due to the small numbers of coauthorships in most categories.

[16] The panel is slightly unbalanced (82 rather than 90 observations) due to merger of some of the firms.

THE ORGANIZATION OF RESEARCH IN DRUG DISCOVERY   177

patterns of association in these data, and we draw no strong inferences about causality. There is clearly potential for joint determination of both research performance and organizational practices and policies by other factors (unobserved differences in human capital for example), as well as for reverse causality between research performance and organizational practice. We make some attempt to control for these difficulties by using lagged rather than contemporaneous realizations of the explanatory variables, and including the stock of past patents (which is essentially a distributed lag on the dependent variable) in some of the regressions. Ideally, of course, we would estimate a structural model of the adoption of organizational practices, coauthoring activity, research expenditure and its allocation across projects, and their impact on research performance, but this requires more data and presents its own problems.

Notwithstanding these concerns, the results reported in Table VI suggest an important relationship between 'connectedness' and research productivity. In Table VI the coefficient on fraction of coauthorships with universities is positive and significant in four of the five regressions reported. In column (2) we control for systematic firm-specific measurement problems, such as variation in the propensity to patent, with firm fixed effects. In column (3) we control explicitly for 'propensity to publish', and other factors internal to the firm which may affect research productivity. In both cases there is little impact on the 'connectedness' coefficient. The positive coefficients on 'pro-publication' and negative coefficients on 'dictator' conform to our priors and to previous results. When we further control for the scope of the firm's research portfolio and its past success (column (4)), the 'connectedness' variable retains significance, though limited variation in the data and multicollinearity between the other explanatory variables weaken some of the other results. The 'connectedness' variable only loses significance in column (5) of the table where we include both a full set of control variables and firm fixed effects, which effectively 'kills' the regression because of multicollinearity. The elasticity of important patents with respect to 'connectedness' is quite large, around 2.2 at the sample mean. The estimated partial effect of publishing more papers is much smaller and statistically insignificant: the estimated elasticity of patents with respect to papers per research dollar is close to zero. This may reflect confounding with other factors, and a variety of other specification problems, but we interpret this as evidence that the *nature* of linkages with public sector research may have a significant effect on private sector research productivity, above and beyond the amount of publishing activity *per se*.

Interestingly, the pro-publication and dictator variables have fairly stable coefficients, consistent with results obtained in previous work with a somewhat different sample. Though coauthoring activity and reportedly pro-publication incentives may only be proxies for 'deeper' internal

TABLE VI
COAUTHORSHIP AND RESEARCH PRODUCTIVITY
OLS REGRESSION. DEPENDENT VARIABLE: LOG(IMPORTANT PATENTS)
SUBSAMPLE OF 10 FIRMS 1980–1988 DATA, 82 OBSERVATIONS

	(1)	(2)	(3)	(4)	(5)
Intercept	−2.556	−1.133	−0.085	−5.633**	−12.17**
	(1.226)	(2.036)	(1.312)	(1.747)	(5.349)
log(research expenditure)	1.307**	0.970*	0.753**	0.686	0.836
	(0.276)	(0.598)	(0.370)	(0.481)	(0.731)
Fraction of coauthorships with universities	11.966**	7.467**	9.050**	7.219**	4.277
	(3.068)	(3.541)	(2.952)	(2.927)	(3.930)
Papers per research dollar			−0.000	0.000	−0.001
			(0.012)	(0.011)	(0.043)
Pro-publication			0.417**	0.259	0.312
			(0.208)	(0.185)	(0.397)
'R&D Dictator'			−0.500**	−0.236	−0.113
			(0.124)	(0.139)	(0.200)
SCOPE: Number of substantial research programs				0.296	0.289
				(0.258)	(0.374)
SCOPE-squared				−0.018	−0.017
				(0.012)	(0.018)
log(Stock of past patents)				0.845**	2.467**
				(0.283)	(1.092)
FIRM DUMMIES		YES			YES
TIME TREND	−0.351**	−0.263**	−0.320**	−0.236**	−0.168
	(0.081)	(0.090)	(0.076)	(0.082)	(0.106)
RMSE	1.711	1.394	1.563	1.350	1.382
R-squared	0.317	0.599	0.452	0.607	0.640

All explanatory variables lagged once with respect to dependent variable.

Standard errors in parentheses
** indicates significant at 5% level
* indicates significant at 10% level

Source: authors' computations.

policies and practices within the firm, or may be the observable consequences of the accumulation of particular types of human capital, there is a hint here that they have statistically distinct relationships with research performance.

A final point to be made about Table VI is that the 'connectedness' variable has explanatory power in both the within and between dimensions of the data. The research productivity of the firms in this sample varies substantially both across firms and within firms over time, as does their measured 'connectedness' and these two effects are systematically related in both dimensions. Our other explanatory variables are more stable over time and thus more easily 'knocked out' by firm fixed effects, and their coefficients more difficult to separately identify. Taken literally the regressions imply an almost contemporaneous impact of changes in 'connectedness' on research productivity. Clearly this is unlikely to be the case, but the somewhat short (and unbalanced) panel we are working with precludes identification of a longer or more complicated lag structure.

By contrast with the coauthoring measure of 'connectedness', presence of a 'star' system is only weakly correlated with research productivity. When included in regressions such as those reported in Table VI, its coefficient is unstable and insignificant. This may simply reflect that this variable, or any similar measure of the presence of 'stars', is necessarily very noisy, and data problems leave us unconvinced that we are measuring it correctly.

We hesitate to over-interpret these results: confounding with aggregate time trends, the small sample imposed by incomplete data, difficulties with lags, causality, and a variety of other measurement problems discussed in previous papers mean that they are not as statistically robust as we would prefer. Furthermore they are offered as descriptive results rather than tests of an underlying behavioral model. Nonetheless they offer support for the hypothesis that the ability to access and interact with public sector research activity may be an important determinant of the productivity of 'downstream' private sector research. This effect is quite large: the estimated coefficients imply differences in research productivity of about 30 percent between the most connected and least connected firms in the sample.

IV. CONCLUSIONS

It is widely accepted that public sector research makes a significant contribution to growth by supplying basic, non-market oriented scientific knowledge that the private sector has weak incentives to produce. The precise mechanisms driving this effect are less well documented, and remain imperfectly understood, but surely are more complex than the simple frictionless transfer of fundamental science into the private sector. Qualitative research has shown that the relationship between public and private sectors is one of reciprocal interaction, and many observers have commented that publicly funded research also has an important role in training researchers and facilitating information flow across a scientific and technical community that transcends organizational boundaries (Nelson [1988], [1990]; Rosenberg and Nelson [1993]; Cohen and Noll [1996]).

In this paper we have drawn on a number of case studies and on a program of qualitative interviews to argue that, at least in the pharmaceutical industry the relationship between the public and private sectors is not well described by a simple 'waterfall' model in which the public sector

180   IAIN M. COCKBURN AND REBECCA M. HENDERSON

produces knowledge that spills over costlessly to downstream researchers. Private sector results sometimes have importance for public sector work, and the transfer of information between public and private sectors can be costly and time consuming.

We have suggested firms wishing to take advantage of public sector research must do more than simply invest in in-house basic research: they must also actively collaborate with their public sector colleagues. Further, we have shown that one measure of the nature and extent of this collaboration, coauthorships across institutions, is correlated with private sector research productivity. While our results must be interpreted with caution, since it is difficult to determine causality and 'connectedness' may be correlated both with a variety of other organizational factors that improve productivity and, plausibly, with other unmeasurable sources of heterogeneity, we believe they reflect substantial differences across firms in their ability to access public knowledge.

We hesitate to generalize from these results, particularly since the biomedical sector is exceptional in terms of the significance of R&D to product market competition, the immediate relevance of fundamental research, the rate of change of its science base, and the magnitude of the public research commitment. But we suspect that they may have relevance to other industries in which incumbent firms are grappling with the power of new technologies. We believe that R&D managers may have much more to learn from academic science than published research results, while scholarly understanding of macroeconomic growth may be enhanced by closer attention to the wider institutional setting of industrial R&D, in particular the nature and extent of linkages to non-profit organizations.

These results suggest a number of directions for future research. On the public policy front, we speculate that the ability to 'do good science' in the private sector may not be supportable in the long run without a close partnership with the institutions of open science. Policies which weaken these institutions, make public sector researchers more market-oriented, or redistribute rents through efforts to increase the appropriability of public research through restrictions in the ways in which public and private sectors work with each other may be therefore be counter-productive in the long run. On the private policy front, we believe that these results highlight the importance of building a richer understanding of the adoption of, and the interaction amongst, key organizational practices. One of the most intriguing aspects of our work is the presence of a multifaceted complex of interrelated organizational drivers of research productivity. The attempt to tease apart these factors and to understand the causes of their diffusion is a principle focus of our ongoing research.

ACCEPTED DECEMBER 1997

THE ORGANIZATION OF RESEARCH IN DRUG DISCOVERY   181

REFERENCES

Aoki, M., 1990, 'Towards An Economic Model of the Japanese firm', *Journal of Economic Literature*, 28(1), pp. 1–27.

Cockburn, I. and Henderson, R., 1994, 'Racing to Invest? The Dynamics of Competition in Ethical Drug Discovery', *Journal of Economics and Management Strategy*, 3(3), pp. 481–519.

Cockburn, I. and Henderson, R., 1996, 'Exploring Inertia: Complementarities and Firm Effects in Pharmaceutical Research', Mimeo, MIT April 1996.

Cohen, W. M. and Levinthal, D. A., 1989, 'Innovation and Learning: The Two Faces of R&D', *Economic Journal*, 99, pp. 569–596.

Cohen, L. and Noll, R., 1996, 'Privatizing Public Research: the New Competitiveness Strategy' in Landau, R., Taylor, T. and Wright, G., (eds.) *The Mosaic of Economic Growth* (Stanford University Press, Stanford).

Comroe, J. and Dripps, R., 1976, 'Scientific Basis for the Support of Biomedical Research', *Science*, 192, p. 105.

Gambardella, A., 1992, 'Competitive Advantage from in-house scientific research: The US pharmaceutical industry in the 1980s', *Research Policy*, XX, pp. 1–17.

Gambardella, A., 1995, *Science and Innovation: the US Pharmaceutical Industry in the 1980s* (Cambridge University Press, Cambridge).

Griliches, Z., 1979, 'Issues in Assessing the Contribution of Research and Development to Productivity Growth', *Bell Journal of Economics*, 10(1), pp. 92–116.

Griliches, Z., 1994, 'Productivity, R&D and the Data Constraint', *American Economic Review*, 84(1), pp. 1–23.

Henderson, R. and Cockburn, I., 1994, 'Measuring Competence? Exploring Firm Effects in Pharmaceutical Research', *Strategic Management Journal*, 15, pp. 63–84.

Henderson, R. and Cockburn, I., 1996, 'Scale, Scope and Spillovers: The Determinants of Research Productivity in Drug Discovery', *RAND Journal of Economics*, 27(1), pp. 32–59.

Hicks, D., 1995, 'Published Papers, Tacit Competencies and Corporate Management of the Public/Private Character of Knowledge', *Industrial and Corporate Change*, 4(2), pp. 401–424.

Jones, C. and Williams, J., 1995, 'Too Much of a Good Thing? The Economics of Investment in R&D' Stanford University Department of Economics Working Paper #96-005.

Koenig, E., 1982, 'A Bibliometric Analysis of Pharmaceutical Research', *Research Policy*, 12(1), pp. 15–36.

Liebskind, J. P., Oliver, A. L., Zucker, L. and Brewer, M., 1995, 'Social Networks, Learning, and Flexibility: Sourcing Scientific Knowledge in New Biotechnology Firms', NBER Working Paper No. 5320, November 1995.

Maxwell, R. A. and Eckhardt, S. B., 1990, *Drug Discovery: A Case Book and Analysis* (Humana Press, Clifton NJ).

Narin, F. and Olivastro, D., 1992, 'Status Report - Linkage Between Technology and Science', *Research Policy*, 21(3), pp. 237–249.

Nelson, R., 1988, 'Institutions Supporting Technical Change in the United States', in Dosi, G. *et al.* (eds.), *Technological Change and Economic Theory* (Pinter, London).

Nelson, R., 1990, 'Capitalism as an Engine of Progress', *Research Policy*, 19(3), pp. 193–214.

182   IAIN M. COCKBURN AND REBECCA M. HENDERSON

Penan, H., 1996, 'R&D Strategy in a Techno-economic network: Alzheimer's disease therapeutic strategies', *Research Policy*, 25, pp. 337–358.

Raiten, D. and Berman, S., 1993, 'Can the Impact of Basic Biomedical Research be Measured? A Case Study Approach'. Working Paper, Life Sciences Research Office, Federation of American Societies for Experimental Biology.

Rosenberg, N. and Nelson, R., 1993, 'American Universities and Technical Advance in Industry', CEPR Working Paper #342, Stanford University.

Stephan, P., 1996, 'An Essay on the Economics of Science', *Journal of Economic Literature*, 34(3), pp. 1191–1235.

Ward, M. and Dranove, D., 1995, 'The Vertical Chain of R&D in the Pharmaceutical Industry'. *Economic Inquiry*, 33, pp. 1–18.

Walsh, V., 1984, 'Invention and Innovation in the Chemical Industry: Demand-pull or Discovery-push?', Research Policy, 13, pp. 211–234.

Wurtman, R. J. and Bettiker, R. L., 1994, 'How to Find a Treatment for Alzheimer's Disease', *Neurobiology of Aging*, 15, pp. S1–S3.

Zucker, L., Darby, M. and Brewer, M., 1997, 'Intellectual Capital and The Birth of U.S. Biotechnology Enterprises', NBER Working Paper No. 4653, February 1994.

Zucker, L., Darby, M. and Armstrong, J., 1994, 'Intellectual Capital and the Firm: the Technology of Geographically Localized Knowledge Spillovers', NBER Working Paper No. 4946, December 1994.

Zucker, L. and Darby, M., 1995, 'Virtuous Circles of Productivity: Star Bioscientists and the Institutional Transformation of Industry', NBER Working Paper No. 5342, November 1995.

[17]

ELSEVIER

Research Policy 26 (1997) 429–446

research
policy

Present at the biotechnological revolution: transformation of technological identity for a large incumbent pharmaceutical firm

Lynne G. Zucker, Michael R. Darby *

University of California, Los Angeles and National Bureau of Economic Research Los Angeles, CA, USA

Abstract

Management of successful incumbent firms experience difficulty in recognizing the need for, and effecting change in the firm's technological identity after an externally generated shift in the industry's technological trajectory. Nonetheless, some large pharmaceutical firms have transformed their technological identity in drug discovery from a chemical/random screening to biological/drug design model. We report how one of the world's most successful incumbents transformed. Technically sophisticated senior management championed the transformation. It was achieved primarily through hiring many new scientists embodying biotechnology; existing personnel acquired the expertise or left. Continual self-transformation is part of the corporate ethos. Some differences in incumbent and entrant technology remain: incumbents use a wider range of techniques consistent with their complementary assets. Publication and incentive compensation policies are driven by the need to attract and retain the best scientists. Professor–firm collaborations are ubiquitous, often non-public, and best identified in quantitative analyses by co-publishing. Collaborations with new biotechnology firms are used primarily to substitute for developing internal expertise judged of marginal value. No drug-discovery collaborations exist with other major incumbents. We identify another seven or eight incumbents similarly transforming as indicated by top scientific talent and patenting success. Published by Elsevier Science B.V.

Keywords: Biotechnology; Collaboration; Incumbent; Pharmaceutical; Transformation

1. Introduction

Technological advance arising within an industry is an effective means of increasing sales and net income of the incumbent firms in an industry, although the gains of the firms leading the innovation may come at some cost in market share and profits of lagging firms. However, when a revolutionary breakthrough in technology originates outside the industry and uses a different set of skills from those required to practice the existing technology, new entrants may replace incumbent firms as a group,

---

* Corresponding author. Anderson School, UCLA, 110 Westwood Plaza, Box 951481, Los Angeles, CA 90095-1481, USA.

0048-7333/97/$17.00 Published by Elsevier Science B.V.
PII S0048-7333(97)00017-6

and even the definition of the industry may be transformed (e.g., from carriages to automobiles).

The wave of innovations in drug discovery associated with the advent of modern biotechnology—beginning in the 1970s, gaining strength in the 1980s, achieving dominance in the 1990s—appears to be an archetypal example of externally generated, incumbent-skill-obsoleting, discontinuous innovation, which the literature predicts leads to replacement of incumbents (pharmaceutical firms) by entrants (new biotechnology firms). There has been an ongoing process of consolidation among the incumbent drug-discovery firms, but a substantial number of incumbent firms surprisingly have flourished. We report here on the experience of one of the largest flourish-

430   *L.G. Zucker, M.R. Darby / Research Policy 26 (1997) 429–446*

ing incumbents to enrich our understanding of the process by which incumbents are or are not supplanted by entrants in the face of an external technological discontinuity.

The paper is organized as follows: Section 1 reviews the literature on technological revolution and displacement of incumbent firms. Section 2 provides a primer on the biotech revolution as it applies to drug discovery. Section 3 presents the goals and methodology of the case study. We turn to Section 4 in reporting what we learned about how this major firm transformed its technological identity to the point that its scientists can claim that there is no difference between the best new biotech firms and itself in how research is done. We next report the firm's approach to collaborations with scientists at universities and at other firms in Section 5. In Section 6, we present some evidence on how generalizable the case study may be by comparing this and other major pharmaceutical firms with the dedicated biotech firms in terms of their access to leading-edge scientists and their genetic-sequence patenting success. Conclusions are drawn in Section 7.

2. Technological revolution and displacement of incumbent firms

The concept of the technological identity of the firm is basic to the issue of displacement of incumbent firms through technological revolution. In the simplest models, all firms have identical production and cost functions. Underlying the enormous population–ecology literature is the idea that each firm has a fixed technological identity, related to the broader concept of entrepreneurial capacity (Friedman, 1976, pp. 106–126), but these identities vary across firms. The entry or exit of the firm depends on how suited its technological identity is to the industry's competitive conditions at any given time (Baum, 1996 provides an excellent recent review of this literature). Kaufman (1971) (pp. 8–23) argues from an organizational point of view that it is extremely difficult for organizations to change. [1] If a firm's technological identity is fixed, then studies of entry or exit can

tell us something about which technology is dominant, for the time being, but transformation for survival is impossible.

Nelson and Winter (1982) elaborate a concept of the firm that is amenable to transformation. They see an organization as separable from the sum of its parts because it embodies particular information about how to do things in a set of task routines that require little direct intervention by management to ensure that the work of the organization gets done. For organizations in technologically-based industries, these task routines and the embodied knowledge they represent will determine the technological identity of the firm. Thus, a firm can have a kind of organizational capital that provides a continuing competitive advantage. Unless conditions change, new entrants cannot eliminate this competitive advantage by hiring away individual employees (although there is occasional litigation over firms hiring a large number of employees in an attempt to replicate these task routines).

Demsetz (1988) sees the identity of the firm determined by the fields for which it has acquired organizational mastery of specialized bodies of knowledge. Individual actors—managers—can choose to enlarge or change the fields for which the organization incorporates mastery (see the review by Tolbert and Zucker, 1996). The technological identity of a great high-technology firm is recognized as the source of its strength, providing persistent high returns (Waring, 1993). It may be difficult to either recognize or effect any substantial change in what has worked so well and, for a time, will continue to produce supranormal earnings in the face of a breakthrough change in technology, which will ultimately convert that strength to an outmoded, uncompetitive technology. The question is when management of an incumbent firm will recognize a new dominant technology, and whether they will then choose to transform the firm's technological identity or gradually withdraw from the industry in the face of the new technology.

As suggested by Dosi (1982), an existing organization can achieve routine excellence in pursuit of a given technological trajectory that carries forward a shared technological paradigm, but this process does not lead to substitution of new technological trajectories and paradigms in the face of opportunities aris-

---

[1] Kaufman does allow for change to occur through personnel turnover.

L.G. Zucker, M.R. Darby / Research Policy 26 (1997) 429–446   431

ing from scientific breakthroughs (see also Klevorick et al., 1995 on the importance of technological opportunity). Lounamaa and March (1987) elucidate some of the difficulties in organizations attempting to learn adaptively in the face of a changing technological environment. (See March and Simon, 1993 for a more nuanced review of the literature on organizational learning.)

Tushman and Anderson (1986) present evidence that the survival of incumbent firms is enhanced by technological discontinuities introduced by incumbent firms, and based on their existing knowledge set while survival is threatened if the discontinuities are external in origin and require mastery of new fields of knowledge. Of course, the choice of exit in the face of new, unfamiliar technologies may be completely in line with rational wealth maximization. Henderson and Clark (1990) and Henderson (1993) elaborate the Tushman and Anderson hypothesis and present further evidence in its support. Reinganum (1983, 1989) also argues that since incumbents have monopoly rents whose value will be lost from radical innovations, such breakthroughs are most likely to come from entrants rather than incumbents.

Kimberly and Quinn (1984) are more optimistic about the possibility of managers taking entrepreneurial actions that transform the organization. Kim (1997) reports numerous case studies in which top management has consciously created crises for organizational subunits to achieve innovative transformation in the rapidly developing Korean economy. In Zucker and Darby (1996a), we report evidence of extensive transformation of most of the world's top-twenty drug-discovery firms by the early 1990s, as evidenced by discovery of new biological entities, genetic sequence patents, and co-publishing with top academic biotech scientists. [2] Therefore, the drug-discovery pharmaceutical industry appears to present a case in which numerous firms have pursued a strategy of transformation of technological identity—adopting the new technological trajectory rather than pursuing 'underinvestment and incompetence as responses to radical innovation'. This surprising success might provide us the basis for better understanding which incumbent firms transform and which die in the face of an external technological discontinuity.

Good fortune—in the form of targeted support from the Alfred P. Sloan Foundation through the NBER Research Program on Industrial Technology and Productivity and personal intervention by NBER President Martin Feldstein—provided us access to the top research and policy management, as well as some archival material, of one of the most successful transforming firms. Section 3 provides a brief primer on biotechnology and its relation to the drug-discovery industry as required for interpreting what we learned about the process of transformation from investigating this firm.

3. The biotechnological revolution in drug discovery

The biotechnological revolution is an outstanding example of the type of technological breakthrough most likely to result in the replacement of previously dominant incumbent firms by newly created firms that encompass the new technology in their technical identity: the new technology's origin is in academic biological sciences, and its practitioners design drugs based on scientific hypotheses, while the drug discovery technology, dominant in the pharmaceutical industry in the 1970s and into the 1980s, was based on chemistry and involved nearly random screening of molecules to discover ones that were effective.

The major pharmaceutical firms of the type we are concerned with in this paper are creators, manufacturers and marketers of human therapeutics, vaccines, and diagnostics. An essential characteristic of this segment of the pharmaceutical industry is that they are involved in discovering new products that are protected for a time by patents. Purely generic manufacturers, whose incomes depend on being low-cost manufacturers and distributors of drugs once they go off patent are not really part of the drug-dis-

---

[2] As discussed below, many drug-discovery firms have pursued a strategy of using new biotechnological techniques to discover small-molecule drugs that can be synthesized by chemical methods rather than produced by living organisms. Thus, all their biotech-based drugs would be classed as new chemical entities, not new biological entities.

432   *L.G. Zucker, M.R. Darby / Research Policy 26 (1997) 429–446*

covery business. [3] The revolution in the biosciences has transformed technologies used in many other industries (including medical supply, chemical, agricultural, food-processing, and brewing), but none so rapidly and dramatically as in drug discovery.

3.1. What is the biotech revolution?

Broadly enough defined, biotechnology has been used as long as people have baked bread and drank wine. Crossbreeding of animals and growing penicillin are other examples of such traditional forms of biotechnology. Today, biotechnology, or biotech, is generally defined more narrowly in terms of using breakthrough technologies such as genetic engineering. An excellent working definition of biotechnology, as put forward by a respondent at the subject firm of the case study, would be as follows:

In discussing biotechnology at [the firm], I use biotechnology to mean the revolutionary breakthroughs in life sciences over the last two decades including especially, the use of recombinant DNA to create living organisms and their cellular, subcellular, and molecular components as a basis for producing both therapeutics and targets for testing and developing therapeutics. Recent developments focus on structural biology, combinatorial chemistry, and gene therapy.

In ongoing research with a number of associates, we have found it useful to date the beginning of the biotech revolution in bioscience with the 1973 discovery by Stanford professor Stanley Cohen and University of California-San Francisco professor Herbert Boyer of the basic technique for recombinant DNA (Cohen et al., 1973). The commercial applications of biotechnology followed quickly in new biotechnology enterprises formed as early as 1975 and 1976. Sindelar (1992, 1993) provides a

useful introduction to these applications in the pharmaceutical industry. [4]

The revolution in bioscience is not completed nor is it entirely clear what will ultimately prove its most important areas of applications in the pharmaceutical industry. It is clear, however, that biotech is a dominant technology for at least some areas of production of biological agents, for creation of targets for screening and evaluating potential pharmaceutical products, and as a methodological base for creating potential pharmaceutical products. Currently, firms in the industry are undergoing a shakeout both as it becomes easier to separate the most effective from the less effective of those firms using the new technologies, and also as government and regulatory initiatives and health-care market restructuring impact anticipated future and current profitability in the pharmaceutical industry. Nonetheless, it is clear that biotech is a dominant technology even as it is unclear what will be the makeup of the industry in which it will be utilized.

3.2. Intellectual human capital and the evolution of industrial biotechnology

In a series of papers recently summarized and extended in Zucker and Darby (1996b), we have shown that the top-producing or 'star' scientists—as measured by frequency of appearance in GenBank,

---

[3] For some purposes, it might be preferable to exclude also diagnostics from the definition of the pharmaceutical industry. However, many new firms aiming to enter the therapeutics and vaccines industry first produce diagnostics as a faster source of revenues utilizing their technologies. The economics of this industry was reviewed extensively by Comanor (1986).

---

[4] Sindelar (1992) (pp. 3–4) notes in reference to pharmaceuticals that modern biotechnological techniques can be divided "into three broad areas…" Recombinant DNA techniques "take identified gene sequences from one organism and place them functionally into another to permit the production of protein medicines such as human insulin, alpha interferon, and colony-stimulating factors. Second, methodologies have been developed for producing monoclonal antibodies, ultrasensitive immune system-derived cells designed to recognize specific substances known as antigens that are uniquely associated with chemicals found in foreign organisms and/or humans. Developments in this field have led to their use as diagnostic agents for laboratory and home use in pregnancy tests and ovulation prediction kits and in the design of site-directed drugs such as OKT-3 for kidney transplant rejection. Finally, the development of technologies to study DND–DNA and DNA–RNA interactions has led to the formation of DNA probes (antisense technology) for a variety of research purposes with potential uses as diagnostics and therapeutics."

L.G. Zucker, M.R. Darby / Research Policy 26 (1997) 429–446   433

the universe of all genetic-sequence discoveries, up to 1990—played a crucial role in determining where and when firms entered biotechnology, and which of them were most successful. Furthermore, during their periods of active publication as or with employees of firms, the stars are more productive (both in terms of articles per year and citations per article) than they are before or after that time. The star effect on firm success is large: an otherwise average firm with five articles co-authored with a local university star is estimated to have 5 more products in development and 3.5 more products on the market by 1991, and 860 more employment growth from 1989 to 1994 (Zucker et al., 1997a,b). The number of universities with top-quality bioscience departments and the number of scientists supported by federal grants in the local region also affect the rate of entry of firms into biotech, but the number of collaborators (co-authors of the stars) does not (unless marginally late in 1980s).

Interestingly, the largest concentrations of the intellectual human capital providing the scientific base for biotechnology was located in California and the Boston area. This distribution was quite different from that of incumbent pharmaceutical firms, although there was another concentration in New York City, which was reasonably close to some of the incumbents. Therefore, a key question for exploration in the case study below is how the firm accessed the scientific base, particularly in view of both geographic impediments, and the ability of star scientists to participate in founding new biotechnology firms, which made many multimillionaires when the new entrants went public.

4. Goals and methodology for the case study

Our earlier empirical work was grounded by case studies of new biotechnology firms in the US and the UK. and fieldwork on four Japanese incumbent firms that had adopted biotechnology (Liebeskind et al., 1996; Darby and Zucker, 1966). In the Japanese fieldwork, we repeatedly heard the view that Japan's system was inhospitable to biotechnology, in part, because of the institutional barriers to founding new biotechnology firms, so that transformation of in-

cumbent firms was the only unsatisfactory alternative. Accordingly, we were delighted by the opportunity to explore the transformation of one of the five largest US pharmaceutical firms, although permission to do so was granted only on condition that the firm not be identified.

This firm is widely and correctly regarded as one of the most successful pharmaceutical companies in the world, with an enviable record of science-based drug discovery and development, outstanding abilities in the management of clinical testing, excellent ability to shepherd a New Drug Application (NDA) through the FDA in as short a time as possible, and a first-rate marketing group to effectively distribute the products when they are approved. It is one of a handful of such firms in the US, each endowed with loyal employees who are both proud of their company, which they believe the best among some outstanding competitors, and very pleased that their personal success is in direct proportion to their ability to contribute to reducing suffering and death among victims of disease.

Given these company attributes, it is easy to see why economists and management strategists would hypothesize that it would prove difficult, if not impossible, to tamper with a proven formula for success. Nonetheless, we found that the ethos of this science-driven company valued innovation and that there was great pride taken in the firm's ability to continuously change how it did its research and development, so that ongoing technological change appeared to be an integral part of the firm's identity. [5]

4.1. Goals

In our experience, lengthy discussions with knowledgeable participants with reinterviews to discuss differences between the information provided and empirical research can lead to better understand-

---

[5] We note that American universities clearly dominate the global market for post-secondary education. This occurs, we believe, precisely because of their success in institutionalizing and rewarding continual self-transformation. Hedberg (1981) argues that such a process of continual learning and unlearning is possible in firms.

434 *L.G. Zucker, M.R. Darby / Research Policy 26 (1997) 429–446*

ing of the underlying processes, institutions and constraints; hence to new hypotheses and better empirical measures for the empirical work. It was necessary, nonetheless, to focus the discussions around some working hypotheses on what determines whether an incumbent firm transforms or dies as the result of an external technological discontinuity. Our first goal is to report what we heard in sufficient detail that readers may come up with their own hypotheses.

The thrust of the literature reviewed in Section 1 is that this sort of externally generated shift in the technological trajectory is most likely to lead to entrants supplanting incumbents. We formalized this null hypothesis as:

H0 immutability: Firms are born with a technological identity, and flourish when that identity has a competitive advantage. In the face of a radical technological breakthrough that makes another technology requiring different human capital dominant, previously successful incumbent firms will be unable to change, and will ultimately be replaced by new firms with the newly dominant technological identity.

With the advantage of hindsight, it was clear that some of the major drug-discovery firms were successfully transforming their technological identity. It seemed to us that preserving the value of organizational capital involved in clinical testing, acquiring regulatory approval, and marketing new drugs provided the incentive for firms to transform the complementary drug-discovery aspect of the business. Thus, we formulated the alternative hypothesis:

H1 Persistent success: In the face of a radical technological breakthrough which makes another technology requiring different human capital dominant, (at least some) successful incumbent firms will change the relevant part of their technological identity, bringing in new human capital, so that the value of their other assets is not wasted.

We knew from our relational database and prior empirical results, as well as the work of others, that the case-study firm would be an example of technological transformation supporting H1 relative to H0. So our practical goals were not to provide a counterexample of H0, which could be done with much less work, but to understand the organizational

mechanisms of transformation, and whether the firm's possession of complementary organizational assets in R&D, as well as in the testing, regulatory, and marketing areas, led to differences in how the firm commercialized the breakthroughs in bioscience, as compared to the dedicated new biotechnology firms born lacking such assets. Since our earlier work had demonstrated that access to and working with top university bioscientists seems to be powerfully linked to success, we were particularly interested in seeing how this firm regarded and used such linkages.

Based on our prior case studies, we isolated three processes, to examine in this case study, that occur sequentially and at any step may lead to either immutability or to persistent success.

1. Detection of a change in the technological environment, with the best predictor the prior investment in R&D (Zucker and Darby, 1996a).

2. Decision to implement the new technology, often initially moving to involve star scientists in company operational decision-making and construction of new scientific teams (often combining new and old employees and thus techniques).

3. Ability to mobilize the necessary resources, either redeploying them from their prior use or raising new resources required to implement the new technology, where resources include scientific personnel, financial resources for product development, and management oversight in the selection of product targets.

We extend resource mobilization theory (developed in the nonprofit context; see McCarthy and Zald, 1977, McCarthy and Wolfson, 1996) to consider both the decision-maker's position within the firm and the availability of necessary resources (e.g., cash flow or external financing).

4.2. Methodology

Our contacts with firm employee's were coordinated by the executive in charge of public policy issues. To familiarize this executive and his colleagues in drug discovery with our interests, we provided first a series of papers reporting our previous research, and then a summary statement of goals (including shedding light on the two hypotheses above) together with a lengthy questionnaire that we

L.G. Zucker, M.R. Darby / Research Policy 26 (1997) 429–446   435

wished to use to guide our interviews during site visits to the main research facilities and the corporate headquarters. A few requests for personnel data were refused due to concerns over privacy of individual employees, but we soon agreed on the basic information to be sought and provided. This approach gave the firm the ability to identify appropriate respondents, and for these respondents to poll colleagues on issues in which they were uncertain. We interviewed a variety of respondents, including the executive vice-president in charge of drug discovery who was more than generous with his time in a sequence of face-to-face and telephone interviews. After each round of interviews, we analyzed what we learned and did preliminary empirical analyses, where indicated, to inform follow-up questions, usually preceded by questionnaires. Limited access to archival data, particularly on publications and presentations, was provided, and we used our copies of these records to test response accuracy. [6]

Upon completing these rounds of interviews, we provided an early draft of this report to our key firm contacts who corrected some misunderstandings, redacted some competitively valuable details of research strategy, and requested rephrasing of identifying passages. Our draft report and tentative findings were presented in a final meeting led by the executive vice-president in charge of drug discovery. We believe that what is reported in Sections 5 and 6 is as accurate an account of the firm's technological transformation as is possible in retrospect, given the proscription on access to personnel data on individual scientists.

5. Transformation as a response to technological revolution

The firm views itself currently as technologically indistinguishable in research and development from any of the best large dedicated biotech firms. The transformation from some involvement to state-of-the-art is seen as occurring between 1985 and 1990. While a few new biotechnology firms were founded

between 1976 and 1979, most were founded in the 1980s, so this transformation might be characterized as lagging the commercial application of the new technology by no more than about five years. In terms of employment, this firm is one of the largest biotech enterprises based on its self-characterization.

5.1. Detection of the technological change

The firm actually was one of the first to market a biotech product, but this product was developed in large part by university scientists who themselves founded new biotechnology firms. Nonetheless, this alliance is evident of management awareness of the commercial importance of biotechnology very early in the 1980s.

5.2. The drivers of the transformation: decision to implement

Firm respondents clearly see the process of transformation as driven by top managers who were technically competent to evaluate the importance of the bioscience breakthroughs to the pharmaceutical industry, and had the vision to devote the resources necessary to ensure that the firm became a world leader in the use of those breakthroughs. [7] These managers included the firm's CEO and the head of the R&D group during the period of transformation. The individual who was the CEO had earlier played a leading role in initiating one of the first biotech collaborations at the firm (see discussion above of detection of the technical change).

5.3. Transformation process: resource mobilization within the firm

In the early 1980s, some research groups at the firm were utilizing biotechnology, but it was not a

---

[6] We are seeking funding to code this archival material in a form suitable for quantitative comparison with similar records acquired from two new biotechnology firms.

[7] In the fieldwork in Japan, we noted that senior research personnel at two major pharmaceutical firms were envious of the ability of American top management to understand and support the necessary technological transformation. In a third such firm, the only Japanese firm to have any star scientists as employees through 1990, senior research personnel attributed the firm's early adoption of biotechnology to the vision and adamant insistence of the CEO that these breakthroughs would transform their industry, if not indeed lead to replacement of the traditional industry.

436   *L.G. Zucker, M.R. Darby / Research Policy 26 (1997) 429–446*

general practice nor one consciously fostered. In 1985, with the appointment, as head of the research group, of a molecular biologist who had the full support of the CEO, the conscious effort to transform how the firm did drug discovery began. Biotechnology was introduced through focused groups, or SWAT teams, at this time. Over a period of three or four years, the firm's scientists generally switched to cloned human targets (receptors, proteins, enzymes, DNA) for initial testing of prospective drugs. [8] The firm's Japanese labs were operated independently and only recently have begun the same transformation under new leadership. By 1990, biotech had permeated the entire research organization, become 'central to the way we do drug discovery,' and the remaining SWAT teams were eliminated, as most of their members had already transferred to research teams focused on particular types of disease. In 1994, the firm hired another of our star scientists to lead developments in the area of gene therapy.

The firm's own biotechnology revolution was accomplished primarily by hiring people knowledgeable in the technology during a period of rapid growth throughout the 1980s and early 1990s:

The strategy was to hire many excellent people to grow our strength in bioscience in the late 1980s. [We] can hire from the best teams because the new biotechnology firms legitimized working in industry. We regularly compete with good university offers in our hiring. As with academic departments, some people already here got excited by what was being done by the new people and adopted the methods as well.

That is, the firm experienced rapidly expanding revenues based on discoveries made using the traditional technologies, and applied large parts of those new resources to acquire the intellectual human capital to replace the very technology that accounted for the current success. We cannot say from one case study how frequently such forward-looking decision-making occurs, but clearly it is possible that a commitment to continuous technological change is a major source of persistent success in high-technology industries. Even in the current period with a relatively stable research budget, the firm is using turnover to hire in targeted areas that go substantially beyond the initial bioscience breakthroughs discussed above.

5.4. Incumbent / entrant differences in applications of biotechnology

There is an ongoing controversy in the pharmaceutical industry as to whether incumbent firms have really adopted biotechnology, or only a part of it. As we have seen above, our respondents say that when one compares how recombinant DNA has been integrated in basic research drug discovery, their firm "is now indistinguishable from the best major biotech companies in how research is done." Biotech company executives sometimes assert that the major pharmaceuticals now use cloned targets to search for the same kind of 'small molecule' drugs they have always produced rather than really using biotechnology to produce 'large molecule' therapies. Major pharmaceutical firms (as seen below) do not accept the factual accuracy of this assertion, saying they do both. In part, this sort of self-conceptualization by advocates of new biotechnology firms can be interpreted as an attempt to define, for the financial markets, a view in which the entrants have a competitive advantage. If the new biotech firms do not dominate on science, few of them could hope to remain competitive with the incumbent firms that have outstanding track records in clinical testing, regulatory affairs, and marketing. One popular scenario for the pharmaceutical industry sees the major firms concentrating on the three latter activities, and increasingly buying their drug discoveries from the new biotechnology firms, thus converting a fixed to a variable cost.

However, this scenario does not recognize the considerable value of integrating testing, regulatory and marketing considerations with decisions on areas

---

[8] A firm scientist explains the advantages of this change: "For example, schizophrenia is a disease involving excess dopamine, and existing drugs operate by suppressing the action of dopamine. This is effective, but results in difficulties with motor function also controlled by dopamine. We can now identify subtypes of dopamine receptors and develop drugs which operate on the relevant subtypes without interfering with the operation of other subtypes. Thus, using biotech permits us to develop effective drugs which are safer and have fewer side effects."

of concentration for research, and on which drug candidates to continue working. [9] Perhaps more importantly, the R&D out-sourcing scenario does not recognize that it is all but impossible for a firm to be an intelligent buyer of research unless the firm has an ongoing capability of doing leading-edge research: [10]

[Our] basic strategy is excellent in-house research. This lets us make better decisions with respect to establishing relationships with new biotechnology firms and, if the right occasion were to arise, purchasing a new biotechnology firm. We have internal evaluators who can adequately assess their research quality, and, therefore, feel that we have effectively turned down deals that were less compelling scientifically.

Nor does the view that this major incumbent firm is indistinguishable in its research technology from the major new biotechnology firms adequately allow for the very real strengths that this sort of firm may have which are not available, without substantial additional investment, to the entrants. Such additional assets could well induce the incumbent firm to use technologies available to it, either in addition to, or as a substitute for those available to the entrants.

We found some evidence in our interviews that there were indeed technologies possessed by the incumbent firm that were viewed by its scientists as providing competitive advantage.

[We have] a competitive advantage as a result of a great history in chemistry. This could have been a disadvantage if there were great resistance to change. It is sometimes suggested, and I believe inappropriately so, that less thought is required [using combinatorial chemistry] than traditional methods. So leadership has been required to support those who adopt the new technology, and to reward those who accelerate the process of drug discovery. [The firm] also has a very significant collection of chemicals that other companies do not have.

Certainly, having more technologies available has the potential to reduce costs. The issue is whether the firm optimizes over the full range of opportunities, and at least this respondent acknowledges the dangers at the root of the immutability hypothesis, and indicates that he believes that they have been overcome. In complementary field work on drug manufacturing (rather than discovery), Pisano (1994) found that large pharmaceutical firms could shorten the process development time by laboratory experimentation for chemical-based drugs, but not for biotechnology-based drugs. This suggests that the manufacturing strengths of these firms are not easily transferred to other technologies.

Research executives at this large incumbent firm state that in fact they try both ends, constructing or identifying targets for drugs and production of drugs by biological processes. They believe that recently "there have not been a lot of successful new proteins; so it looks like the 'low-hanging fruit' was picked early. The remaining areas of application, like septic shock, have proved to be very complex." Indeed, they would argue that "the biotechs have themselves become less optimistic about proteins as therapeutics, and thus, have moved away from proteins and set up combinatorial chemistry groups of their own."

At this point, we see no yardstick to measure the differences in research strategy between any entrant/incumbent pair much less a typical one. We do believe that the case illustrates that the differences may be smaller than popularly believed, and that there is no necessary presumption that any differences favor the new entrants over the incumbents.

5.5. Incumbent / entrant differences in providing information to financial markets

Discussions with firm executives suggested that the Zucker et al. (1997a,b) finding—that significantly higher numbers of products in development are reported by entrants relative to incumbents, other things equal—may reflect different financial reporting approaches conditioned by different financial circumstances:

... biotech firms seem to announce drugs in development earlier. We want to avoid raising hopes that are

---

[9] See also Aghion and Tirole (1994) for an illuminating analysis of the advantages and risks of integrated vs. out-sourced R&D.

[10] Arora and Gambardella (1994) consider the case of biotechnology and point to differences among incumbents in ability to evaluate information and profit from collaborations as a significant competitive factor.

438   *L.G. Zucker, M.R. Darby / Research Policy 26 (1997) 429–446*

very often disappointed and so release relatively little information until the principle is proven in humans near the end of Phase II clinicals. Publications by scientists early on are fine, but as a company we try not to raise expectations since we neither want to raise patients' hopes without being able to deliver nor to violate our fiduciary responsibilities to not make unfounded claims. It also is something of a competitive advantage not to discuss at an early stage what seems to be promising...

And in another context:

We don't do a lot of public relations aimed at the current price of our stock. We believe that if we deliver the fruits of excellent research the stock price will take care of itself. The strategic orientation is rather different when you aren't pressed to raise money to cover the burn rate... [11] On the other hand, because of the importance of breakthroughs in the area of AIDS, we broke our usual policy and have been disclosing information about our AIDS drug development earlier than normal...

This sort of reported difference in announcement policies suggest that caution is warranted with respect to interpreting products-in-development differences as due to greater research productivity by the entrant new biotechnology firms.

5.6. Human resource policies during the transformation

Our previous research (Zucker and Darby, 1996b) indicates that star scientists combining genius and knowledge of emergent technologies are the gold deposits around which firms and their success were built subsequent to the biotech breakthroughs. These

scientists had the ability to become a founder of a new biotechnology firm, earning in some cases literally hundreds, if not thousands of millions of dollars when the firm was taken public. In addition, star scientists may be pursuing personal goals of scientific achievement, including perhaps the Nobel prize. Japanese respondents point to factors making it impossible there to either start one's own firm, or to pursue scientific achievement outside the university as factors holding back their country's commercialization of biotechnology (Darby and Zucker, 1966). The ability of university-based scientists to start their own firms in the 1980s, and continue affiliation with their university and active scientific publication, provided opportunities that could not be offered by any major pharmaceutical company. However, the ability of a biotech star to break the bank declined dramatically in the latter half of the 1980s as the techniques diffused more widely.

By 1985, when this incumbent firm launched its effort to transform its technological identity, it could offer an overall employment package that was attractive to a number of the best scientists.

5.7. Publication policy

Unlike the reported case for Japan, this firm—and they believe much the same is true at other major pharmaceutical firms—follows a very liberal policy on scientific publication. Beyond a possible brief delay to prepare patent filings, the firm encourages publication of research results. [12] The policy is rationalized as follows:

We see some danger of losing our competitive advantage by publishing, but a much greater danger if we do anything that deters the best scientists from coming here. Further, we need for our scientists to have great reputations in order to bring others like them to [the firm]. We are the beneficiaries of worldwide scientific research, and thus we also need to contribute to this pool of scientific knowledge, creating a public good.... Relative to new biotechnology

---

[11] The 'burn rate' is the term used by analysts specializing in new biotechnology firms for the [negative] "sum of the net cash flow from operating activities per month, plus net cash flows from investing activities per month, plus capital spending per month." (Lee and Burrill, 1994; p. 54) The survival index is the "burn rate divided into existing cash, cash equivalents, short-term investments, and long-term marketable securities. This calculation reflects the number of months a company can survive at its existing net burn rate" in the absence of off-book resources or commitments, regulatory approvals which can dramatically alter operating cash flows, or sales of fixed assets or debt or equity.

[12] "Sometimes there is a delay due to patenting, but when I was in academe, I observed a tendency to delay and skim the cream using new discoveries before publishing them."

L.G. Zucker, M.R. Darby / Research Policy 26 (1997) 429–446   439

firms, [we] may believe more strongly in the commonality of research tools because we have a wider array of methodologies and products.

Zucker et al. (1997a,b) show that star scientists affiliated with firms, particularly those with patented discoveries, are typically much more highly cited by other scientists than stars working in universities; [13] so the ability to pursue and publish scientifically interesting and important research would not appear to disadvantage this firm relative to new biotechnology firms, where, arguably, the scientist–entrepreneur must devote more of his or her time to management activities.

One of the firms we studied in Japan, as well as some other large pharmaceutical firms, have attempted with some success to attract top scientists by establishing a quasi-independent, almost academic bioscience research institute that has some interaction with the separate applied R & D group. Separate basic and applied groups had existed for a time at the subject firm, but were integrated prior to and independently of the decision to adopt biotechnological methods company-wide. Our respondents viewed the independent institute model as a less productive approach, and reported no conflict between the integrated research group approach and recruitment of top scientists.

5.8. Incentives

A large pharmaceutical firm like the one studied here can offer a very attractive compensation and working-conditions package for outstanding scientists. Research teams, organized around a specific target and led by a champion, operate internally as 'mini-companies;' so the leader can enjoy much of the research independence experienced by the scientist–entrepreneur in a firm without the same risk, initial sacrifice, and pressures.

The manager with operational responsibility for research is certainly supportive of individual initiative:

An important part of my job is to avoid bureaucracy at all costs so as to keep the science productive. The research teams in effect are many 'small companies.' My job is to nurture these small units, identify leaders, try to add in the areas that will be important, and to convince people to stop projects that aren't going anywhere without squelching creativity.

Further the firm has the ability to very quickly shift substantial resources to support promising ideas.

While the firm certainly does not offer the upside potential one enjoys in one's own firm, neither is there the downside, and there are very substantial financial incentives for top scientists. The corporate compensation philosophy is to reward for success. "There is a single basic measure of research productivity: Are you finding therapeutic compound candidates?" For those who are successful, incentive compensation in the form of bonuses and stock options can form a very substantial portion of the total package.

A key incentive plan for research scientists was instituted in 1985 in addition to the corporate-wide plans. Under this drug-discoverer system, several scientists, who played a major role in identifying a new drug candidate, are granted stock options that vest at specific mileposts in the development of the new drug candidate. The option price of the stock is the market price at time of issue. Recently, the program has been expanded to include other team members who have made important contributions to the discovery and development of the compound. These options would also vest in line with the milepost schedule.

Economists cannot help noting that under such a stock option system, managers profit from lower current stock prices so long as favorable news eventually emerges and is reflected in the stock price at the time they exercise their options. Thus, except for employees about to leave the firm, the incentive system is consistent with the policy of not discussing products in development (except in scientific journals) until they have been successfully proven in humans.

---

[13] The ratio is 6.5 times as many citations comparing firm scientists with patents to university scientists without patents, where, in each case, the comparison is restricted to the elite group of star scientists.

440 *L.G. Zucker, M.R. Darby / Research Policy 26 (1997) 429–446*

6. Collaborations of the firm with external scientists

6.1. Collaborations involving university faculty and students

Collaborations with university professors, their students, and their departments are common, often quite informal, and rarely publicly acknowledged:

Nearly every research program [here] has at least one university collaboration. Our scientists are told that its their job to find out what is important in their field worldwide and bring it into [the firm]. That is, scientists [here] should think about themselves as running the research for the whole world, and then bring in those other people who are needed to do that research.

...

Co-publishing is about as good an indicator as you can get of commonality of interests between [the company] and an academic collaborator. Although formal relationships are on a publicly available list, many relationships are not publicly acknowledged. We focus on a group of major universities which we support and whose students we actively recruit, so recruitment of students would not generally indicate collaboration with their professor. We don't hire collaborators just because they were collaborators.

In this and other fieldwork, we have repeatedly validated the usefulness of linking academic scientists to firms by bibliometric research on patterns of co-publication. As indicated in Section 2, this concept of linkage is powerfully predictive of firm success when academic star scientists are involved.

The company provides support for students, junior faculty, and relevant departments, as well as entering into direct collaborations with particular professors. The collaborations may involve a little more than informal exchanges of reagents needed in each other's research, to more elaborate and long-lasting efforts with particular therapeutic goals. The firm's expertise in knocking out particular genes through recombinant DNA, and in using drugs that knock out their effects, tools which get at gene expression, makes the firm a particularly attractive collaborator for university scientists.

6.2. Collaborations with other firms

While involved in multiple marketing arrangements with large and small firms and basic-research collaborations with small firms, the firm generally does not collaborate on basic research with other large firms. This lack of collaboration with large firms is not a matter of policy, but rather reflects the difficulties involved in working out complicated issues on marketing rights and other terms that are not so difficult with the small biotechs.

With respect to research collaborations with small new biotechnology firms, the firm is especially interested in collaborations where the particular expertise held by others is needed for a particular project, but is not thought worthwhile to build up internally. Sometimes successful collaboration reverses that judgement and leads the firm to undertake acquisition of the capability internally. Collaborations are not seen as shortcuts to acquiring new technologies for internal use. As discussed above, since the firm's strategy emphasizes excellent in-house research, collaborations with other firms do not play a central part in their effort to identify new drugs.

7. Evidence on generalizability of the transformation experience

We have seen that the incumbent firm started a bit late in the biotech revolution, but then devoted enough resources to transform its research technology to state-of-the-art. There is a natural question as to whether this is a peculiarity of the particular firm which we studied, or whether this pattern might have been followed more generally by major pharmaceutical companies. A definitive analysis of these issues is beyond the scope of the present paper, but it is possible to shed some light on the generalizability, as well as provide some useful information by considering patterns of affiliation and linkage of stars with incumbents and entrants, and also patterns of patenting of genetic sequences by the different types of firms.

7.1. Patterns of affiliation and linkage of star scientists

Zucker et al. (1997a,b) validated a method of measuring the strength of connection between star scientists and commercial enterprises by counting the number of publications written by a star giving the particular firm as an affiliation—or, if the star lists another affiliation—written by a star with a co-author who gives the firm as his or her affiliation (in which case the star is said to be linked to the firm). There is some evidence that scientists who are nearby are likely to be more involved with the firm than those farther away, so we classify linked stars by whether they are affiliated with an organization which is located in the same region (functional economic area as defined by the US Bureau of Economic Analysis) as the firm, in another US region, or in a foreign country. Firms with access to leading-edge science as evidenced by such affiliations and linkages perform significantly better than the vast majority of enterprises that lack such access.

Table 1 reports the history of such affiliations and linkages of stars to particular firms classified as dedicated biotech firms (entrants), major pharmaceutical firms, and the remaining incumbents for the periods 1976–1980, 1981–1985, 1986–1990. As we see, during the first five years of the biotech revolution, only one well-known entrant had the intellectual human capital that we are measuring here. In the second five years, 17 firms had demonstrated substantial access to intellectual human capital of which almost 24% were major pharmaceutical firms, and the remainder were entrants. Quantitatively, the pharmaceuticals lagged further, however, with all 97 articles by affiliated stars being published by stars affiliated with entrants, and only 19% of 52 linked articles linked to pharmaceutical firms. In the third five-year period 1986–1990, there appears to be evidence of a general catch-up effort by pharmaceutical firms. Pharmaceutical firms (including the subject of our case study) begin to have star scientists publishing as their employees (11%) and their share of linked articles rises to 24% (excluding the nascent

Table 1
Publications by star bioscientists affiliated with or linked to US firms

Variables	No. of firms	Publication counts of stars			
		Affiliated stars	Linked in region	Linked in other US	Linked foreign
1976–1980					
Dedicated biotech firms	1	9	0	0	0
Major pharmaceutical firms	0	0	0	0	0
Other incumbent subunits	0	0	0	0	0
Total for all firms	1	9	0	0	0
1981–1985					
Dedicated biotech firms	13	97	20	12	10
Major pharmaceutical firms	4	0	2	7	1
Other incumbent subunits	0	0	0	0	0
Total for all firms	17	97	22	19	11
1986–1990					
Dedicated biotech firms	19	68	16	30	6
Major pharmaceutical firms	8	8	3	9	4
Other incumbent subunits	3	0	2	2	0
Total for all firms	30	76	21	41	10
1976–1990					
Dedicated biotech firms	22	174	36	42	16
Major pharmaceutical firms	9	8	5	16	5
Other incumbent subunits	3	0	2	2	0
Total for all firms	34	182	43	60	21

442   *L.G. Zucker, M.R. Darby / Research Policy 26 (1997) 429–446*

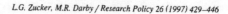

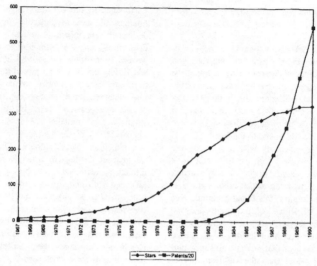

Fig. 1. Cumulative number of stars and genetic-sequence patents granted in the world. Source: GenBank™, Release 81.0, February 15, 1994, and calculations by the authors.

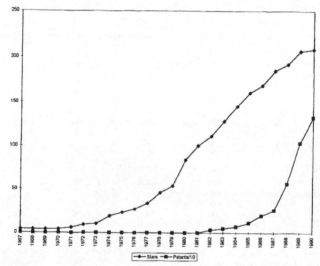

Fig. 2. Cumulative number of US stars and US genetic-sequence patents granted. Source: GenBank™, Release 81.0, February 15, 1994, and calculations by the authors.

L.G. Zucker, M.R. Darby / Research Policy 26 (1997) 429–446

Table 2
Patents granted to US firms with affiliated or linked stars, 1980–1990

Period	Major pharmaceutical firms	Dedicated biotech firms	All firms
1980	0	2	2
1981	0	4	4
1982	4	17	21
1983	0	9	9
1984	1	29	30
1985	4	34	38
1986	3	49	52
1987	10	18	28
1988	45	101	146
1989	43	152	195
1990	7	79	86
1980–85	9	95	104
1986–90	108	399	507
1980–90	117	494	611

group of incumbent firms in other industries with significant scientific capital).

7.2. Patterns of patent production

Fig. 1 indicates that, while the science diffused rapidly in the late 1970s in terms of initial publications of stars, patenting of genetic sequences did not boom until the mid-1980s. [14] Fig. 2 provides the data indicating a similar pattern when the quantities are limited to US values only. GenBank has data on 3353 patents granted through the end of 1990, of which we were able to link 611, or 18.2% of the world total, to 21 of the 34 firms examined in Table 1. [15] Table 2 provides annual data for total numbers of genetic-sequence patents granted to the major pharmaceuticals with ties to stars, to the corresponding entrants, and their sum. For convenience, the

---

[14] The availability of patent protection for genetically engineered organisms was doubtful until the US Supreme Court's 1980 decision in Diamond v. Chakrabarty; see Eisenberg (1987).

[15] We matched genetic-sequence patents to 8 of the 9 major pharmaceutical firms, 13 of the 22 dedicated biotech firms, but found no genetic sequence patents for the 3 other incumbents. Of course, the latter group were late on the scene and may appear in patent data after the 1990 cutoff in the data which we have so far analyzed. The case-study firm was in the middle of its group in terms of frequency and onset of patenting.

sums are provided for 1980–1985, 1986–1990, and 1980–1990. Again, we find that the major pharmaceuticals lagged behind the dedicated biotech firms but then began catching up quickly in the late 1980s: they had only 8.7% of total patents for 1980–1985, but this rose to 21.3% in 1986–1990. Given an average lag of perhaps two or three years between application and granting of the patent, this performance is even more remarkable.

7.3. Tentative conclusions on generalizability

In the early years of the biotech revolution, a few great scientists, who were also great entrepreneurs, recognized the value in the pharmaceutical industry of the scientific breakthroughs being made. Few if any outsiders could adequately judge whether their vision was right and probably none would or could pay a conventional compensation package that would match their true worth when the tacit knowledge essential to biotechnology was held in very few hands and brains. Some of these scientists proved right in their vision and became multimillionaires or billionaires. Others, although perhaps equally able and visionary, proved unlucky in their choice of problems or approaches and did not do as well, although we have found few star scientists who became principals of firms before 1986 and are not, by now, multimillionaires. Certainly, some very good scientists who do not have the record of achievement of our stars were also lucky and are very rich men and women today, but their odds were considerably diminished from those of equally situated stars.

After 1985, the science diffused rapidly and such extraordinary returns do not appear to have been there for the star scientists, and empirically, Zucker, Darby and Brewer (1997) report that the stars no longer played such a key role in determining the location of new entrants using biotechnology. It is certainly reasonable that the combination of technological successes and the more affordable compensation demands of top biotech scientists made this period an attractive one for the firm in our case study, and the other firms highlighted in Tables 1 and 2 to begin a wrenching and still expensive transformation of their technological identities.

Clearly not all major pharmaceutical firms have transformed their technological identity. However, the firm we studied has, and we find quantitative evidence that a number of such firms have followed a similar path in terms of both timing and success. Cookson (1995) reports that "[t]oday, genetic engineering is used daily as a laboratory tool by every research-based pharmaceutical and biotech company" and quotes Dr. Francois L'Eplattanier, head of R&D for Ciba of Switzerland: "Genetic engineering is absolutely essential for us. If we were not active in genetic engineering, we would be out of the game entirely by the beginning of the next century." Of course, recognition of the competitive necessity to transform the firm's technological identity is not the same thing as achieving that transformation in an effective way, so that the firm's superior performance is maintained. In future research, we shall test possible determinants of ability to transform suggested by the case study, such as top management's technical expertise.

8. Conclusions

We can draw a number of significant conclusions from the case study reported here.

(a) We have evidence that in one major incumbent firm, the biotechnology revolution fundamentally changed the firm's technology identity, a counter-example to the hypothesis that these identities are immutable; and consistent with the hypothesis that persistently successful firms maximize their wealth by transforming their technological identity as required to remain competitive in the face of technological revolutions.

(b) Senior management with the scientific ability to assess the breakthroughs championed the technological transformation.

(c) The technological transformation was achieved primarily through hiring new personnel embodying the new technology and incorporating them into the existing structure. Special subunits played only a transitional role, and collaborations and joint ventures with university scientists and new biotechnology firms were used primarily to augment internal expertise with explicit decision-making on the issue of whether this expertise was worth developing internally.

(d) There is some evidence that biotechnology applications in the incumbent firm are more likely to be used in combination with other technologies than in entrants which tend to use biotechnology for both discovery and production of new therapeutic entities. This difference in emphasis may result in value-enhancing synergies for the incumbent firm because of the wealth of related knowledge that makes for more effective, possibly different, applications of the new technologies.

(e) University–firm collaborations are ubiquitous, often non-public, and best identified in quantitative analyses by co-publishing. Hiring is not significantly related to such collaborations.

(f) The firm is capable of recruiting star scientists with an overall working-conditions/employment package which includes, for those with identifiable contributions to drug discovery, stock options which vest as the drug candidate progresses through clinical trials and FDA approval.

(g) While not all incumbent major pharmaceutical companies have changed their technological identities, we were able to identify another seven or eight such firms that seem to be following a similar path, both in terms of involving top scientific talent and in terms of patenting success.

Acknowledgements

This research has been supported by grants from the Alfred P. Sloan Foundation through the NBER Research Program on Industrial Technology and Productivity, the National Science Foundation (SES 9012925), the University of California Systemwide Biotechnology Research and Education Program, the University of California Systemwide Pacific Rim Research Program, the UCLA Center for American Politics and Public Policy, and the UCLA Institute of Industrial Relations. It would not have been possible without the assistance of Martin Feldstein in gaining access to the firm studied and the interest and efforts of the respondents there. Useful comments on earlier drafts have been received from Douglas L. Cocks, William Comanor, Timothy L. Hunt, Michael Intriligator, Jeffrey L. Tarlowe, and other participants in presentations to the Ad Hoc Working Group on the Economics of the Pharmaceutical Industry and the joint Pharmaceuticals Workshop of the UCLA Departments of Health Services and Economics. Princi-

L.G. Zucker, M.R. Darby / Research Policy 26 (1997) 429–446   445

pal research assistance for this paper was provided by Maximo Torero and Jeff Armstrong. We are indebted to a remarkably talented team of post-doctoral fellows Zhong Deng, Julia Liebeskind, and Yusheng Peng and research assistants Paul J. Alapat, Jeff Armstrong, Cherie Barba, Lynda J. Kim, Kerry Knight, Edmundo Murrugara, Amalya Oliver, Alan Paul, Erika Rick, Maximo Torero, Alan Wang, and Mavis Wu. This paper is a part of the NBER's research program in Productivity. Any opinions expressed are those of the authors and not those of the National Bureau of Economic Research.

References

Aghion, P., Tirole, J., 1994. The management of innovation. Q. J. Econ. 109, 1185–1209.

Arora, A., Gambardella, A., 1994. Evaluating technological information and utilizing it: scientific knowledge, technological capability, and external linkages in biotechnology. J. Econ. Behav. Organization 24, 91–114.

Baum, J.A.C., 1996. Organizational ecology. In: Clegg, S.R., Hardy, C., Nord, W.R. (Eds.), Handbook of Organization Studies. Sage Publications, London, 77–114.

Cohen, S., Chang, A., Boyer, H., Helling, R., 1973. Construction of biologically functional bacterial plasmids in vitro. Proc. Natl. Acad. Sci. 70, 3240–3244.

Comanor, W.S., 1986. The political economy of the pharmaceutical industry. J. Econ. Literature 24, 1178–1217.

Cookson, C., 1995. Essential engineering: how different is biotechnology? Financial Times, April 25, p. 10.

Darby, M.R., L.G. Zucker, 1966. Star scientists, institutions, and the entry of Japanese biotechnology enterprises. Natl. Bureau of Economic Res. Working Paper No. 5795.

Demsetz, H., 1988. The theory of the firm revisited. J. Law, Econ. Organization 4, 141–161.

Dosi, G., 1982. Technological paradigms and technological trajectories: a suggested interpretation of the determinants and directions of technical change. Res. Policy 11, 147–162.

Eisenberg, R.S., 1987. Proprietary rights and the norms of science in biotechnology research. Yale Law J. 97, 177–231.

Friedman, M., 1976. Price Theory. Aldine Publishing, Chicago.

Hedberg, B., 1981. How organizations learn and unlearn. In: Nystrom, P.C., Starbuck, W.H. (Eds.), Handbook of Organizational Design, Vol. 1. Oxford Univ. Press, Oxford, pp. 3–27.

Henderson, R., 1993. Underinvestment and incompetence as responses to radical innovation: evidence from the photolithographic alignment industry. Rand J. Econ. 24, 248–270.

Henderson, R., Clark, K.B., 1990. Architectural innovation: the reconfiguration of existing product technologies and the failure of established firms. Administrative Sci. Q. 35, 9–30.

Kaufman, H., 1971. The Limits of Organizational Change. Univ. of Alabama Press, University, AL.

Kim, L., 1997. Imitation to Innovation: The Dynamics of Korea's Technological Learning. Harvard Business School Press, Boston.

Kimberly, J.R., Quinn, R.E., 1984. The challenge of transition management. In: Kimberly, J.R., Quinn, R.E. (Eds.), Managing Organizational Transitions. Richard D. Irwin, Homewood, IL.

Klevorick, A.K., Levin, R.C., Nelson, R.R., Winter, S.G., 1995. On the sources and significance of interindustry differences in technological opportunities. Res. Policy 24, 185–205.

Lee, K.B., Jr., Burrill, G.S., 1994. Biotech 95: Reform, Restructure, Renewal. The Ernst and Young Ninth Annual Report on the Biotechnology Industry. Ernst and Young, Palo Alto, CA.

Liebeskind, J.P., Oliver, A.L., Zucker, L.G., Brewer, M.B., 1996. Social networks, learning, and flexibility: sourcing scientific knowledge in new biotechnology firms. Organization Sci. 7, 428–443.

Lounamaa, P.H., March, J.G., 1987. Adaptive coordination of a learning team. Manage. Sci. 33, 107–123.

March, J.G., Simon, H.A., 1993. Organizations, 2nd edn., Blackwell, Cambridge, MA.

McCarthy, J.D., Wolfson, M., 1996. Resource mobilization by local social movement organizations: agency, strategy, and organization in the movement against drinking and driving. Am. Sociological Rev. 61, 1070–1088.

McCarthy, J.D., Zald, M.N., 1977. Resource mobilization and social movements: a partial theory. Am. J. Sociology 82, 1212–1241.

Nelson, R.R., Winter, S.G., 1982. An Evolutionary Theory of Economic Change. Harvard Univ. Press, Cambridge, MA.

Pisano, G.P., 1994. Knowledge, integration, and the locus of learning: an empirical analysis of process development. Strategic Manage. J. 15, 85–100.

Reinganum, J.F., 1983. Uncertain innovation and the persistence of monopoly. Am. Econ. Rev. 73, 741–748.

Reinganum, J.F., 1989. The timing of innovation: research. development, and diffusion. In: Schmalensee, R., Willig, R.D. (Eds.), Handbook of Industrial Organization, Vol. 1. North-Holland, New York, 849–908.

Sindelar, R.D., 1992. Overview/preview of current and future recombinant DNA-produced pharmaceuticals. Drug Topics, Suppl., April 20, 3–16.

Sindelar, R.D., 1993. The pharmacy of the future. Drug Topics 137 (9), 66–84.

Tolbert, P.S., L.G. Zucker, 1996. The institutionalization of institutional theory. In: Clegg, S.R., Hardy, C., Nord, W.R. (Eds.), Handbook of Organization Studies. Sage Publications, London, pp. 175–190.

Tushman, M.L., Anderson, P., 1986. Technological discontinuities and organizational environments. Administrative Sci. Q. 31, 439–465.

Waring, G.F., 1993. Industry differences in the persistence of competitive advantage. unpublished doctoral dissertation. Univ. of California, Los Angeles.

Zucker, L.G., Darby, M.R., 1996a. Costly information: firm transformation, exit, or persistent failure. Am. Behavioral Scientist 39, 959–974.

Zucker, L.G., Darby, M.R., 1996b. Star scientists and institutional transformation: patterns of invention and innovation in the formation of the biotechnology industry. Proc. Natl. Acad. Sci. 93, 12709–12716.

Zucker, L.G., Darby, M.R., Armstrong, J., 1997. Intellectual human capital and the firm: the technology of geographically localized knowledge spillovers. Economic Inquiry, in press.

Zucker, L.G., Darby, M.R., Brewer, M.B., 1997. Intellectual human capital and the birth of US biotechnology enterprises. Am. Econ. Rev., in press.

[18]

ELSEVIER

Research Policy 31 (2002) 381–398

research
policy

www.elsevier.com/locate/econbase

Who is interested in biotech? R&D strategies, knowledge base and market sales of Indian biopharmaceutical firms

Shyama V. Ramani*

*Department of Economics and Social Sciences, Institut of National de la Recherche Agronomique (INRA),
Universite Pierre Mendes France, BP 47, 38070 Grenoble Cedex 9, France*

Received 14 April 2000; received in revised form 7 November 2000; accepted 18 January 2001

Abstract

This paper addresses three main questions on Indian pharmaceutical firms that have integrated biotechnology in their marketing, production or research activities: (i) What kind of labour stocks of the knowledge base have an impact on market sales? (ii) Which components of the R&D strategy are strategic substitutes and which are strategic complements? (iii) What are the distinguishing features of firms that have already integrated biotechnology in their research activities? The paper shows that market sales are an increasing function of qualified labour stocks. Internal R&D and foreign collaborations are strategic substitutes, while patents and publications are strategic complements. Firms that are active in biotechnology research are likely to be younger and implementing more aggressive learning strategies. © 2002 Elsevier Science B.V. All rights reserved.

Keywords: India; Biotechnology; Pharmaceutical sector; R&D strategies

1. Introduction

The biopharmaceutical sector refers to pharmaceutical firms that have integrated modern biotechnology in their research, production or marketing activities. Modern biotechnology pertains to a set of techniques that involve manipulation or change of the genetic patrimony of living organisms. Since from 1980s, modern biotechnology has been integrated in a number of industries such as pharmaceuticals, chemicals, agribusiness, agriculture and environment. In the pharmaceutical sector, advances in modern biotechnology have initiated a radical change in the nature of the search processes for the creation of new drugs (i.e. creation by rational design rather than by trial and error methods). They have also led to the creation of radical

and incremental product innovations and brought down the costs of production of pharmaceutical products (OTA, 1991). At present, the 10 top selling biopharmaceutical drugs have an annual world wide sales of more than US$ 6 billion and all of them have been created and are being marketed by American or Western European firms (Ernst and Young, 1998). It is clear that biotechnology will have an increasing influence on the evolution of the global pharmaceutical industry and that the bulk of the investment in biotechnology will continue to be in the pharmaceutical sector.

The Indian pharmaceutical industry is the 12th largest in the world accounting for a market of about US$ 2.5 billion. The supply side is highly fragmented with at least 3000 firms in the "organised" sector and at least 13,000 firms in the "unorganised" small-scale sector (CMIE, 1996). However, only about 48 pharmaceutical firms have been listed in government directories as being active in the biopharmaceutical field.

* Tel.: +33-4-7682-5412; fax: +33-4-7682-5455.
E-mail address: ramani@grenoble.inra.fr (S.V. Ramani).

382   *S.V. Ramani / Research Policy 31 (2002) 381–398*

At present, the firms in the biotechnology bandwagon are of two types. The first type is marketing a foreign product or producing a biotech-based product using a license without undertaking research in biotechnology (though it may be doing research in some other discipline). The second type is integrating biotechnology in its research activities. Given these two types of biopharmaceutical firms, this paper tries to address three central questions: (i) What kind of labour stocks forming the knowledge base have an impact on market sales? In particular, do labour stocks allocated to R&D have a positive influence on market sales? (ii) Which components of the R&D strategy are strategic substitutes and which are strategic complements? (iii) What are the distinguishing features of firms that have already integrated biotechnology in their research activities?

Though, very small, it is important to study the nature and impact of the R&D strategies of this set of firms for three reasons. Firstly, they are the only Indian firms, which, in the future, may be able to offer cheaper local equivalents of the biopharmaceutical products presently being sold in India by Western firms (either directly or through an Indian marketing partner). Secondly, time is running out to catch up with Western firms in this field because once India implements the WTO–GATT agreement (in 2005) Indian firms will be effectively barred from replicating innovations patented in Western countries. Thirdly, the success of these biopharmaceutical firms will determine whether this sector is likely to grow in the future.

By studying the Indian biopharmaceutical sector, the present article attempts to make two types of contributions. The first is to the existing literature on the R&D activities of Indian firms and the second is to the literature on the integration of biotechnology in Indian firms. With respect to the former, the paper re-examines the issues raised in the context of the biopharmaceutical industry. Furthermore, it considers the knowledge base of firms in terms of their labour stocks and expands the definition of an R&D strategy to include a vector of actions and examines their relations with market sales. Finally, it distinguishes between firms that are active in biotechnology research, as different from firms that are only marketing or producing a biotechnology-based product, and identifies their specific characteristics. These contributions taken together attempt to provide some

insight on the integration of biotechnology in the Indian pharmaceutical sector in more concrete, quantitative terms than existing studies, which are mainly of a historical or institutional nature (Acharya (1995); Sasson (1993); Ramani and Visalakshi (2001)).The principal results of the paper can be summarised as follows. In the Indian biopharmaceutical sector, R&D expenditure intensity is not linked to firm size, but to research orientation. Market sales are positively correlated to the knowledge base of firms as embodied in their qualified personnel outside of their R&D department. An R&D strategy of these firms is given by a three dimensional vector related to the acquisition of knowledge, disclosure of knowledge and internal creation of knowledge. Either knowledge can be acquired in-house through employing more people in the R&D department and spending more on R&D or it can be acquired from abroad through foreign collaborations. Either new knowledge can be disclosed in the form of patents and publications or it can be kept within the firm by having more qualified people in the R&D department. Thirdly, firms can choose to create knowledge throughout the firm by recruiting qualified personnel or focus on creating knowledge through the R&D department by allocating more qualified personnel exclusively to the R&D department. Finally, firms that are doing research in biotechnology are likely to be young, with a higher R&D expenditure intensity, a higher proportion of qualified employees and a higher proportion of employees in the R&D department.

If India can be taken as a case study of an emerging economy, then it means that in such countries, the participation in the biotechnology revolution (with respect to the pharmaceutical sector) is limited to a very small fraction of local firms. Research is mainly undertaken by younger, small or medium sized firms. Large firms serve to provide market partnerships for foreign multinationals. The major differences with biopharmaceutical firms in developed countries are that patents and publications play an insignificant role as strategic tools for creating value or market signalling and R&D collaborations between local agents (firms or public laboratories) have a very marginal impact on the creation of innovations.

This article is organised as follows. Section 2 presents the background of the context being studied. It contains a brief note on the evolution of the pharmaceutical sector and biotechnology in India. It then reviews the literature on the R&D activities of Indian manufacturing firms. Section 3 presents the hypotheses, the construction of the database and the variables considered. Section 4 details the methodology, the statistical results obtained and a discussion of the results. Finally Section 4 concludes with policy implications that can be inferred from the analysis.

2. Background of the context studied

2.1. Evolution of the Indian pharmaceutical industry[1]

When India attained its independence in 1947 it had a pharmaceutical industry of a very modest size with a market of about US$ 28.5 million (Ahmad, 1988). There were several Indian-owned firms in the field but their operations were on a much smaller scale than those of the foreign multinationals or MNCs. The production of pharmaceuticals involves two phases: the manufacture of basic ingredients that are called "bulk drugs" and their subsequent "formulation" for final use by consumers, in the form of tablets, capsules, syrups, injectibles, drops and sprays. No Indian company was a major factor in either field at the time of independence and there was heavy dependence on imported foreign drugs which were marketed either by MNCs already established in India or by local agents of other MNCs that did not have a local presence. In order to reduce the dependence on imports and on Western MNCs, at least for vitally needed antibiotics, the government of India undertook large investments to establish a network of public sector enterprises (Singh, 1985). The most important among these were Hindustan Antibiotics Limited (HAL) and Indian Drugs and Pharmaceuticals Limited (IDPL). The move was useful and timely but it was not a comprehensive response to the country's healthcare needs.

The foreign multinationals formulated their drugs in India, importing the bulk drugs from their home countries. It was their contention that the locally available bulk drugs were not of the desired quality. This led to drug prices that were regarded as being too high

by the consumers as well as by the government. Thus, in 1965 the government pegged drug prices at levels that prevailed as on 1 April 1963. The "drug price control" order of 1970 brought under price control a number of bulk drugs and selected formulations and also set a ceiling on the overall profits of companies in the pharmaceutical sector. The control regime was continually opposed by both MNCs and fledgling Indian companies. They argued that high import duties were largely responsible for pushing up prices and that price controls discouraged the flow of investment into the industry by depressing the earnings of companies. Discouraged by what they regarded as low margins that could be made under the price control regime, MNCs became disinclined to increase their investment in their Indian subsidiaries or expand their manufacturing activities significantly. They evinced little interest in developing R&D activities based in India.

In order to develop the indigenous pharmaceutical industry at a much faster pace, the Indian government enacted the Indian Patents Law in 1972. The act ensured patent protection only to production processes and not to the products themselves. The provision left the way open for Indian companies to develop and market substitutes for MNC products by simply evolving some process variations. This expedient was not something invented by the government of India. Japan, for instance had such a provision in place for several years in order to promote its own indigenous pharmaceutical industry (Probert, 1994). The communist countries did not respect Western patents either. That the government of India made its move a quarter of a century after the country attained its freedom testifies to its inadequate awareness and appreciation during earlier years of what countries like Japan were doing and of what Indian private enterprise might be capable of achieving in the pharmaceutical industry.

Initially the multinationals did not see the new patent act as a threat to their market position as they assumed that it would be beyond the technological competence of the Indian pharmaceutical companies to do "reverse engineering" and formulate products equivalent to those of the MNCs (Redwood, 1994). The immediate impact was slight. However, the patent act opened up opportunities which in time some alert and aggressive Indian companies equipped themselves to exploit. Those that were unimaginative and timid were left behind. The MNCs had underestimated

---

[1] This section on the historical evolution of the Indian pharmaceutical sector is largely drawn from Ramani and Venkataramani (2001)

the capability of Indian technologists and the entrepreneurial skills of the Indian businessmen, and overestimated the appeal of their brand names for the price conscious consumer. The consumer was quite willing to go for a lower priced Indian product with its own brand name.

In 1976, among the top 20 firms which held 57.19% of the pharmaceutical market, there were only 4 Indian firms. However, by 1995 only 7 MNCs (including their subsidiaries) figured among the top 20 pharmaceutical companies in India and together they could claim only 15.1% of the total market. Indian companies that had won a place in the 1995–1996 list ranked in order of their market share were: Ranbaxy, Lupin, Cipla, Dabur, SOL Pharma, Sarabhai, Torrent, Dr. Reddy's, Allembic, Kopran, Ipca and Cadilla. In addition, there were 38 other Indian owned pharmaceutical companies that were among the top 50 in terms of sales (US$ 22 million or more) during 1995–1996. Only 12 MNCs figured in the list, of whom only 3 made it to the top 10: Glaxo, Hoechst, and Pfizer.

Clearly, in order to compete against entrenched and popular MNC brands, the Indian substitutes had to become of comparable quality and cheaper in price. These requirements made it incumbent on Indian engineers and managers to pay continuing attention to cost reduction and quality control. Many of the companies in the top 100, recognising the opportunities afforded by the Indian Patent Law, made modest investments in R&D activities resulting in an enhancement of their technical capabilities in working out processes for the production of selected drugs identified by them as having good commercial prospects. However, R&D expenditures as a percentage of sales still remained quite low compared to figures in the advanced countries and companies generally tended to raise it to just the point needed for the production of the identified drugs. Most importantly, while successful Indian companies had demonstrated their capabilities in bringing out very satisfactory substitutes for a number of patented Western products and expanded their sales in India and in the overseas markets through lower prices, none of them had come up with a significant innovation in the form of a new drug based on indigenous R&D. The Indian pharmaceutical firms had their knowledge base firmly embedded in organic and synthetic chemistry. They had not made any efforts to integrate other scientific disciplines to create or re-engineer innovations.

These firms were then confronted with biotechnology during the 1980s, a set of techniques based on recent developments in the life sciences that was new, different and much more complex to integrate requiring a multi-disciplinary team to create a product.

2.2. A note on biotechnology in India[2]

Biotechnology in India emerged largely due to the key role played by scientists turned administrators in the government ministries. In this note, we briefly outline the strategy and role of the Indian government in the initiation of the biotechnology sectors. The strategy of the Indian government can be grouped into three stages: initiation, building scientific competence, and reaching out to the private sector.

2.2.1. Initiation [1981–1986]

In India the push to develop the biotechnology sectors came from reputed scientists who had been brought into the administration. In 1982, Dr. S. Varadarajan, then secretary of the Department of Science and Technology headed the National Biotechnology Board (NBTB). Its objectives were to: (a) identify priority areas in biotechnology; (b) identify infrastructural needs and (c) implement a co-ordinated programme to realise certain national objectives. To this end, a number of pilot programmes were proposed in the 6th (1981–1985) and 7th (1986–1990) five year plans. However, during its 4 years of existence from 1982 to 1986, the NBTB seems to have achieved only objectives (a) and (b).

In 1986, the NBTB was replaced by a separate government department called the Department of Biotechnology (DBT, 1993). It functioned under the aegis of the Ministry of Science and Technology. The main reason for this evolution seems to have been the realisation that biotechnology is a generic technology whose progress requires the development of a variety of competencies in a variety of scientific disciplines. In order to achieve this co-ordinated development, an agency working together in tandem with the Ministry of Science and Technology was deemed necessary. It set out to implement the objectives of the earlier body such as the development of scientific competence

[2] The note 2.2 on government strategy for biotechnology in India is based on Ramani and Visalakshi (2001).

S.V. Ramani / Research Policy 31 (2002) 381–398   385

in selected non-capital intensive disciplines (genetic engineering, vaccines, food production, edible oils). The establishment of DBT served as a signal that the government considered biotechnology to be a priority area for development. It was welcomed by academics, national laboratories as well as industrialists.

2.2.2. Creation of scientific competence [1986–1990]

The first target was to create a core of researchers competent in biotechnology. Grants were given to the network of research institutions and university departments to undertake biotechnology related projects. Grants were also provided to selected teaching and research institutes partially supported by the government, such as the Indian Institute of Sciences, Indian Institutes of Technology, All India Institute of Medical Sciences, National Chemical Laboratory, Tata Institute of Energy Research, Tata Institute of Fundamental Research etc. The DBT also participated in the creation of new institutions such as the National Institute of Immunology, Centre for Cellular and Molecular Biology, National Facility for Animal Tissue and Cell culture, and International Centre for Genetic Engineering (in collaboration with UNIDO).

2.2.3. Reaching out to the private sector [since 1990]

In India, as in most developing countries, the number of financial institutions that invest in a new technology is extremely limited and even then they tend to be risk averse and bureaucratic in their approval process. The government of India tried to remedy this problem through the creation of the Biotechnology Consortium of India Ltd. (Biotech Consortium Ltd., 1994) or BCIL as a public company in 1990. It was set up jointly by the DBT (1993) government sponsored financial institutions like the Industrial Development Bank of India, the Industrial Credit and Investment Corporation of India and "about 30 industries, mainly in the private sector". It was to fulfil the same functions as the venture capital companies in the US, i.e. promote the creation of firms by not only providing venture capital but also complementary competencies required by scientists to set up firms. Thus, it was to guide start-ups, arrange technology transfers and support their efforts to find financing. As of 1997, they had been involved in fund syndication for 3 companies, technology scale up of 1 project, packaging

technology for 3 projects, and transfer of technology from laboratories for 6 companies.[3] BCIL's main activity seems to be conducting techno-economic feasibility studies and monitoring activities for its institutional shareholders like the ICICI and government bodies like the Department of Science and Technology. In short, the impact of BCIL, both in the creation of new firms and new products has been rather limited. A few other venture capital fund companies have also set been set up since then by the government.

2.3. Impact of the government strategy on the integration of biotechnology in the pharmaceutical sector

The strategy of the Indian government focused on the two ends of the commercialisation spectrum: public research networks and final markets. It funded public research and regulated the final market. Its weak point was the link. It did not have a strategy for the efficient transformation of research into usable technology. There were no well thought out practical goals or plans made for the effective utilisation of competent manpower. While this indispensable intermediate exercise to transform scientific competence into technological competence was largely skipped, the government concerned itself with the final product markets and fiscal measures such as price control and distribution measures to benefit the masses.

From the period of initiation of biotechnology, the pharmaceutical industry did not figure high in the thinking of the National Board (NBTB), and the non-association of any competent scientist or industrialist from the pharmaceutical sector in its deliberations, had its own consequences. Afterwards, no grants were available from the DBT for the modest R&D establishments that were being set up by some pharmaceutical majors. The thrust of government strategy was on agriculture rather than healthcare, because of the former's intrinsic importance to the economy and the existence of a good record of indigenous research accomplishment. The meagre research output of pharmaceutical enterprises and the minor role of pharmaceuticals related research in the large government supported research establishment had

---

[3] BCIL—a profile, New Delhi, Biotechnology Consortium of India Ltd., 1996.

386 *S.V. Ramani / Research Policy 31 (2002) 381–398*

their inevitable impact on the resources made available to the pharmaceutical industry. There were also none from the industry itself to make the point that a determined effort should be made for developing new drugs through biotechnological techniques even for the major diseases afflicting the people of the third world.

However, the efforts of the Indian government created substantial awareness of the implications of biotechnology for firms. A number of large firms in the pharmaceutical industry began to invest in biotechnology. A few new dedicated biotechnology firms were created by public laboratory researchers or industrial scientists. At present there are about 100–150 firms active in biotechnology in India (i.e. they have integrated biotechnology techniques in either their research or production or are marketing biotech-based products). About one-third of these firms are active in the pharmaceutical sector. A small proportion of the biotech firms are newly created firms (about 10–15%) and a smaller proportion of them have been created by scientists from public laboratories. With respect to the pharmaceutical industry, the biotech industry has currently well developed strengths in the following areas: vaccine technology, antibiotic fermentation, enzyme fermentation, rDNA technology for R&D, diagnostic probes for tropical diseases, screening of plant and microbial extracts for molecules and clinical testing.

2.4. Review of the literature on R&D in the Indian manufacturing sectors

The literature on the R&D activities of Indian firms has mainly focused on three issues: (i) the impact of R&D expenditure on factor productivity; (ii) the relationship between R&D expenditure intensity and firm size; and (iii) the relationship between R&D expenditure intensity and foreign collaborations, where R&D expenditure intensity is the ratio of R&D expenditure to market sales.

2.4.1. R&D expenditure intensity and factor productivity

Several empirical studies have examined the impact of R&D strategies on the knowledge base of Indian firms through econometric estimations of the production function (Basant and Fikkert (1996); Raut (1995); Ferrantino (1992)). Basant and Fikkert (1996) find that

factor productivity has increased in the scientific industries (chemicals, drugs and electrical appliances) by about 2%, and decreased in the non-scientific industries due to R&D expenditure. On the other hand, Raut (1995) concludes that while in-house R&D of firms have not had any significant effect on firm productivity, firms have gained from the industry wide R&D spillovers resulting from the R&D efforts of other firms in the industry. Finally, Ferrantino (1992) asserts that factor productivity of Indian firms has stagnated while there has been a substantial increase in the qualification of personnel.

We do not examine the impact of R&D strategies on a representative production function because clearly the firms in our sample set have become more productive in the sense that they have diversified into a new field. Furthermore, this issue cannot be studied unless the functional form of the production function is assumed to remain constant which is very unlikely to be the case for an emerging sector.

2.4.2. R&D expenditure intensity and size of the firm

A number of authors have studied the relationship between firm size and R&D expenditure intensity in the Indian manufacturing sectors. Desai (1980) and Kumar and Saqib (1996) find that R&D intensity is an increasing function of firm size because a firm needs to be of a minimum size in order to be able to invest in R&D. Having established an R&D unit, it then enjoys increasing returns to scale. However, Katrak (1989, 1994), Siddharthan and Agarwal (1992) show that R&D intensity is a decreasing function of firm size. Their argument is that returns to R&D do not proportionately increase with increase in size and therefore large firms tend to have lower research intensity. Furthermore, large firms have established market niches and the required technological competence to ensure products of quality and hence do not perceive any need to engage in R&D. Still others like Siddharthan (1988) and Nath (1993) propose a U-shaped relation between R&D intensity and size. Nath (1993) argues that large firms engage in R&D to conceive major innovations to create a competitive advantage in the long run, while small firms spend on R&D to create minor innovations to maintain their competitive advantage in the short run but this relation is influenced by the structure of the industry being considered.

S.V. Ramani / Research Policy 31 (2002) 381–398   387

2.4.3. R&D expenditure intensity and foreign collaborations

In the literature, the relation between internal R&D and foreign collaborations remains an ongoing debate. Some economists assert that internal R&D is a substitute for import of technology. Desai (1980, 1988) argues that Indian R&D given its limited sources can only focus on short term projects and therefore it is more economical to buy rather than make technology that requires medium to long term investment in knowledge generation. Basant and Fikkert (1996) find that the stock of technology imports is always significantly negatively related to in-house R&D. They argue that since returns to technology imports are greater than to internal R&D and since both are substitutes in knowledge production, firms buy from abroad when they can. Spillovers from abroad on the other hand are significantly positively related to in-house R&D indicating that such spillovers are complements to in-house R&D.

Others however assert that technology imports are a complement to internal R&D (Katrak, 1985, 1989, 1994; Deolalikar and Evenson, 1989). Here the basic assumption fuelling the analysis is that Indian R&D is mainly adaptive rather than innovative. Therefore, in order to be efficient in identifying and adapting useful information, processes, or products obtained from Western firms it is necessary to maintain a sufficient level of knowledge through engaging in internal R&D. Siddharthan (1988) further notes that this complementarily is a decreasing function of the technological sophistication of the sector concerned. However, Siddharthan and Agarwal (1992) find that when other firm characteristics like past successes or expenditure on skilled personnel are taken into account, R&D intensity ceases to have any relationship with technology imports. Kumar and Saqib (1996) call for a fresh look at this debate, as they find no significant relation between technology imports and R&D intensities.

Thus, there is no consensus on any of the three issues raised in the literature on the R&D activities of Indian firms. Such diverse results on the impact of R&D strategies could stem from the fact that they analyse different databases and they consider different indicators of R&D strategies. In what follows, we will examine the second and third issues, i.e. the relationship between R&D expenditure intensity and firm size and the relationship between R&D expenditure intensity and foreign collaborations with respect to the Indian biopharmaceutical sector.

3. Formulation of hypotheses, database and variables

In this section, we define the notion of "knowledge base" and "R&D strategy" as used in this paper. Then we present the construction of the database, the variables considered and the sample set of firms.

3.1. Knowledge base embodied in labour stocks and R&D strategy vectors

According to traditional economic theory, the technical knowledge of a firm about the production process is given by its production function, that indicates the maximum output that can be produced from a given combination of tangible inputs, say capital and labour. When a multi-product firm is considered, the production function is replaced by a production possibilities set, which gives the set or combinations of maximum outputs that can be produced from a set of inputs. In both cases, this production technology of the firm is considered to be fixed and constant over time. However, it is now commonly acknowledged that the productivity of factor inputs can change as the firm learns more about the production process or acquires "knowledge stocks". (Grilliches, 1979, 1995).

Thus, a firm starts with four elements: (i) a production possibilities set giving the technology blueprint available to the firm; (ii) an initial knowledge base embodied in its labour stocks; (iii) non-labour stocks and (iv) a mode of governance including an R&D strategy. As a function of these four elements, a firm produces in each time period, new knowledge stocks and final commodity bundles. The final commodity bundles generate market sales which are used to maintain the resources of the firm. The new knowledge stocks change the production possibility set of the firm and may lead to quality improvement, cost reduction or increase in the variety of products produced. Furthermore, if we assume strong market competition and price taking firms, then market performance can be given by market sales.

388   S.V. Ramani / Research Policy 31 (2002) 381–398

Why do we evoke this simple scheme, when there exists so many sophisticated theories of the firm? This is for two reasons. The first is to focus on the need to examine the relation between the composition of labour stocks and market sales. In high tech sectors, where human ingenuity is the key to the creation of innovations and increase in market shares, it seems likely that the composition of the knowledge base of the firm as embodied in its labour stocks is crucial to its market performance. Labour stocks can consist of qualified or non-qualified labour and it can be allocated to R&D or non-R&D tasks. This gives rise to four types of stocks (qualified R&D, non-qualified R&D, qualified non-R&D, non-qualified non-R&D) whose values and proportions are likely to impact the market performance. In any sector, where knowledge generation is important for market performance, the latter is likely to depend on the stocks of labour in the R&D department or at least the total stock of qualified labour in the firm.

The second notion that is sought to be promoted is that an R&D strategy is actually a vector of possible actions rather than being identified with simply R&D expenditure. Let us define an R&D strategy as the vector of decisions related to the acquisition and disclosure of knowledge.

With respect to the acquisition of knowledge, two kinds of decisions can be considered, labour allocations and technology transactions. Labour allocations refer to the quantity, quality and distribution of labour within the firm and have an impact on the "learning by doing" of employees. Knowledge can also be acquired from outside of the firm through technology transactions. Technology transactions refer to technology purchases (i.e. R&D expenditure on capital stocks) and technology alliances. The latter can be either with public laboratories or with other firms.

Once new knowledge is created, a firm has to decide how much of the knowledge should reside within a firm and how much should be disclosed. Disclosure can take two forms: with protection in the form of patents or without protection in the form of publications.

What is likely to be the relation between the different components of the R&D strategy? A priori all the components of the R&D strategy would seem to be strategic complements. In reality, it would depend on the context studied and the constraints of the firms considered.

In the pharmaceutical sector of developed countries, Cockburn et al. (1999) assert that "science-driven", or "rational" drug discovery is both a technology for discovering new drugs and a set of managerial practises for organising and motivating research workers. They point out that firms invest in leading edge research because it increases the efficiency of the knowledge production process (Henderson and Cockburn, 1994; Gambardella, 1995). They confirm that in the pharmaceutical sector some firms have become active participants in the creation of scientific knowledge instead of being only passive users. These firms publish and patent actively and have extensive research networks with research laboratories. Investment in patenting in the biopharmaceutical sectors has also been noted to be a means by which firms can mark their territory in limited technological space to gain future rent. If the situation is similar in developing countries we should find the same pattern. In other words, we should find that all parameters of the R&D strategy take on higher values for firms which undertake biotech research as compared to firms which are simply marketing or producing a biotech-based product without doing research.

The central questions of the paper as may be recalled are: (i) What kind of labour stocks forming the knowledge base have an impact on market sales? (ii) Which components of the R&D strategy are strategic substitutes and which are strategic complements? (iii) What are the distinguishing features of firms that have already integrated biotechnology in their research activities? From the arguments developed in this section, we can propose the following three hypotheses as initial responses to the above questions to be tested with data.

H1: market sales are an increasing function of the stocks of qualified labour or an increasing function of labour stocks allocated to R&D activities.

H2: all parameters of an R&D strategy are strategic complements.

H3: all parameters of an R&D strategy will take on higher values for firms which are active in biotechnology research (as opposed to firms which are only marketing or producing a biotech product but not doing biotech research).

S.V. Ramani / Research Policy 31 (2002) 381–398 389

Table 1
Variables considered

Variables considered

Market performance
  Market sales

Firm characteristics
  Age
  Technological orientation = (0 if not doing biopharmaceutical R&D, 1 if doing biopharmaceutical R&D)

Knowledge base
  Size/total number of personnel
  Qualified personnel
  R&D personnel
  Qualified personnel in R&D

R&D strategies
  R&D expenditure
  R&D expenditure intensity = R&D expenditure/total sales
  R&D employment intensity = employees involved in R&D/total number of employees
  Qualification intensity = number of employees with a masters or Ph.D. degree/total number of employees
  R&D qualification intensity = number of employees with a masters or Ph.D. degree in R&D/total number of employees in R&D
  Academic collaborations = (number of technology agreements since 1970)
  Foreign collaborations = (number of technology agreements since 1970)
  Publications = (between 1970 and 1994)
  Patents = (granted between 1970 and 1994)

3.2. Construction of the database

We first compiled a list of firms active in the biopharmaceutical sector from three documentary sources published by the Ministry of Science and Technology of the government of India: (i) reports of the Department of biotechnology; (ii) the "Directory of biotechnology industries and institutions in India" and (iii) the directory on the "Research profile of biotechnology activities in India". They yielded 48 pharmaceutical firms as being active in the biotechnology sectors. From the various reports of the department of biotechnology we were able to compile information on the different variables for 24 of the 48 firms. We were able to interview the CEOs of 8 more firms and obtain information directly on these also. The information on labour allocations, patents, publications and R&D expenditures are not normally published in company reports. Both the reports of the department of biotechnology and our information were based on answers to questionnaires. Thus, all information were voluntary disclosures by the firms themselves including data on whether they were undertaking biotech research or simply marketing or

producing a biotech product without doing in-house biotech research. Information on patents were also obtained from the responses to questionnaires and referred to patents actually obtained by the firm between 1970 and 1994.[4] Pooling these two sources of data we obtained information on 32 of the 48 firms. The information collected pertained to the year 1994–1995.

3.3. Variables considered

Four types of variables were considered market sales, firm characteristics, initial knowledge base as embodied in labour stocks and R&D strategies. They are given in Table 1. Two kinds of firm characteristics were noted: age and technological orientation. The latter was a dummy variable that associated value 1 with a firm conducting biopharmaceutical research and value 0 with a firm that was not conducting biopharmaceutical research at the time of the data collection. This distinction was made to

---

[4] This period was considered by the department of biotechnology in their reports and we had to use the same to be consistent.

390 S.V. Ramani / Research Policy 31 (2002) 381–398

differentiate the firms active in biopharmaceuticals through marketing a product for a multinational or producing a biotech-based product on the basis of a license without undertaking their own biotech research.

The knowledge base of the firm embodied in its labour stocks was classified as follows. Labour within in a firm can be either qualified (with a Masters degree or more) or non-qualified. These two kinds of labour can be allocated to R&D or non-R&D tasks. This gives the four kinds of labour stocks: total personnel, total qualified personnel, total personnel in R&D and total qualified personnel in R&D.

R&D strategies were considered in terms of labour allocations, technology transactions and disclosed knowledge. This gave us 9 indicators of the R&D strategy of a firm as shown in Table 1. R&D expenditures were taken into account both as a stock and as an intensity variable because larger firms usually spend more in absolute amounts on R&D, but this does not mean that such large firms are pursuing a more aggressive R&D strategy. They might in fact be re-investing a lower proportion of their sales revenue in R&D activity or having a lower rate of new knowledge creation. Two kinds of technology alliances were noted: alliances with public laboratories or with foreign firms. These were technology agreements with or without equity participation of the foreign partner. There were no alliances between Indian firms themselves in our data set.

3.4. Firms in the sample set

Among the 32 firms in our sample set, there were 3 Indian subsidiaries of MNCs, 2 government or public sector firms and 27 private sector firms. Out of the 32 firms, 26 were established firms that had diversified into biotechnology and 6 were new founded dedicated biotechnology firms. All the firms in our data set were either marketing, producing or doing research on a biotech-based product. About 19 firms, were only marketing or producing a biotech-based product without doing biotech research. In terms of size, 11 firms were very large firms with more than 1000 employees, 15 were medium sized firms with between 100 and 1000 employees and all the remaining 6 firms were new dedicated biotechnology firms with less than 100 employees.

4. Methodology and results

4.1. Methodology

To identify the relation between market sales, knowledge base in the form of labour stocks and R&D strategies, a correlation matrix was computed. Next a model of market sales as a linear function of the labour stocks forming the knowledge base was estimated using the method of "step wise linear regressions". There was also an attempt to model R&D intensity as a function of market sales, firm characteristics and other R&D strategies but this did not yield results that were more definite than the correlation analysis. Next, to understand the relation between the different R&D strategies, a principal component analysis (PCA) was conducted on the R&D strategies.[5] Finally, the distinctive features of firms active in biopharmaceutical research were identified through an ANOVA analysis.

4.2. Relation between market sales, knowledge base and R&D strategies

The correlation matrix between all the quantitative variables (i.e. all except biotech research) is given in Table 2. Evidently it could throw light only on the first two hypotheses because the firms doing biotech research were not distinguished from others.

- The first hypothesis H1 is strongly supported as the market sales are significantly correlated to knowledge stocks in the form of total personnel, total qualified personnel and total R&D personnel.
- The second hypothesis H2 is weakly supported. All the different components of an R&D strategy are not strategic complements. R&D expenditure intensity is significantly positively correlated with R&D employment intensity. Not surprisingly, firms with a higher proportion of R&D personnel spend a higher proportion of their sales revenue on R&D activities. Such firms are likely to be small and young firms. Foreign collaborations are significantly negatively correlated with qualification intensity, but they are positively correlated with the stock of qualified personnel in R&D. Patents and publications are strate-

---

[5] R&D expenditure in absolute terms was dropped out of the PCA and ANOVA analysis as we wanted to examine the relations between all the intensity measures.

S.V. Ramani/Research Policy 31 (2002) 381–398

Table 2
Descriptive statistics on sample firms[a]

Variable	1	2	3	4	5	6	7	8	9	10	11	12	13	14	15	Mean	S.D.
Market sales	1.0															40.11	54.47
Firm characteristics																	
Age	0.23	1.0														27.16	18.80
Knowledge base																	
Total personnel/size	0.69**	0.34	1.0													1459.75	1876.49
Qualified personnel	0.8**	0.35	0.86**	1.0												817.62	1184.01
R&D personnel	0.50**	0.15	0.65**	0.42*	1.0											64.31	58.74
Qualified personnel in R&D	0.8	0.21	0.6**	0.25	0.63**	1.0										32.32	34.27
R&D strategies																	
R&D expenditure	0.78**	-0.03	0.43	0.61*	0.40	-0.06	1.0									1.23	2.39
R&D expenditure intensity	0.12	-0.36*	-0.21	-0.17	-0.22	-0.13	0.05	1.0								0.04	0.07
R&D employment intensity	-0.2	-0.37*	-0.42*	-0.42*	-0.06	-0.22	-0.07	0.64**	1.0							0.12	0.14
Qualification intensity	0.13	-0.02	-0.02	0.25	-0.30	-0.39*	0.28	0.03	-0.16	1.0						0.57	0.20
R&D qualification intensity	-0.31	-0.15	-0.01	-0.09	-0.36*	0.36*	-0.33	0.04	-0.24	0.06	1.0					0.58	0.33
Academic collaborations	-0.21	-0.37*	-0.16	-0.22	0.01	-0.17	-0.14	0.09	0.21	0.30	-0.18	1.0				1.69	1.38
Foreign collaborations	-0.02	0.03	0.15	0.03	0.31	0.38*	-0.08	-0.34	-0.20	0.21	0.08	-0.22	1.0			2.69	3.34
Publications	0.2	0.06	0.30	0.02	0.41*	0.30	0.07	-0.14	-0.06	-0.54**	-0.15	-0.05	0.1	1.0		5.20	12.17
Patents	0.4*	0.14	0.22	0.24	0.24	-0.05	0.17	-0.07	-0.16	0.10	-0.23	0.02	-0.08	0.51**	1.0	3.47	10.31

[a] Units: US$ million.
* Pearson's correlation coefficient is significant at 1%.
** Pearson's correlation coefficient is significant at 5%.

Table 3
Model of market sales

Explanatory variable	Coefficient (β)	t-Value
Qualified personnel in non-R&D	0.102	−3.19*
Non-qualified personnel in R&D	1.15	6.38*
Constant	−4.487	3.64
$R^2 = 0.78$	$F = 38.7$*	

* $p \leq 0.005$.

gic complements but they are not significantly correlated to any other R&D strategies. It is noteworthy that publications are positively correlated to R&D personnel while patents are positively correlated to market sales. This reveals that patents are of some importance to market performance.

Two other points of interest concern academic collaborations. Younger firms seem to be entering into more collaborations with public laboratories than older ones. The mean of number of academic collaborations is less than that of foreign collaborations, implying the Indian firms tend to initiate more collaborations with foreign firms than with research centres in their own country.

With respect to the issues raised in the literature on Indian R&D, it can be inferred that R&D expenditure intensity is likely to depend on the characteristics of the firm such as age rather than size. In fact, R&D expenditure intensity is negatively correlated to firm size and market sales though this is not statistically significant. It is more difficult to draw conclusions on the relationship between foreign collaborations and internal R&D. While foreign collaborations are significantly negatively correlated with qualification intensity, they are positively correlated to the absolute stock of qualified personnel in the R&D department.

To further identify the labour stock that impact market sales, a stepwise regression was run and the result is given in Table 3. It indicates that market sales increases with an increase in the qualified labour outside of the R&D department or the non-qualified labour in the R&D department.

We also tried to estimate a model for R&D intensity by means of a stepwise regression but the only significant coefficient in the model was that of R&D expenditure intensity and age as already revealed by the correlation matrix. Younger firms or firms with a

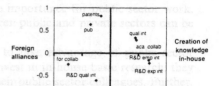

Fig. 1. Relations between R&D strategies; for. collab: foreign collaborations; pub.: publications; qual. int.: qualifications intensity; aca. collab: academic collaborations; R&D emp. int.: R&D employment intensity; R&D exp. int.: R&D expenditure intensity; R&D qual. int.: R&D qualifications intensity.

higher ratio of personnel allocated to R&D activities exhibit a higher R&D expenditure intensity.

4.3. Relations between the different R&D strategies

In order to understand the relations of substitutability or complementarily between the different R&D strategies, a principal component analysis was conducted. The analysis yielded three factors that accounted for about 60% of the total information (or variance) contained in the sample. Fig. 1 shows the variables mapped along the first and second factors and Fig. 2 shows the variables mapped along the first

Fig. 2. Relations between R&D strategies: for. collab: foreign collaborations; pub.: publications; qual. int.: qualifications intensity; aca. collab: academic collaborations; R&D emp. int.: R&D employment intensity; R&D exp. int.: R&D expenditure intensity; R&D qual. int.: R&D qualifications intensity.

S.V. Ramani/Research Policy 31 (2002) 381–398   393

Table 4
Distinctive features of firms active in biotechnology research

Variable	Mean of variable for firms not active in biotech research	Mean of variable for firms active in biotech research
Market sales	165.84	102.52
Firm characteristics		
Age**	42.56	21.13
Knowledge base		
Size/total personnel	2121.00	1201.00
Qualified personnel	861.86	803.54
R&D personnel	86.33	55.69
Qualified personnel in R&D	60.50	22.52
R&D strategies		
R&D expenditure intensity**	1.38	5.29
R&D employment intensity**	0.056	0.14
Qualification intensity**	0.45	0.61
R&D qualification intensity	0.64	0.56
Academic collaborations	1.11	1.91
Foreign collaborations	4.67	1.91
Publications	7.0	4.54
Patents	4.50	3.09

** t-Test for distinct means significant at 5% or less.

and third factors. The three factors seem to embody the following three axes of R&D strategy:

- acquisition of knowledge (as defined by R&D employment intensity, R&D expenditure intensity, foreign collaborations);
- disclosure of knowledge (as defined by patents, publications and R&D qualification intensity);
- creation of knowledge within the firm (as defined by R&D employment intensity and qualification intensity).

Fig. 1 shows that in the acquisition of knowledge there are two possible substitutable strategies. Either knowledge can be created by the Indian firms through internal R&D or be acquired from abroad. Internal R&D is given by R&D expenditure intensity and R&D employment intensity. Foreign acquisitions are given by foreign collaborations. It is interesting that while collaborations with public laboratories are complements to internal R&D (both variables being on the same side of the horizontal axis), foreign collaborations are substitutes to internal R&D (both variables being on opposite sides of the horizontal axis). Thus, foreign collaborations and internal R&D seem to be substitutes and not complements.

Let us now come to the disclosure of knowledge. Fig. 1 shows patents and publications to be on the opposite side of R&D qualification intensity in terms of the horizontal axis. This indicates that either the firms can disclose knowledge in the form of patents or publications (which are complements) or let the knowledge reside within the firm by increasing the proportion of qualified people in the R&D department.

Finally, with respect to the creation of knowledge, in Fig. 2, the two furthermost variables on the vertical axis defining the third factor, are qualification intensity and R&D employment intensity. These are also aligned on opposite sides of the horizontal axis. This indicates that the firm can either choose to create knowledge throughout the firm by recruiting more qualified people and distributing them throughout firm or it can focus on creating knowledge through the R&D department through allocating more personnel to this department.

4.3.1. Distinguishing features of biotech firms

The distinguishing features of biotech firms were identified through an ANOVA analysis, the results of which are given in Table 4. The distinguishing features that are statistically significant are age and learning

394   *S.V. Ramani / Research Policy 31 (2002) 381–398*

strategies. Firms active in biotechnology research are younger, allocate a higher proportion of personnel to R&D, reinvest more of their sales revenue on R&D and have a higher proportion of qualified personnel on their payrolls. It may be recalled that according to hypothesis H3, all parameters of R&D strategy are expected to take on higher values for firms undertaking research in biotechnology. This is clearly not the case for R&D qualification intensity, foreign collaborations, publications and patents.

4.4. Discussion of results

Two issues that have been examined in the literature of Indian R&D and that are of relevance to the subject at hand are the relationship between firm size and R&D expenditure intensity and the relationship between foreign collaborations and R&D expenditure intensity. Our results support the view that R&D expenditure intensity decreases with size of the firm and is a strategic substitute for foreign collaborations.

As in the developed countries, the Indian pharmaceutical sector comprises three kinds of companies: large incumbent firms, small and medium sized incumbent firms and new dedicated biotechnology firms (usually very small). The firms that are active in biopharmaceuticals are mainly medium sized and big companies and new dedicated biotechnology firms. The medium sized and big firms have the resources to diversify into a new field. The new firms are created through commercialisation of a specific biotech-based knowledge. However, many of these firms are simply marketing the product of a Western multinational or producing a biotech-based product using a license. Only some of them are trying to build their own knowledge base through investment in biotechnology research. We would expect such firms, which are investing in diversifying their knowledge to have a higher R&D intensity than others. Thus, R&D intensity would not be dependent on the size of the firm but rather on the research orientation of the firm.

Foreign collaborations and R&D expenditure intensity could be strategic substitutes because of two possible reasons. Firstly, Indian firms have a knowledge retard with respect to biotechnology. Therefore, they may not yet have the absorptive capacity to use foreign technology as a complement to their own knowledge base. Secondly, it could be due to the difficulties

and uncertainties of international technology transfer. Often, a pertinent knowledge transfer does not occur. Thus, foreign collaborations are sought only for technology that cannot be developed economically in-house.

We now discuss the results that were unexpected or counter intuitive. The first was with respect to the determinants of market sales. It was hardly strange that market sales increased with the qualified labour outside of the R&D department. However, it was surprising that market sales increased with the non-qualified personnel in the R&D department. The latter relation is very counter intuitive and could be due to two possible reasons. Either the qualified personnel in the R&D departments are mis-managed and contribute little to market performance or this is due to the nature of our sample. In our sample, firms with a high R&D qualification intensity are younger, smaller firms, which have lower market sales and this could be leading to a positive relation between market sales and the non-qualified personnel in the R&D departments.

It has been mentioned that patents and publications are strategic tools for a firm to improve its market position in the biopharmaceutical sectors of developed countries. In contrast, in the Indian case, while patents and publications are clearly strategic complements, they were not correlated to any indicators of R&D strategy. Moreover, they are aligned in opposition to R&D qualification intensity, implying that firms with a high proportion of qualified people in the R&D department do not seek to patent or publish. This is quite counter intuitive.

This result may be due to the state of the patenting bureaucracy in developing countries including in India. In the Indian pharmaceutical sector, Redwood (1994) and Lanjouw (1998) assert that it is not common to publish or patent because most of the research is on the engineering side and therefore the knowledge created is tacit residing in an individual or sets of individuals. It is difficult to translate such knowledge into writing and therefore there is no incentive to apply for patents. They also note that India at present does not have the infrastructure to ensure an efficient patenting process. The patent offices are very poorly staffed, they have very limited resources, there are not many patent lawyers and there are not many people who know both the science and the law. This makes patent application a time consuming and costly affair

S.V. Ramani / Research Policy 31 (2002) 381–398   395

within India and too costly an investment outside of India. Finally, patent protection is not effective because patent litigation usually costs more than out of court settlements. This is surely going to change but it will take time. Thus, patents and publications may be more the result of a firm specific managerial orientation rather than being correlated to other R&D strategies.

Another counter-intuitive result is that firms that are active in biotech research are less interested in publications and patents and they are less connected to international networks. A number of explanations are possible. Such firms could be working on projects close to being commercialised and therefore have no interest in publishing their work. The lack of concern for patenting may be because they do not perceive its benefits to be sufficiently high, or because patenting is not a routine that firms are forced to think about, since the patenting bureaucracy is not yet well developed in India. Finally, it could be due to the nature of the sample set. Firms that are active in research are younger and they may not have accumulated a large number of patents or publications yet.

The fact that firms undertaking biotech research are less inserted in international networks could be due to the strategic foundations of international collaboration. Many of the firms undertaking biotech research are small dedicated biotechnology companies with a strong knowledge base in biotechnology. They often focus on creating drugs and diagnostics for diseases prevalent in India such as leprosy, malaria, filariasis, etc. that are not of interest to the Western multinationals. Western firms also prefer to collaborate with large rather than small Indian firms since they perceive the large firms to pose less of a market risk.

5. Conclusions

Strategic positioning of firms for the integration of new technology in emerging economies is not frequently studied because the set of firms engaged in such activity is small and data is often not available. Most of the existing works have looked at the Indian manufacturing sector at large using data pertaining to the pre-liberalisation era. In contrast, this paper has focused on one sector, namely the biopharmaceutical sector, and its objective was to examine the impact of

knowledge stocks and the nature of R&D strategies of firms in this sector.

Our analysis showed that market performance is positively correlated with the knowledge base of the firm as embodied in its qualified labour outside of the R&D department. The three factors defining the R&D strategies were acquisition of knowledge, disclosure of knowledge and internal creation of knowledge. New technology could be acquired from abroad or created within Indian firms through increasing the R&D personnel. Knowledge could be disclosed in the form of patents or publications or remain as tacit knowledge within the firm. Knowledge could be created in the firm by increasing the R&D department personnel or in a diffused manner throughout the firm by increasing the qualified personnel.

Finally, the analysis revealed that firms that are likely to make inroads into the biopharmaceuticals sector have to be identified by their technology strategy and not their resources. It showed that firms that are producing biotechnology products are likely to have a strong research base. They may not be into publishing or patenting but they allocate a high proportion of their labour to R&D activities and employ a substantial number of qualified personnel for conducting R&D.

5.1. Recommendations for foreign firms, small Indian firms, and large Indian firms

The study indicates that Indian firms seeking foreign technological collaborations in the biopharmaceuticals sector are likely to be not doing research in biotechnology and are likely to have a lower proportion of qualified personnel. These firms buy technology because it is more economical to buy than to create internally. In the same market, there are new and small firms, which are research-intensive and which allocate a substantial proportion of their employees to R&D activity. Therefore, a potential exists for forming "research contracts" or technological collaborations between these small R&D intensive Indian firms and Western firms, as has happened between Indian and Western firms in the field of micro-electronics. These have to be initiated by Western firms since the small Indian firms are usually not searching to collaborate with foreign firms. At the same time, if small firms seek foreign collaborations, then they have to become

396   S.V. Ramani / Research Policy 31 (2002) 381–398

more visible on the international scene through publications and patents.

The negative correlation between R&D strategy indicators and market sales indicates that employing educated engineers and technicians who have acquired sufficient knowledge in the universities largely satisfies the needs of large firms. Large firms also tend to have fewer academic collaborations. However, since knowledge stocks clearly have a positive impact on the market performance of large firms, they might well reconsider whether increasing these parameters would be better for maintaining their competitive advantage in the long run.

An often highlighted feature of biopharmaceutical firms in the developed countries is their complex web of strategic alliances with other firms and public research laboratories (Orsenigo, 1989; Pisano, 1991). It is usually proposed that knowledge creation through internal R&D and external strategic alliances are strategic complements. In the case of large firms, it has been shown that larger the investment in internal R&D or internal learning, larger the number of external strategic alliances (Arora and Gambardella, 1990). Such a phenomenon is completely absent in the Indian pharmaceutical sector.

Most of the interfirm collaborations in biotechnology in developed countries occur at a pre-competitive stage, i.e. they are on projects that are not close to being commercialised. They are also initiated when the R&D costs or the R&D risks are too high to be supported by a single firm. Thus, one plausible reason for the non-initiation of interfirm collaborations between Indian firms could be due to the fact that there has been intense competition only to develop and commercialise already patented drugs. In such cases, the R&D costs are not high, the R&D risks are not high and the product can be immediately commercialised, which leaves little incentive for interfirm co-operation. In the technology races, which occur periodically in the Indian bulk drugs market, "a winner takes most" game is set into motion leaving little scope for inter-firm co-operation. However, as production in the pharmaceutical sector becomes more and more knowledge intensive, in order to compete in the international arena, firms can do better by initiating co-operative alliances with research laboratories and other firms on projects at a pre-competitive stage.

5.2. Policy recommendations

In terms of policy formulation, the two most striking features in need of reform are the lack of interfirm co-operation between Indian firms and the low impact of public research laboratories on the market sales or research strategies of Indian firms. Given the paucity of resources to which all emerging economies are subject to, and in order not to aggravate the north–south gap, it is necessary to maximise the economic returns from existing investment in public research. Thus, conventions have to promoted for the transfer of knowledge from public laboratories to private firms and then for its transformation into commercialisable technology. Ramani and Visalakshi (2001) have argued that with respect to biotechnology, the Indian policy so far has tried to emulate the American model to some degree, whereby public research is funded and promoted and the market is expected to generate new firms and new innovations. In the American context, there is conversion of knowledge into technology by the market itself, because of actively functioning networks between the different agents of the innovation system, such as public laboratories, pharmaceutical firms, new biotechnology firms, government and financiers. Such networks are already less active in Europe and even more dormant in emerging economies like India. India has been successfully able to develop the nuclear bomb, satellites and the super computer because such projects involved a group of scientists who were given directives under a "mission mode", i.e. under a clearly defined system of milestone targets and associated rewards. This route cannot be pursued in the integration of biotechnology because biotechnology involves a variety of techniques with multi-sectoral applications. A variety of agents have to mobilised in order to integrate biotechnology in any particular sector. Thus, it may be worthwhile for the Indian government to consider more intervention in the creation of networks between Indian firms and between public laboratories and private firms themselves through national programs, as some European countries such as France have successfully done, in order to accelerate the integration of biotechnology and generally the creation of innovations in the Indian pharmaceutical sector.

S.V. Ramani/Research Policy 31 (2002) 381–398   397

5.3. Limitations and suggestions for further study

Our primary problem was to obtain comprehensive data on the biopharmaceutical firms. There is not much data available on the R&D activities or technology related purchases of firms. Often data from different sources are contradictory and a considerable time has to be spent in identifying the correct information. Telephone interviews or direct interviews are necessary to obtain relevant data on many private limited firms. Any extension of the present work can thus envisage the amelioration of the database used. Secondly, different measures of market sales such as net profits can be considered in the place of sales if such data can be obtained. Case studies can also be conducted to open "the black box" of international strategic alliances in order to identify the conditions favourable to the initiation and success of technology collaborations with Indian firms.

Acknowledgements

I gratefully acknowledge financial assistance from INRA under the programme "Theory of Contracts" that made the research for this paper possible. I would especially like to thank Lionel Nesta for his extremely useful comments. Many thanks also go to Olivier Taramasco, Paolo Saviotti and the participants of the CSIR/RADMA conference (December 1999, New Delhi) for their useful suggestions. I am also very grateful for the insightful comments of the two referees. I remain solely responsible for any errors.

References

Acharya, R., 1995. The impact of New Technologies on Economic Growth and Trade: A case Study of Biotechnology. Ph.D. Thesis, University of Maastricht.

Ahmad, H., 1988. Technological Development in Drugs and Pharmaceutical Industry in India. Navrang, New Delhi.

Arora, A., Gambardella, A., 1990. Complementarity and external linkages: the strategies of large firms in biotechnology. Journal of Industrial Economics 4, 361–379.

Basant, R., Fikkert, B., 1996. The effects of R&D, foreign technology purchase, and domestic and international spillovers on productivity in Indian firms. Review of Economics and Statistics, pp. 187–199.

Biotech Consortium Ltd., 1994. Directory of Biotechnology Industries and Institutions in India. New Delhi, India.

Centre for Monitoring Indian Economy (CMIE), 1996. The Indian Corporate Sector. Bombay.

Cockburn, I., Henderson, R., Stern, S., 1999. The diffusion of science driven drug discovery: organizational change in pharmaceutical research. Working paper 7359. National Bureau of Economic Research, 1050 Massachusetts Avenue, Cambridge, MA 02138.

Deolalikar, A.B., Evenson, R.E., 1989. Technology production and technology purchase in Indian industry: an econometric analysis. Review of Economics and Statistics, pp. 687–692.

Department of Biotechnology, Government of India, 1993. Research Profile of Biotechnology Activities in India. New Delhi, India.

Desai, A.V., 1980. The origin and direction of industrial R&D in India. Research Policy 9, 74–96.

Desai, A.V., 1988. Technology acquisition and application: interpretations of the Indian experience. In: Lucas, R.E., Papanek, G.F. (Eds.), The Indian economy. Oxford University Press, Oxford.

Ernst, Young, L.L.P., 1998. New Directions 98, The 12th Biotechnology Industry Annual Report, CA.

Ferrantino, M.J., 1992. Technology expenditures, factor intensity and efficiency in Indian manufacturing. The Review of Economics and Statistics, pp. 689–699.

Gambardella, A., 1995. Science and Innovation in the US Pharmaceutical Industry. Cambridge University Press, Cambridge.

Grilliches, Z., 1979. Issues in assessing the contribution of research and development to productivity growth. Bell Journal of Economics 10, 92–116.

Grilliches, Z., 1995. R&D and productivity. Handbook of the Economics of Innovation and Technological Change. Oxford University Press, UK and Cambridge University Press, USA, Blackwell Scientific Publications, Oxford, pp. 52–89.

Henderson, R., Cockburn, I., 1994. Measuring competence? Exploring firm effects in pharmaceutical research. Strategic Management Journal 15, 63–84.

Katrak, H., 1985. Imported technology, enterprise size and R&D in a newly industrializing country: The Indian experience. Oxford Bulletin of Economics and Statistics 47, 213–229.

Katrak, H., 1989. Imported technologies and R&D in a newly industrialising country. Journal of development economics 31, 123–139.

Katrak, H., 1994. Imports of technology, enterprise size and R&D-based production in a newly industrializing country: the evidence from Indian enterprises. World Development 22, 1599–1608.

Kumar, N., Saqib, M., 1996. Firm size, opportunities for adaptation and in-house R&D activity in developing countries: the case of Indian manufacturing. Research Policy 25, 713–722.

Lanjouw, J.O., 1998. The introduction of pharmaceutical patents in Indian: heartless exploitation of the poor and suffering? National Bureau of Economic Research, Working paper 6366.

Nath, P., 1993. Firm size and in-house R&D: the Indian experience revisited. The Developing Economies 3, 329–344.

Office of Technology Assessment, O.T.A., 1991. Biotechnology in a global economy. US congress office of Technology Assessment, US Government Printing Office, Washington DC.

398   *S.V. Ramani / Research Policy 31 (2002) 381–398*

Orsenigo, L., 1989. The Emergence of Biotechnology. Pinter, London.

Pisano, G.P., 1991. The governance of innovation: vertical integration and collaborative arrangements in the biotechnology industry. Research Policy 20, 237–249.

Probert, J., 1994. Japanese Pharmaceutical Firms. Players in the European Market. In: Helmut Schutle (Ed.), The Global Competitiveness of the Asian Firm. St. Martins Press, New York.

Ramani, S.V., Venkataramani, M.S., 2001. Rising to the technological challenge: possibilities for integration of biotechnology in the Indian pharmaceutical industry. The International Journal of Biotechnology, 3, no 1/2.

Ramani, S.V., Visalakshi, S., 2001. The chicken or the egg problem revisited: the role of resources and incentives in the integration of biotechnology techniques. The International Journal of Biotechnology, in press.

Raut, L.R., 1995. R&D spillover and productivity growth: evidence from Indian private firms. Journal of Development Economics 48, 1–23.

Redwood, H., 1994. New Horizons in India. Oldwicks Press Limited, Suffolk, England.

Sasson, A., 1993. Biotechnologies in Developing Countries, Present and Future, UNESCO, Paris.

Siddharthan, N.S., 1988. In-house R&D, imported technology, and firm size: lessons from Indian experience. The Developing Economies 3, 212–221.

Siddharthan, N.S., Agarwal, R.N., 1992. Determinants of R&D decisions: a cross-section study of Indian private corporate firms. Economic Innovation and New Technology 2, 103–110.

Singh, S., 1985. Multinational Corporations and the Indian Drug Industry. Criterion Publications, New Delhi.

Name Index

The International Library of Critical Writings in Economics

the capability of Indian technologists and the entrepreneurial skills of the Indian businessmen, and overestimated the appeal of their brand names for the price conscious consumer. The consumer was quite willing to go for a lower-priced Indian product with its own brand name.

In 1976, among the top 20 firms which held 57.05% of the pharmaceutical market, there were only 4 Indian firms. However, by 1995 only 7 MNCs (including their subsidiaries) figured among the top 20 pharmaceutical companies in India and together they could claim only 15.1% of the total market. Indian companies that had won a place in the 1995-1996 list, ranked in order of their market share were: Ranbaxy, Lupin, Cipla, Dabur, SOL Pharma, Sarabhai, Torrent, Dr. Reddy's, Alkembie Kopran, Ipca and Cadilla. In addition, there were 38 other Indian owned pharmaceutical companies that were among the top 50 in terms of sales (US$ 22 million or more) during 1995-1996. Only 12 MNCs figured in the list, of whom only 3 made it to the top 10: Glaxo, Hoechst, and Pfizer.

Clearly, in order to compete against entrenched and popular MNC brands, the Indian substitutes had to become of comparable quality and cheaper in price. These requirements made it incumbent on Indian engineers and managers to pay continuing attention to cost reduction and quality control. Many of the companies in the top 100, recognising the opportunities afforded by the Indian Patent Law, made modest investments in R&D activities resulting in an enhancement of their technical capabilities in working out processes for the production of selected drugs identified by them as having good commercial prospects. However, R&D expenditures as a percentage of sales still remained quite low compared to figures in the advanced countries and companies generally tended to raise it to just the point needed for the production of the identified drugs. Most importantly, while successful Indian companies had demonstrated their capabilities in bringing out very satisfactory substitutes for a number of patented Western products and expanded their sales in India and in the overseas markets through lower prices, none of them had come up with a significant innovation in the form of a new drug based on indigenous R&D. The Indian pharmaceutical firms had their knowledge base firmly embedded in organic and synthetic chemistry. They had not made any efforts to integrate other scientific disciplines to create or re-engineer innovations.

These firms were then confronted with biotechnology during the 1990s, a set of techniques based on recent developments in the life sciences that was new, different and much more complex to integrate requiring a multi-disciplinary team to create a product.

2.2. Strategy for biotechnology in India

Biotechnology in India emerged largely due to the key role played by scientists turned administrators in the government machinery. In this section, we introduce the strategy and role of the Indian government in the initiation of the biotechnology sectors. The strategy of the Indian government can be grouped into three stages: initiation, building scientific competence, and reaching out to the private sector.

2.2.1. Initiation, 1982-[1986]

In India, the push to develop the biotechnology sectors came from reputed scientists who had been brought into the administration. In 1982, Dr. S. Varadarajan, then Secretary of the Department of Science and Technology headed the National Biotechnology Board (NBTB). Its objectives were to (a) identify priority areas in biotechnology, (b) identify infrastructural needs, and (c) implement a co-ordinated programme to realise certain national objectives. To this end, a number of pilot programmes were proposed in the 6th (1980-1985) and 7th (1985-1990) five-year plans. However, during its 4 years of existence, from 1982 to 1986, the NBTB seems to have achieved only objectives (a) and (b).

In 1986, the NBTB was replaced by a separate government department, namely a Department of Biotechnology (DBT, 1986). It was placed under the aegis of the Ministry of Science and Technology. The main reason for this evolution seems to have been the realisation that biotechnology is a generic technology whose progress requires the development of a variety of competencies in a variety of scientific disciplines. In order to achieve this co-ordinated development, an agency working together in unison with the Ministry of Science and Technology was deemed necessary. It was set up to implement the objectives of the earlier body such as the development of scientific competence

[1] The 2001 R&D government strategy for biotechnology in China is reviewed (Rangan and Guillemin, 2004).

336   *S. b. Kumar/Research Policy 31 (2002) 341-346*

Why do we evoke this simple scheme, when there exists so many sophisticated theories of the firm? This is for two reasons. The first is to focus on the need to examine the relation between the composition of labour stocks and market sales. In high tech sectors, where human ingenuity is the key to the creation of innovations and increase in market shares, it seems likely that the composition of the knowledge base of the firm as embodied in its labour stocks is crucial to its market performance. Labour stocks can consist of qualified or non-qualified labour and it can be allocated to R&D or non-R&D tasks. This gives rise to four types of stocks (qualified R&D, non-qualified R&D, qualified non-R&D, non-qualified non-R&D) whose values and proportions are likely to impact the market performance. In any sector, where knowledge generation is important for market performance the latter is likely to depend on the stocks of labour in the R&D department, or at least the total stock of qualified labour in the firm.

The second notion that is sought to be promoted is that an R&D strategy is actually a vector of possible actions rather than being identified with simply R&D expenditure. Let us define an R&D strategy as the vector of decisions related to the acquisition and disclosure of knowledge.

With respect to the acquisition of knowledge, two kinds of decisions can be considered, labour allocations and technology transactions. Labour allocations refer to the quantity, quality and distribution of labour within the firm and have an impact on the learning by doing of employees. Knowledge can also be acquired from outside of the firm through technology transactions. Technology transactions refer to technology purchases (i.e. R&D expenditure on current stock) and technology alliances. The latter can be either with public laboratories or with other firms.

Once new knowledge is created, a firm has to decide how much of the knowledge should reside within a firm and how much should be disclosed. Disclosure can take two forms: with protection in the form of patents or without protection in the form of publications.

What is likely to be the relation between the different components of the R&D strategy? A priori all the components of the R&D strategy would seem to

be strategic complements. In reality, it would depend on the context studied and the constraints of the firms considered.

In the pharmaceutical sector of developed countries, Cockburn et al. (1999) assert that "science-driven", or "rational" drug discovery is both a technology for discovering new drugs and a set of managerial practises for organising and motivating research workers. They point out that firms invest in leading edge research because it increases the efficiency of the knowledge production process (Henderson and Cockburn, 1994; Gambardella, 1995). They confirm that in the pharmaceutical sector some firms have become active participants in the creation of scientific knowledge instead of being only passive users. These firms publish and patent actively and have extensive research networks with research laboratories. Investment in patenting in the biopharmaceutical sectors has also been noted to be a means by which firms can mark their territory in limited technological space to gain future rent. If the situation is similar in developing countries we should find the same pattern. In other words, we should find that all parameters of the R&D strategy take on higher values for firms which undertake biotech research as compared to firms which are simply marketing or producing a biotech-based product without doing research.

The central questions of the paper as may be recalled are: (i) What kind of labour stocks forming the knowledge base have an impact on market sales? (ii) Which components of the R&D strategy are strategic substitutes and which are strategic complements? (iii) What are the distinguishing features of firms that have already integrated biotechnology in their research activities? From the arguments developed in this section, we can propose the following three hypotheses as initial responses to the above questions to be tested with data.

H1: market sales are an increasing function of the stocks of qualified labour or an increasing function of labour stocks allocated to R&D activities.

H2: all parameters of an R&D strategy are strategic complements.

H3: all parameters of an R&D strategy will take on higher values for firms which are active in biotechnology research (as opposed to firms which are only marketing or producing a biotech product but not doing biotech research).